AF269610

The People of Palomas

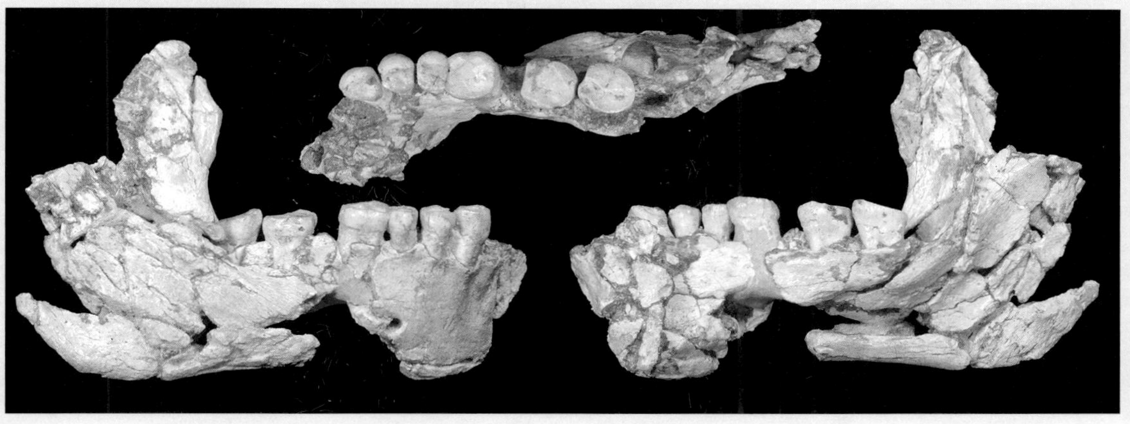

THE PEOPLE OF PALOMAS

Neandertals from the
Sima de las Palomas del Cabezo Gordo,
Southeastern Spain

EDITED BY ERIK TRINKAUS AND
MICHAEL J. WALKER

Texas A&M University Press
College Station, Texas

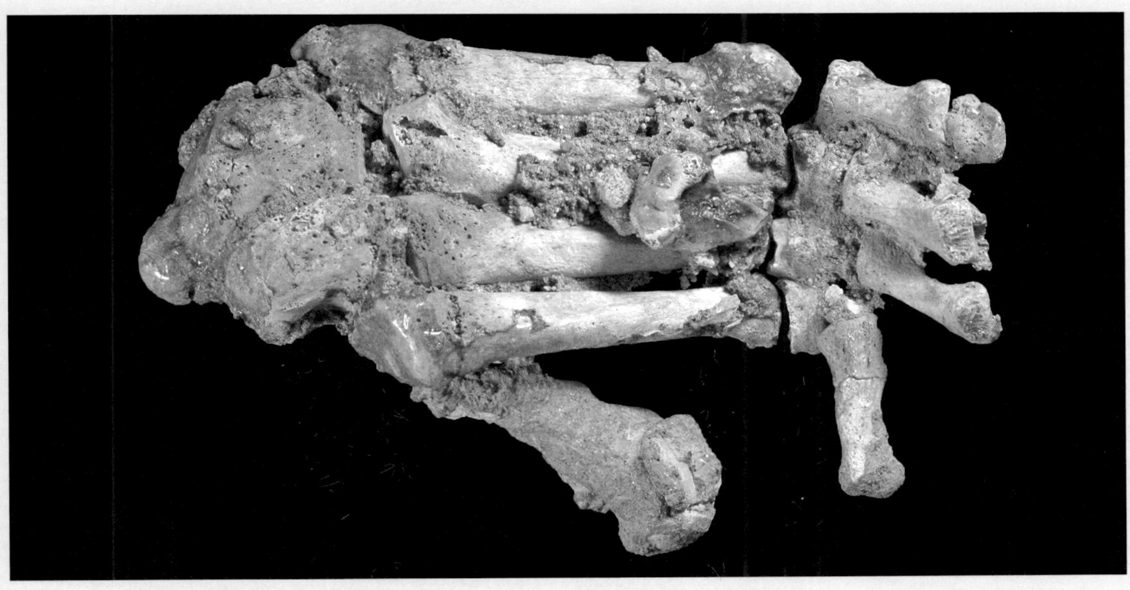

Library of Congress Cataloging-in-Publication Data

Names: Trinkaus, Erik, editor. | Walker, Michael J.,
    editor. | Trinkaus, Erik. Neandertals from the Sima de
    las Palomas.
Title: The people of Palomas : Neandertals from the
    Sima de las Palomas del Cabezo Gordo, southeastern
    Spain / edited by Erik Trinkaus and Michael J. Walker.
Other titles: Texas A & M University anthropology
    series ; no. 19.
Description: First edition. | College Station : Texas
    A&M University Press, [2017] | Series: Texas A&M
    University anthropology series ; number nineteen |
    Includes bibliographical references and index.
Identifiers: LCCN 2016039265| ISBN 9781623494797
    (hardcover (printed case) : alk. paper) |
    ISBN 9781623494803 (e-book : alk. paper)
Subjects: LCSH: Neanderthals—Spain—Murcia
    (Region). | Prehistoric peoples—Spain—Murcia
    (Region). | Murcia (Spain : Region)—Antiquities. |
    Antiquities, Prehistoric—Spain—Murcia (Region). |
Human growth. |
    Teeth—Growth. | Dental anthropology.
Classification: LCC GN285 .P46 2017 | DDC
    569.9/86094677—dc23 LC record available at
    https://lccn.loc.gov/2016039265

Contents

Preface

FOR MOST OF THE HISTORY of human paleontology, the paleontological monographic descriptions of substantial samples of human fossil remains have been undertaken by single individuals or, at the most, pairs of individuals. These efforts have produced a series of now classic monographs describing the remains from Spy, Krapina, La Chapelle-aux-Saints, Oberkassel, La Quina, Předmostí, Skhul and Tabun, Zhoukoudian, Olduvai Gorge, La Ferrassie, Qafzeh, and Shanidar, among others. Yet during the past few decades, it has become apparent that the proper presentation and analysis of a large sample (or even small sets of isolated bones and teeth) requires the collaboration of multiple individuals with different areas of knowledge and technical expertise as well as diverse perspectives. It is with this recent approach that this treatment of the human remains from the Sima de las Palomas has emerged.

A decade ago, when Michael J. Walker invited Erik Trinkaus to undertake the process of bringing the Palomas Neandertal remains to the larger profession, it was readily apparent that any such effort would involve a substantial set of individuals. Some of these people had already been involved with the Palomas remains, especially A. Vincent Lombardi through his years of excavating at the Sima de las Palomas and his early involvement with the human remains. Other individuals—Josefina Zapata, Alejandro Pérez-Pérez, and Beatriz Pinilla—had begun their studies of the large sample of dental remains. At the same time, Jon Ortega had been involved with the excavation and cleaning of the human remains (especially Palomas 92) and was increasingly involved in their analysis.

Once we, along with the late Josep Gibert, had sorted out the basics of the project and had begun the process of publishing short notes on aspects of the collection, it was then appropriate to bring in other individuals—ones with appropriate and complementary paleoanthropological expertises. It was therefore as a result of these different individuals' work, separately and together, that the contents of this volume have emerged and come together. All of these individuals are authors or coauthors of the chapters in this volume. In addition, a substantial number of individuals have furnished expertise and analyses in the field and the laboratory to enable placing the Palomas human remains in their geological and chronological context; most of them are mentioned in chapter 2. Through this process, João Zilhão has provided helpful feedback and perspectives.

Any project such as this one is of course dependent on substantial financial and logistical support for the fieldwork, the descriptions of the Palomas human fossils, the collection of comparative data, and the analysis of the resultant data. The following institutions have provided economic, material, and personnel support to Michael J. Walker and colleagues for the Sima de las Palomas excavations and laboratory analysis: the Spanish Government for research grants CGL2005/02410/BTE, BOS/2002/02375, PB/98/0405, and PB/92/0971 and Anglo-Spanish Integrated Actions HB/1993/0114 and HB/1995/002; the Consejería de Cultura of the Comunidad Autónoma de la Región de Murcia for excavation and research grants CTC/DGC/SPH/063/2001, CCE/DGC/IPH/SAR0/1998, CCE/DGC/IPH/SAR/1997, CE/DGC/IPH/SAR/011/1996, CCE/DGC/IPH/SAR/1995, CCE/DGC/IPH/SAR/1994, and PSH/93/52 and the Fundación Séneca of the Comunidad Autónoma de la Región de Murcia for research grant 05584/ARQ/07 for the infrastructure at the site; the town council of Torre Pacheco and its Civic Centre at Dolores de Pacheco for hosting the Murcian Association for the Study of Palaeoanthropology and the Quaternary (MUPANTQUAT) annual Field School (with the Earthwatch Institute, 1994–2001); Universidad de Murcia for research grant 2009/12441 and collaboration from both its official research group for *Quaternary Paleoecology, Palaeoanthropology and Technology* and its Departamento de Zoología y Antropología Física; and the Institut Paleontològic "Dr. M. Crusafont" of the Diputació de Barcelona at Sabadell.

Funding for dental analyses to A. Pérez-Pérez came from MEC Spanish Government grants CCGL2007–60802, CCGL2010–15340, and CGL2011–22999; Generalitat de Catalunya grant 2009SGR884; and the Leakey Foundation. B. Pinilla was also supported by MEC Spanish Government grants CCGL2007–60802 and CGL2011–22999 plus a predoctoral grant (AP2006–01274) from the same institution. The Leakey Foundation and Washington University provided support for S. A. Lacy and J. C. Willman. The transportation of the micro-CT scanner to Murcia by P. Bayle and K. A. Robson Brown was funded by the LabEx des Sciences Archéologiques de Bordeaux (ANR program of prospective investments, ANR-10-LABX-52), and the University of Bristol Institute for Advanced Studies, with additional funding from IdEx Bordeaux/CNRS (ANR-10-IDEX-03–02). Visits for Erik Trinkaus to Murcia were supported by Washington University and Murcia University, with funding from the National Science, Leakey, and Wenner-Gren Foundations; the Centre National de la Recherche Scientifique (CNRS); and B. I. Gorokhoff for the collection of comparative data. In addition, a large number of curators and colleagues have made comparative fossil remains available to a number of us, and many of our colleagues have provided constructive discussions concerning aspects of the Palomas Neandertal remains and their analysis. Sheela Athreya and Hélène Rougier in particular have provided feedback on the volume as it has come together. To all of these institutions, funding agencies, and individuals, we are immensely grateful.

It should be mentioned here that the decision to bring together the various data on the Palomas remains at this time, but to minimally include data from the Palomas 96 and 97 associated skeletons, was based on one of the constraints resulting from the nature of the Sima de las Palomas. Portions of the sediment column within the shaft are relatively soft, and fossils can be cleaned easily from those levels. However, the associated skeletons derive from extremely hard, brecciated sediments. The bones are rather soft, and the process of extracting the remains from their sediment is long and arduous. Palomas 92, having been found much earlier, is largely clean, sufficiently extracted to permit its detailed analysis. Palomas 96 and 97 remain a long way from being adequately cleaned for analysis, much of which will be done virtually. It was therefore decided, given the work that had already been done on the isolated remains plus Palomas 92, that it was appropriate to proceed at this time with a volume on these fossils, with plans for a separate volume on the Palomas 96 and 97 remains when they have sufficiently emerged from their entombment.

It is therefore our hope that these presentations and analyses of the Palomas Neandertal remains will provide a better window on the nature of Neandertal biology, on its variation, and on its presence in southeastern Iberia.

—ERIK TRINKAUS, Saint Louis
MICHAEL J. WALKER, Murcia

The People of Palomas

Introduction

The Neandertals from the Sima de las Palomas

ERIK TRINKAUS AND MICHAEL J. WALKER

HUMAN REMAINS REFERRED to the Neandertals have been known from Iberia since the nineteenth-century beginnings of the paleontological considerations of human origins, but they have only come to figure substantially in considerations of Neandertal paleobiology and population relationships in recent decades. The Forbes' Quarry (Gibraltar) 1 cranium, found in 1848, was the second Neandertal fossil to be discovered (after Engis 2) and the second one to be recognized as an archaic human (after Feldhofer 1). Subsequently, the Banyoles (Bañolas) 1 mandible, discovered in 1887, was one of the earliest Neandertal mandibles to be found and the first largely complete one, second to La Naulette 1 and close in time to Spy 1 and 2. Yet until recently Iberia has contributed only peripherally to ongoing issues of Neandertal paleontology, since subsequent discoveries through much of the twentieth century consisted of isolated teeth and fragmentary skeletal elements (Garralda 2006).

However, in this context, a series of recent discoveries has greatly increased the Iberian Neandertal paleontological presence (Quam et al. 2001; Lorenzo and Montes 2001; Arsuaga et al. 2001a, 2007; Barroso-Ruíz et al. 2003; Daura et al. 2005; Rosas et al. 2006; Sarrión 2006; Trinkaus et al. 2007; Walker et al. 2008, 2011b; Willman et al. 2012; Garralda et al. 2014). These discoveries have helped, with the persistence of the Middle Paleolithic in the region (Zilhão 2006; Rodríguez-Vidal et al. 2014), to establish the late presence of the Neandertals south of the Pyrenees (Trinkaus et al. 2007; Walker et al. 2008). Furthermore, analyses of their remains have increasingly documented aspects of their paleobiology (Lalueza et al. 1993a; Arsuaga et al. 2001b; Rosas et al. 2006; Walker et al. 2011a, 2011c; Estalrrich et al. 2011; Estalrrich and Rosas 2013) and have raised issues regarding Neandertal variability (Rosas et al. 2006; Walker et al. 2008, 2010; Alcázar de Velasco et al. 2011). At the same time, Neandertal predecessors are well documented at the Sima de los Huesos, Atapuerca (Arsuaga et al. 2014, 2015), and there is fossil evidence of their earlier Upper Paleolithic modern human successors (Zilhão and Trinkaus 2002a; Arsuaga et al. 2002).

This growing knowledge of the Neandertals in Iberia is relevant to broader issues of Neandertal biology and their place in Late Pleistocene human population dynamics, given the cul-de-sac nature of the Iberian Peninsula (assuming that the Straits of Gibraltar were indeed an impediment to human movement and communication). Not only is Europe (at least west of the Dardanelles) a cul-de-sac within Eurasia, but Iberia represents the extreme of any isolation by distance that would have occurred relative to the rest of the Late Pleistocene global human population. Moreover, those Neandertal populations south of the Pyrenees may well have been further ecologically separated from the remainder of the European Neandertals (Zilhão 2006).

At the same time as these discoveries of, and

renewed interest in, Iberian Neandertals, there has been continued focus on more general issues regarding the Neandertals. The century-old debate regarding the role of the Neandertals in modern human emergence in western Eurasia appears to be reaching a consensus. The dominant view has become one in which modern humans emerged in eastern Africa in the late Middle Pleistocene, temporarily spread into southwestern Asia and apparently more permanently into southeastern Asia, but only dispersed permanently through western Eurasia (the principal Neandertal region) after ≈50,000 years ago (Trinkaus 2013; Liu et al. 2015). In the final stage, the paleontological evidence (Trinkaus 2007; Cartmill and Smith 2009) indicates modest interbreeding between dispersing modern humans and resident Neandertals, resulting in early modern human populations that were basically modern in their biology but retained a variety of archaic and/or Neandertal features (Trinkaus and Zilhão 2013) that gradually waned over the subsequent tens of millennia (Trinkaus 2007; Fu et al. 2015). Moreover, such a scenario accords well with both the plethora of shared cultural attributes of the two human groups (Villa and Roebroeks 2014) and the documented porousness of inferred reproductive boundaries between closely related mammalian (including primate) species (Jolly 2003; Holliday 2006; Holliday et al. 2014).

At the same time, a variety of analyses of Neandertal paleobiology related to activity levels, paleopathology, life history parameters, and so on, have narrowed the previously perceived gaps between the Neandertals and early modern humans in terms of biological reflections of adaptive behavior (Trinkaus 2013). Yes, the Neandertals were different from modern humans morphologically, and there were developmental patterns that led to these anatomical contrasts. However, when the Neandertals are appropriately compared to their contemporaneous modern humans (the Middle Paleolithic ones from eastern Africa and southwest Asia) and to their immediate successors (the Early/Mid Upper Paleolithic modern humans), most of the suggested differences are either reduced in effect or disappear fully. The more common comparisons to Holocene humans often confuse the biological differences between Neandertals and early modern humans with those between Pleistocene modern humans and late Holocene sedentary and mechanized modern humans.

These considerations of the dynamic process of Neandertal paleoanthropology, a process that continues to evolve after a century and a half (Trinkaus and Shipman 1993), provide a context for the assessment here of the Neandertal remains from the Sima de las Palomas. These variably fragmentary remains of Neandertals, recovered since 1991 within and upon the slopes of the Cabezo Gordo in southeastern Iberia, are not likely to resolve these ever evolving aspects of Neandertal inquiry. Yet these remains provide considerable additions to the overall European Neandertal sample. They do so especially for Iberian Neandertals, which tend to be fragmentary and/or widely scattered despite the number of sites that have yielded them recently (El Sidrón being the one real exception).

With these considerations in mind, we have brought together a variety of analyses of the Palomas human remains, analyses that provide a basic paleontological description, furnish standard descriptive data on them, and assess aspects of their paleobiology. These analyses are arranged in a traditional anatomical framework, from crania and mandibles to teeth and to postcrania. Given preservation and the variety of approaches available for their analysis, the majority of the chapters are focused on the dental remains, those in alveoli and especially the isolated teeth. One of the specimens, Palomas 92, consists of associated postcrania, albeit in various states of preservation and cementation. The remainder of the elements are isolated and currently cannot be associated by individual, although it is likely that the teeth in particular represent far fewer individuals than the total number of specimens enumerated. They are nonetheless described and assessed individually. Note that two of the associated skeletons from Palomas, Palomas 96 and 97

(Walker et al. 2011b, 2012) are not included here, although references to some aspects of their morphology and comparative data, as available, are included. They are currently emerging from breccia, and they will be considered in detail when that process is sufficiently advanced.

It is therefore hoped that this volume will provide a sufficiently thorough description and analysis of the Palomas Neandertal remains to permit their integration into the Late Pleistocene human fossil record. A few aspects of their morphology and paleobiology have been presented (e.g. Walker et al. 1998, 2008, 2010, 2011a, 2011c), but the more complete assessment of the sample and considerations of a diversity of aspects of these elements provided here should further this integrative process.

2 The Context of the Sima de las Palomas Neandertals

MICHAEL J. WALKER, MARIANO V. LÓPEZ, MARIA HABER, AND ERIK TRINKAUS

ON THE COASTAL PLAIN OF MURCIA, in southeastern Spain, there is an isolated hill of Permo-Triassic marble, Cabezo Gordo (Big Hill), that rises 312 m above sea level. It lies in the municipality of Torre Pacheco, between Murcia city and the Mediterranean saltwater lagoon of the Mar Menor. It is moderately north of Cartagena and the Cabo de Palos along the Mediterranean Coast (Plates 2.1 to 2.3). Cabezo Gordo has several natural karstic cavities, including vertical shafts containing variably brecciated Quaternary deposits. One shaft, on the southern face, is the Sima de las Palomas (Dove or Pigeon Hole), at 37°47'59" N, 0°53'45" W (37.793508, −1.859436). The top of the Sima de las Palomas is 125 m above sea level, and the cave system has a maximum depth of 31 m.

The Sima de las Palomas (Figs. 2.1 and 2.2) was filled with sediments during the late Middle Pleistocene and earlier Late Pleistocene, apparently from the time of Marine Isotope Stage (MIS) 6 to the middle of MIS 3 (see dating discussion, below). Only a column of brecciated sediment plastered against the rear wall of the main shaft and the main chamber remained after most of the fill was removed by nineteenth-century miners. They exploited veins of the iron ore magnetite, which were the object of fifteen mining concessions on Cabezo Gordo. Needing water to wash iron ore, they broke into the cave from the

hillside, apparently seeking water (the cave is dry today), by widening an 11-m-deep natural cleft that opens beside a trivial superficial vein of magnetite (no veins exist inside the cave). In this cleft, the miners built a dry-stone revetment as a landing for their ladders above and below it (Walker et al. 2012). They later dynamited a horizontal tunnel from the hillside through into the main chamber, leaving behind considerable rubble there and on the hillside around the tunnel entrance (Plate 2.4).

In 1991, a local wildlife enthusiast and speleologist, Juan Carlos Blanco, was abseiling down the shaft when, 3 m below its rock overhang, he noticed a fossil, which he removed from brecciated sediment. When he showed it to Michael J. Walker and Josep Gibert, they recognized it as being the crushed, anatomically connected mandible and maxillae of a Neandertal with an almost complete permanent dentition; it was called Cabezo Gordo 1 (CG1), later renamed Palomas 1 (SP1) (Plate 5.1).

Work at Sima de las Palomas

Sieving the mine rubble began in 1992 with the recovery of late Middle and early-middle Late Pleistocene faunal remains (Table 2.1), Middle Paleolithic artifacts, and more Neandertal fossils (Gibert et al. 1994). The latter included the

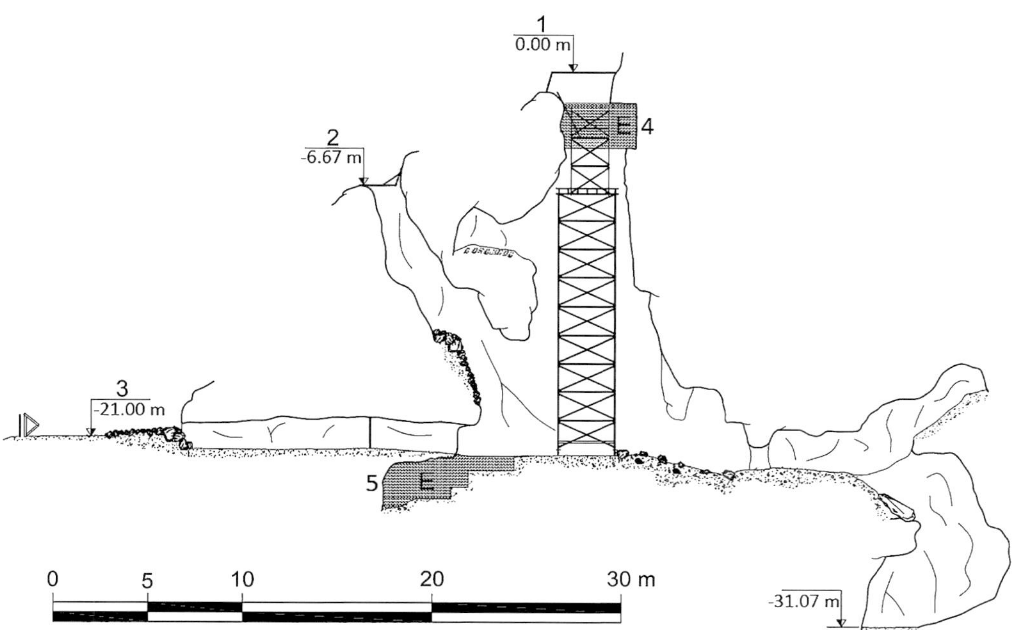

FIG. 2.1. Elevation of the Sima de las Palomas del Cabezo Gordo. *1*, Top of main shaft (0.0 datum). *2*, Top of the cleft containing the dry-stone revetment made by miners. *3*, Entrance to horizontal tunnel dynamited by miners from the hillside through to the main chamber about 1900. The terrace at the entrance has been built up on the 45° sloping marble bedrock of the hillside using mine rubble that has been sieved in order to retrieve fossils and archeological finds thrown out by mining activity. *E*, Areas of the cave where archeological excavation has taken place. *4*, The Upper Cutting excavation area. *5*, The Lower Cutting excavation area. (We thank Ignacio Nicolás-Vázquez, director of the Escuela Murciana de Espeleología de la Federación de Espeleología de la Región de Murcia and members of the Grupo de Excursionistas de la Villa de Alcantarilla for carrying out the topographic survey of the cave and generating the diagrams.)

FIG. 2.2. Plan of the Sima de las Palomas del Cabezo Gordo. *E*, Areas of the cave where archeological excavation has taken place. *1*, Top of main shaft (0.0 datum) and Upper Cutting excavation. *2*, Top of the cleft containing the dry-stone revetment made by miners. *3*, Entrance to horizontal tunnel dynamited by miners from the hillside through to the main chamber about 1900. The terrace at the entrance has been built up on the 45° sloping marble bedrock of the hillside using mine rubble. (Plan courtesy of Ignacio Nicolás-Vázquez, director of the Escuela Murciana de Espeleología de la Federación de Espeleología de la Región de Murcia, and members of the Grupo de Excursionistas de la Villa de Alcantarilla.)

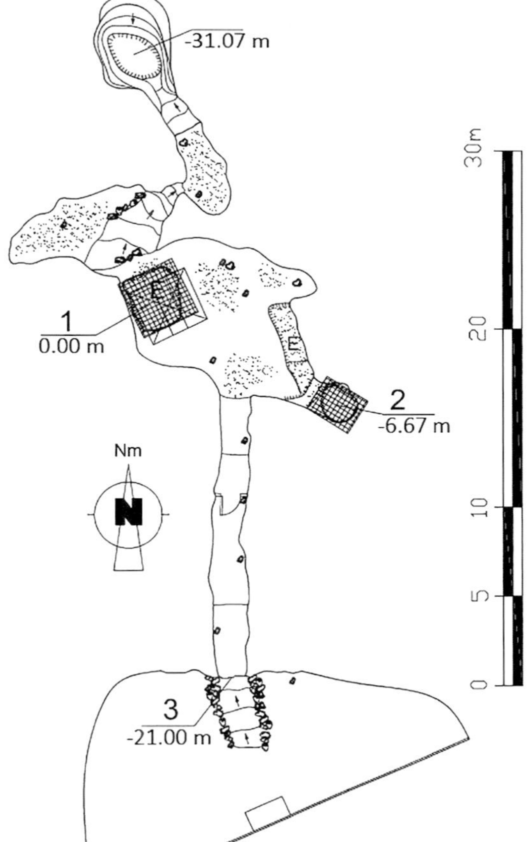

Palomas 2 to 7, 10, and 11 cranial and mandibular remains, among them the morphologically diagnostic Palomas 6 mandible and Palomas 11 supraorbital torus. Taken for cleaning and study to the Institut Paleontológic "Dr. M. Crusafont" at Sabadell near Barcelona, the fossils are now at Murcia, together with subsequently excavated finds, in a facility provided by the Murcian Regional Authority (Comunidad Autónoma de la Región de Murcia) until its newly built Museum for Paleontology and Human Evolution at Cabezo Gordo is ready to house them. The importance of the initial discoveries among the mine rubble, backed up by initial electron spin resonance (ESR) age determinations on three animal bone fragments, led the regional government to install protective gates and grilles and make available scaffolding to enable systematic excavation, from above downwards, of the variably brecciated sedimentary column in the main shaft (Figs. 2.1 and 2.2; Plates 2.5 and 2.6).

Excavation began in 1994 in the Upper Cutting, which occupies an L-shaped area around the open shaft. As of 2015, the excavation reached a depth of 5 m below the overhanging rock and ≈2 m below where Palomas 1 was found. Excavation also was undertaken in sediments below the floor of the main chamber (the Lower Cutting). A 2 m depth of mine rubble was found there to cover a 3 m depth of almost sterile Pleistocene sediment. The excavation was suspended on encountering a steeply sloping calcrete flowstone that cemented large stones.

Until the death of Josep Gibert in 2007, excavation at Sima de las Palomas was codirected by Walker and Gibert, with research funding channeled through Murcia University; the codirectors now are Walker, Mariano V. López, and Maria Haber, and excavation is under the auspices of the Murcian Association for the Study of Palaeoanthropology and the Quaternary (MUPANTQUAT) at its yearly summer field school. Since 1992, a constant feature is the extensive collaboration of colleagues, graduate students, and undergraduates. Meriting special mention is the longstanding collaboration of A. Vincent Lombardi with both the description and analysis of the site's

abundant Neandertal teeth and assistance in the field. In 2003, Walker invited Josefina Zapata to join in studying the dental remains. In 2006, he invited Trinkaus, at the suggestion of João Zilhão, to oversee and coordinate the analysis of the human fossil collection. At the same time, the dental remains were made available to Alejandro Pérez-Pérez and Beatriz Pinilla at Universitat de Barcelona for microscope analysis of postcanine wear. In 2012, Katharine Robson Brown and Priscilla Bayle brought the University of Bristol's micro-CT scanner to Murcia and generated micro-CT scans of all the teeth and those bones small enough to fit within the scanner. At the same time, Sarah Lacy and John Willman analyzed the teeth and mandibles for oral pathology and anterior dental wear, respectively. Jon Ortega, assisted by Klára Karaková, undertook the cleaning and inventorying of the skeletal remains (especially of the associated Palomas 92, 96, and 97 partial skeletons), and Christopher Zollikofer, Marcia Ponce de León, Amalia Agut-Giménez, and Marta Soler-Laguía have given Jon Ortega help with using the Murcia University Veterinary Hospital CT scanner to generate preliminary CT scans of the associated skeletons (Walker et al. 2013). Additionally, Amanda Henry and Domingo Salazar-García identified phytoliths in dental calculus of Sima de las Palomas Neandertals (Salazar-García et al. 2013). All of the aforementioned individuals have been involved directly with the study of the Palomas human remains, which is the focus of this volume. There have also been a substantial number of geologists, palynologists, paleontologists, analytical chemists, geochronologists, archeologists, mineralogists, microbiologists, and anthracologists, as well as numerous field assistants, without whose invaluable and enthusiastic collaboration research at the Sima de las Palomas could never have borne fruit; we thank them all (for more details, see Walker et al. 1999, 2008, 2012; Walker 2001). The work has also been funded by a diversity of institutions during the quarter-century of excavation and analysis (see Preface).

The Sima de las Palomas human fossils were published initially in preliminary inventories and descriptions (Gibert et al. 1994; Walker et al. 1998, 1999; Walker 2001). Subsequently, those remains from the uppermost sedimentary layers of the Upper Cutting were reassessed in the context of an updated inventory and revised chronological framework (Walker et al. 2008), followed by descriptions of the mandibles and postcrania (Walker et al. 2010, 2011a), the carious lesions of Palomas 25 and 59 (Walker et al. 2011c), and a preliminary description of the Palomas 96 skeleton (Walker et al. 2011b). Data from the Palomas human remains have also been included in various considerations of Neandertal/Pleistocene *Homo* paleobiology.

The Upper Cutting of the Sima de las Palomas

All of the human remains with stratigraphic contexts come from sediments in the Upper Cutting (Fig. 2.3). The nature, stratigraphy, and chronology of the Upper Cutting deposits are central to the context of these human remains, and a brief account follows (a more detailed account superseding older publications is given in Walker et al. 2012).

The west side of the Upper Cutting contains partly cemented tumbled rocks and stones forming an éboulis that is Conglomerate A (Plate 2.7) sloping down at 30–40° from the entrance, where remnants are seen cemented onto a sill of marble bedrock. Abutting Conglomerate A are later sediments, bedded almost horizontally, which are unconsolidated, coarse, gritty, light beige, and interspersed with several small angular stone clasts; they filled the inner, eastern part of the natural cavity up to its marble roof. These uppermost sediments contained only Middle Paleolithic artifacts and Neandertal fossils without later material (Walker et al. 2008). Vertical thin sinters of $CaCO_3$ formed in the sediments after they had accumulated, and minor condensations of it formed around or on bones within the infilling. Cementation of Conglomerate A likely occurred at the same time, because its slope had neither a weathered surface nor a sloping calcrete flowstone covering (initially it was thought, mistakenly, that Conglomerate A

corresponded to a rock fall postdating the uppermost sediments disrupted by a violent fall). We do not know how much time may have elapsed during and between sedimentation of Conglomerate A and the sediments that later accumulated around it. Finally, a thin calcrete flowstone sealed the entire fill of the cave. Presumably, the uppermost sediments inside the cavity and the fossil and archeological remains within them entered before the top of Conglomerate A became so consolidated as to block up the cave entrance completely.

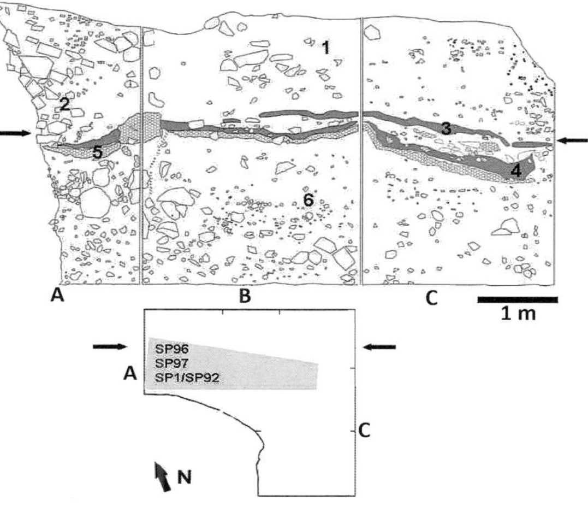

FIG. 2.3. Current stratigraphic profiles (*above*) of the Upper Cutting and horizontal projection (below) of the area under excavation. The A, B, and C sections in the upper diagram are the approximately west, north, and east profiles of the excavated area, and they match the A, B, and C sides of the lower diagram. 1, the uppermost beige sediment that contained Neandertal teeth and bones and Middle Paleolithic technology. 2, the partly cemented rock tumble or éboulis, Conglomerate A, containing associated human remains. 3, Upper Gray Layer. 4, Lower Gray Layer. 5, Conglomerate B. 6, Deeper sediments (currently under excavation) containing Mousterian implements and abundant burnt animal bone fragments. The shaded area in the bottom part of the figure is a horizontal projection at the level of the arrows in the top part, corresponding approximately to where the Palomas 92, 96, and 97 associated skeletal remains and the Palomas 1 facial remains lay. Although shown schematically as horizontally adjacent to each other in the lower diagram, Palomas 96 was stratigraphically above and Palomas 1 and 92 were below.

About 1 m down, the uppermost horizontally bedded beige sediments were partly interrupted by a variable layer of dark gray sediment (the Upper Gray Layer) that abutted onto part of Conglomerate A, reaching a thickness of 0.25 m in the northeast corner of the Upper Cutting and thinning out to the northwest and southeast. This Upper Gray Layer provided burnt finds and chemical signals of combustion (Walker et al. 2012). If human access to this part of the cave occurred when the Upper Gray Layer accumulated, reopening of the main shaft by mining has destroyed archeological evidence of it. This is particularly unfortunate, because Conglomerate A contained important Neanderthal remains. Moreover, because it "did not offer a uniformly altered surface, it might be wondered whether there was no great temporal separation from the upper grey layer of sediment" (Walker et al. 2012).

Both Conglomerate A and those later sediments abutting it lay on a very extensive pale gray sediment with weathered stones, the Lower Gray Layer. Combustion played a minor part at most in producing this layer, which seems to have been a result mainly of processes involving waterlogging (and perhaps microorganisms) that probably were favored by the impermeability of the underlying, very heavily cemented, angular small stone chips of Conglomerate B, which also contained a few fragments of bone and flint. Rock hard, it corresponds to a cemented cryoclastic scree, utterly different from Conglomerate A (see Walker et al. 2012 for further details). Sloping downward from northwest to southeast, Conglomerate B occupied the entire area of the cavity as a thin band (0.15–0.25 m thick). Conglomerate B in turn covers deeper sediments, containing abundant burnt faunal remains and Mousterian artifacts, in which ongoing excavation is 5 m below the roof of the cavity.

The foregoing account implies four different phases of human impingement. First, there was food preparation at the cave mouth, responsible for finds in sediments below Conglomerate B. Later, there were deposited the Palomas 92, 97, and 96 articulated skeletons excavated from Conglomerate A. Later still, there was activity associated with the Upper Gray Layer. Finally, there was activity around the cave mouth responsible for the remains that derived from the beige sediment above the Upper Gray Layer.

Because initial excavation of the uppermost beige sediment was conducted in arbitrary horizontal levels (spits), it was some time before stones on one side of the excavation area were recognized as being the tip of what was designated Conglomerate A. Consequently, quasi-stratigraphic designations of both the excavated human remains (see the inventories in the descriptive chapters) and the dated samples corresponded in some cases to horizontal levels cutting across both the uppermost beige sediment and Conglomerate A. Therefore, assigning finds and samples correctly to one or the other, and thereby ascertaining their relative chronology, demands paying as much detailed attention to their relative horizontal coordinates as to their vertical ones. Failure to do so underlies misunderstandings of the probable ages of some remains by Santamaría and de la Rasilla (2013) and Wood et al. (2013).

The Chronological Framework for the Sima de las Palomas

The geological ages of the levels within the Sima de las Palomas have presented a series of challenges. The Neandertal fossils may well come from times close to, or beyond, the limits of radiocarbon dating. There are concerns inherent in applying other radiometric techniques to a stratigraphic column that has suffered from ravages of mining. However, diverse forms of evidence, in the context of the depositional sequence of the stratigraphy, provide a framework for approximating the ages of the human remains.

Faunal Chronological Indications

The initial indications of the geological age of the Sima de las Palomas stratigraphic column came from the identifications of faunal taxa, albeit primarily for materials out of stratigraphic context. The faunal list (Table 2.1) includes a number of taxa that are present through the Late Pleistocene and Holocene of Europe and therefore of minimal chronological value. However, other represented taxa (for example, *Hippopotamus, Crocuta, Panthera*) have been extinct within southwestern Europe since at least the late last glacial and are well represented in Late Pleistocene Middle Paleolithic assemblages. There are also remains of *Stephanorhinus* identified from the mine rubble, indicating a late Middle Pleistocene or initial Late Pleistocene age (Gibert et al. 1994). Additionally, a skull of *Panthera pardus* cf. *lunellensis* (also indicating a late Middle Pleistocene age) was removed in 1991 from 2 m above the floor of the main chamber, and it could have been derived from higher in the stratigraphic column. The faunal profile, therefore, largely constrains the fossiliferous portion of the stratigraphic column to the late Middle Pleistocene through the Late Pleistocene.

Electron Spin Resonance Dating

Before excavation began, bone fragments with breccia cement, found in mine rubble, were sent by M. J. Walker to P. J. Pomery at Queensland University for electron spin resonance (ESR) dating. The samples gave ages of ≈83, ≈146, and ≈532 ka cal BP based on an assumed background dose rate of 1 Gy/ka. If the rate was assumed to be 2 Gy/ka, the ages would become ≈41.5, ≈73, and ≈266 ka cal BP, respectively (Gibert et al. 1994). Subsequently, a background dose rate of 1.32 ± 0.06 Gy/ka was estimated for the uppermost beige-colored sediments in the Upper Cutting by J. L. Schwenninger (in Walker et al. 2008; see optically stimulated luminescence dating, below). Using that value, the ESR paleodoses provide ages of ≈63, ≈111, and ≈403 ka cal BP, with unknown confidence intervals. The first two ages agree with the paleontological indications of an early–middle Late Pleistocene age. Because earlier Middle Pleistocene fauna have yet to be found at the site, the substantially older third date may be erroneous.

Uranium-Series Dating of Sediment Sample

Before systematic excavation commenced, two carbonate samples were collected at 2–3 m above the base of the breccia column and from

slightly below where Palomas 1 had been found; a third sample (M5) was collected from either Conglomerate B or immediately underneath it (Sánchez-Cabeza et al. 1999). The samples were dated at the Autonomous University of Barcelona using uranium-thorium dating "based on the determination of the ^{230}Th/^{234}U disequilibrium produced during formation of the carbonates" (Sánchez-Cabeza et al. 1999, 261). The two deeply lying samples (M1 and M2) provided statistically identical ages in the initial Late Pleistocene (118 +20/−16 ka cal BP and 124 +20/−15 ka cal BP, respectively). Sample M5 gave an age of 56 +13/−10 ka cal BP, pointing towards an early MIS 3 age for the Upper Cutting, although the large standard error precludes further precision.

Uranium-Series Dating—Bone

Uranium-series determinations on bones excavated in the Upper Cutting have been made using the diffusion-adsorption (D-A) model to account for uranium uptake and laser ablation multicollector plasma mass spectrometry (LA-ICP-MS) at Bristol University (A. Pike in Walker et al. 2008). A determination was made on a metacarpal (APSLP1; E in Fig. 2.4) of the Palomas 96 skeleton, whose head rested against both hands raised up from flexed elbows, giving it a relatively high position (level 2c) within Conglomerate A. Two additional ones were on unburnt faunal bones (APSLP4 and APSLP6; C and D in Fig. 2.4). The three samples provided ages of 54.1 ± 7.7 ka, 43.8 ± 1.5 ka, and 51.0 ± 2.5 ka, respectively (2σ errors). The 95% confidence intervals are therefore 61.6–46.6, 45.3–42.3, and 53.5–48.5 ka, respectively. Although excavated in the uppermost coarse beige-colored sediment (level 2l), APSLP6 lay close to Conglomerate A, from which it may well have been displaced in antiquity (given the incomplete consolidation of Conglomerate A). Despite having been excavated at a higher level (2i) in the same sediment, at a position 0.15–0.2 m above APSLP6, APSLP4 lay further away from Conglomerate A, and the APSLP4 determination is consistent with that sediment being later than Conglomerate A. It is reasonable to view APSLP1 and APSLP6 as indicating the age of

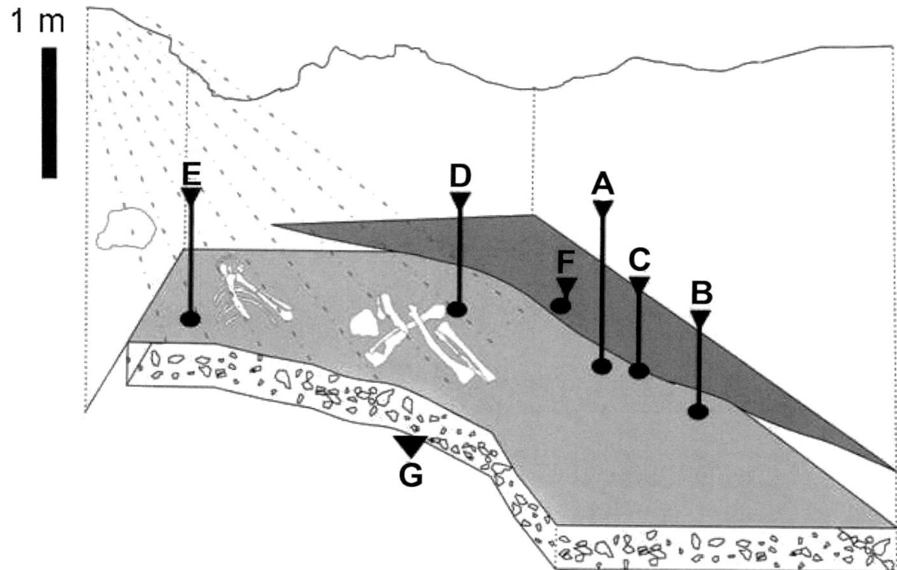

FIG. 2.4. Diagram to show provenience of samples on which radiometric estimates were determined. ● = horizontal positions of the samples; ▼ = relative vertical positions of the samples. The stratigraphy of the Conglomerate A slopes downwards from upper left to lower right and is indicated by the dotted lines (see the slope of the stones drawn in Fig. 2.3). The Upper Gray Layer is in dark gray, and the Lower Gray Layer is shown in light gray. A, OxA-10666 radiocarbon sample; B, OxA-15423 radiocarbon sample; C, APSLP4 U-series on bone sample; D, APSLP6 U-series on bone sample; E, APSLP1 U-series on Palomas 96 metacarpal; F, X2509 OSL on Upper Gray Layer sediment; G, M5 U-series on carbonate sample. The schematic skeletal remains in the light gray represent Palomas 92 (left) and Palomas 96 (right).

Conglomerate A. Nevertheless, as noted by Pike (17), "these dates must be treated as of unknown accuracy, since a change in the geochemistry of the burial environment will result in the bone rapidly re-equilibrating, which may include further uptake of uranium (leading to underestimated apparent dates) or the loss of uranium (leading to older apparent dates)." With that important qualification, the U-series determinations indicate an age for the specimens in the earlier MIS 3.

Optically Stimulated Luminescence Dating

A bloc sediment sample (X2509) of ≈1.5 kg was taken from the Upper Gray Layer by a horizontal bore into the archaeological section or profile of the Upper Cutting (F in Fig. 2.4). It was treated for optically stimulated luminescence (OSL) dating of sedimentary quartz (J. L. Schwenninger in Walker et al. 2008), following sample treatment procedures at the Luminescence Dating Unit of the Oxford University Research Laboratory for Archaeology and the History of Art (RLAHA). Six aliquots were measured, providing a mean weighted paleodose of 72.36 ± 4.99 Gy. To determine the dose rate, the results of *in situ* γ-ray spectroscopy measurements were combined with

elemental analysis by inductively coupled plasma mass spectroscopy (ICP-MS) using a fusion sample preparation method. The calculated total dose rate was 1.32 ± 0.06 Gy/ka, and the calculated OSL age estimate for the X2509 sample was therefore 54.7 ± 4.7 ka (1σ error). Given the stratigraphic position of the sample in the Upper Gray Layer, it therefore provides a maximum age for the horizontal coarse sediments in the west side of the Upper Cutting and a minimum age for the antecedent Conglomerate A, albeit with 95% confidence interval from ≈64 ka to ≈45 ka. It should be pointed out, however, that the X2509 and APSLP4 were excavated at positions in unit 1 that were near to each other, both horizontally and vertically, and their 95% confidence intervals overlap at ≈45 ka.

Radiocarbon Dating

The two radiocarbon estimates from the Upper Cutting require detailed consideration. They were determined at the Oxford Radiocarbon Accelerator Unit (ORAU) of the RLAHA. The results were originally (T. Higham in Walker et al. 2008) accepted as finite dates based on their chemistry, with the expected minimal qualifications that apply to virtually all ^{14}C determinations in the time range of the results (>30 ka ^{14}C BP). Subsequent rejection of the dates by Wood et al. (2013), without reconsideration of their paleochemistry, was based on a misrepresentation of the stratigraphy of the Upper Cutting, the combining of dates from Conglomerate A and the later uppermost horizontal sediment deposited against it, and the ignoring of the large confidence intervals of the U-series and OSL determinations (see also Galván et al. 2014). The absence of an ultrafiltration step in the sample preparation (Ramsey et al. 2004), deemed by Wood et al. (2013) to be important, is irrelevant; the two samples were of burnt bone and were therefore treated as charcoal samples and processed accordingly.

Both bone samples came from the horizontally bedded coarse sediments. OxA-10666 was determined on a nondiagnostic burnt faunal specimen (740 mg) that was cemented to the medial surface of the unburnt Palomas 59 mandibular corpus (A in Fig. 2.4). Although not chemically dating Palomas 59, OxA-10666 should be regarded as dating this Neandertal mandible. If the dated bone fragment became cemented to Palomas 59 following deposition of the latter, the two bones might differ slightly in absolute age. However, possible vertical displacements in the coarse beige sediment notwithstanding, both bones lay within it before their cementation (plausibly by the same process responsible for its vertical $CaCO_3$ sinters), and small differences in their absolute ages could be encompassed by the error margins of the radiocarbon determination. The second radiocarbon sample (OxA-15423) was charred rabbit bone (257 mg) found in the Upper Gray Layer (B in Fig. 2.4); it was at a lateral distance from Conglomerate A even greater than that of OxA-10666, because the OxA-15423 bone was discovered in the horizontal sedimentary bore that provided the X2509 OSL sample when it was opened at RLAHA.

Both samples were processed with the standard ORAU treatment for charcoal (acid-base-acid, or ABA), although diagnostics normally used for collagen were also applied. The OxA-10666 sample had a 5.3% yield, and the OxA-15423 one a 2.5% yield, with %C values of 8.0% and 40.9%,

respectively. The $\delta^{13}C$ values were −21.0‰ and −22.3‰, as would be expected for collagen in Pleistocene Europe; contra Higham (in Walker et al. 2008), these values are moderately low but within the range of $\delta^{13}C$ values documented for a variety of European Late Pleistocene fauna (for example, Stevens and Hedges 2004; Trinkaus and Richards 2013) as well as recent northern hemisphere fauna (Robu et al. 2013). The $\delta^{13}C$ values do not by themselves indicate the absorption of non-collagenous carbon. The C:N atomic ratios of 6.9 and 5.8 are high for fresh collagen (DeNiro 1985), but they are as expected for the mixture of charcoal, collagen, and burnt bone, the charcoal augmenting the carbon in the collagen without additional nitrogen (T. Higham in Walker et al. 2008).

The radiocarbon dates are 34,450 ± 600 ^{14}C BP for OxA-10666 and 35,030 ± 270 ^{14}C BP for OxA-15423 (1σ errors). These determinations provide 95% confidence intervals of 40,950–37,662 cal BP and 40,986–38,850 cal BP, respectively (T. Higham in Walker et al. 2008). With respect to these determinations, Higham (13) commented: "Taken together, the results provide increased confidence in the results being finite and not underestimates of age. The consistency in the two ages is a further measure of the probable finite nature of these dates." They may nonetheless "fall closer to the upper limits of the 95% confidence intervals" (Walker et al. 2008). In other words, although one can never be entirely certain that radiocarbon ages in the time range of the Palomas samples are not minimum ages, because small amounts of intrusive younger carbon would have a disproportionate effect on the results, the natures of the samples and their chemistry provided no clear reason to doubt that these were finite determinations.

Paleoclimatic Considerations

Taken together, the several radiometric determinations provide the general chronological range from Conglomerate B and the overlying Conglomerate A to later horizontally bedded sediments: between ≈64 ka cal BP and ≈38 ka cal BP, thereby spanning much of MIS 3b. The horizontally bedded sediments have given a palynological record indicating mild, moist climatic conditions (Carrión et al. 2003); moreover, cold-climate mammals are conspicuous by their absence (Table 2.1). Deeper in the deposits, below Conglomerate B, there is evidence of *Hystrix javanica*, further supporting a warm climatic regime through the sequence (Rhodes et al. 2013).

The flora and fauna from the Upper Cutting imply that the deposits formed during one or more of the relatively warm stages, the Greenland Interstadials (GIs), part of the known sequence of global climatic fluctuations during MIS 3 (Wang et al. 2001; Svensson et al. 2008; Fleitmann et al. 2009). The deposits did not form during particularly cold phases, although the underlying Conglomerate B may correspond to a cold, dry period. Following Svensson et al. (2008; see also Hemming 2004), therefore, none of the other Palomas levels are likely to date to the cold Heinrich Event 4 (HE4; ≈40 to ≈38.5 ka) or Heinrich Event 5 (HE5; ≈47.7 to ≈48.8 ka). There were multiple warm phases around these cold peaks, including GI-8 (≈38 ka) immediately after HE4, GI-9 to GI-12 (≈40 to ≈47 ka) between HE5 and HE4, and GI-13 (≈49 ka) plus earlier ones ≥54 ka prior to HE5 (Svensson et al. 2008; Wolff et al. 2010). However, the effect of the cold peaks at Sima de las Palomas may have been minor because the uppermost horizontally bedded sediments contain pollen of thermophyll species (*Maytenus senegalensis europea, Withania frutescens, Osyris quadripartita, Periploca laevigata angustifolia*) that do not regenerate readily after prolonged severe frosts but presumably had survived locally throughout MIS 3 (Carrión et al. 2003, 2005). The paleoclimatic indicators from the Upper Cutting therefore are compatible with the accumulations in the Sima de las Palomas correlating with multiple periods through the maximum time of sedimentation indicated by the various radiometric determinations and the faunal remains.

Paleolithic Considerations

All of the Paleolithic artifacts that have been excavated in the Sima de las Palomas are either directly diagnostic of Middle Paleolithic (Mousterian) technology or fall comfortably within its range of variation (Gibert et al. 1994; Walker et al. 1999, 2012; Walker 2001; see discussion below). Securely dated Middle Paleolithic assemblages are known from southeastern Iberia dating to as recently as 38–36.5 ka (Zilhão et al. 2010; Angelucci et al. 2013), within GI-8. Similarly recent Middle Paleolithic assemblages are known from elsewhere in southern Iberia (Hoffmann et al. 2013; Rodríguez-Vidal et al. 2014). Recent attempts to discount a relatively late survival of the Middle Paleolithic in the region (Wood et al. 2013; Galván et al. 2014) have only documented that some of the Middle Paleolithic sites in the region do not have late Middle Paleolithic assemblages. The range of radiometric determinations for the Upper Cutting, therefore, falls comfortably within the known range of dates for similar archeological assemblages.

Chronological Summary

It should be evident from the foregoing data and considerations that there are inherent constraints on determining, with accuracy and precision, the absolute ages of different parts of the Upper Cutting sequence. In addition to the methodological and technical complexity, the geological context and the time period in question make it difficult to arrive at precise ages for the various stratigraphic features of the Upper Cutting. The uranium-series (and related ESR and OSL) dating applications are consistent in giving ages between ≈56 ka and ≈44 ka from Conglomerate B to the Upper Gray Layer. Nevertheless, confidence intervals for these determinations admit a wide range from ≈64 ka to ≈41 ka. The two radiocarbon dates imply ages in the vicinity of 40 ka for samples from the horizontally bedded sediments abutting on Conglomerate A; given paleoclimatic indicators and MIS global climatic fluctuations, this might indicate that some of those sediments were deposited within GI-8 after HE4. Other parts of the horizontally bedded sediments might have been deposited slightly earlier, within GI-12 to GI-9, between HE5 and HE4, given the overlap at

≈45 ka of the 95% confidence intervals of X2509 and APSLP4.

It is not unthinkable that the Upper Gray Layer could be the result of fires lit inside the cave by Neandertals ≈45 ka, after Conglomerate A had formed, whereas overlying sediments and their Neandertal fossils are the outcome of later deposition from above ≈40 ka. This scenario implies that Conglomerate A and the Palomas 92, 96, and 97 partial skeletons entombed within it were deposited ≈50 ka. It should be kept in mind, moreover, that the unconsolidated nature of the rock tumble could have permitted sediment and small fossils within it to become displaced outwards by natural processes or human agency, contributing variably to the lower part of sediments accumulating around it. It should not be forgotten that excavation below Conglomerate B has removed a depth of some 2 m of sediments containing abundant Mousterian and burnt animal bones, indicating yet another manner in which Neandertals left their signature in the cave.

Notwithstanding the foregoing conjecture, because there are no paleochemistry criteria for regarding the two radiocarbon dates OxA-10666 and OxA-15423 as being more than moderately too young, there are no clear reasons to consider the uppermost sediments as being substantially older than the confidence intervals of OxA-10666 and OxA-15423, taking into consideration the stratigraphic position of the two dates with regard to the earlier dates from the Upper Cutting and the large confidence intervals of the uranium-series and OSL estimates. Furthermore, dates of between 45 and 35 ka come from other Middle Paleolithic sites in southeastern Spain. Taken together, these associations and chronological assessments place the *in situ* Palomas Neandertal remains within the earlier MIS 3, and the scattered remains excavated in the uppermost sediments appear to be relatively late within the classic Middle Paleolithic.

The Palomas Middle Paleolithic Assemblage

As noted above, all of the diagnostic artifacts from the Sima de las Palomas can be included comfortably within the Middle Paleolithic technology. The full assemblage is summarized here in terms of raw material and general categories of implements.

Lithic Raw Materials

The majority of the retouched artifacts are side scrapers, and of those, most were flaked from chert, even though quartz was the most abundant raw material (Tables 2.2 and 2.3).

Much of the chert ranges in hue from white through light gray and bluish gray to dark gray. No chert outcrops exist on Cabezo Gordo, and it is therefore not local. The Sima de las Palomas has provided only 29 chert cores and rare flakes with cortex or patina; initial reduction of cores therefore largely took place elsewhere. The predominant rock of the Cabezo Gordo is a light gray marble containing frequent milky masses of calcite, with poorly developed crystalline structure when not amorphous and crisscrossed by metal veins (particularly of magnetite). In addition, schist beds just below the summit include small nodules of colorless or milky quartz. In the Sierra de Carrascoy, more than 20 km west, there are small, worn nodules of brittle, tabular chert (mainly of pale hues) that were formerly incorporated into Tertiary beds, ultimately deriving from the Jurassic mountains. The location of a place on the flank of the Sierra de Carrascoy near Corvera where chert, presumably of better quality, was taken for wooden threshing sledges is currently unknown.

A few items excavated at Palomas are of reddish brown and chocolate brown chert, recalling that of the La Crisoleja hilltop outcrop of hydrothermal chert about 25 km south, near La Unión, which also contains a gray blue chert similar to pieces from Palomas. A few excavated pieces of jasperlike chert are similar to raw material from the coastal spit La Manga del Mar Menor in the neighborhood of Punta de la Raja, Cerro del Calnegre, and Isla del Ciervo, about 25 km east of Palomas.

Pink quartzite cobbles occur in gravels at the foot of Cabezo Gordo, and some quartzite artifacts have been found at Palomas. It is likely that the small, smooth, subspherical cobbles of marble excavated at the site came from these gravels or from ones in the more distant Rambla de Albujón or other watercourses in the coastal plain (Campo de Cartagena) below Cabezo Gordo; some of these cobbles could have been hammerstones or pestles.

Because the innumerable excavated angular fragments of fractured marble seem mostly to be detritus from the hillside, only fragments that bore clear conchoidal fractures or adherent breccia were inventoried; additional marble fragments lacking ostensible signs of knapping could have been used but are not classified as artifacts. By contrast, there may be an overrepresentation of quartz, given its source near the summit of the hill, even though many pieces bear no signs of knapping. Calcite is ubiquitous in the hillside marble as both veins and masses with poorly developed crystalline structure when not amorphous, and many excavated pieces with possible, albeit often dubious, signs of knapping were inventoried. Some excavated nodules and fragments of a cream- or fawn-colored, marly limestone (or calcrete) are of unknown but possibly local origin; they often show concave fracture surfaces and occasionally signs of use or secondary knapping. The resultant distribution of catalogued artifacts by raw material is in Table 2.2.

The Palomas Artifacts

Of the 2,624 lithic items inventoried as of 2015, 1,448 were excavated in units 1, 2, and 3; 17 in conglomerate B (unit 4); and 1,080 in layers 5 and 6. A further 79 artifacts were recovered out of context. Whereas only 29 cores were identified, 1,478 items are fragments or knapping spalls, and 865 are flakes without retouch and mostly with plane striking platforms. An additional 252 items have retouch that consisted usually of simple abrupt or semiabrupt working (Plate 2.8; Table 2.4). Only rarely is there additional secondary knapping of the kind that gives rise to stepped or scalar retouch.

For the most part, retouch was limited to a single row of removals at 40–50° to the ventral (bulbar) surface of a flake, although steeper flaking

producing a more abruptly knapped edge is not uncommon. Only rarely was still further knapping applied that gave rise to a stepped effect. The majority of the classifiable retouched tools are side-scrapers (78.8%). They were knapped predominantly on chert (Table 2.3).

Denticulate, toothed, and notched artifacts with one or more notches of variable size are present at Palomas on chert and quartz and, to a lesser extent, calcite and quartzite. (Included here are keeled artifacts with deep or steep notches and edges converging in a tip or point, so-called Tayac points). Sima de las Palomas has also produced triangular points on flat flakes, both with and without neat secondary knapping along the edges of the flakes. Frequently a faceted striking platform of such flakes had been prepared by the Levallois technique before their removal from the original core, giving rise to symmetrical triangular Levallois points. Yet triangular points can also be produced by other techniques (so-called pseudo-Levallois points), and examples of both types occur at Palomas. Levallois cores have not been found at Palomas, and therefore the points were likely made elsewhere. In addition, several flakes and flake blades have simple striking platforms that are of very small size, suggesting indirect percussion for their removal.

At the Sima de las Palomas, the only examples of worked or utilized organic materials are a hippopotamus incisor with an artificial groove (found by sieving mine rubble) and part of a smooth bone spatulate artifact with a rounded end (excavated deep in layer 5 in 2011). The bone resembles Upper Paleolithic implements called lissoirs that have been identified in two Middle Palaeolithic sites (Soressi et al. 2013). Additional bone elements may have been used, but that has yet to be confirmed microscopically.

The Contexts of the Human Remains

All the Palomas Neandertal remains that have secure stratigraphic contexts were excavated in the Upper Cutting. The three partial skeletons, the Palomas 92 and 96 young adults and the

Palomas 97 juvenile, were excavated in the brecciated éboulis of Conglomerate A. Some large marble blocks in it weighed over 50 kg. It is unsurprising that the skeletons underwent some displacement from strict anatomical position during the processes of sedimentary consolidation and fossilization. They were variably crushed, distorted, and then cemented in brecciated sediment, which affected especially the pelvis and femora of Palomas 92 and the skulls of Palomas 96 and 97. Likewise, the Palomas 1 maxillae and mandible, discovered in 1991, had experienced a similar crushing and distortion. Because Palomas 1 was reported to have been found close to where the undiscovered head of Palomas 92 might have been, one of us (Walker) has referred to the fossils as SP92/SP1 for simplicity, whereas the other (Trinkaus) has pointed out that the Palomas 1 dental attrition implies an individual who died at an age older than that which is inferred from inspection of the Palomas 92 postcrania (chapter 13), so therefore SP92 and SP1 should belong to different individuals.

Scattered teeth and bone fragments of other Neandertals were excavated in the uppermost sediments that accumulated against the slope of Conglomerate A. Some of them (for example, Palomas 59) were embedded in a sedimentary matrix with precipitation of $CaCO_3$, and most of them are broken (other than phalanges and teeth). None of these isolated remains show the kind of crushing that is seen in the skeletons from Conglomerate A, although their surrounding sediments did not contain large stones or rocks. The isolated human remains from these sediments therefore experienced a different taphonomic history from that of the three associated partial skeletons, reflecting differences both in site formation processes and undoubtedly in use of the site by Neandertals.

One of us has inferred based on anatomical positions and contexts (Walker et al. 2012) that the skeletons entered the cave as intentional burials, whereas the isolated elements most likely came from decomposed cadavers or disarticulated skeletons accumulated in the cave. Similar combinations of articulated and disassociated human remains are known from other Middle Paleolithic sites (for example, Amud, La Chapelle-aux-Saints, Kebara, La Quina, Shanidar, Skhul, and Tabun) as well as Upper Paleolithic ones (see discussion in Trinkaus et al. 2014a), albeit not necessarily from the same stratigraphic levels.

In support of the notion that intentional arrangement of cadavers had taken place, and that they were covered with stones (perhaps to deter scavengers), are the following observations: First, Palomas 96, 97, and 92 lay immediately on top of each other, head uppermost to the west, feet downwards to the east (a position observed at La Ferrassie and other sites), and many of their skeletal remains were found in appropriate articulated anatomical relations, subsequent crushing and distortion notwithstanding. Second, lying on their sides, with knees and elbows flexed, Palomas 96 and 97 had both hands raised against the face, as would happen were the bodies placed with the elbows flexed and the arms against the sides; similar positions have been noted for several Neandertals buried on their sides (Defleur 1993). Third, whereas the site has provided remains of scavengers (leopard, hyena, porcupine), the retention of skeletal order shows that the cadavers escaped scavenging despite the presence of large carnivore remains in close proximity; there were two leopard (*Panthera pardus*) metacarpal bones beside the skull of Palomas 97 and two leopard paws in anatomical connection slightly further away. In addition, beside Palomas 97 were the calcanei, astragali, and cuboid bones of two *Equus ferus*; one group was extensively burned and cemented to the skull of Palomas 97, whereas the other set, unburnt and including the distal tibia, was immediately behind the juvenile's trunk. Fourth, close to Palomas 92 there were excavated, in apparently undisturbed sediment, 9 retouched Middle Paleolithic artifacts, 12 unretouched flakes, and more than 100 knapping spalls and fragments of flint, calcite, and quartz.

Numerous human remains were found when sieving the rubble left behind by miners in the Sima de las Palomas tunnel and on the hillside around its entrance. It would be tempting to ascribe all those human remains to an origin in the Upper Cutting levels, being mainly fragments comparable to those from the uppermost sediments of the Upper Cutting, were it not for the fact that, unlike human remains from those sediments, several show signs of burning: namely, the Palomas 2, 11, 12, 30, and 62 cranial fragments; the Palomas 6 and 23 mandibles; and the Palomas 9 and 17 postcrania. The burnt pieces may have come from sediments below Conglomerate B, which has provided many burnt animal bones and two human teeth to date. Alternatively, together with other *ex situ* elements they could have been altered on the hillside above the shaft and subsequently washed into the deposits, later to be scattered on the hillside by the miners.

Given their excavated contexts in the Upper Cutting, Palomas 1, 92, 96, and 97 date from somewhere between ≈60 and ≈45 ka cal BP, possibly from ≈50 ka cal BP. The isolated remains from the uppermost sediments date from between ≈45 and ≈38 ka cal BP, allowing for a possible modest underestimation of the ages of the radiocarbon samples. The materials from the hillside rubble, if mostly from the sediments below Conglomerate B, should date from between ≈65 and ≈55 ka cal BP; otherwise, they could date from anywhere from the late Middle Pleistocene to the middle of MIS 3.

Summary

The fieldwork at the Sima de las Palomas, from the initial discovery to the cleaning of the site and collection of disturbed remains and to the excavation of the Upper Cutting, has given us a considerable number of human fossils, Late Pleistocene faunal remains, and Middle Paleolithic artifacts. The human remains consist of three samples: those found out of context in mine rubble in the main chamber and mine tunnel and around the tunnel's entrance on the hillside; those from the Conglomerate A breccia; and those from the sediments that accumulated subsequently around Conglomerate A. Stratigraphic relationships between the first sample and the other two samples will remain unknown until reliable nondestructive dating techniques can be applied directly to the human remains. The second sample is stratigraphically older than the third one, but the time gap between them remains unclear, given the large analytical errors associated especially with the uranium-series and OSL dates. Two of the associated skeletons from the second sample, Palomas 96 and 97, are not of direct concern here, although we include limited data about them. The mandibular and dental remains of Palomas 1 and the postcrania of Palomas 92 are described and compared along with the other Palomas remains. Because the comparative sample of Neandertals from western Eurasia spans far more time than the Palomas material, the age range at Palomas should have little effect on what follows, namely, morphological and paleobiological assessments of these Neandertal remains from southeastern Iberia.

CLASS	ORDER	FAMILY	GENUS, SPECIES
Mammalia	Carnivora	Felidae	*Panthera pardus* (leopard)
			Felis (*Lynx*) cf. *lynx* (lynx)
			Felis cf. *sylvestris* (wild cat)
		Hyaenidae	*Crocuta crocuta spelaea* (cave hyena)
		Canidae	*Canis* cf. *lupus* (wolf)
			Vulpes vulpes (fox)
		Mustelidae	*Meles* cf. *meles* (badger)
	Perissodactyla	Equidae	*Equus ferus* (horse)
		Rhinocerotidae	*Stephanorhinus* sp. (rhinoceros)
	Artiodactyla	Hippopotamidae	*Hippopotamus amphibius* (hippopotamus)
		Bovidae	*Bos* cf. *primigenius* (aurochs)
			Capra sp. (*C. ibex*; *C. pyrenaica*)
			Cervus elaphus (red deer)
			Dama sp. (fallow deer)
	Rodentia	Hystricidae	*Hystrix javanica* (porcupine)
		Leporidae	*Oryctolagus cuniculus* (European rabbit)
			indet. (possibly *Lepus* hare)
	Insectivora	Erinaceidae	*Erinaceus* sp. (hedgehog)
	Chiroptera	Vespertillionidae	*Myotis* sp. (bat)
Reptilia	Testudinida	Testudinidae	*Testudo hermanni* (Mediterranean tortoise)
	Squamata	Lacertidae	indet. (lizard)
Aves*	Falconiformes	Falconidae	*Falco tinnunculus*[b] (Eurasian kestrel)
			Falco naumanni[b] (lesser kestrel)
	(Galliformes)	Phasianidae	*Alectoris rufa*[b] (red-legged partridge)
	Columbiformes	Columbidae	*Columba livia*[a] (rock dove/rock pigeon)
	Strigiformes	Strigidae	*Athene noctua*[b] (little owl)
	Passeriformes	Alaudidae	*Galerida cristata/theklae*[b] (crested/thekla's lark)
		Turdidae	*Saxicola torquata*[b] (African stonechat))
			Monticola solitarius[b] (blue rock-thrush)
		Corvidae	*Pyrrhocorax graculus*[c] (yellow-billed chough)
			Pyrrhocorax pyrrhocorax[a, c] (red-billed chough)
			Corvus corone[b] (carrion crow)
		Passeridae	*Passer domesticus*[b] (house sparrow)
		Emberizidae	*Emberiza sp.*[c] (bunting)

Note: Most of the extinct species are of fossils excavated in the Upper Cutting, although the list also includes some known only from the mine rubble (for example, *Crocuta crocuta spelaea*, *Hippopotamus amphibious*, and *Stephanorhinus*).
*Although some bones of *C. livia* were excavated in the uppermost sediments of the Upper Cutting, it is likely that most of the bird bones shown in the table are modern contaminants; bird species marked [a] come from the Upper Cutting, those marked [b] come from the Lower Cutting, and those marked [c] come from mine rubble sieved on the hillside.

TABLE 2.2. Raw material distribution for the Sima de las Palomas flaked stone artifacts

RAW MATERIAL	N	%
quartz	1,234	47.0
chert	886	33.8
marble	10	0.4
quartzite	32	1.2
calcite and limestone	419	16.0
uncertain owing to breccia coating	43	1.6
TOTAL	2,624	100.0

TABLE 2.3. Lithic raw material distribution for the side-scrapers from Sima de las Palomas

RAW MATERIAL	N	%
chert	61	74.4
calcite and limestone	7	8.5
quartz	8	9.8
quartzite	5	6.1
marble	1	1.2
TOTAL	82	100.0

TABLE 2.4. Distribution of retouched lithic artifacts from the Sima de las Palomas

ARTIFACT TYPE	N	%
gravers/burins	3	1.2
denticulate/toothed/notched pieces	3	1.2
slugs/limaces	2	0.8
"Levallois" and "pseudo-Levallois" triangular flat points	5	2.0
Mousterian points	3	1.2
side-scrapers	82	32.6
steep scrapers	6	2.4
flakes/pieces with some secondary knapping/retouch	148	58.7
TOTAL	252	100.0

The Palomas Neandertal Sample 3

ERIK TRINKAUS

THE SAMPLE OF NEANDERTAL remains from the Sima de las Palomas consists of pieces of neurocranium, fragmentary maxillary alveoli adherent to teeth, partial mandibles providing variable amounts of information, a long series of isolated teeth plus dental remains in the maxillary and mandibular alveoli, a number of isolated and mostly partial postcranial elements, and an associated partial postcranial skeleton (Palomas 92). The sample thus includes 13 cranial vault pieces (1 immature), 8 partial mandibles (4 immature), 17 isolated postcranial elements (5 immature), 96 teeth (36 in maxillae or mandibles, 15 deciduous, and 5 partially formed), and 14 partial dental roots (11 in mandibles). In addition, the Palomas 92 partial skeleton retains 70 bones, of which 42 derive from the hands and left foot. Therefore, the Palomas Neandertal sample consists of 218 bones and teeth. Two additional associated partial skeletons, Palomas 96 and 97 (Figs. 3.1 and 3.2), are not directly considered here (see Walker et al. 2011b, 2012). However, the available data for them are included as appropriate in the respective descriptive and comparative chapters.

The preservation, ages at death (or general maturity), descriptive morphology, and morphometrics are presented in the following chapters, organized anatomically. The craniomandibular and postcranial remains are largely considered in individual chapters by anatomical region. The dental remains, given their abundance in the Palomas sample and the diversity of approaches to their morphology and paleobiology, are presented

in chapter 6 and then compared in that chapter and subsequent ones.

Inventories of the remains with basic contextual information are provided with each of the primary descriptive chapters. The specimen numbers are based on the dates of discovery of

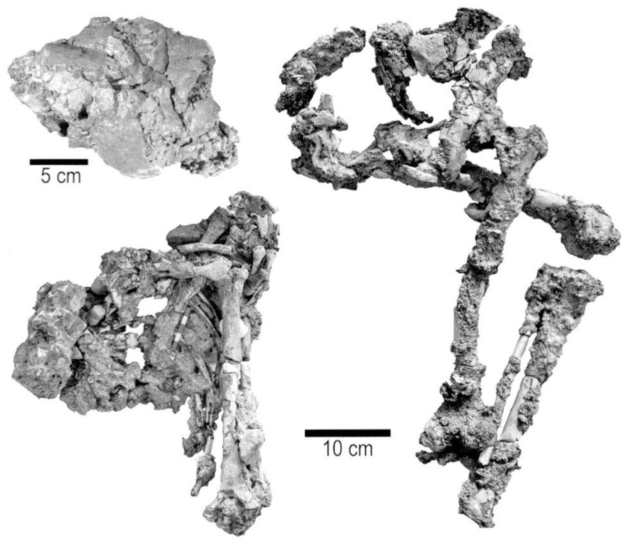

FIG. 3.1. The more complete sections preserved for the Palomas 96 young adult partial skeleton. *Upper left*, the cranium and mandible in right lateral view. *Lower left*, the right superior thorax, shoulder, arm, and hands in dorsal view, with the elbow hyperflexed and the hand remains below the lateral clavicle and scapula and the proximal humerus. *Right*, the pelvis, both femora, and the left tibia and fibula. The left arm and hand remains are preserved separately, as are the sacrum, the caudal cervical vertebrae, and the more cranial thoracic vertebrae (cf. Walker et al. 2011b).

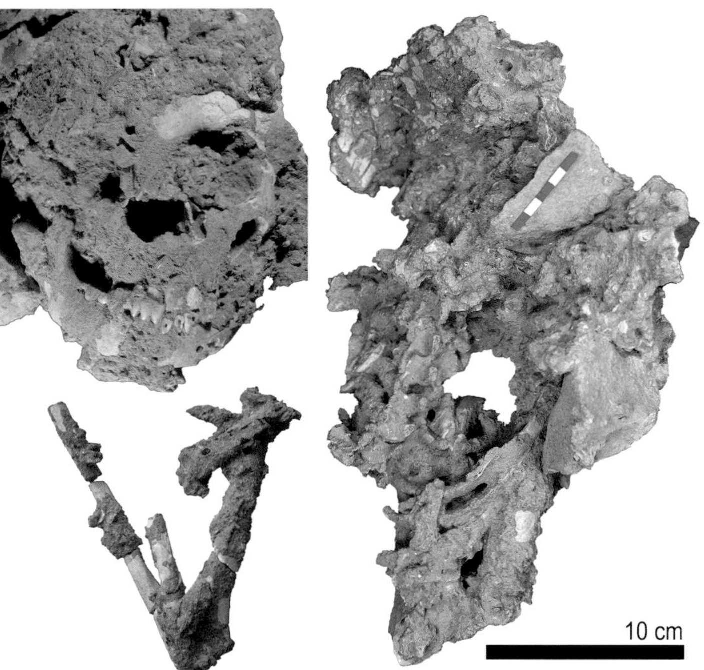

FIG. 3.2. The more complete portions of the Palomas 97 juvenile partial skeleton. *Upper left*, anterior view of the craniofacial block. In addition to the facial portions exposed, it retains most of the neurocranium (crushed posteriorly), the complete developing dentition *in situ*, and portions of both hand skeletons. *Lower left*, the left humerus, ulna, and radius. *Right*, the thoracic to crural block, retaining 11 vertebrae, ≥6 ribs, both proximal femora, and portions of the pelvis.

the specimens, whether *ex situ* or *in situ*, and they are therefore not sequential within each of the chapter inventories. The inventory tables derive from that presented in Walker et al. 2008, with updates for some of the specimens, especially for additional elements of Palomas 92 that have emerged from its encasing breccia and reidentification of some teeth following reassessment and virtual cleaning. The Palomas 96 and 97 partial skeletons (Walker et al. 2011b, 2012) are only occasionally included in these considerations, since they are still emerging from breccia and limited morphological aspects of them are available for comparative analysis (Walker et al. 2011b). Data for Palomas 96 are included particularly as they relate to aspects preserved in Palomas 92 and the isolated pieces.

The Palomas 92 partial postcrania represent one individual (as do each of Palomas 96 and 97). However, it is not possible to associate the other remains. There are a couple of apparent pairs of antimeric teeth, although additional pairs may exist in the sample. The degree of breakage of most of the isolated skeletal elements prevents articular or bilateral association. The issue is further complicated by the stratigraphic separation of some of the elements found *in situ* and by the lack of stratigraphic context for those found *ex situ* on the hillside or at the base of the pit. The remains are therefore considered as individual specimens, as is reflected in the separate Palomas (SP) numbers, bearing in mind that the original number of individuals was undoubtedly smaller than the 100 numbered specimens.

The Palomas Sample as Neandertals

Late Pleistocene human remains from the Sima de las Palomas have been referred since the initial discoveries in 1991 to the Neandertal sample. The site has only yielded Middle Paleolithic artifacts *in situ*, and it is near the end of the southwestern cul-de-sac of Europe. It is therefore to be expected that the individuals whose remains have survived should be referred to the Neandertals, given that Neandertals are the only humans known to have been associated with the Middle Paleolithic in Europe and that other Iberian Middle Paleolithic humans are generally considered to be Neandertals.

There is little question that the Palomas humans were nonmodern, late archaic humans. This is evident in a variety of aspects of their skeletal remains, including supraorbital tori, mandibular symphyses, maxillary incisors, terminal manual phalanges, and femoral diaphyseal shape. These aspects and others have been considered previously (Walker et al. 2008, 2010, 2011a, 2011b) and are detailed here. What has remained to be assessed is the extent to which the Palomas human fossils have their affinities specifically with the other fossils referred to the Neandertals *sensu stricto*, as opposed to being merely late archaic humans. In other words, to what extent do they exhibit clear, uniquely derived features of the Neandertals, many of which are found occasionally elsewhere (Trinkaus 2006a; Wu et al. 2014)?

The Palomas humans will be nonetheless referred to as Neandertals throughout this volume. This appellation is appropriate given their place in time and space and previous considerations of their morphology. Yet whether the Palomas sample conforms entirely to the Neandertal pattern needs to be reconsidered in light of the distribution of morphological configurations in the sample (see chapter 16).

The Comparative Framework

Assessment of the Palomas human remains, in terms of morphology and paleobiology, requires placing them in an appropriate human paleontological context. The obvious comparative sample is the one of western Eurasian late archaic humans, the Neandertals *sensu stricto*. The majority of these remains date to the early last glacial, MIS 4 and MIS 3b, but there are occasional ones that derive from MIS 5. Almost all of them are associated with the Middle Paleolithic, but there are a few from initial Upper Paleolithic contexts in western Europe. Included with them is the initial Late Pleistocene (MIS 6/5) large sample from Krapina as well as the associated skeleton (Tabun 1) and several isolated remains from Tabun (the Tabun 2 mandible is likely earlier and is included in the Middle Pleistocene sample). In most of the comparisons, these Neandertals are pooled together; given differential sample sizes for a couple of the comparisons, the Krapina specimens are identified separately, but they are nonetheless considered as part of the larger Neandertal sample. The sample is limited geographically to Eurasian remains from central Asia to the Atlantic coast. Specimens from further east in Asia that have been referred to the Neandertal sample based on ancient DNA sequences are included separately if they provide relevant morphological data.

In order to bracket this Neandertal sample, data are included, as available, for western Eurasian Middle Pleistocene fossils and for early modern humans. The Middle Pleistocene sample derives principally from central and western Europe, and it is dominated by the large sample from the Sima de los Huesos at Atapuerca (Atapuerca-SH). These remains are generally considered to represent Neandertal ancestors within western Eurasia (Stefan and Trinkaus 1998; Arsuaga et al. 2014), and they are included within that framework. In a few of the comparisons, in which only a few late Middle Pleistocene specimens provide comparative data, they are included with the Late Pleistocene Neandertals.

The early modern human samples consist of an MIS 5 southwest Asian sample from the sites of Qafzeh and Skhul and an earlier Upper Paleolithic sample from western Eurasia. The former is Middle Paleolithic in context, and it is therefore referred to as the Middle Paleolithic modern humans (MPMH). The latter sample includes remains from the Early Upper Paleolithic (EUP) and especially Mid Upper Paleolithic (MUP), all within MIS 3. It is variably referred to as earlier Upper Paleolithic or Early/Mid Upper Paleolithic and abbreviated as E/MUP. In a few cases for which E/MUP sample sizes are very small, data from through the Upper Paleolithic (Early, Mid, and Late Upper Paleolithic) are pooled; such cases are indicated as such.

In addition, to provide a general reference, data are occasionally included for samples (separate or pooled) of late Holocene ("recent") humans. They are not intended to be exhaustive, but in some cases they provide a context with respect to levels and patterns of variation for which the paleontological samples are inadequate.

A consensus has not been reached (and may never be) regarding the number and appropriate designations of formal taxonomic categories within the genus *Homo*. As the human fossil record improves through the Pleistocene and across the Old World, it is increasingly difficult to designate morphological boundaries between groups, the primary one that appears to be consistent being between archaic and modern humans. For these reasons, formal taxonomic species designations are not employed.

The data employed for the quantitative comparisons (morphometric and discrete traits) derive principally from the authors' analyses of the original remains and/or high-resolution replicas (principally of dental crowns). These observations have been supplemented by data from published descriptions of the fossils. Additional comparative data, especially for Upper Paleolithic and Middle Pleistocene humans, have been made available by S. E. Bailey, S. E. Churchill, T. W. Holliday, B. Holt, C. Lorenzo and F. H. Smith.

Quantitative Assessment

The description and evaluation of the Palomas human remains employ a variety of approaches, including verbal description, standard osteometrics (Bräuer 1988), cross-sectional geometry, dental tissue proportions, and morphological, attritional, and paleopathological ordinal scales. Many of these were scored on the fossils or measured with digital calipers. Cross-sectional subperiosteal contours were either transferred for digitizing with polysiloxane molding putty or used scaled photographs of fossilization breaks. Internal configurations of the teeth and many of the smaller skeletal elements were accessed through micro-CT scanning. The specimens were CT scanned using a Skyscan 1172 system located at the University of Bristol and transported to Murcia by Katharine Robson Brown and Priscilla Bayle for the Palomas fossils. It was set at 100 kV and 100 µA with an Al/Cu filter to provide images with an isometric voxel resolution of 35 µm. Analysis of the resulting reconstructed image files was achieved using CTAn (v. 1.9.2.5, Skyscan, Kontich, Belgium) and Avizo (version 6) software. To the extents necessary, the details of the techniques as applied to aspects of the remains are provided in the individual chapters.

Issues of Maturity

The Palomas Neandertal sample contains a substantial number of remains of both immature and mature individuals. The age-at-death range of the sample extends from infancy to middle adulthood (see especially chapters 5, 6, and 13). While there are a number of mature specimens, most explicitly mandibles with M_3s in occlusion, obviously immature skeletal specimens are present among the cranial fragments (Palomas 5), the mandibles (Palomas 7, 49, 80, and 88), and the postcrania (Palomas 8, 14, 32, 66, and 86). Immature remains are, however, best represented by the isolated teeth and the mandibles. Of the 57 isolated teeth and mandibles for which an age at death can be approximated, 80.7% are immature (Table 15.1). If the three associated skeletons, Palomas 92, 96, and 97 are added to them (the first two are mature; the third is juvenile), the sample increases to 60, of which 78.3% are immature. Counting truly minimal MNIs (minimum numbers of individuals), not taking into account details of antimeric morphology or detailed levels of calcification, 64.2% of the 14 MNIs are immature (58.8% with the three partial skeletons).

In addition, older adults appear to be absent from the sample. The only specimen that shows moderately advanced dental wear is Palomas 1; its molar wear scores range from 4a to 5c (see chapter 6), which indicate substantial occlusal enamel despite variable amounts of exposed dentin (Smith 1984). In comparison to late archaic and earlier Upper Paleolithic humans, for whom nondental age indicators are preserved, this degree of postcanine occlusal wear would indicate an age at death in the fourth or possibly the fifth decade (Trinkaus 1995, 2011a; Hillson et al. 2006; Trinkaus et al. 2014a). The Palomas dental sample (including teeth in alveoli) therefore reflects a young age distribution—from infants to prime-age adults.

Among the postcrania, excluding the obviously immature specimens, one can ask to what extent they derive from fully mature (third decade or older) adults as opposed to a mix of adolescents and adults. The lack of ventral body fusion in the Palomas 92 sacrum (discussed in chapter 13) indicates an early-third-decade age at death for that individual, as it does for Palomas 96 (Walker et al. 2011b). For the other postcrania, the primary criterion is epiphyseal fusion. For several of the

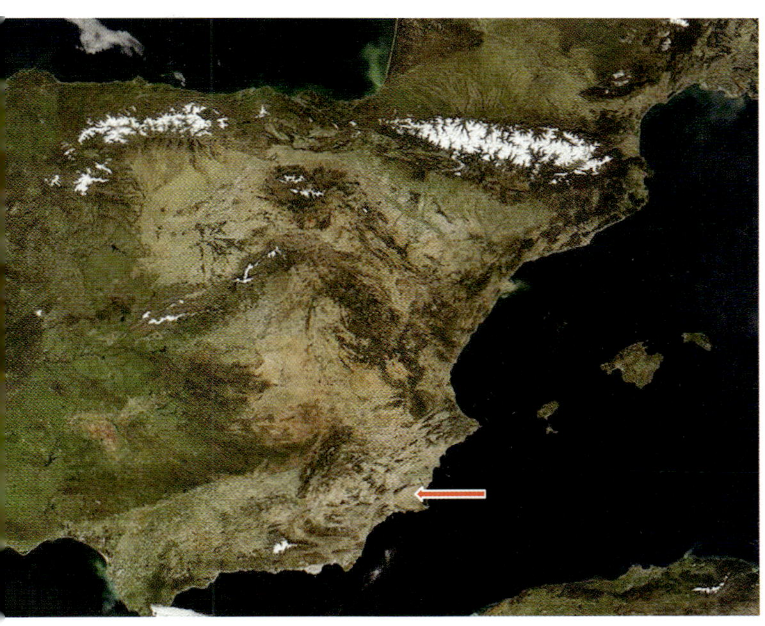

PLATE 2.1. Satellite image of Iberia, with the position of Cabezo Gordo and the Sima de las Palomas indicated by the arrow in southeastern Iberia. (From NASA/Jeff Schmaltz, LANCE/EOSDIS MODIS Rapid Response Team at NASA GSFC, http://commons.wikimedia.org/wiki/File:Iberia_Europe_satfoto_2014067.jpg)

PLATE 2.2. Satellite views of the southeastern Spain coastal plain, with Cabezo Gordo (*above*) and the Sima de las Palomas (*below*). (Image © Google Earth / Landsat / US Geological Survey)

PLATE 2.3. View of the Cabezo Gordo from the south, with the location of the Sima de las Palomas indicated.

PLATE 2.4. Entrances to the Sima de las Palomas, including the entrance to the main shaft (*left arrow*) and the one to the cleft through which the nineteenth-century miners initially entered the cave (*right arrow*).

PLATE 2.5. The main chamber shaft of the Sima de las Palomas with the scaffolding erected in 1994 to permit excavation of the Upper Cutting.

PLATE 2.6. Views of the excavations of the Upper Cutting. *A*, the entrance above; *B* and *C*, excavation from the top of the scaffolding; *D*, the surface of Conglomerate B after cleaning.

PLATE 2.7. Excavation of Conglomerate A in the Sima de las Palomas.

PLATE 2.8. Select Middle Paleolithic artifacts from the Sima de las Palomas.

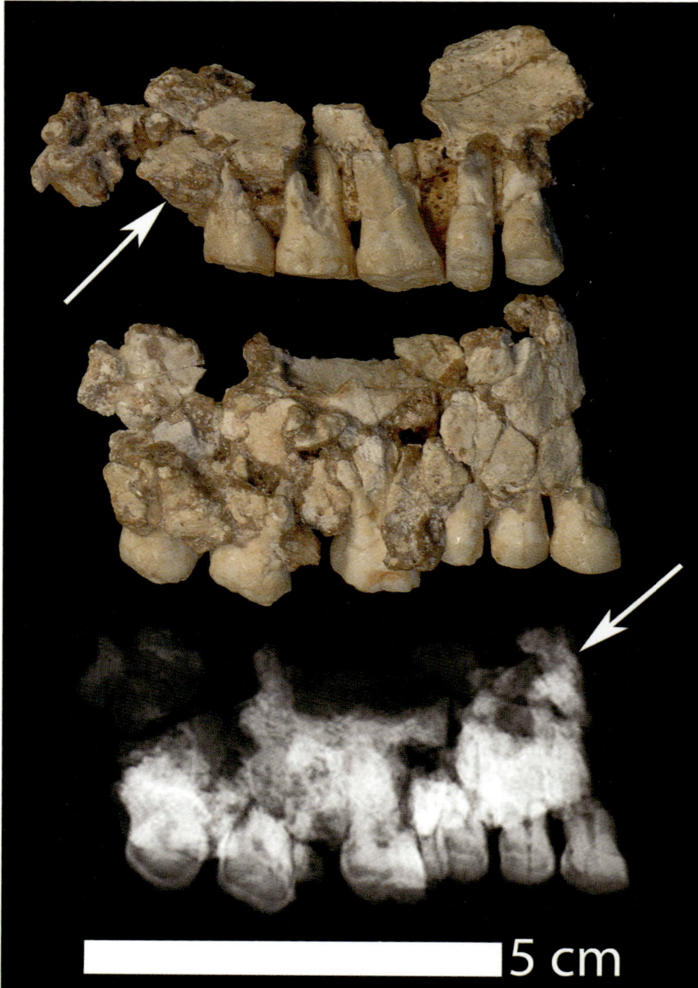

PLATE 4.1. Medial view of the Palomas 1 left maxilla (*above*), lateral view of the right maxilla (*middle*), and lateral-medial radiograph of the right maxilla (below). The arrows point to the retromolar alveolar bone (*left above*) and the radiodensity inferred to be the right nasal floor (*right below*). See also views in Plate 5.1.

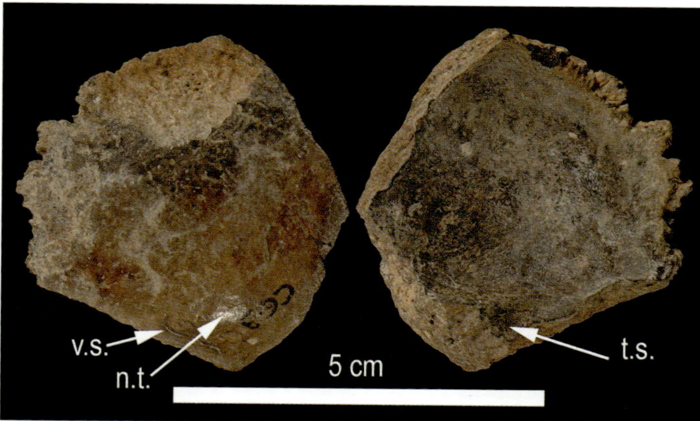

PLATE 4.2. External (*left*) and internal (*right*) views of the Palomas 2 left occipital bone, with arrows pointing to t.s.: the transverse sinus sulcus; n.t.: a probable portion of a small nuchal torus; v.s.: vascular sulcus.

PLATE 4.3. The Palomas 3 right parietal piece, in external (*left*) and internal (*right*) views, showing st: stephanion; c.s.: coronal suture.

PLATE 4.4. Exterior (*left*) and interior (*right*) views of the Palomas 4 left posteromedial parietal bone. Double arrow shows the position of lambda.

PLATE 4.5. Exocranial (*left or above*) and endocranial (*right or below*) views of the Palomas 5 anterior, probably right parietal bone; the Palomas 10 sutural bone; and the Palomas 56 parietal piece.

PLATE 4.6. The Palomas 11 anterolateral right frontal bone in superior (*left*), anterior (*above right*), and orbital/endocranial (*below right*) views.

PLATE 4.7. Anterior (*above*), medial (*left*), and lateral (*right*) views of the Palomas 12 left supraorbital torus piece.

PLATE 4.8. Exocranial (*above*) and endocranial (*below*) views of the Palomas 30 right posteroinferior parietal bone; ≈ast.: approximate position of asterion; t.s.: transverse sinus sulcus.

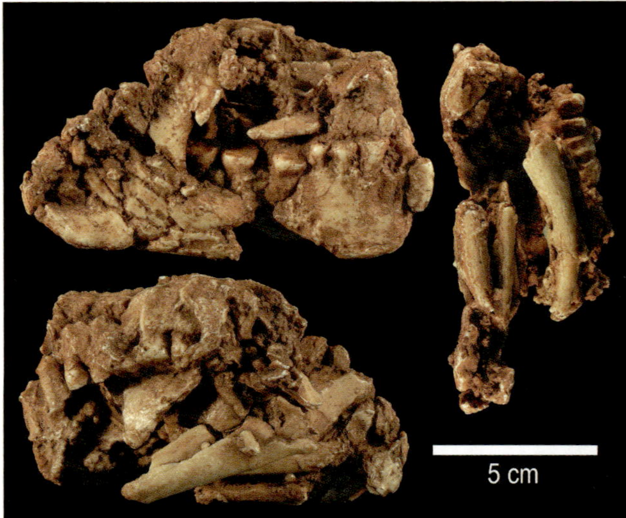

PLATE 5.1. Views of a cast of the Palomas 1 maxilla, mandible, and dentition as discovered. *Upper left*: right lateral view; *lower left*: left inferolateral view; *right*: inferior view.

PLATE 5.3. Lateral (*above*) and medial (*below*) views of the Palomas 1 left mandibular corpus.

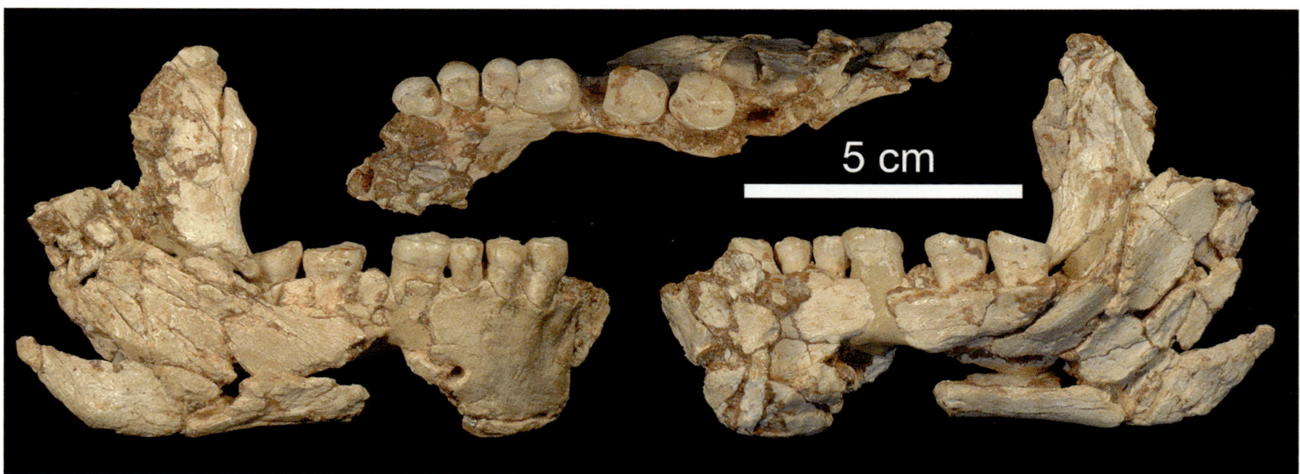

PLATE 5.2. Lateral (*left*), medial (*right*), and superior (*above*) views of the Palomas 1 right mandibular corpus and ramus.

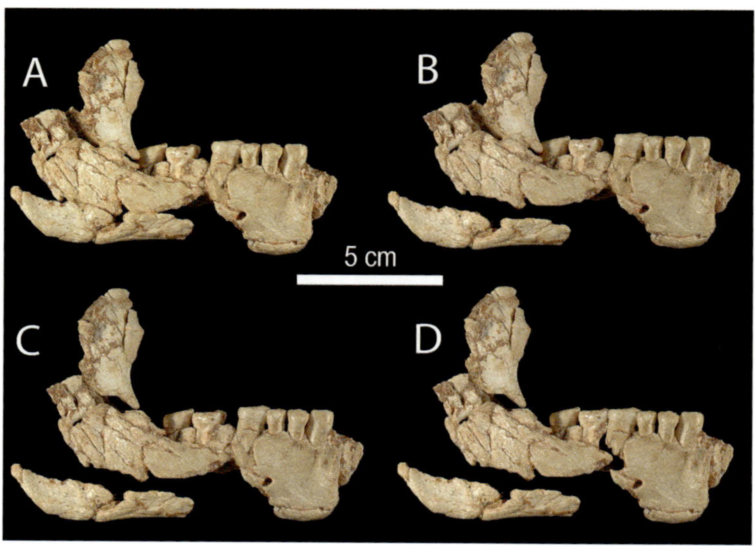

PLATE 5.4. Image sequence showing the photographic repositioning of portions of the Palomas 1 right mandible. *A*, Lateral view of the right mandibular piece as currently conserved; see also Plates 5.2 and 5.3. *B*, Posteroinferior repositioning of the posteroinferior corpus and inferior ramus using the corpus height derived from the left corpus. *C*, Posterosuperior displacement of the coronoid process, using the preserved contour posterior to the M3 and the similarly indicated anteroinferior ramus margin preserved on the left side. *D*, Closure of the gap between the right M1 and M2.

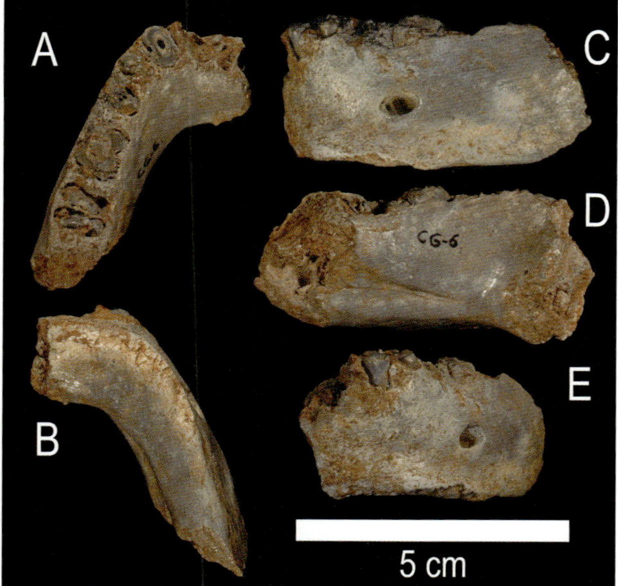

PLATE 5.5. Views of the Palomas 6 left mandibular corpus. *A*, superior (occlusal); *B*, inferior; *C*, anterolateral; *D*, posteromedial; *E*, anterior. The C and D views are in the plane of the lateral corpus.

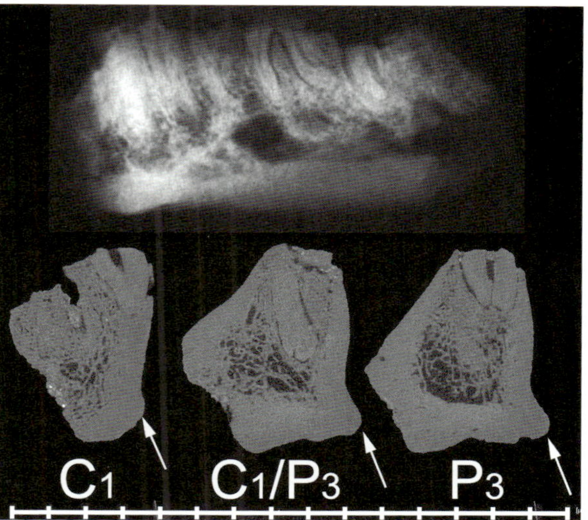

PLATE 5.6. Lateromedial radiograph of the Palomas 6 mandibular corpus (*above*) and micro-CT slices of the corpus at the mid-canine cervix (C1), the C1/P3 septum at the alveolar margin, and the mid-P3 cervix (P3) (*below*). Lateral is to the right in the micro-CT slices. The arrows point to the inferolateral protrusion of the corpus.

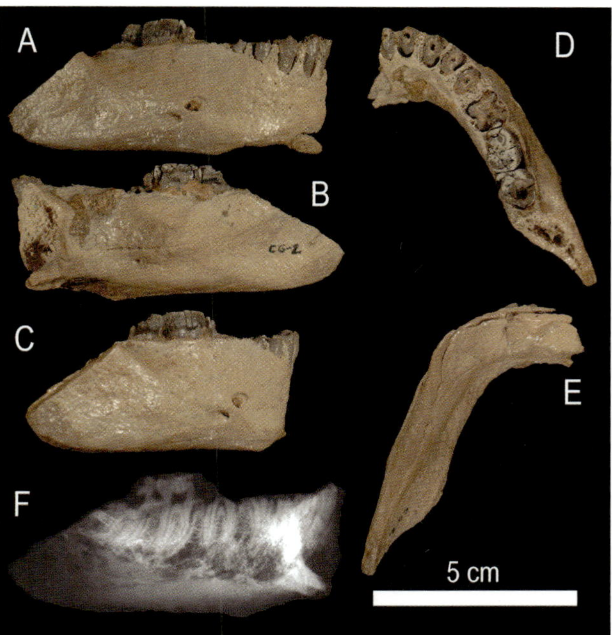

PLATE 5.7. Views of the Palomas 23 right mandibular corpus. *A* and *B*, anterolateral and posteromedial views in the plane of the lateral corpus; *C*, lateral view in *norma lateralis*; *D* and *E*, occlusal and inferior views; *F*, lateral-medial radiograph in the plane of the lateral corpus.

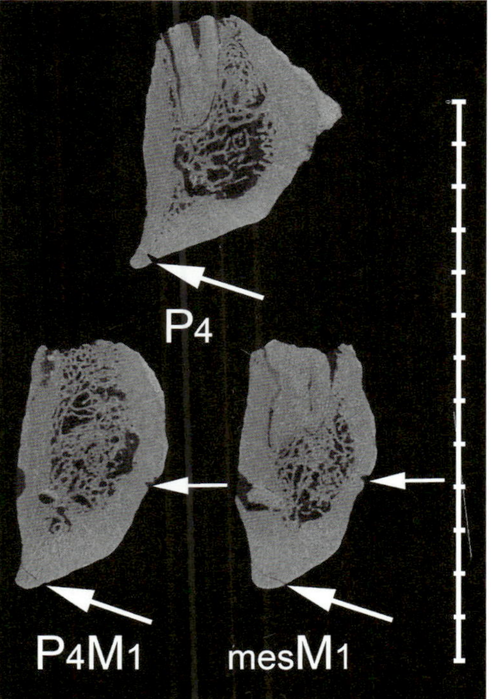

PLATE 5.8. Coronal plane micro-CT slices through the Palomas 23 mandibular corpus at the levels of the middle of the P4, the P4/M1 interdental septum, and the mesial roots of the M1. Lateral is to the left. The separation lines between the primary inferior corpus cortical bone and the inferior edge are indicated by the larger oblique arrows. The small surface crack on the medial side is indicated by the smaller horizontal arrows. The mesial end of the primary mental foramen is evident in the P4/M1 slice, and the double nature of the mental foramen is evident in the mesial M1 slice (see also Plate 5.7A).

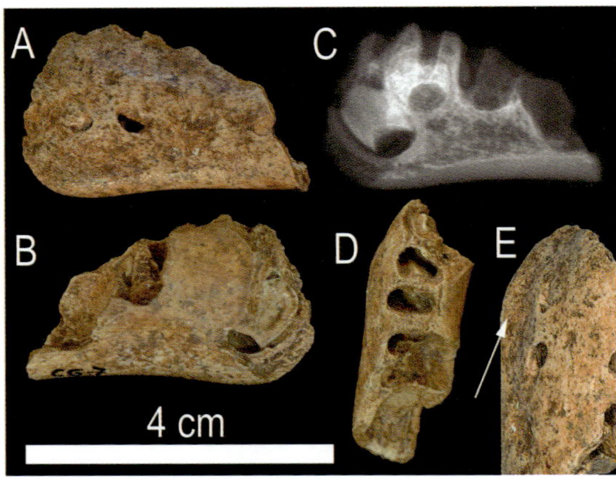

PLATE 5.9. Views of the Palomas 7 immature mandible. A, lateral; B, medial; C, lateral-medial radiograph; and D, occlusal. In E, the superolateral view of the anterolateral corpus (enlarged relative to the other views), the arrow indicates the inferolateral projection of the corpus (anterior marginal tubercle); the detailed image is intentionally cropped.

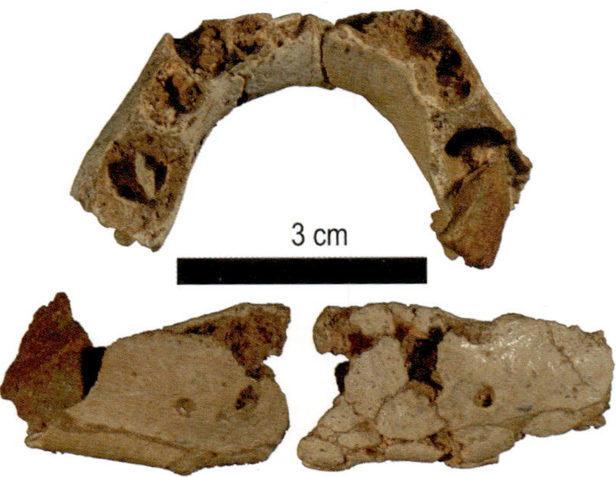

PLATE 5.10. Superior/occlusal (*above*), right lateral (*below left*), and left lateral (*below right*) views of the Palomas 49 immature mandible. For the superior view, the two halves of the corpus are positioned along the symphyseal break using the lingual contour. A piece of faunal bone is adherent to the right dm2 alveolus.

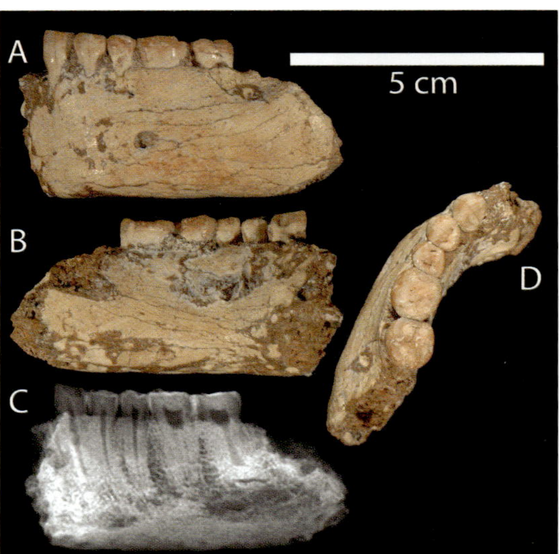

PLATE 5.11. Views of the Palomas 59 left mandibular corpus. A and B, anterolateral and posteromedial views in the plane of the lateral corpus; C, lateral-medial radiograph in the plane of the lateral corpus; D, occlusal view.

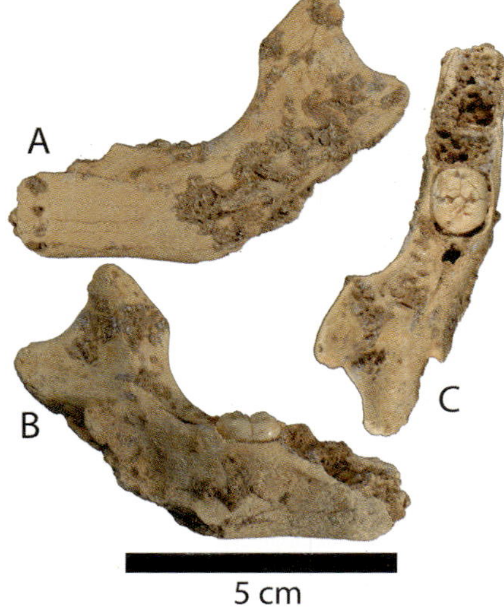

Plate 5.12. Views of the Palomas 80 immature left mandibular corpus and ramus. A and B, anterolateral and posteromedial views in the plane of the lateral corpus; C, occlusal/superior view.

elements, the metaphyseal regions are absent, and others normally fuse in midadolescence (Scheuer et al. 2000). These bones are treated as though they are fully mature, but one can none-theless query whether further growth might have taken place had they all lived to full adulthood.

As a result of this mortality distribution of the Palomas sample, clearly immature elements are compared to the (limited) available Pleistocene immature remains, which is most possible for the mandible. The other bones are assessed relative to the available adult remains, bearing in mind that any differences related to skeletal hypertro-phy might be reflecting ages at death as well as differential loading regimes.

Some Notes on the Human Inventories

In 2006, when all of the human remains were brought together in Murcia, it was decided to inventory the sample again and to renumber the specimens according to their dates of discovery. Given the presence of other karstic sinkholes with Pleistocene remains in the Cabezo Gordo (for example, the nearby Sima des las Arañas, which has yielded a Neandertal incisor), it was decided no longer to use the Cabezo Gordo (CG) designation for the fossils but to refer to them as Palomas, abbreviated as SP, with the appro-priate number (see Walker et al. 2008 and the following primarily descriptive chapters). This is the numbering system employed here, although many of the specimens found prior to 2006 are labeled with CG numbers (evident in the photo-graphs). The CG numbers, as well as field exca-vation numbers for more recently found remains, are provided in the inventory tables. It should be mentioned that the original CG inventories con-tain some elements subsequently determined to be nonhuman or too incomplete to assess; they have been deleted from the Palomas Neandertal sample and are not considered further.

After the Palomas 1 facial remains were discov-ered, they were taken to Sabadell, to the paleonto-logical laboratory of Josep Gibert, for separation, cleaning, and partial reassembly. In the process, a few of the anterior teeth became separated from the remainder of the specimen. Three of these teeth were inventoried separately as Palomas 20 to 22, but they most probably derive from Palo-mas 1, through comparison to a cast of the spec-imen as originally discovered (Plate 5.1). These teeth are numbered and described separately, but they are included with the other Palomas 1 teeth, particularly for the assessment of dental propor-tions along the arcade (see Fig. 7.6).

During the early years of work at Palomas, and especially with the collection of materials from the miners' debris, the paleontological remains were taken to Gibert's laboratory in Sabadell. Gibert carried most of those fossils to Murcia in 2006, but after his untimely death in 2007, sev-eral of the elements still in his laboratory could not be located. Little information is available for the missing skeletal pieces, but measurements are available for most of the missing teeth.

4   The Palomas Cranial Remains

ERIK TRINKAUS

GIVEN THE ABUNDANCE of individual elements, and especially teeth, from the Sima de las Palomas, and the distinctiveness of human cranial remains, Neandertal cranial remains are surprisingly rare and fragmentary. There are the largely complete but extensively damaged crania in the Palomas 96 and 97 associated skeletons (not considered here; see Walker et al. 2011b, 2012), but the other cranial material consists of the maxillary alveolar fragments with the Palomas 1 mandible and dental arcades, ten pieces of neurocranium, and fragments of alveoli adherent to isolated teeth (see chapter 6). The first was found *in situ*, but the remainder of the neurocranial pieces were recognized from the material scattered by miners on the hillside (Table 4.1). Only the alveolar fragments on the Palomas 51 and 68 maxillary teeth have more precise stratigraphic positions. This dearth of cranial remains is all the more surprising given their overabundance among isolated Neandertal remains, the number of pieces into which a broken cranium can descend, and their high level of paleontological distinctiveness.

Some of these elements provide little more than documentation of their presence, but a few of them furnish comparative paleontological data for the Palomas sample. The preserved elements are described, and, to the extent possible, salient features are compared to Late Pleistocene human variation.

The Palomas Cranial Remains

The Palomas 1 Maxillae

From the crushed and cemented mass that was Palomas 1 *in situ* (Plate 5.1), it has been possible to extract most of the maxillary dental arcades and adherent pieces of alveolar bone (Plates 4.1, 6.1, and 6.2). The right maxilla retains external bone up to above the level of the nasal sill above C^1 and P^3 for a mesiodistal distance of 20.6 mm, and there are crushed fragments across the M^1 to M^3 buccal roots for 14.5 mm mesiodistally. The lingual alveolar bone is retained up to the palatal roof above C^1 to P^4 for 20.4 mm mesiodistally, and then there is a section of lateral palatal angle above M^1 and M^2. The floor of the maxillary sinus is preserved for 21.5 mm, primarily above the M^1 roots.

On the left side there are buccal interproximal fragments at C^1/P^3, P^3/P^4, and P^4/M^1 plus lateral alveolar bone across the M^3 roots extending back into crushed bone in the retromolar region. Lingually on the left maxilla there is a piece of superiorly displaced bone that was across the P^3 and P^4 roots, 18.7 mm mesiodistal. There are then fragments adjacent to the P^4 to M^1 and the M^2 to M^3.

Morphologically, the Palomas 1 maxillae provide little comparative evidence. The left side, especially laterally, does provide evidence of a substantial maxillary retromolar space, ≈10 mm mesiodistal. In this respect, it conforms to the presence of a mandibular retromolar space indicated by the rearranged mandibular elements (chapter 5). Similar maxillary retromolar spaces are present in the Amud 1, La Ferrassie 1, and Shanidar 1, 2, and 5 Neandertals; other Neandertal crania do not sufficiently preserve the fragile maxillary retromolar alveolar bone. There is also a large maxillary one on Skhul 5, but earlier Upper Paleolithic crania retaining the retromolar region have small to nonexistent retromolar spaces. Given the fragility of the region, overall frequencies of maxillary retromolar spaces are not known, but mandibular ones are variably present in Late Pleistocene samples (Table 5.8; see also Wu and Trinkaus 2014).

The region of the right lateral nasal margin is crushed and brecciated internally in the region above the C^1. However, the lateral bone is largely continuous from the alveolar margin to the superior break, and the lateral radiograph of the piece shows what is interpreted as a posteriorly descending nasal floor above the distal I^2 and the C^1 (Plate 4.1). If correctly interpreted, this would provide Palomas 1 with a bilevel nasal floor. Even if the anterior nasal floor piece has been distorted, its position relative to the nasal floor (and palate) is sufficiently elevated to indicate a bilevel nasal floor. A marked bilevel nasal floor is also present on Palomas 96. Bilevel nasal floors, as opposed to sloping or level nasal floors (Franciscus 2003), are the dominant form among the Neandertals, occurring in 75.0% of Neandertal crania (n = 25 not including Palomas 1 and 96; 81.5% including the Palomas maxillae); a level floor is present in only one of them, and the other four are sloping. Bilevel nasal floors are also the dominant configuration among eastern Eurasian archaic humans (Wu et al. 2012). Among early modern humans, bilevel nasal floors are present in one-quarter of a Middle Paleolithic modern human (MPMH) sample (n = 8) and in only one (Sunghir 1; 4.0%, n =

25) of Early/Mid Upper Paleolithic (E/MUP) modern humans (Trinkaus et al. 2014a). The other MPMH nasal floors are equally divided between level and sloping, and all but one of the other E/MUP ones are level. In a pooled recent human sample, level and sloping nasal floors occur about equally (47.4% and 41.3%), but bilevel ones are considerably less common (11.3%, n = 638) (Franciscus 2003; Wu et al. 2012).

Palomas 2 Occipital Bone

Palomas 2 (Plate 4.2) consists of a lateral piece of the left occipital bone with the lateral cerebral occipital lobe fossa, the superior edge of the lateral transverse sinus sulcus, ≈34 mm of the lateral lambdoid suture, and 12.5–13.0 mm of concave suture towards asterion. The concave nature of the suture indicates the former presence of a sutural ossicle in the lambdoid suture at this level. However, this area is not for articulation with the Palomas 10 sutural ossicle (Plate 4.5), since the sutures do not fit.

On the inferior medial margin of the piece there is a thinning and exterior concavity of the bone. This may represent the superior lateral corner of a suprainiac fossa, but the bone surface remains smooth and is insufficient to confirm the presence of such a feature. Further laterally and inferiorly there is a transverse swelling (Plate 4.2), which should represent a portion of a modest nuchal torus. On the lateral portion of the apparent nuchal torus is a distinct vascular sulcus (Plate 4.2). Similar vascular sulci are evident in this region on 60% of the Neandertal occipital bones, and they appear to reflect a branch of the occipital artery crossing the nuchal torus (Rougier 2003). They are also variably evident on Middle Pleistocene and recent human remains (Rougier 2003).

Palomas 3 Parietal Bone

Palomas 3 (Plate 4.3) represents the anterior portion of a right parietal bone, with 20.5 mm of the coronal suture, 27.0 mm of the anterior meningeal vessel sulcus running parallel to the coronal suture with three faint parallel branches coming off of the vertical sulcus, 46.0 mm of a faint temporal line with stephanion at the superior end of the preserved coronal suture, and 31.3 mm of the anterior squamous (temporal and/or sphenoidal) suture broken at both ends of the preserved section. Pterion is absent.

The piece retains few features. The anterior meningeal vessel sulcus is 2.7 mm wide in its inferior portion, becoming 8.4 mm wide superiorly. The temporal line is faintly marked, and it does not rise above the even curve of the exocranial surface. In *norma occipitalis*, the bone has an even convexity. This suggests a *forme en bombe* (Boule 1911–13), but the bone is not sufficient to confirm its presence.

Palomas 4 Parietal Bone

The Palomas 4 left parietal bone (Plate 4.4) preserves a posterior section of the sagittal suture (≈18 mm long), lambda, and a slightly longer (≈27 mm) section of the left lambdoid suture. The external surface is encrusted and has lost some surface cortical bone along the lambdoid suture. There is a prominent meningeal vessel sulcus leading to the sagittal suture, which runs adjacent to a distinct parietal foramen. There is evidence of a vascular sulcus leading posteromedially from the parietal foramen, as well as another small foramen (evident endocranially) closer to lambda. The foramen is not sufficiently large to be an "enlarged parietal foramen," a rare variant nonetheless known among late archaic humans (Hauser and DeStefano 1989; Wu et al. 2013), but its position and vascular sulcus suggest a small version of a similar structure. The courses of the sutures near lambda indicate that a lambdoidal sutural ossicle, if present, would have been entirely occipital.

Palomas 5 Parietal Bone

Palomas 5 is a thin piece of anterior parietal bone (Plate 4.5), with 14 mm of a shallow sulcus along the coronal suture and 31.5 mm of the more distinct anterior meningeal vessel sulcus running parallel to the coronal suture. It is probably a right parietal bone. In the broken edges of the bone, the tables are relatively thick and there is little diploë, indicating an immature individual. This is supported by the thickness measures of 3.8 to 4.5 mm across the bone. Comparative Late Pleistocene parietal thicknesses at the eminence are 7.9 ± 1.3 mm (n = 12) for Neandertals, 8.0 ± 2.6 mm (n = 5) for Middle Paleolithic modern humans, and 6.1 ± 1.8 mm (n = 7) for earlier Upper Paleolithic humans.

Palomas 10 Sutural Ossicle

Palomas 10 is a complete sutural ossicle (Plate 4.5). Exocranially, its perpendicular maximum diameters are 25.0 and 20.0 mm. Endocranially, its perpendicular maximum diameters are 19.0 and 13.3 mm. The ossicle has on one side a fine and beveled suture that is not lambdoid and probably not sagittal. There is a coarser suture around the other side, perpendicular to the exocranial surface. It is probably therefore coronal or bregmatic. It does not articulate with any of the other preserved cranial sutures.

Palomas 11 Frontal Bone

Palomas 11 (Plate 4.6) consists of an anterolateral piece of the right frontal bone, with the lateral supraorbital torus retaining lateral endocranial surface (31 mm wide and 21 mm high), the adjacent orbital roof surface (22 mm deep and 25.5 mm wide), 12 mm mediolaterally of the anterior supraorbital torus, all of the frontozygomatic suture, the temporal crest around its concave curve, and 19.3 mm of the lateral supratoral sulcus from the temporal crest. The frontozygomatic suture is complete and unfused. There is no trace of the frontal sinus, but the medial extent of the bone is probably too lateral to show the sinus, given the restrictions of Neandertal frontal sinuses to the medial halves of their supraorbital tori (Vlček 1967; Zollikofer et al.

2008). The anterior surface of the middle of the supraorbital torus was exfoliated, such that only the lateral extent retains the full parasagittal toral contour.

Palomas 11 has a clear supraorbital torus (as opposed to separate superciliary arch and lateral trigone). To the extent preserved, the orbital and supratoral surfaces are largely parallel, the thickness decreases only modestly laterally, and the anterior surface remains rounded (as opposed to angled) as it reaches the frontozygomatic suture. It is accompanied by a clear supratoral sulcus, at least as far medially as midorbit. The frontal squamous profile, relative to the plane of the orbital roof, is also receding from the supratoral sulcus. Although some early modern humans without supraorbital tori have similarly thick or thicker lateral orbital margins, none of them possess this combination of anterolateral frontal features.

The supraorbital torus is nonetheless relatively thin, with a lateral thickness of 7.0 mm. This value is below the smallest ones in a comparative Neandertal sample (11.0 mm: Krapina 35.7 and Saint-Césaire 1), but it is close to that of Palomas 96 (7.5 mm). As such it is 3.2 standard deviations below a Neandertal mean (12.1 ± 1.6 mm, n = 24; 11.9 ± 1.8 mm, n = 25 with Palomas 96). It is also among the thinner of the lateral superciliary arches of an E/MUP sample (8.8 ± 1.4 mm, n = 10, exceeding only those of Dolní Věstonice 3 (6.0 mm) and Mladeč 2 (6.6 mm) (comparative data from Smith and Ranyard 1980; F. H. Smith pers. comm.).

The supraorbital torus does not project markedly from the endocranial surface either, since its midorbit projection is estimated at ≈18 mm (between 17 and 19 mm). This value is ≈2.8 standard deviations below a Neandertal mean (23.9 ± 2.1 mm, n = 20), but it is similar to those of a small E/MUP sample (19.2 ± 3.1 mm, n = 6) (comparative data from F. H. Smith, pers. comm.). It is unclear whether these modest values for the Palomas 11 supraorbital torus thickness and projection reflect a reduced facial robusticity and/ or size or, as with Palomas 92, 96, and other

postcrania, an overall small facial and body size (Walker et al. 2011b; chapter 14).

The modest supraorbital torus of Palomas 11 is nonetheless associated with a large frontozygomatic suture, 7.5 mm mediolaterally wide and 16.5 mm anteroposteriorly long. As has been noted with respect to Neandertals and other archaic humans (Smith 1976, 1993), this appears to be a general feature of archaic, as opposed to modern, humans.

The frontal bone also exhibits a prominent temporal crest, extending posteriorly from the frontozygomatic suture around the superior temporal fossa. The full posterior extent of the crest, as opposed to the temporal line, is not known, since it was broken off well anterior of the coronal suture.

Palomas 12 Frontal Bone

The small Palomas 12 element (Plate 4.7) consists of a piece of a left supraorbital torus with portions of the superior, anterior, and orbital surfaces plus dense trabeculae in the middle. It extends from the medial orbit to the vicinity of the midsupraorbital torus. On the medial side of the anterior surface is the beveling associated with the supraorbital notch. It does not continue sufficiently posteriorly to preserve any of the supratoral sulcus. There are dense trabeculae with no trace of a frontal sinus. Given the presence of large frontal sinuses in all Neandertal frontal bones to mid-orbit, this is unusual.

The one comparative aspect is its midorbital thickness of 9.2 mm. This value is below a Neandertal sample mean (10.8 ± 1.9 mm, n = 31) but well within its range of variation. It is above those of an E/MUP sample (5.7 ± 1.1 mm, n = 10), most closely approached by that of Mladeč 5 (7.7 mm) (Smith and Ranyard 1980; Trinkaus 1983; F. H. Smith pers. comm.).

Palomas 30 Parietal Bone

Palomas 30 is a posteroinferior corner of a right parietal bone with 14 mm of the posterior squamous suture, 34 mm of the parietomastoid suture, and 15 mm of the lambdoid suture (measurements assuming that asterion is at the corner of the bone) (Plate 4.8). The exocranial and especially the endocranial surfaces are well preserved.

Assuming that there are no ossicles near asterion, the parietomastoid suture appears to be overall horizontal but with an anterior portion that turns inferiorly into the parietal notch (for entomion). There is a meningeal vessel sulcus arcing across from entomion to the lower lambdoid suture. Furthermore, there is a distinct sulcus, several millimeters wide, extending across the parietomastoid suture onto the endocranial parietal surface and fading as it goes posteriorly to the lambdoid suture. It is bordered superiorly by a raised swelling that is a continuation of the petrous ridge of the temporal bone. This should be the anterolateral extent of the transverse sinus, as it arcs inferiorly to the sigmoid sinus. The fading out posteriorly is commonly found in the lateral transverse sinus, whether on the posterior parietal bone or the anterolateral occipital bone. In contrast to many archaic humans, but as in most modern humans (Grimaud-Hervé 1997; Verna 2006; Rougier and Trinkaus 2013), it therefore courses across the posteroinferior parietal bone rather than crossing directly from the occipital bone to the temporal bone across the mastooccipital suture. Variation in the osteological position of the sulcus is ultimately related to the size of the posterior cranial fossa (and its contents) and hence the position of the tentorium cerebelli.

Palomas 56 Parietal Bone

This element is a middle portion of a relatively thick cranial vault bone with no sutures (Plate 4.5). There is a distinct meningeal vessel groove arcing across the endocranial surface, with a bifurcation on the wider end of the piece. It is therefore identified as a piece of parietal bone. The thickness in the middle is 8.5 mm, modestly above a Neandertal midparietal thickness mean (7.9 ± 1.3 mm, n = 12).

Palomas 62 Frontal Bone

A burnt section of the right lateral supraorbital region, with a portion of the trigone, was found on the hillside in 1998. Its whereabouts are currently unknown.

Palomas 51 and 68

The Palomas 51 right M^3 and the Palomas 68 right P^3 and P^4 have substantial pieces of buccal alveolar bone adherent to their roots (Plates 6.41 and 6.52). Both have lateral matrix encrustations. They only provide indications of intact buccal alveoli without any indications of alveolar degenerations.

Summary

The Palomas cranial remains are few, small, and fragmentary, but they provide a few aspects of comparative interest. Most of them appear to be mature, or at least adolescent, given size and thickness, although Palomas 5 was probably immature. Palomas 12 and especially Palomas 11 have small supraorbital tori, albeit ones that conform to the Neandertal pattern of full toral development. Palomas 11 accompanies the small torus with a large frontozygomatic suture and a prominent temporal crest. There is a suggestion of *forme en bombe* vault shape for Palomas 3. Palomas 2 should have had a lateral lambdoidal sutural ossicle; Palomas 10 is a sutural ossicle, probably from the coronal suture. The lateral course of the transverse sinus is unclear in Palomas 2, but it was parietal in Palomas 30. The Palomas 56 parietal bone is moderately thick. Palomas 1 exhibits a maxillary retromolar space and a bilevel nasal floor.

It is of note that none of these Palomas cranial fragments present any lesions. Minor lesions to the cranial vault are ubiquitous among Pleistocene humans (Wu et al. 2011), and they are evident on even small pieces of cranial vault similar to the Palomas ones (Wu and Trinkaus 2015). Their absence may be due in part to the generally young age at death of the Palomas sample.

SP NO.	OLD NO.	IDENTIFICATION	MATURITY	PROVENIENCE	DISCOVERY DATE	COMMENTS
1a, 1b	CG-1	Maxillary alveoli right and left	mature	Shaft upper breccia	1991	crushed
2	CG-3	Occipital squamous left	mature	Hillside rubble	1992	burnt
3	CG-4	Parietal anterior right	mature	Hillside rubble	1992	—
4	CG-5	Parietal posterior medial left	mature	Hillside rubble	1992	—
5	CG-16	Parietal anterior right?	immature	Hillside rubble	1992	—
10	CG-10	Sutural ossicle (coronal?)	mature	Mine level rubble	1993	burnt
11	CG-15	Frontal supraorbital torus right trigone	mature	Main chamber rubble	1993	burnt
12	CG-14	Frontal supraorbital torus left	mature	Hillside rubble	1994	—
30	CG-13	Parietal posterior inferior right	mature	Hillside rubble	8 July 1995	burnt
51	CG-43	Maxillary alveolus on M^3 right	mature	Upper cutting level 1	9 August 1996	—
56	CG-68	Parietal		Hillside rubble	17 August 1997	—
62	CG-74	Frontal supraorbital torus right trigone	mature	Hillside rubble	1998	burnt
68	CG-92	Maxillary alveoli on P^3–P^4 right	mature?	Upper cutting level 1A	29 July 2001	—

5 The Palomas Mandibles

ERIK TRINKAUS, JOSEFINA ZAPATA, PRISCILLA BAYLE, KATHARINE A. ROBSON BROWN, MIGUEL ALCARAZ, A. VINCENT LOMBARDI, AND MICHAEL J. WALKER

THE PALOMAS NEANDERTAL sample preserves a series of partial mandibles, including the incomplete and distorted Palomas 1 one; the partial mature corpori of Palomas 6, 23, and 59; and the partial immature Palomas 7, 49, 80, and 88 mandibles (Table 5.1). Palomas 1 was found in the upper shaft breccia, and Palomas 6, 7, and 23 were found out of context on the hillside or at the bottom of the mine shaft. The remainder were discovered during excavations in the Upper Cutting. Although none of them provide overall mandibular dimensions, all of them furnish data on discrete mandibular features, the mature mandibles permit corpus measurements, and the Palomas 49 immature corpus provides arcade breadths. As such, they add substantially to considerations of later Pleistocene mandibular morphology, for both adults and immature individuals (see Mallegni and Trinkaus 1997; Stefan and Trinkaus 1998; Coqueugniot 1999; Rosas 2001; Richards et al. 2003; Verna et al. 2012; Wu and Trinkaus 2014). In addition, there are incomplete mandibular rami present with the crushed Palomas 96 cranium and most of a mandible with Palomas 97 (the latter almost entirely encased in breccia).

These mandibular remains were described previously (Walker et al. 2010), and several aspects of their description have been updated with additional assessments of the remains and an expanding later Pleistocene comparative data set. For the details on the immature ages at death and the associated dental remains, see chapter 6.

The Palomas 1 Mandible

Preservation

This badly fragmented, distorted, and cemented lower facial fossil was found in the upper shaft breccia in 1991 (Plate 5.1). It retains much of the mandible and alveolar portions of the maxillae. The pieces were separated into several elements, and various distorted portions have been reassembled as best as they can be. Currently, Palomas 1 consists of four sections: the right and left maxillae (chapter 4) and the right and left mandibles. There is inevitable distortion and movement of sections of bone relative to each other, separation between teeth that were in interproximal contact, and variable surface erosion. However, some basic information, particularly on the mandibular discrete morphology, is possible.

The right mandible (Palomas 1c) retains bone from the symphysis to the middle of the ramus, and it includes the C_1 to the M_3 (Plate 5.2). The C_1 to M_1 are close to their original occlusal positions. However, there is a gap between the M_1 and M_2, and the M_3 is displaced superiorly relative to the M_2. The right mandibular maximum preserved length is 113 mm; the maximum height at the coronoid process is 75 mm, which is an underestimate of the original height. Given the cementing of the bone fragments and the softness of the bone itself, it is not possible to separate out the individual pieces to produce a restored right mandible. Moreover, the edges of the fragments are too eroded for secure fits between many of them.

The right corpus is largely intact from the I_1/I_2 interdental septum to the P_4/M_1 interdental septum. The original alveolar margin is present by the C_1 and P_3. The basilar margin is intact below the I_2 to P_3, but the margin itself has been reattached. The mental foramen lacks its posterior margin. It is located below the P_4, even though it appears to be below the P_4/M_1 interdental septum, given distortion of the mandible. The medial corpus is intact from C_1/P_3 to P_4/M_1 along the superior half of the corpus. There are additional but distorted bone pieces down to the medial side of the basilar margin.

Fragments of bone extend across the symphysis, but they are distorted and pushed anteriorly. The superior 16 mm of the midline with the mesial I_1 alveolus is present, but it has been pushed left and twisted. The lingual symphysis was crushed lingually and fragmented.

Posteriorly, there is a gap of bone below the M_1. Between the M_1 and M_2 there are bone fragments in the alveolar process, but it has resulted in a separation of the M_2 and M_1 crowns of several millimeters. From the mesial M_2 to the retromolar alveolar bone process, the superior half of the corpus is largely intact and contains the M_2 and M_3, plus a clear open retromolar space. Joined to this superior corpus is the basilar margin from below M_1/M_2 to around most of the rounded gonial angle. However, the contact of the basilar margin with the superior corpus is not the original one, and the corpus height was reduced significantly based on the measurements from the left side. The piece of inferior bone needs to be displaced inferiorly, leaving a gap between it and the lateral corpus by the eminence.

The anterior ramus up to the coronoid process appears to be largely intact, with surface and break erosion. However, the piece of anterior ramus is positioned too far anteriorly and inferiorly relative to the posterior corpus, such that it eliminates any suggestion of a retromolar space. This is an artifact of fossilization and assembly, since the less complete but undistorted left side shows a substantial retromolar space. Posteriorly on the ramus, the bone descends into crushed and distorted bone. The mandibular foramen, the inferior mandibular notch (all except on the coronoid process), the posterior ramus, the condylar neck, and the condyle are absent. It is therefore not possible to determine the configurations of the mandibular foramen, the medial pterygoid insertion, and the condyle position. The gonial angle, or curve, is present and attached to the piece of posterior inferior corpus. Along with the inferior margin, it needs to be displaced inferiorly and posteriorly from its preserved position.

The left mandible (Palomas 1d; Plate 5.3) preserves several pieces of the buccal corpus below the M_1 to M_3, for a mesiodistal length of 31 mm and height of 21 mm below the M_1 and 10 mm below the M_3. To it is attached a piece of bone, originally separate, that provides the basilar margin from below the P_4 to distal of the M_3. The posterior margin of the mental foramen is located along the anterior break of the lateral corpus, indicating that its middle was just mesial of the M_1, or below the P_4/M_1 interdental septum. Medially, the superior one-third to one-half of the corpus is present from the middle of the M_1 to the distal retromolar space. The mylohyoid line is evident below the M_3. The three molars are in place in the alveolar bone, but they are at different heights, with the M_2 below the M_1 and M_3. The overall dimensions of the piece are 64 mm anteroposterior (inferiorly) and 37.5 mm high without the teeth.

Given these distortions and gaps in the Palomas 1 mandible, and in order to provide a better visual assessment of the Palomas 1 mandible, the principal pieces of the right side (the anterior corpus, the posterior inferior corpus and ramus, the posterior superior corpus and anterior ramus, and the coronoid process) have been separated with image editing software and repositioned closer to their original positions (Plate 5.4). The posterior corpus height is based on the more complete left side, and the other arrangements were made by matching edges and preserved contours. No measurements have been taken from the repositioned image.

Age at Death

The dentition has had its occlusal surfaces planed off with reductions in crown heights, but there is only modest dentin exposure on most of the teeth (see chapter 6). The degree of dentin exposure can be subsumed within Smith's (1984) stage 4 wear for all of the teeth except the M_1s and the left M_3, which are a better fit to stage 5. Compared to other Neandertals (as well as other Late Pleistocene humans and recent foragers on native diets; Davies and Pedersen 1955; Moorrees 1957; Trinkaus 1995; Hillson et al. 2006), these degrees of wear suggest an age at death in the fourth decade postnatal.

Morphology

The distorted nature of the Palomas 1 mandible only permits some corpus height and breadth measurements, combining both sides (Table 5.2), as well as observations of several discrete traits. As preserved and through the visual reconstruction of the right corpus and ramus using observations from the left corpus (Plates 5.2 to 5.4), an overall impression of its proportions can be obtained.

The symphysis is sufficiently intact and undistorted to the I_1/I_2 interdental septum to indicate an anterior buccal depression below the right I_1 to the I_2/C_1, up to 16.7 mm from the alveolar margin. The symphysis is largely vertical or slightly retreating relative to the alveolar plane, although damage makes any such assessment tenuous; it was not markedly retreating. It exhibits a gentle swelling, but the mental trigone area is not preserved. It is conservatively scored as mentum osseum rank 2–3 (retreating [2] or vertical [3], with a clear but nonprojecting mental trigone [Dobson and Trinkaus 2002]).

The right mental foramen is single, opens laterally and slightly posteroinferiorly, and is located slightly below the vertical middle of the corpus. The left one is indicated by its posterior margin.

The right one is below the P_4, and the left one is below the P_4/M_1 interdental septum, assuming that the foramen was round and single. The lateral eminence begins below the M_2, but it is not prominent. The gonial angle is evenly rounded, and there is a smooth and rounded concavity in lateral view between its anterior extent and the basilar margin of the corpus. It is not possible to assess whether it was inverted, as are the majority of the Neandertal gonial angles (68.2% inverted, 27.3% straight, n = 22; Wu and Trinkaus 2014).

There is a distinct retromolar space on the left side, ≈10 mm from the distal M_3 to the anterior ramus as preserved. Visual placement of the right anterior ramus to an anatomically correct position based on bone contours (Plate 5.4) provides a similar retromolar space. The right coronoid process is high and prominent, with a marked endocoronoid buttress. The inferior two-thirds of its anterior margin is concave, and the superior portion (damaged) would have been anteriorly projecting relative to the lower portion. The shape of the mandibular notch cannot be directly assessed. However, the coronoid process is high; the current posterior height (basilar margin to coronoid tip) is 73.5 mm, and photographic reconstruction places it ≈80 mm above the inferior corpus margin. It would require an exceptionally high condyle to make the resultant notch evenly rounded, so it was probably asymmetrical with a high coronoid process.

The Palomas 6 Mandible

Preservation

The Palomas 6 mandible (Plate 5.5), found *ex situ* on the hillside, retains a left mandibular corpus from the symphysis to the middle of the M_3 alveolus. The I_1 and I_2 alveoli are largely absent, but the remainder of the symphyseal midline is present (if irregular) across the break. The alveolar bone is largely present, but the buccal margin is intact only by the distal M_2 root and lingually from C_1/P_3 to M_1/M_2. The interdental septa are all damaged, and therefore all of the corpus heights involve minor estimation. Only the roots of the C_1 to M_2

are present. The mesial surface of the mesial M_3 root socket is preserved, but the remainder of the M_3 socket is absent. At the posterior end of the corpus, the internal surface bone is present only to the middle of the distal M_2 root, whereas externally, it is preserved to the apex of the mesial M_3 root socket. Maximum preserved (oblique) length is $\approx$64 mm.

The whole bone was heated, and the loss of the tooth crowns was probably due to that burning. The bone surface is a hard gray-brown with reddish sediment filling fine depressions, and the tooth crowns were all lost with angular breaks in the remaining roots. These changes are commensurate with exposure of the mandible to fire. However, the bone morphology does not appear to have been altered by the burning. The external surfaces and the exposed trabeculae all appear as normal bone with none of the exfoliation or cracking associated with burning. It remains possible that some features, such as the size of the mental foramen, may have been altered by the heating, but the radiograph of the bone indicates that the whole of the inferior alveolar nerve canal is large (Plate 5.6). Therefore, it does not support an interpretation of distortion through heating.

In this context, the anterior and lateral margin of the inferior corpus has an irregular surface, 5 mm wide anteriorly and anterolaterally, narrowing to 3 mm wide where it ends below the P_4/M_1 interdental septum (Plate 5.5B). This area superficially resembles a muscle insertion rugosity, even though it would be very unusual for any of the muscles (for example, the anterior belly of digastric) normally inserting in the region. Detailed inspection, however, indicates that the original surface bone along this "rugosity" was exfoliated in small chips, probably from a combination of the heating and the displacement of the bone onto the hillside in the nineteenth century.

Age at Death

Palomas 6 was fully mature at death, as indicated by the mesial side of the M_3 socket exposed at the posterior break. The loss of the tooth crowns prevents further assessment of its age at death.

Morphology

Palomas 6 retains the left symphysis and lateral corpus (Plate 5.5; Table 5.2). The anterior symphyseal surface retreats relative to the alveolar plane; the symphysis is insufficiently complete at midline to measure its anterior angle directly, but extrapolation from the preserved bone provides an anterior angle (infradentale to pogonion versus the alveolar plane) of $\approx$70°. The lateral extent of the mental trigone is evident below the distal I_1 and the I_2 alveoli. The trigone is $\approx$5 mm high at the break slightly lateral of the midline, and it ends laterally below the I_2/C_1 interdental septum. In combination with the symphyseal retreat, this provides a mentum osseum rank of 2. There is an anterior buccal depression below the I_2/C_1, 11.5 mm high and 7.2 mm wide, but damage precludes determining if it extended below the I_1.

The anterior dental arcade appears to angle strongly around the C_1, with a more transverse incisal profile. However, this impression may be exaggerated by the loss of the incisors and their associated alveolar bone, especially labially. The lingual symphysis has a modest planum alveolare, which descends smoothly to a thickening of the symphysis. The thickening produces an even convexity from the planum alveolare to the basilar margin, so it is hard to describe it as either a superior or an inferior transverse torus, rather than just a transverse torus. The effect is a thick symphysis that spans the height of the bone. There are two small genioglossal tubercles, projecting less than 1 mm, and there is a small accessory foramen above the superior tubercle (Plate 5.5D).

As a result of the bone erosion on the anterior margin of the inferior symphysis, the normal insertion area for the anterior belly of the digastric muscle appears to be absent. An alternative assessment of the inferior symphysis is that the depression, extending from the midline $\approx$17 mm laterally across the middle of the inferior symphysis, is the left digastric fossa. The fossa has rounded and indistinct margins, especially anteriorly. It lacks the teardrop contour of most Neandertal (and recent human) digastric impressions, including those of Palomas 23 (Plate 5.7) and the immature Palomas 49. The surface of this fossa faces posteroinferiorly.

Along the anterior and lateral margins of the modestly eroded anterior anterolateral margin of the inferior surface, there is a small anterolateral eversion of the external corpus. It extends from the symphyseal midline around the curve below the canine and then along the lateral corpus. It is most prominent from below the C_I to the P_4, decreases to the midline symphyseal break, and tapers off as it goes posteriorly to below approximately the M_2. It is evident from the shallow sulcus along the inferior external corpus, producing a distinct hollow below the C_I and P_3 (Plate 5.5C and E). If the inferior fossa just lateral of midline is the digastric fossa, then the ≈5 mm of bone anterior of that fossa would be part of this anterior and lateral lip of bone. The micro-CT slices through the corpus show that the cortical bone of the lip is continuous with the inferior corporeal cortical bone (Plate 5.6), in contrast to that of Palomas 23 (see below).

This lip of bone is unusual, among both Neandertals and recent humans, since the associated corpus margin is usually rounded from lateral to inferior with a variably pronounced anterior marginal tubercle (Rosas 2001). Moreover, the projecting lip was probably larger prior to the postmortem erosion that removed the external edge of the inferior surface. If it represents the remains of an elongated anterior marginal tubercle, then it would be an unusual development of that feature in terms of its elongation from near midline to the M_2. It would also be unusual in having its greatest prominence below C_I to P_3; among Neandertals, anterior marginal tubercles are centered from P_3/P_4 to M_1 (Rosas 2001). If it is distinct from an anterior marginal tubercle, then its significance is unclear.

The lateral corpus retains the thickness present in the symphysis, with its corpus breadths exceeding those of Palomas 1 despite a shorter corporeal height (Table 5.2). Its external surface has a prominent lateral eminence below the M_2, which appears more as a rounded tubercle, ≈12 mm anteroposterior and ≈5 mm high, than as the usual swelling for the eminence. The posterior end of this swelling is separated by a slight depression from the anteroinferior end of the ramal margin.

Below the P_4/M_1 interdental septum, but extending below both the P_4 and the mesial M_1 root, is a very large mental foramen. Its midpoint is midway between the alveolar and the basilar margins. It is ellipsoid in shape, opens directly laterally, and has the orientation of its maximum dimension sloping slightly anteroinferior to posterosuperior. Radiographically (Plate 5.6), it is associated with a large inferior alveolar nerve canal, larger in vertical dimension than the foramen, suggesting that the mental foramen dimension is likely to be an anatomical variant and not the result of the burning. Medially, there is an angle along the mylohyoid line, with a posteriorly widening concavity below it. The inferomedial surface is smooth.

It is not possible to determine directly whether the mandible had a retromolar space. However, if the line of the anterior ramal margin, as preserved below the distal M_2 and the mesial M_3, is extended in an even arc posterosuperiorly, and the original M_3 crown is given a conservative mesiodistal diameter of ≈10 mm (Late Pleistocene pooled sample: 11.6 ± 0.8 mm, N = 66), the line of the anterior ramal margin would cross the distal portion of that M_3 crown in *norma lateralis*. Palomas 6 therefore almost certainly lacked a retromolar space.

The Palomas 7 Immature Mandible

Preservation

The Palomas 7 left mandibular corpus (Plate 5.9) retains the corpus from the distal dc_I socket with the C_I germ below it to the mesial side of the M_1 crypt. The basilar margin is complete from the C_I crypt to the distal M_1 crypt. The alveolar process is only complete along the buccal dm_1 and mesial dm_2, the lingual dm_1, and the septa between the dm_1 roots and between the dm_1 and the dm_2. The lateral surface is largely intact from the canine to the middle of the M_1 crypt, but the internal surface is present principally below the dm_1. The maximum preserved length is 38 mm; maximum preserved height is 23 mm.

Age at Death

By the morphology of the root sockets, the dc_1, dm_1 and dm_2 were fully erupted and in occlusion (Plate 5.9). This is especially evident for the two deciduous molars. The C_1 germ is formed close to the cervix. This degree of dental development suggests an age at death of three and one-half to four and one-half years postnatal (chapter 6). At that developmental age, the M_1 should be just beginning its root formation (between R_i and $R_{1/4}$). The mesial side of the M_1 alveolus therefore reflects primarily the crypt for the M_1 crown, with little root formation.

Morphology

The lateral corpus appears to have some angulation around the dc_1/dm_1 interdental septum, but the anterior breakage prevents determining how pronounced was the angle (Plate 5.9; Table 5.3). Inferiorly, this is accentuated by a rounded margin of bone along the external basilar margin.

The rounded external margin begins mesially below the dc_1, is most prominent below the mesial dm_1, and then gradually fades away by the mesial M_1 crypt. It is part of a continuous rounding with the inferior margin of the corpus, with no evidence of a digastric insertion or rugosity. There is a shallow sulcus along the superior margin of the external rounding, which separates it from the lateral surface of the corpus. This feature resembles an immature anterior marginal tubercle or what could be a juvenile manifestation of the more elongated external inferior lip of bone on Palomas 6 (Plate 5.5).

There are two mental foramina: a slightly larger one below the dm_1/dm_2, which opens posteriorly, and a smaller one below the mesial dm_1 root, which opens anteriorly.

The Palomas 23 Mandible

Preservation

Palomas 23 (Plate 5.7) consists of a right mandibular corpus with the exposed roots of the right I_1 to M_1, the crown of the M_2, and the mesial third of the crown of the M_3. It was found *ex situ* on the hillside. The anterior midline break is largely vertical but with a piece of the inferior margin extending ≈ 10 mm onto the left side with the medial half of the left digastric fossa. Posteriorly, the bone is broken obliquely from the retromolar surface just distal of the M_3 to the middle of the gonial angle. There is a thin layer of matrix adherent to the internal surface of the corpus between the I_2 and the middle of the M_1. Its maximum oblique length is 83 mm.

The original alveolar margin is present only around the M_2 and M_3. More mesially, all of the alveolar margins and interdental septa are broken. The only teeth retaining any of their crowns are the M_1 to M_3 (see chapter 6). However, the M_1 has only a tiny piece of the midcrown with no cervix, the M_2 lacks the lingual enamel, and the M_3 retains only root and pieces of the mesial crown. The roots for the I_1 to P_4 were broken off below their cervices.

The bone was burnt after burial, with the tooth crowns largely or completely breaking off on the I_1 to M_1 and becoming gray and cracked on the M_2 and M_3 (Plates 6.21 and 6.22). The bone is a uniform gray-brown. The burning appears to have produced a couple of cracks in the surface bone. There is a thin crack on the medial side of the corpus, below the premolars and M_1 (Plate 5.7B). It is evident in the P_4 to M_1 micro-CT slices (Plate 5.8), and it only penetrated through the external half of the cortical bone. More pronounced is a separation of the inferior margin of the mandible from the symphyseal midline to the anterior end of the gonial curve (Plates 5.7A and E and Plate 5.8). The separation is most pronounced below the incisors, where there is a gap between the digastric fossae anterior margins and the separated bone, evident in inferior view. The portion toward the midline is then bent inferiorly. As it extends distally from the area of the C_1 to the distal molars, it is apparent as thin separation lines inferomedially and inferolaterally. In the P_4 to M_1 micro-CT slices of the corpus (Plate 5.8), it can be seen that the separation continues across the inferior corpus, indicating that it is not just a surface crack but a longitudinal piece of bone along the basilar corporeal margin that is distinct from the inferior cortical bone of the corpus. What remains unclear is whether this marginal bone represents a strip of the primary corporeal cortical bone that was separated across the basilar margin from the burning, or whether it is an extra strip of bone that was incompletely fused to the inferior corpus and then became partially separated due to the heating of the mandible.

These considerations are relevant for the interpretation of the inferior corporeal morphology of Palomas 23. However, these postmortem changes do not appear to have affected the bone's morphology beyond the areas of the cracks. The bone is therefore taken to represent its perimortem morphology.

Age at Death

The M_3 was fully occlusal, and therefore Palomas 23 was mature. The occlusal wear (following Smith 1984) was stage 4c on the M_2. Comparisons to the occlusal attrition of other Neandertals aged by alternative techniques (Trinkaus 1995) suggest an age at death in the middle of the third decade.

Morphology

The symphysis of Palomas 23 (Plate 5.7; Table 5.2) retreats (anterior symphyseal angle: infradentale-pogonion relative to the alveolar plane: 76°), but it is vertically straight in *norma lateralis* right. There is a hint of a mental trigone, and therefore it has a mentum osseum rank of 2. Lingually, there is a prominent superior transverse symphyseal torus but no inferior transverse torus. There does not appear to be a planum alveolare, although one might construe the superior portion of the superior transverse torus to constitute a planum alveolare. There are no anterior buccal depressions. The genioglossal tubercles are small, rounded, and nonprojecting. There are two accessory foramina lateral to the superior genioglossal tubercle, one below the middle of the M_1 and one smaller below the mesial M_2.

The digastric fossae are delineated anteriorly, medially, posteriorly on their medial halves, and to some extent posterolaterally. They comprise the normal teardrop shape with the apex distally on the mandible and a midline separation with a triangular raised area anteriorly (Plate 5.7E). The fossae are directed slightly posteriorly of directly inferior.

Along the external (anterolateral/buccal) margin of the more complete right digastric fossa is the thin lip of bone that became partially separated postmortem (see above). The degree of projection of this lip of bone is uncertain below the incisors, due to the postmortem separation. However, distally it forms a prominent, laterally directed crest that is most prominent below the canine and premolars, especially below the P_4 (Plates 5.7D and E and 5.8). It then diminishes by the M_1 and becomes the rounded margin of bone that extends to the distal corpus, to just anterior of where the gonial curve would produce a downward curving of the basilar margin in lateral view. As noted above, this strip of bone, as a laterally directed crest more mesially and as an inferior rounding distally, has a separation seam between it and the corporeal cortical bone that extends from medial to lateral.

It is unclear whether the partially separated inferior margin of bone and its associated external lip are (1) a normal inferior margin of the mandibular corpus that became separated by the burning along normal longitudinal fibers of the mandible, or (2) an extra layer of bone that formed along the inferior corpus and became fused to it (only to partially separate due to the postmortem burning). If it were removed, the remaining inferior corpus contour would closely resemble those of other Neandertal mandibles, with a generally straight inferior edge and then an inferior flaring of the rounded gonial region (also evident in Palomas 1 [Plate 5.2]). As suggested for a similar inferolateral lip on the Palomas 6 corpus, it could be an elongated anterior marginal tubercle; the maximum projection of it in the region of the P_4 conforms to the locations of anterior marginal tubercles among the Neandertals (Rosas 2001). Yet as an anterior marginal tubercle it would be unusual in its prolongation around the anterior symphysis. It would also be exceptional in being on an apparently additional strip of bone that was largely, but not entirely, fused to the inferior corpus prior to the postmortem heating. Certainly the early development of an anterior marginal tubercle in Neandertal mandibles does not consist of a separate ossification (Crevecoeur et al. 2010; see also Palomas 7 and 49). If this aspect of the Palomas 23 mandible is not related to an anterior marginal tubercle, its significance is unclear.

The lateral corpus has parallel alveolar and basilar margins, with the variation in corpus heights (Table 5.2) due to alveolar variation. It is minimally taller than Palomas 6 and a few millimeters shorter than Palomas 1. The mental foramen is double, with the larger foramen positioned more anterosuperiorly and opening in that direction, and the smaller one positioned and opening posteroinferiorly. The larger foramen is also double within the lateral opening (Plates 5.7A and 5.8). The larger foramen is under the M_1, and the smaller one is below the M_1/M_2 interdental septum. The lateral eminence is below the M_3 and is small. Its position is indicated mostly by the beginning of the anterior ramal root. Internally the mylohyoid line is weak with little angulation. It is also well below the alveolar plane at the M_3.

The anterior ramal root is $\approx$3 mm posterior of the M_3 at the alveolar plane, and there is the beginning of a retromolar alveolar surface preserved distal of the M_3. The two features combine to indicate the presence of at least a small retromolar space.

Only a portion of the gonial angle is preserved; it is evenly rounded, continuing the line of the basilar margin. If one visually deletes the extra lip of bone back to the region of the M_3 along the join between it and the corpus, the Palomas 23 mandible would have had a modest concavity in the inferior margin just anterior of the rounding for the gonial angle. The gonial angle appears to be slightly inverted, as with 68.2% (n = 22) of the Neandertals.

Only the inferior portion of the medial pterygoid insertion is preserved, but it is remarkable for the absence of rugosity or tubercles. It is not possible to determine whether a prominent superior medial tubercle was present.

The Palomas 49 Immature Mandible

Preservation

The Palomas 49 immature mandible (Plate 5.10) preserves the corpus from the right dm_2 crypt to the left dm_2/M_1 interdental septum. It is in two pieces that articulate cleanly along the lingual symphysis opposite the right di_1 socket, despite minimal rounding of the join edges. The anterior symphysis is present only along its inferior margin, but the symphysis is complete from the lingual di_1/di_1 alveolar margin around lingually to the anteroinferior edge of the deciduous incisor sockets. The lingual corpus is complete between the two distal dm_2 crypts with damage to the right distal dm_2 crypt. The right external corpus is intact from the middle of the dc_1 socket to the mesial end of the dm_2 crypt, with the full height preserved principally by the dm_1 socket. The left external corpus is similarly preserved, but the inferior two-thirds of the surface adjacent to the dm_2 crypt are present. The inferior surface is complete from the middle of the right dm_2 around the symphysis to the left dm_1/dm_2.

The mandible preserves the damaged alveoli of both sides of the lower dentition from di_1 to dm_2. There is matrix in all of the sockets/crypts, and the labial alveolar bone of the deciduous incisors is missing. There are fragments of the left dc_1 and dm_1, and the partially formed crowns of the left I_2 and C_1 are contained within the corpus.

Age at Death

The age at death of Palomas 49 based on its dentition is assessed in detail in chapter 6. The presence of the germs of both the left I_2 ($\approx Cr_{3/4}$) and the left C_1 ($\approx Cr_{1/2}$) in their crypts and the degrees of formation of the dm_1 and dm_2 sockets are consistent in providing an age at death of approximately two and one-half years.

Morphology

The anterior symphysis is damaged superiorly, and it is best considered slightly retreating to vertical, depending upon how broad one makes the deciduous incisors and hence the estimated position of infradentale (Plate 5.10; Table 5.3). Given bone loss and crushing in the region of the anterior symphysis, it is not possible to determine whether a trigone was present.

The lingual symphysis presents a clear planum alveolare, although it may be due to the forming permanent incisor crowns. Those incisor crowns should have reached at least half of their crown formation by the age at death of Palomas 49, as is indicated by the approximately three-quarters formation of the left I_2 crown in its crypt (Plate 6.39). There is no evidence of a transverse torus. The digastric fossae are present on the inferior surface of the symphysis, with distinct concavities but no rugosity plus a prominent midline peak. The better preserved right one is 2.8 mm antero-posterior and 4.9 mm long (anteromedially to posterolaterally).

The mental foramina are single on both sides, and each one is below the middle of the dm_1. The right one is slightly smaller than the left one.

On both sides of the corpus there appears to be a gentle swelling of bone along the inferolateral margin. It is continuous on the right side and ends posteriorly in a small tubercle; however, the postmortem separation of it from the more superior lateral corpus makes it clear that it is a fossilization artifact. A similar swelling is evident for ≈5 mm below the mental foramen on the left side, with a slight sulcus above it. It is likely to be an immature form of an anterior marginal tubercle.

The Palomas 59 Mandible

Preservation

The Palomas 59 mandible was found *in situ* and consists of the left corpus from the damaged symphysis and I_2 alveolus to the region of the M_3 and the lateral eminence (Plate 5.11). There are small areas of thin matrix on the bone, and the lingual side by the M_3 region is chipped away on the superior half of the surface. Otherwise, the surfaces present are largely intact. The mandible retains the C_1 to the M_2.

Age at Death

The individual had an age at death of a young adult, with full occlusal eruption of the preserved tooth crowns and modest wear on them. The M_2 presents more advanced wear (stage 4 of Smith 1984) than the M_1 (stage 3) or the more mesial teeth (stage 3 for the C_1 and P_4, stage 2 for the P_3), but it was anomalous with respect to its occlusal wear (see discussion in chapter 6 and especially chapter 10). There is nonetheless substantial interproximal attrition between the P_4/M_1 and M_1/M_2, but not at the C_1/P_3 or the P_3/P_4. The M_3 is absent. There is what resembles the mesial surface of the socket for a mesial root of the M_3 on the broken posterior edge of the corpus, but it is unclear whether it was such a bone surface. However, on the distal M_2 there is a distinct interproximal facet, 2.7 mm high and 3.6 mm wide. The mandible therefore derives from a young adult with anomalous dental wear.

Morphology

The Palomas 59 (Plate 5.11; Table 5.2) symphyseal morphology assessment is approximate, given the absence of most of the region. However, the anterior midline appears to have been relatively vertical with a modest retreat. There is no evidence of the trigone on the preserved portion below the I_2, but it is likely to have been mentum osseum rank 3, vertical without projection of the trigone; it is conservatively scored as 2–3. Lingually, there is no planum alveolare. There is a lingual inferior swelling more laterally, so there may have been a small inferior transverse torus.

The single mental foramen has its middle at the mesial edge of the mesial M_1 root. It is therefore best considered as intermediate between the M_1 and P_4/M_1 designations. The corpus has parallel alveolar and basilar planes with a pronounced basilar margin. There is no evidence of an inferolateral eversion or lip similar to those on Palomas 6 and 23. The modest lateral eminence is below M_2 and M_3. The mylohyoid line is rounded, and it was well below the M_3 alveolar margin. The mylohyoid line to alveolus distance cannot be accurately measured at the M_3, since both ends must be estimated; continuing the lines of the mylohyoid line and the alveolar plane provides a height of ≈6 mm. The region of the retromolar space is not preserved, but one can estimate the mesiodistal diameter of the M_3 crown at ≈10 mm (see above) and continue the curving line from the lateral eminence. This suggests that a small retromolar space was present.

The Palomas 80 Immature Mandible

Preservation

Palomas 80 is a left mandibular posterior corpus and ramus, found *in situ*, with the M_2 at alveolar eruption and the M_3 germ completely within its crypt, plus damaged alveolar bone for the P_4 and M_1 (Plate 5.12). Anteriorly the corpus is broken through the lower portion of the P_3/P_4 interdental septum. The external corpus has an irregular vertical break at the anterior end, but the internal corpus is broken anteroinferiorly to posterosuperiorly, reaching the alveolar margin only at the mesial M_2. The sockets for the P_4 and M_1 are largely absent within the corpus, and the space is filled with hard matrix. However, the basilar margin is largely complete from the anterior break to directly below the M_2/M_3 interdental (or intercryptal) septum. The alveolar margin is intact only by the buccal M_2, the distal M_2 crypt, and around the M_3 crypt. There is an opening 4.5 mm anteroposterior and 2.5 mm mediolateral into the M_3 crypt.

The ramus retains a complete coronoid process, with the superior margin of the mandibular notch from the coronoid tip to its lowest point anterior of the condyle; the latter is indicated by the notch margin becoming horizontal just anterior of the posterior break. Medially, the ramal surface is intact down to the complete lingula, but the mandibular foramen is broken away posteroinferiorly and the inferior alveolar nerve canal is exposed along the posteroinferior break. The gonial angle is completely absent, as are the condyle and the condylar neck. There is nonetheless an indication of a downturning of the basilar margin below the M_2 for the beginning of a gonial

flare. The external ramal surface has a series of matrix encrustations parallel to the oblique break of the posteroinferior ramus. The maximum length of the mandible is 75.5 mm; its maximum height to the coronoid process is 54 mm.

Age at Death

The erupting M_2 is at about $R_{3/4}$, and the M_3 is at about $Cr_{3/4}$, providing an age at death of 10 to 12 years postnatal for Palomas 80 (Smith 1991; AlQahtani et al. 2010; chapter 6).

Morphology

The remaining immature lateral corpus (Plate 5.12; Table 5.3) is evenly rounded superoinferiorly, especially by the M_2, and there is no discrete lateral eminence. It is evenly concave medially below the M_1. There is no clear mylohyoid line, but there is a modest mylohyoid angle 4 mm below the alveolar plane at the M_1/M_2. The basilar margin is parallel to the alveolar plane, with slight concavities below the P_4 and M_2. Posteriorly, there is slight downward turn of the basilar corpus margin, suggesting that the gonial region extended inferiorly of the plane of the basilar margin.

The mental foramen is not preserved on the mandible, but the canal for the inferior alveolar nerve is evident in the anterior corpus break. This indicates that the mental foramen was at least as mesial as the middle of the P_3. Given that it would have migrated distally through facial growth had Palomas 80 lived to maturity (Coqueugniot 1999), it is placed conservatively at from P_3/P_4 to P_4 for adult comparisons (Table 5.6). It may well have been more mesial.

The anterior margin of the ramus curves up from the distal M_2 and is concave anteriorly, with the small opening for the M_3 crypt occurring within the lower portion of that curve. The curve then rises up to become a vertical anterior margin relative to the alveolar plane. There is an angle of ≈136° between the concave lower portion and the straight upper portion of the anterior ramus. The coronoid process then angles posteriorly, to reach a peak ≈29 mm above the alveolar plane. The anterior mandibular notch margin has an even curve down to its lowest point, $\approx$14 mm posterior of the coronoid tip, $\approx$16 mm anterior of the lingula, $\approx$26 mm posterior of the anterior ramus, and $\approx$35 mm posterior of the distal M_2, all measurements parallel to the alveolar plane. The mandibular notch is close to parallel to the alveolar plane at its lowest point posteriorly, which should have been close to the condylar neck, hence indicating an asymmetrical mandibular notch.

There is modest concavity to the ramus laterally below the anterior mandibular notch. Medially, the endocoronoid crest is well developed. Although the posteroinferior mandibular foramen is not preserved, the lingula is complete, including its posteroinferior edge. The lingula turns strongly medially, and there was no bridging of the mandibular foramen. The tip of the lingula is 23–24 mm posterior of the distal M_2.

The Palomas 88 Immature Mandibular Fragment

Preservation

The Palomas 88 specimen (Plate 6.68) consists of a lateral corpus fragment preserving the buccal distal dm_1 socket, the buccal dm_2 socket, and bone around the posterosuperior mental foramen. It contains the left dm_2 lacking only the tip of the mesial root. Maximum length and height of the bone are 22.5 and 15.5 mm.

Age at Death

The dm_2 crown retains thin matrix, so it is not possible to determine if it was unworn. However, any occlusal and mesial interproximal wear was little more than a polishing of the enamel (stage 1–2 of Smith 1984). The distal root apex is closed. The tooth lacks a distal interproximal facet. These observations suggest a tooth recently in full occlusion, or three to four years old at death (Friedlaender and Bailit 1969; Lunt and Law 1974; Liversidge and Molleson 2004).

Morphology

The only mandibular feature discernible on the Palomas 88 fragment is the position of its mental foramen (Plate 6.68). It is below the distal root of the dm_1, with its center 11 mm from the alveolar margin.

Comparative Assessment of the Palomas Mandibles

The overall proportions and a number of the details of the human mandible are the result of its developmental, functional, and spatial conformation to the cranial and pharyngeal regions, in addition to maintaining biomechanical effectiveness for mastication, deglutition, and respiration (Enlow and Hans 1996). In this context, many, if not most, of the quantitative (both morphometric and discrete trait) attributes employed to assess fossil *Homo* mandibles are clearly secondary, if not tertiary, reflections of spatial, developmental, and functional constraints and demands upon the bone (see, for example, Trinkaus 1993; Franciscus and Trinkaus 1995; Stefan and Trinkaus 1998; Rosas 2001). Yet thorough assessment of these aspects in the Middle and Late Pleistocene requires associated, mature, nonpathological facial skeletons essentially complete on at least one side; the current *Homo* fossil record provides less than a dozen minimally adequate specimens prior to the Upper Paleolithic. Given the partial preservation and anatomical isolation of the majority of the Late Pleistocene human mandibles, including those from the Sima de las Palomas, the Palomas mandibles are assessed using linear, angular, and cross-sectional morphometrics plus discrete traits, bearing in mind that multiple factors may well determine the affinities of these traits.

The Symphysis

The mandibular symphysis is preserved for Palomas 23 and is partially present on Palomas 6, and Palomas 1 and 59 retain lateral portions of it (Plates 5.2, 5.5, 5.7, and 5.11). Palomas 6 and 23 exhibit mentum osseum rank 2 (sloping with a hint of a mental trigone), and the two others exhibit either 2 or 3 (Dobson and Trinkaus 2002). It is unlikely that any of them had clear projection of the tuber symphyseos, and hence they are aligned with other Neandertal mandibles and distinct from those of all but two of the early modern human mandibles (Table 5.4).

Only one of the Palomas mandibles (Palomas 23) provides an anterior symphyseal angle (76°), but the angle for Palomas 6 can be estimated to be ≈70° (Table 5.5). Comparison of the former to later Pleistocene human distributions (Fig. 5.1) places it below the median of the Neandertal sample and separate from all of the early modern humans. (Only Předmostí 3, with its marked alveolar prognathism, has an angle <89° among the pooled early modern humans [n = 18].) The estimate for the Palomas 6 angle places it similarly among the Neandertals, even further from the early modern humans.

Comparison of the cross-sectional major axis angle (Fig. 5.1) provides similar results. The two Palomas angles are well below the Neandertal median, and only Palomas 23 (72°) is matched by an early modern human (Skhul 7). A pan–Old World Middle Pleistocene sample has even lower angles on average, indicating more retreating symphyses than most of the Neandertals.

The sizes and proportions of the symphysis can be assessed for Palomas 6 and 23 using geometric parameters of their symphyseal cross-sections (Table 5.5). The cross-sectional parameters were determined by digitizing the symphyseal subperiosteal contour with a Summagraphics 1812 tablet, using the alveolar plane for orientation. The measures were computed with SLICE/SLCOMM (Nagurka and Hayes 1980; Eschman 1992), modeling the symphysis as a solid beam. The comparison of the maximum (generally anterosuperior to posteroinferior) second moment of area (I_{max}) to its perpendicular (I_{min}) (Fig. 5.2) provides no meaningful separation of the comparative samples; the high Neandertal size outlier is Kebara 2. The values for Palomas 6 relative to those of Palomas 23 show its greater overall cross-sectional size

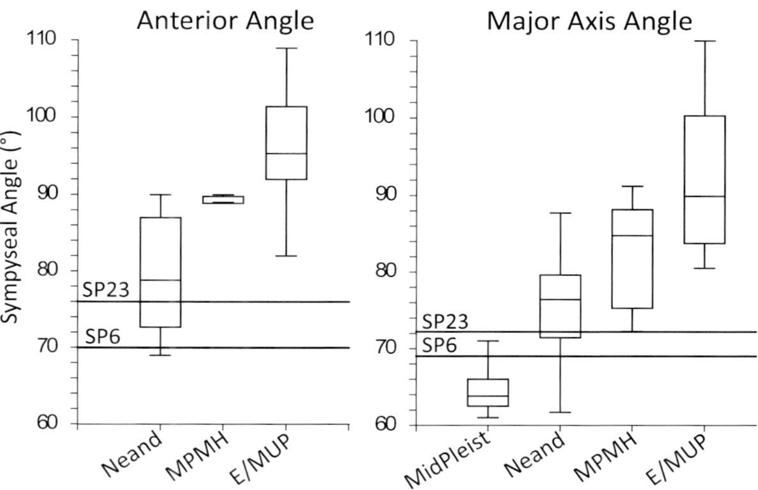

FIG. 5.1. Box plots of (*left*) the anterior symphyseal angle (infradentale-pogonion versus alveolar plane) and (*right*) the cross-sectional major axis symphyseal angle (major axis versus alveolar plane) for Palomas 6 and 23 and comparative samples. Given anteroinferior rounding, anterior symphyseal angles are not available for the Middle Pleistocene sample. The Middle Pleistocene cross-sectional angles include non-European specimens. Anterior angle sample sizes are Neandertals (Neand), 22; Middle Paleolithic modern humans (MPMH), 4; Early/Mid Upper Paleolithic modern humans (E/MUP), 15. Cross-sectional angle samples are Middle Pleistocene (MidPleist), 9; Neandertals, 24; Middle Paleolithic modern humans, 5; Early/Mid Upper Paleolithic modern humans, 13.

(total subperiosteal area of ≈365 mm² versus 290 mm²), but both of the Palomas mandibles cluster along the shorter and thicker portion of the Pleistocene distribution. They are nonetheless close to mandibles from all four comparative samples.

It is also possible to scale the symphyseal vertical second moment of area (I_x) against the dental arcade breadth (Fig. 5.3). The anteroposterior second moment of area should be scaled against mandibular length (Daegling 2001), which neither of the Palomas mandibles preserves. There is little variation in the I_x values across the comparative samples; it is principally the wider dental arcades of the Middle and Late Pleistocene archaic humans that make them appear more gracile. Palomas 23 is among the mandibles with the shorter symphyses and wider arcades, exceeded

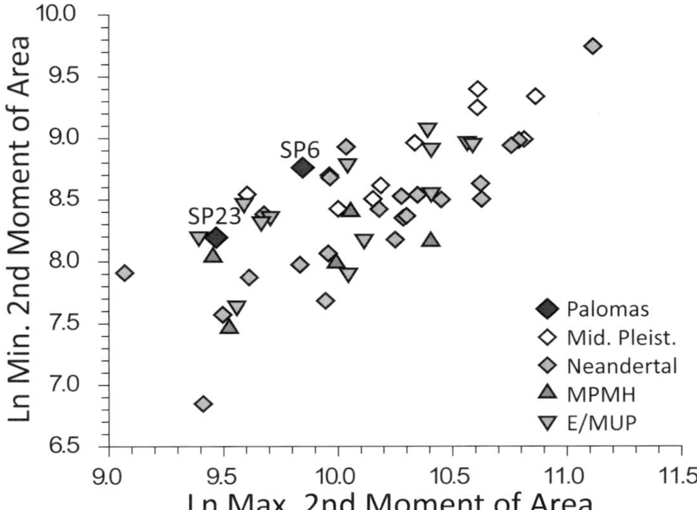

FIG. 5.2. Comparison of the minimum versus maximum second moments of area of the mandibular symphyseal cross-section, modeled as a solid beam. Mid. Pleist.: Middle Pleistocene; MPMH: Middle Paleolithic modern humans; E/MUP: Early and Mid Upper Paleolithic modern humans.

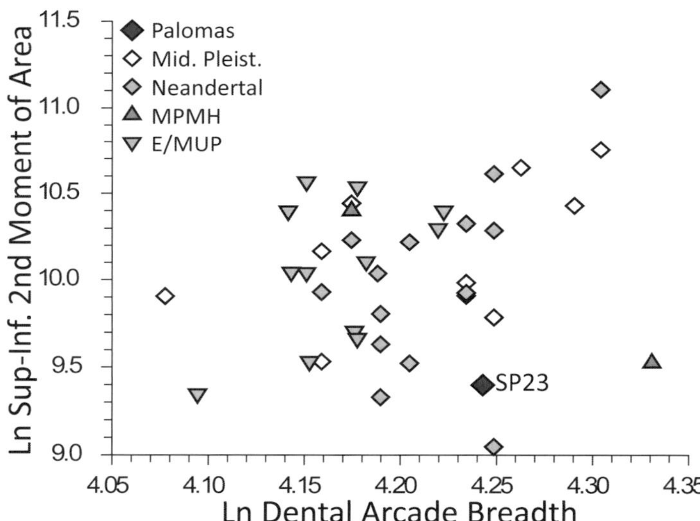

FIG. 5.3. Comparison of the symphyseal vertical second moment of area versus dental arcade breadth for Palomas 23 and comparative samples. Mid. Pleist.: Middle Pleistocene; MPMH: Middle Paleolithic modern humans; E/MUP: Early and Mid Upper Paleolithic modern humans.

in this aspect only by Qafzeh 7 and Vindija 226, bearing in the mind the estimated dental arcade breadths for both Palomas mandibles.

The Lateral Corpus

The lateral corpori of the Palomas mandibles exhibit generally parallel alveolar and basilar margins, nonprominent lateral eminences, and otherwise smooth surfaces. The adult mental foramina are largely below P_4/M_1 and M_1 (all except for the right Palomas 1 foramen and the more distal secondary one of Palomas 23). In this they are similar to most of the Neandertal and Middle Pleistocene mandibles (Table 5.6). They overlap the ranges of variation of the early modern human samples, although they are the less common configurations for those samples.

The three very young Palomas mandibles (Palomas 7, 49, and 88) have mental foramina in vicinity of the dm_1, the most common position for the mental foramen among Neandertals less than six years old (Coqueugniot 1999). Young early modern humans have similar positions for the mental foramen (Trinkaus 2002). However, the older, early adolescent Palomas 80 mandible has its mental foramen mesial of the P_3/P_4 interdental septum; among immature Neandertals greater than six years old, all have the mental foramen at (n = 1) or distal (n = 9) of the P_3/P_4 septum (Coqueugniot 1999; Quam et al. 2001). Similarly, all but one (Les Rois 1) of the early modern human mandibles (N = 12) in this age range have the mental foramen distal of the P_3 (or dm_1). Palomas 80 therefore has a rather mesial position for this foramen for either a Neandertal or an early modern human.

Assessment of adult lateral corpus thickness, using height and breadth at the mental foramen (Fig. 5.4) places all four Palomas mandibles at the low end of the later Pleistocene range of variation in height. The Palomas 6 corpus height measurement may be slightly underestimated, given postmortem damage to the bone (see above), but increases of its corpus height to a maximum of 30 mm would still maintain it among the shorter Neandertal mandibles. There is no significant difference in height across the comparative samples (Kruskal-Wallis p = 0.104), but they are significantly different in breadth (p < 0.001), with a consistent decrease from the Middle Pleistocene to the Neandertals to the Middle Paleolithic and then to Upper Paleolithic modern humans. Only one of the Neandertals has a breadth measurement <14.0 mm (Subalyuk 1 at ≈12 mm, although it lacks the basilar portion near the mental foramen), and most of the early modern human corpus breadths fall below that value. Only Pataud 1 and Qafzeh 9, plus the northeast African Nazlet Khater 2 and the east Asian Zhiren 3 (included respectively with the Upper Paleolithic and Middle Paleolithic modern humans), have corpus breadths >14 mm. Two of the Palomas mandibles, 1 and 6, are well within the overall Neandertal range. Palomas 23 is at the lower margin of most of that distribution. Palomas 59, however, has one of thinnest of these Neandertal mandibular corpori, exceeding only Subalyuk 1, and it clusters with the early modern human mandibles in this feature.

Palomas 59 does have among the smallest of the known Neandertal teeth; its M_1 crown breadth (9.5 mm) is 2.80 standard deviations from a Neandertal sample mean (10.9 ± 0.5 mm, n = 51) and its M_2 breadth (9.3 mm) is 2.25 standard deviations from a similar sample mean (11.0 ± 0.8 mm, n = 44). However, among Late Pleistocene humans, the individual with the largest molars (Oase 1) has a modest corpus breadth (Trinkaus and Rougier 2013; Trinkaus et al. 2013), and the mandible with the thickest corpus (Kebara 2) has modest molars (Tillier et al. 1989). Least squares regressions of mental foramen corpus breadth against M_1 and M_2 crown breadths across the pooled Late Pleistocene sample, not including the Palomas specimens, produces negative slopes of −0.904 (n = 36) and −0.034 (n = 35), respectively, neither of which is significantly different from zero (p = 0.203 and p = 0.954, respectively). The modest corpus breadth of Palomas 59 is therefore not the result of its small teeth.

Later Pleistocene immature mandibles can be assessed by plotting lateral corpus dimensions against age, given expected increases with age

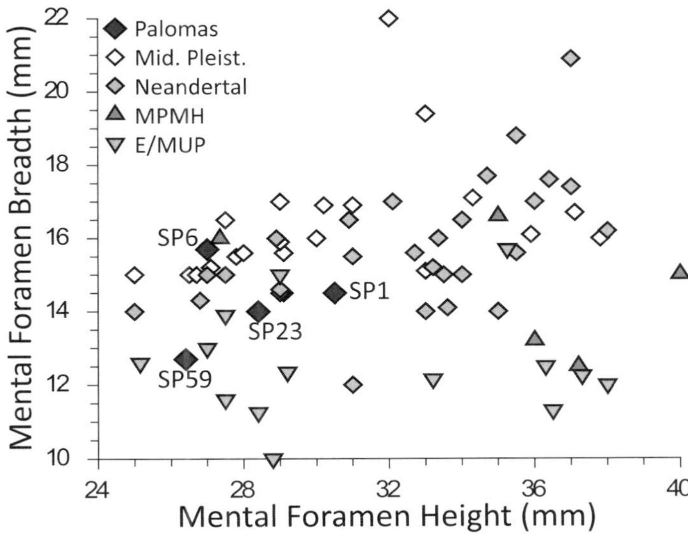

FIG. 5.4. Comparison of mandibular corpus breadth versus height at the mental foramen for the Palomas and comparative sample mature mandibles. Mid. Pleist.: Middle Pleistocene; MPMH: Middle Paleolithic modern humans; E/MUP: Early and Mid Upper Paleolithic modern humans. The high outliers are Arago 2 and Mauer 1 (Middle Pleistocene) and Kebara 2 (Neandertal).

despite the presence of forming tooth crowns within them. The samples are generally similar in corpus height at the dm_1/dm_2 (Fig. 5.5), a measure that increases similarly with age in all of the samples. The similarity parallels the nonsignificant difference in corpus height among mature specimens. However, there is less of a clear increase in breadth with age, probably due to the formation of the tooth crowns prior to the roots (influencing corpus breadth earlier in development than corpus height). The younger Neandertal immature mandibles tend to be thicker than similarly aged early modern human ones (Fig. 5.5). The two Palomas mandibles providing both of these corpus measurements, Palomas 7 and 49, fall among these younger Neandertals, separate from the Upper Paleolithic ones, especially in corpus breadth. The older Palomas 80 mandible does not quite extend mesially with the full corpus to the dm_1/dm_2 (or P_3/P_4), but it provides a corpus breadth at dm_2/M_1. Given the similarity of breadth measurements between these two locations among the mature

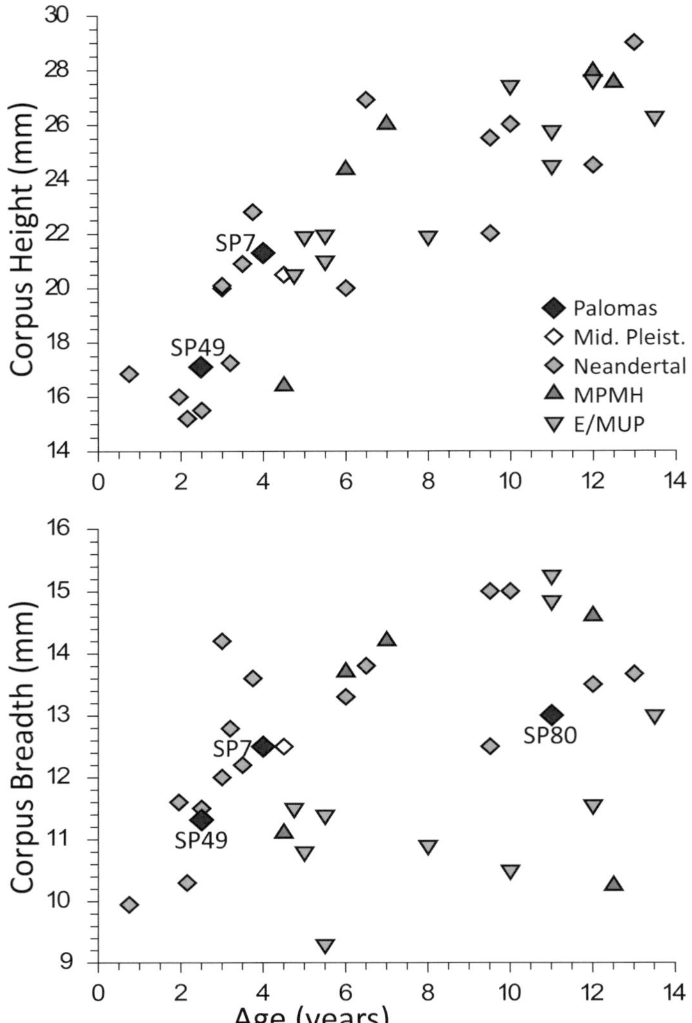

FIG. 5.5. Comparisons of corpus dimensions at the dm$_1$/dm$_2$ (or P$_3$/P$_4$) for immature mandibles versus modal age at death based on their dentitions, for height (*above*) and breadth (*below*). The breadth for Palomas 80 is estimated from its more distal (P$_4$/M$_1$) breadth. Mid. Pleist.: Middle Pleistocene; MPMH: Middle Paleolithic modern humans; E/MUP: Early and Mid Upper Paleolithic modern humans.

Palomas mandibles (Table 5.2), its breadth of 13 mm is included at its average age estimate (Fig. 5.5). This value places it among the narrower Neandertal ones and in the middle of the early modern human values for that early-second-decade age. Consequently, the age-relative lateral

corpus dimensions of these immature mandibles provide little separation across the later Pleistocene samples in height and only a partial one in breadth, and the Palomas ones fall with the other immature Neandertal specimens.

The Inferolateral Corporeal Margins

Associated with these corpus characteristics are the inferolateral protrusions of the corpori present on Palomas 6 and 23, with a smaller and more restrictive projection on Palomas 7 (Plates 5.5 to 5.9). Palomas 59 lacks one, Palomas 1 and 80 appear not to have them, and their presence on Palomas 49 is equivocal. In a sample of 24 late adolescent and adult Neandertal mandibles, many of them (62.5%) have a gently rounded swelling along the inferolateral mandibular corpus, occasionally with a prominent anterior marginal tubercle, principally below the premolars and mesial molars, with a shallow longitudinal sulcus above it (see also Rosas 2001). Yet all of them have normal digastric impressions, none of them possess a lateral crest or extended swelling (as in Palomas 6, 7, and 23), and none of them have what may be a secondary edge of bone partially fused onto the usual inferior corpus (as in Palomas 23). The prominent part of the projection on Palomas 23 below the P$_4$ may be an anterior marginal tubercle, but the remainder of the extension and its partial separation from the inferior corpus are unknown among other mature Neandertals. It is also possible that the swellings on Palomas 6 and 7 share an etiology with anterior marginal tubercles.

Among the eight Neandertal mandibles between the ages of two and five years, six (Dederiyeh 1 and 2, Devil's Tower 1, Roc-de-Marsal 1, Molare 1, Pech-de-l'Azé 1) lack the ridge present on Palomas 7. However, one (Archi 1) has a similar bony formation (Mallegni and Trinkaus 1997), and Dederiyeh 1 and Spy 6 have distinct anterior marginal tubercles (Crevecoeur et al. 2010; Fukase et al. 2015). There is also a very small crest in the same position on the late Middle Pleistocene immature La Chaise 13 mandible

(Tillier and Genet-Varcin 1980). None of the immature or mature early modern human mandibles known to us exhibit an extended inferolateral projection along the anterior and lateral corpus.

The Dental Arcade

Breadths are provided for the Palomas 23 dental arcade (Table 5.7) based on a doubling of the distance from the midline to the right side, the midline having been determined from the anterior and posterior symphyseal surfaces. The measurements are therefore approximate. The resultant breadth at the M_1/M_2 ($\approx$65 mm) is in the middle of the later Pleistocene ranges of variation (Neandertals: 67.3 ± 3.5 mm, n = 15; pooled early modern human sample: 64.7 ± 3.6 mm, n = 14).

The bi-dc_1 and bi-dm_2 arcade breadths can be used to calculate an index for immature mandibles that is then plotted against developmental age (Fig. 5.6). The value for Palomas 49 (62.9) is in the middle of the Neandertal variation for younger mandibles, joined by the one Middle Pleistocene immature mandible and the early modern human ones. They all have anterior arcade breadths about two-thirds of their more distal breadths. The older, early-second-decade mandibles are relatively wider anteriorly, perhaps related to the full eruption of the permanent incisors (the high outlier is Sunghir 3). There is little difference across the samples, and Palomas 49, as one of the youngest of them providing both measurements, is among the other specimens. The Neandertal (and early modern human) arcade breadths have previously been shown to be large compared to those of recent humans (Mallegni and Trinkaus 1997), and the Neandertal ones appear to be large relative to lateral corpus height (Verna et al. 2012), but the anterior to posterior proportions remain similar across these later Pleistocene immature specimens. Moreover, even though some recent human infants have relatively narrow anterior dental arcades, the later Pleistocene specimens all fall well within recent variation (see also Mallegni and Trinkaus 1997).

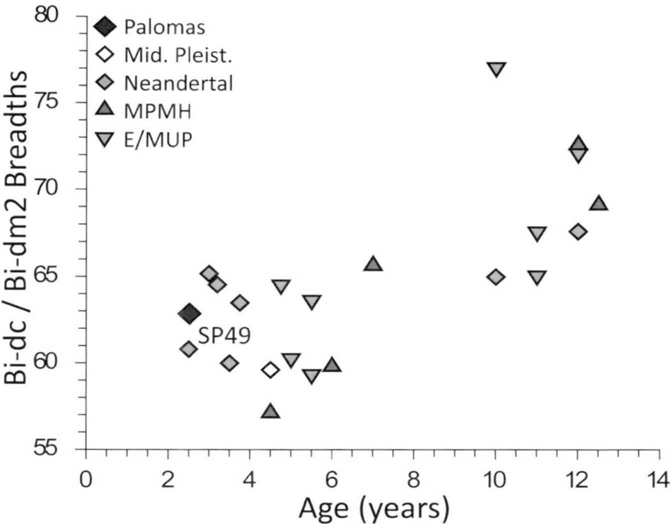

FIG. 5.6. Comparison of the index of bi-dc_1 versus bi-dm_2 arcade breadth against modal age at death for Palomas 49 and immature later Pleistocene mandibles. Mid. Pleist.: Middle Pleistocene; MPMH: Middle Paleolithic modern humans; E/MUP: Early and Mid Upper Paleolithic modern humans.

The Ramus

None of the Palomas mandibles retain a complete ramus, but four of them give indications of their retromolar proportions, and two mandibles provide discrete traits on their rami (Table 5.8). There are also a few features evident on the Palomas 96 mandible.

Palomas 1 exhibits a large retromolar space, especially on the less complete but undistorted left side. Palomas 23 retains the anteroinferior portion of one. It is likely that Palomas 59 possessed a small retromolar space. However, a conservative reconstruction of the distal M_3 and anterior ramal positions of Palomas 6 indicate that it lacked such a gap in *norma lateralis*. The resultant frequency of 75% is essentially the same as that of the larger Neandertal sample, similar to that of the Middle Pleistocene Europeans, and distinct from the Upper Paleolithic early modern human distribution. The Middle Paleolithic modern humans have an intermediate frequency.

However, as noted (Franciscus and Trinkaus 1995; Rosas and Bastir 2004), retromolar space presence is a secondary reflection of mandible length, dental arcade length, and ramus breadth proportions and not a trait per se.

Neandertal rami have been characterized as having relatively asymmetrical mandibular notches, with the lowest point adjacent to the condylar neck and a frequent bridging of the mandibular foramen (Kallay 1970; F. H. Smith 1978; Stefan and Trinkaus 1998; Jidoi et al. 2000; Wu and Trinkaus 2014). Almost all early modern humans have symmetrical mandibular notches, the overwhelmingly dominant pattern among recent humans, whereas about two-thirds of the Neandertals and half of the Middle Pleistocene specimens have the other arrangement. Palomas 1 and 80, although lacking the condylar portion of the notch, are very likely to have had the asymmetrical arrangement, and Palomas 96 has a distinctly asymmetrical mandibular notch. Palomas 80 lacks bridging of the mandibular foramen (horizontal-oval form, lingular bridging), given its projecting lingula, the most frequent configuration in all of the samples despite an elevated occurrence of bridging among the Neandertals. The Palomas 96 ramus also exhibits a prominent superior medial pterygoid tubercle, a common Neandertal feature but one that is not unique to them (Richards et al. 2003; Wu and Trinkaus 2014). Its notch crest also appears to have been toward the middle of the condyle. Therefore, in these discrete traits of the ramus, the Palomas mandibles are comfortably within the Neandertal range of variation.

Summary

The Palomas mandibles therefore fall within the morphological range of variation of Neandertal mandibles, and they are distinct from those of early modern humans in a number of characteristics. The principal features that confirm this morphological alignment are their anterior symphyseal configurations, their symphyseal angles (both anterior and cross-sectional), the relatively distal positions of (most of) the mental foramina, the thick lateral corpori of three of the mandibles, the presence of a retromolar space in three of the specimens, and the asymmetrical mandibular notches of Palomas 1 and 80 (plus Palomas 96).

At the same time, Palomas 59 has one of the thinnest lateral mandibular corpori known for a Neandertal, with only Subalyuk 1 providing a lower corpus breadth. Palomas 80 has an unusually mesial position of the mental foramen, matched among Neandertal and early modern human late juvenile and adolescent mandibles only by the late Aurignacian early modern human Les Rois 1. Mental foramen position, as with retromolar space presence, is a complex product of total facial length, dental arcade length, and inferior alveolar nerve canal length (Trinkaus 1993). Since most Neandertal facial lengths are similar to, or shorter than, those of earlier *Homo* (Trinkaus 2003), and their dental arcade lengths are similar to those of most other Pleistocene *Homo* (Franciscus and Trinkaus 1995), the position of most Neandertal mental foramina below more distal teeth suggests that they share a shortened inferior alveolar canal length with modern humans.

The more mesial position of the mental foramen in Palomas 80 could therefore reflect either a retained ancestral long canal or, as suggested by Rosas et al. (2006), a reflection of a foreshortened facial length among southern European Neandertals. Palomas 80 is not sufficiently complete to assess its overall dimensions, but visual comparison of it to similarly aged Neandertal mandibles (especially Malarnaud 1 and Teshik-Tash 1) does not imply any reduction in mandibular length. As noted above, the mandible length of Palomas 1 cannot be measured reliably, but general assessment from rearranging the pieces of the right side (Plate 5.4) suggests a superior mandible length (infradentale to midcondyle) in the vicinity of 110 mm, well within the Neandertal range of ≈100 to ≈122 mm (see chapter 14). It is also possible to measure the minimum ramus breadth of Palomas 96 (40.0 mm), which is close to the middle of a Neandertal range of variation

(41.2 ± 2.8 mm, n = 16). Given preservation, it is therefore unclear whether the Palomas mandibles exhibit a trend toward facial shortening as suggested by Rosas et al. (2006) for more southern Neandertals. Nonetheless, all of them have relatively low corpus heights, and the small size or apparent absence of retromolar spaces in Palomas 6 and 59 (despite small teeth in the latter) imply a similar pattern for these Mediterranean Neandertals.

In this context, the significance of the inferolateral projections on Palomas 6, 7, and 23 (plus Archi 1 and La Chaise 13) is unclear. They do not appear to be pathological. The only muscles inserting in the region are digastric anteroinferiorly and platysma laterally, neither one of which is likely to produce the more posterior portions of the flanges. They may represent variants of anterior marginal tubercles, especially the immature ones. It is also possible that they represent a genetically relevant trait (*sensu* Hauser and DeStefano 1989), in which case they might indicate a close population relationship among the three Palomas individuals.

At the same time, there is considerable variation in the mandibles from the Sima de las Palomas in discrete traits, corpus breadths, and symphyseal morphology. Even though this level of variation is evident in the overall Neandertal sample, the Palomas mandibles emphasize the degree of variation present within, as well as across, site-specific samples of Neandertals.

TABLE 5.1. Inventory of the Palomas human mandibular remains

SP NO.	OLD NO.	IDENTIFICATION	MATURITY	PROVENIENCE	DISCOVERY DATE
1c	CG-1[a]	Mandible right with I_1, C_1 to M_3	mature	Shaft upper breccia	1991
1d	CG-1[b]	Mandible left with C_1, M_1 to M_3	mature	Shaft upper breccia	1991
6	CG-6	Mandible corpus left with C_1 to M_2 roots	mature	Hillside rubble	1993
7	CG-7	Mandible mesial corpus left C_1 and P_3 germs	immature	Main chamber mine rubble	1993
23	CG-2	Mandible right from left I_1 to gonion with right I_1 to M_3 roots or teeth	mature	Hillside rubble	2 July 1995
49	CG-51; CG-49	Mandible corpus with left dc_1 and dm_1 fragments and left I_2 and C_1 germs	immature	Upper cutting level IA	2 August 1996
59	CG-70	Mandibular body left from I_2 to M_2 with C_1 to M_3	mature	Upper cutting level 2f	1 August 1998
80	CG-105	Mandible corpus/ramus left with M_2 and M_3 germ	immature	Upper cutting level 2d	5 August 2003
88	(Code 15)	Mandibular left lateral corpus and dm_2	immature	Upper cutting level 2g	23 July 2004

[a]The original cast (Fig. 5.1) includes a right I_1; it is now missing, and its description in chapter 6 is based on the original cast.
[b]The original cast (Fig. 5.1) includes the left I_1 to P_3; the I_1, I_2, and P_3 are likely the teeth inventoried as Palomas 20 to 22. The C_1 is missing, and its description is based on the cast (see chapter 6).

TABLE 5.2. Mandibular corpus measurements of the mature Palomas mandibles, in millimeters and degrees

MEASUREMENT	PALOMAS 1	PALOMAS 6	PALOMAS 23	PALOMAS 59
Symphyseal height	—	((29.0))[a]	25.4	—
Symphyseal breadth	—	16.5	14.2	—
Anterior symphyseal angle	—	((70°))[b]	76°	—
Height @ mental foramen	(30.5) rt	(27.0)	28.4	26.4
Breadth @ mental foramen	(15.0) rt	15.7	14.0	12.7
Height @ P_3	30.6 rt	28.0	(25.9)	—
Breadth @ P_3	—	15.9	15.0	—
Height @ P_3/P_4	30.7 rt	(27.5)	26.3	25.3
Breadth @ P_3/P_4	(14.5) rt	15.6[c]	14.2[c]	13.0
Height @ P_4/M_1	((31.5)) rt	(26.6)	27.8	26.4
Breadth @ P_4/M_1	—	15.6	14.0	12.7
Height @ M_1/M_2	32.7 lt	26.1	28.8	26.7
Breadth @ M_1/M_2	—	15.6	13.6	14.3
Height @ M_2/M_3	32.2 lt	26.5	27.7	26.0
Breadth @ M_2/M_3	—	(15.7)	14.0	15.6
Height @ M_3	35.2 lt	—	27.1	—
Breadth @ M_3	—	—	14.4	—
Mental foramen number	1	1	2	1
Mental foramen position[d]	P_4, P_4/M_1	P_4/M_1	M_1, M_1/M_2	P_4/M_1
Mental foramen ant-post diameter[e]	—	6.6	5.7	4.0
Mental foramen sup-inf diameter	3.0 rt	4.0	3.5	2.7
Mental foramen to alveolar plane[d]	17.1, (18.0)	(15.0)	17.0	14.5
Mylohyoid line to M_3 alveolus	9.2 lt	—	10.8	≈6.0

Note: Single parentheses indicate minor estimation; double parentheses indicate more substantial estimation.
[a]Estimated alveolar plane position, hence more than minimal estimation.
[b]The Palomas 6 anterior symphyseal angle was previously estimated at 75°–80°; reassessment of the missing infradentale position suggests a value closer to 70°.
[c]Corpus breadth without the inferolateral crests; the maximum breadth is 16.1 mm for Palomas 6 and 15.0 mm for Palomas 23.
[d]Mental foramen position and distance to the alveolar plane for Palomas 1 are provided for the right and then the left sides. For Palomas 23 the two locations reflect the presence of two foramina; the foramen diameters and the single distance to the alveolar plane refer to the larger, more anterosuperior foramen.
[e]Perpendicular diameters of the mental foramen. The maximum mental foramen diameters for Palomas 6, on the lateral corpus, are 7.9 and 5.2 mm for height and breadth. Dimensions for Palomas 23 are those of its larger, more mesial foramen.

TABLE 5.3. Mandibular corpus measurements of the Palomas immature mandibles, in millimeters

MEASUREMENT	PALOMAS 7	PALOMAS 49		PALOMAS 80	PALOMAS 88
	Left	Right	Left	Left	Left
Symphyseal height	—	(21.0)		—	—
Height @ dc_1/dm_1	—	21.2	20.2	—	—
Breadth @ dc_1/dm_1	—	11.4	11.5	—	—
Height @ mental foramen	21.3	19.2	—	—	—
Breadth @ mental foramen	12.7	11.3	11.3	—	—
Height @ dm_1	22.8	—	(19.2)	—	—
Height @ dm_1/dm_2	21.3	17.1	—	—	—
Breadth @ dm_1/dm_2	12.5	11.3	11.3	—	—
Height @ dm_2	(18.5)	—	—	—	—
Breadth @ dm_2/M_1	—	—	—	13.0	—
Height @ M_1/M_2	—	—	—	21.7	—
Breadth @ M_1/M_2	—	—	—	14.2	—
Height @ M_2/M_3	—	—	—	(24.0)	—
Breadth @ M_2/M_3	—	—	—	13.9	—
Mental foramen number	2	1	1	—	1
Mental foramen position	dm_1, dm_1/dm_2	dm_1	dm_1	—	dm_1
Mental for. to alveolar plane	—	—	—	—	(11.0)

Note: Single parentheses indicate minor estimation.

TABLE 5.4. Anterior mandibular symphysis mentum osseum ranks

	MENTUM OSSEUM RANK					
	1	2	3	4	5	N
Palomas 1		(X)				
Palomas 6		X				
Palomas 23		X				
Palomas 59		(X)				
Middle Pleistocene	40.0%	60.0%	—	—	—	10
Neandertals	18.5%	59.3%	22.2%	—	—	27
MPMH	—	14.3%	—	85.7%	—	7
E/MUP	—	—	2.6%	71.1%	26.3%	38

Note: The mentum osseum ranks follow Dobson and Trinkaus (2002). The Palomas 1 and 59 ranks are in parentheses, given damage to their mandibular symphyses. Middle Pleistocene = European Middle Pleistocene archaic humans; MPMH = Middle Paleolithic modern humans; E/MUP = Early and Mid Upper Paleolithic modern humans.

TABLE 5.5. Palomas 6 and 23 mandibular symphyseal cross-sectional parameters, modeling the symphysis as a solid beam

PARAMETER	PALOMAS 6[a]	PALOMAS 23[b]
Total area (mm²)	(365)	290
Maximum 2nd moment of area (I_{max}) (mm⁴)	(18891)	12930
Minimum 2nd moment of area (I_{min}) (mm⁴)	(6360)	3613
Anteroposterior 2nd moment of area (I_y) (mm⁴)	(8008)	4489
Superoinferior 2nd moment of area (I_x) (mm⁴)	(17342)	12053
Theta (orientation of I_{max} relative to the alveolar plane)	(69°)	72°

[a]The Palomas 6 cross-section is from a scaled photograph of the fossilization break, and its superior portion is located slightly lateral of the symphyseal midline. Its values are placed in parentheses, since the superolabial corner was visually interpolated following the preserved contours and estimated alveolar thickness.
[b]The Palomas 23 cross-section has been transferred using polysiloxane putty and represents the midline cross-section without the displaced anteroinferior digastric flange.

TABLE 5.6. Mental foramen position relative to the dentition for mature mandibles, counting each side as 0.5 in cases of asymmetry

	$P_3 + P_3/P_4$ [a]	P_4	P_4/M_1	M_1	N
Palomas 1[b]		X	X		
Palomas 6			X		
Palomas 23				X	
Palomas 59[c]			X		
Palomas 80 imm[d]	(X)				
Middle Pleistocene	4.5%	—	25.0%	70.5%	22
Neandertals	—	11.4%	34.3%	54.3%	35
MPMH	—	57.1%	28.6%	14.3%	7
E/MUP	10.8%	68.9%	20.3%	—	37

Note: Middle Pleistocene = European Middle Pleistocene archaic humans; MPMH = Middle Paleolithic modern humans; E/MUP = Early and Mid Upper Paleolithic modern humans.
[a]$P_3 + P_3/P_4$ pools together the numbers of mental foramina located mesial of the P_4.
[b]The two values for Palomas 1 reflect its right and left sides.
[c]The intermediate position for Palomas 59 represents its position on the edge of the mesial M_1 root.
[d]The Palomas 80 inferior alveolar nerve canal extends mesially to at least the level of the P_3, but the final adult position of the mental foramen may well have been more distal. It is therefore placed conservatively at P_3/P_4 to P_4. This ambiguity is indicated by placing its location in parentheses.

TABLE 5.7. External dental arcade breadths across the buccal interdental septa for the Palomas 23 adult mandible and the Palomas 49 immature mandible, in millimeters

	PALOMAS 23	PALOMAS 49
at the C_1/P_3, dc_1/dm_1	(35.8)	35.2
at the P_3/P_4, dm_1/dm_2	(44.4)	41.8
at the P_4/M_1, dm_2/M_1	(52.6)	(56.0)
at the M_1/M_2	(64.8)	—
at the M_2/M_3	(69.6)	—
at the distal M_3	(75.5)	—

Note: Estimated values are in parentheses; they are two times the distance from the midline (as indicated by the symphysis) to the preserved side.

TABLE 5.8. Lateral mandibular discrete traits, counting presence as the purportedly Neandertal configuration of the trait; percent present (N)

INDIVIDUAL OR GROUP	RETROMOLAR SPACE PRESENCE	MANDIBULAR FORAMEN BRIDGING	MANDIBULAR NOTCH ASYMMETRICAL
Palomas 1	Present	—	((Present))
Palomas 6	(Absent)	—	—
Palomas 23	Present	—	—
Palomas 59	(Present)	—	—
Palomas 80 (imm.)	—	Absent	Present
Palomas 96	—	—	Present
Middle Pleistocene	66.7% (21)	7.7% (13)	50.0% (11)
Neandertals	73.1% (26)	28.5% (26)	66.7% (12)
MPMH	42.9% (7)	0.0% (5)	0.0% (4)
E/MUP	13.8% (29)	1.9% (26)	3.7% (27)

Note: Middle Pleistocene = European Middle Pleistocene archaic humans; MPMH = Middle Paleolithic modern humans; E/MUP = Early and Mid Upper Paleolithic modern humans. The single parentheses indicate minor estimation of the trait presence; the double parentheses refer to the more ambigious indications of the Palomas 1 mandibular notch configuration.

6 The Palomas Dental Remains

Preservation, Wear, and Morphology

JOSEFINA ZAPATA, PRISCILLA BAYLE, A. VINCENT LOMBARDI,
ALEJANDRO PÉREZ-PÉREZ, AND ERIK TRINKAUS

THE DENTAL REMAINS from the Sima de las Palomas consist of a largely complete dentition in the fragmentary maxillae and mandible of Palomas 1, a C_I to M_2 series preserved in the partial left hemimandible of Palomas 59, broken roots in the Palomas 6 mandible, incompletely formed tooth crowns in the Palomas 7 and 80 immature mandibles, fire-damaged roots and three partial molar crowns in the Palomas 23 mandible, two premolars in the Palomas 68 maxillary fragment, a deciduous molar in the Palomas 88 mandibular fragment, and a large series of isolated deciduous and permanent teeth (Table 6.1). These dental remains are described and illustrated here individually, with notes on their preservation, current condition, occlusal and interproximal attrition, probable ages at death, and considerations of the occlusal and radicular morphologies of the individual teeth. Dental pathological lesions are noted, but more detailed considerations of them and assessments of oral abnormalities are provided in chapter 12. These tooth-by-tooth descriptions are followed by a summary and comparison of crown discrete traits.

Given the hard, brecciated sediment of much of the Sima de las Palomas stratigraphic column, a number of the teeth have hard, adhering matrix on them. Most of the matrix has been removed from the specimens. However, for a number of them it has not been possible to remove all of the matrix (mechanically or chemically) without risking damage to the tooth. Yet since the teeth were micro-CT scanned (chapter 3), it has been possible to virtually clean a number of the teeth to reveal the occlusal morphology as well as some internal structures. As is appropriate, the resultant "cleaned" images are provided, together with photographs of the teeth in their current conditions.

Unfortunately, through the peregrinations of a portion of the Palomas human remains, several isolated teeth that were originally inventoried— Palomas 27, 38, 41, 43, and 47 (Walker 2001; Walker et al. 2008)—are currently missing. Data on them are included from Walker et al. (1998) as available (especially dental metrics; see chapter 7), but those data are less complete than the data assembled from the currently available remains.

The assessments of dental age at death for immature specimens are based on tooth calcification, following Lunt and Law (1974), Smith (1991), Hillson (1996), and AlQahtani et al. (2010), using when appropriate the calcification stages of Moorrees et al. (1963) (for example, $Cr_{3/4}$ for the crown three-quarters complete, Cr_c for the crown just complete, $R_{3/4}$ for the root three-quarters complete, and so on). Dental eruption schedules follow Friedlaender and Bailit (1969), Wolpoff (1979), Smith (1991), and Liversidge and Molleson (2004).

The estimations of the ages at death for the

mature specimens, plus the probably immature teeth that are nonetheless completely formed, are based on degrees of overall occlusal wear. The actual rates of wear in the Palomas sample are unknown, and too few specimens provide associated dentitions to use the sequential aging technique of Miles (1962). It is assumed that the experienced rates of occlusal abrasion were similar to those of other Late Pleistocene human samples, especially those that are associated with dental calcification and nondental skeletal indicators of ages at death (Wolpoff 1979; Trinkaus 1983; Hillson et al. 2006; Trinkaus et al. 2014a). Moreover, these rates of overall dental wear appear to have been generally similar across these Late Pleistocene samples as well as a substantial proportion of preindustrial recent humans (Wolpoff 1979; Trinkaus 1983; Hillson 1996; Hillson et al. 2006). For reference, the occlusal wear is scored using the stages of Smith (1984). For teeth whose wear matches one of the images within a row subsumed by a single stage, they are designated a to c, left to right in Smith's diagrams.

The occlusal morphology of the teeth is described as available given variable amounts of antemortem abrasion, postmortem damage, and adhering hard matrix on the teeth. For traits for which grades, or stages of development, of the feature have been defined through the Arizona State University Dental Anthropology System (ASUDAS) (Turner et al. 1991; Scott and Turner 1997) and/or its expansion for Pleistocene humans (Bailey 2002a, 2002b, 2006; see also Martinón-Torres et al. 2012), those grades are indicated in the individual tooth descriptions, whether they come from the primary, recent human–oriented, ASUDAS or from its expansion by Bailey.

The Palomas frequencies for a select set of these traits, ones that are commonly recorded and have been assessed for substantial samples of Pleistocene western Eurasian humans (Bailey 2006; Martinón-Torres et al. 2012; Bailey and Zubov 2014), are compared to the comparative data sets for permanent (Tables 6.6 and 6.7) and deciduous teeth (Table 6.9). It should be kept in mind that these grades are ordinal scores of what are continuous variations, and that fine distinctions between adjacent grades are often hard to make, especially on worn, damaged, or obscured crowns. For this reason, following Bailey (2006), the frequency comparisons provided in Tables 6.6 and 6.7 pool together the higher scores for most features into a "presence" category.

Finally, all of the Palomas teeth, except for those securely in alveoli, are numbered and treated as isolated teeth. A few of them (Palomas 20 to 22) probably derive from the Palomas 1 maxillae and mandible, based on visual matching with the cast of the original (before cleaning) specimen (Plate 5.1); they are described separately but are also considered with the remainder of Palomas 1 (see chapters 7 and 11). There are also a few sets of anterior teeth that may well go together based on similar wear and hypoplasias (see below and chapter 11). These possible associations, as well as a few that do not work, are mentioned in the individual descriptions. The specimens nonetheless remain numbered separately.

The Palomas 1 Dentition

The Palomas 1 maxilla and mandible were discovered badly crushed, distorted, and twisted as a cemented mass (Plate 5.1). The four main elements were cast and separated, and then the related pieces were put together as well as could be done given distortion and missing portions (Plate 5.4). The teeth are retained in the four separate pieces (two maxillae and two hemimandibles). The maxillary teeth consist of the right C^1 and the right and left P^3 to M^3 (Plate 6.1). The mandibular ones are the right C_1 to M_3 and the left M_1 to M_3 (Plate 6.8). In addition, there are four teeth that were separated from the maxillae and mandibular sections: the left C^1, the left C_1, and the right and left I_1s (also evident on the cast). The left I_1 may well be the left I_1 inventoried as Palomas 21, and the left I_2 may well be the left I_2 inventoried as Palomas 20; they are described below. The left P_3 inventoried as Palomas 22 may represent the missing left P_3 of Palomas 1, but it is also inventoried and described below.

The preserved teeth in their alveoli are worn

but generally well preserved. There are signs of mild chemical erosion to most of them, as well as to the adjacent bone, but it has had little effect on the macroscopic morphology of the teeth.

Maxillary C¹ Right

The crown is largely complete (Plate 6.2), with a chip of enamel (1.5 × 5.7 mm) lost postmortem from the labial side. In addition, the mesial side was eroded, removing the mesial half of the root and leaving the pulp canal open lengthwise. In lingual view, the tooth exhibits a vertical fracture running along its length in a zigzag shape. The crown is worn (stage 4–5) with exposed dentin. There is a mesial interproximal facet (2.4 × 3.9 mm). The degree of the occlusal wear makes it difficult to assess the development of the marginal ridges. There is slight cingulum development (grade 1).

Maxillary C¹ Left

The tooth retains most of its crown, and the root is missing, with only 4.5 mm of its lingual surface preserved (Plate 6.2). It was separated from the maxilla during cleaning but is evident next to the left P³ in the precleaning cast (Plate 5.1). The tooth was eroded postmortem, and it shows several pits. In labial view, the crown exhibits two vertical fractures running along its length near the interproximal margins. The crown is worn (stage 4), with exposed dentin. There is both a mesial interproximal facet (2.4 × 3.9 mm) and a distal one (2.7 × 4.2 mm). This tooth shows a little less wear of the marginal ridges than the right canine. A slight distal marginal ridge blends into a slightly moderated cingulum. The mesial marginal ridge is fused to tuberculum dentale, but lingual wear obscures the grade of development (grade ≥2).

Maxillary P³ Right

In buccal view, this tooth lost a chip (2.5 × 2.2 mm) of root postmortem (Plate 6.3). The crown exhibits two microfractures, vertically oriented,

one bisecting the paracone buccally and the other bisecting the mesial margin followed by the mesial interproximal surface. The crown presents occlusal wear with the cusps worn off (stage 4b–5a) and exposed dentin. There is a mesial interproximal facet 5.6 mm wide and a distal one 5.3 mm wide.

The degree of the occlusal wear makes it impossible to evaluate the development of the essential crests and the marginal ridges. The shape of lingual contour indicates that the protocone was situated centrally. The occlusal surface shows the remains of a weak mesiodistal fissure separating the paracone and the protocone, which suggests the absence of a transverse crest.

Maxillary P³ Left

This tooth (Plate 6.3) retains the crown and root, with two chips of the root lost postmortem, a buccal chip 0.3 × 2.7 mm and a lingual chip 3.9 × 3.4 mm. The tooth shows a transverse, zigzag fracture on the upper third of the root. Buccally, the crown presents a vertical microfracture, which bisects the paracone. The distal interproximal facet measures 1.5 × 4.2 mm. The crown presents occlusal wear with the cusps worn off (stage 4b–5a) and exposed dentin.

The degree of the occlusal wear makes it impossible to evaluate the development of the essential crests and the marginal ridges. The shape of lingual contour suggests that the protocone was situated more mesially than the one on the right P³. The occlusal surface shows that the paracone and the protocone are separated by a well-defined, uninterrupted sagittal sulcus. This sulcus is confined to the central portion of the tooth and does not extend to either the mesial or distal margins; hence, there was an absence of a developmental groove.

Maxillary P⁴ Right

In buccal view, this tooth (Plate 6.4) presents a microfracture, vertically oriented, that bisects the paracone and the root. The crown has occlusal wear with the cusps worn off and exposure of the

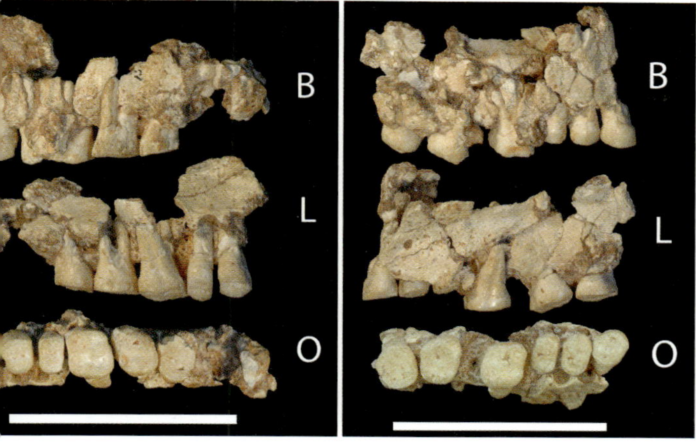

PLATE 6.1. The maxillary dentition of Palomas 1, in buccal (B), lingual (L) and occlusal (O) views. Scales: 5 cm

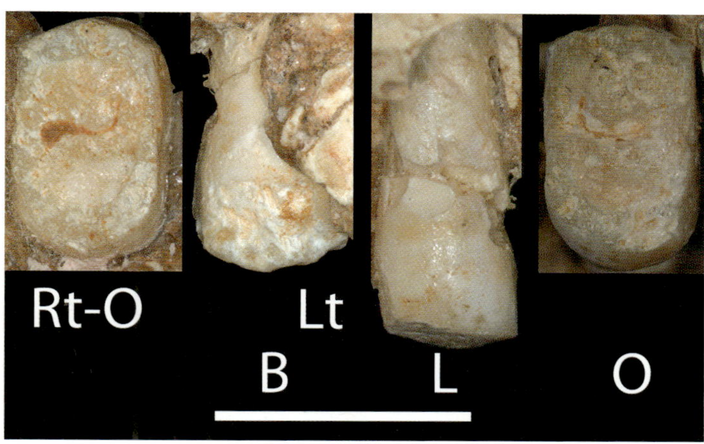

PLATE 6.4. The Palomas 1 right P⁴ in occlusal (O) view and the left P⁴ in buccal (B), lingual (L), and occlusal (O) views. Scale: 1 cm

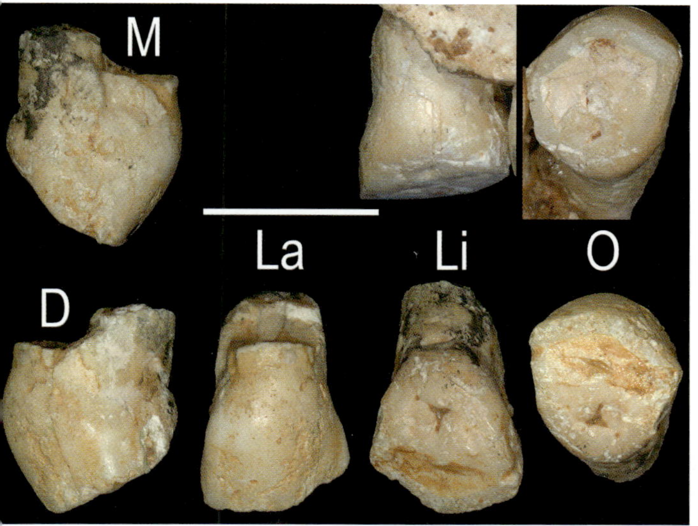

PLATE 6.2. The Palomas 1 right and left C¹'s. The in situ right C¹ (*above right*) is in lingual (Li) and occlusal (O) views, whereas the separated *left* C¹ (*below*) is in mesial (M), distal (D), labial (La), lingual (Li), and occlusal (O) views. Scale: 1 cm

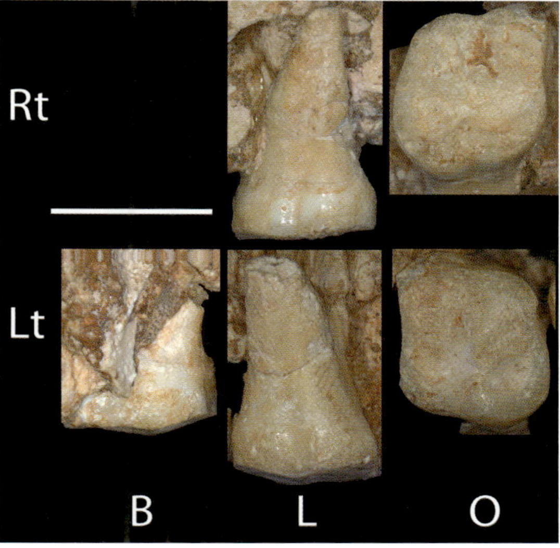

PLATE 6.5. The Palomas 1 right M¹ in lingual (L) and occlusal (O) views and the left M¹ in buccal (B), lingual (L), and occlusal (O) views. Scale: 1 cm

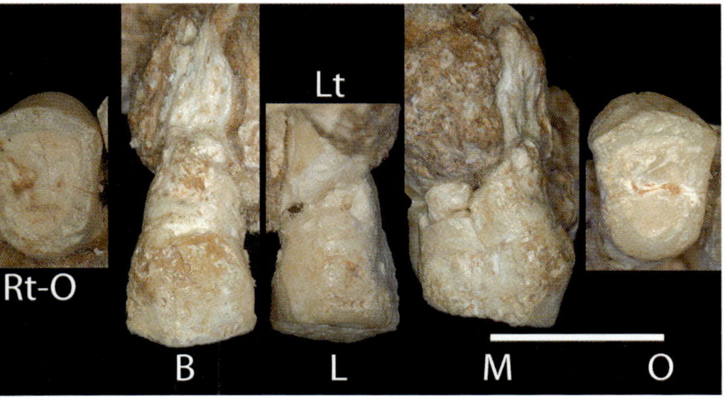

PLATE 6.3. The Palomas 1 right P³ in occlusal (O) view and the left P³ in buccal (B), lingual (L), mesial (M), and occlusal (O) views. Scale: 1 cm

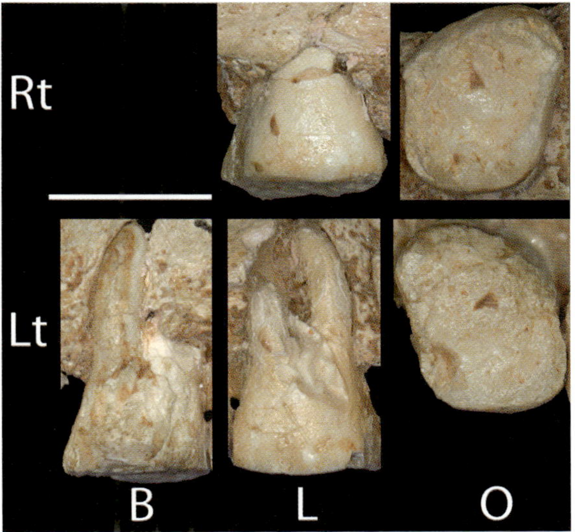

PLATE 6.6. The Palomas 1 right M² in lingual (L) and occlusal (O) views and the left M² in buccal (B), lingual (L), and occlusal (O) views. Scale: 1 cm

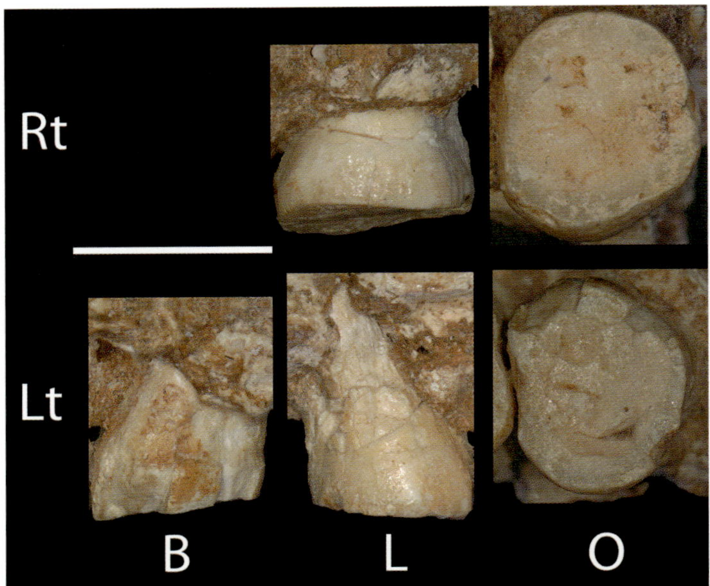

PLATE 6.7. The Palomas 1 right M³ in lingual (L) and occlusal (O) views and the left M³ in buccal (B), lingual (L), and occlusal (O) views. Scale: 1 cm

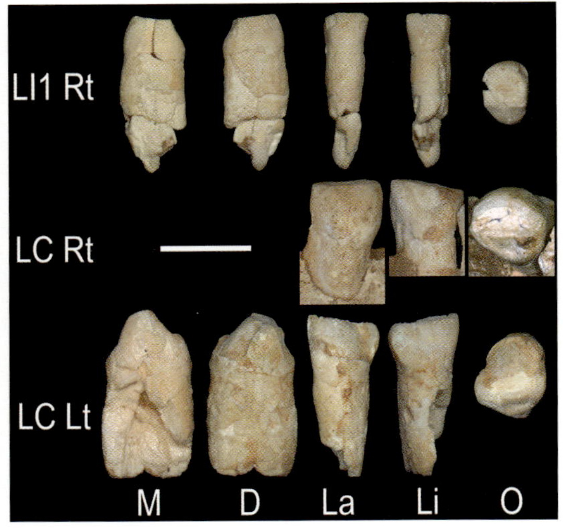

PLATE 6.9. The Palomas 1 right I1 (*above*), right C₁ (*middle*), and *left* C₁ (below) in mesial (M), distal (D), labial (La), lingual (Li), and occlusal (O) views. Scale: 1 cm

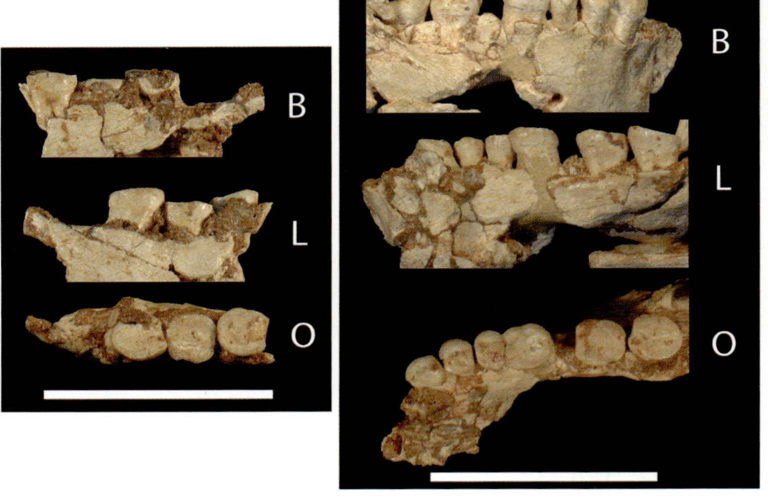

PLATE 6.8. The mandibular dentition of Palomas 1, in buccal (B), lingual (L), and occlusal (O) views. Scales: 5 cm

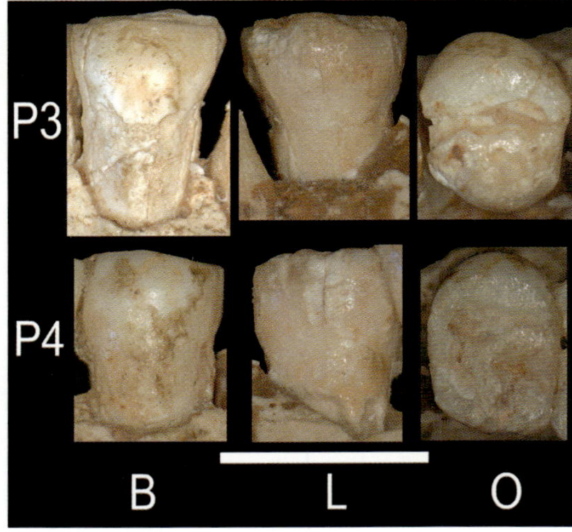

PLATE 6.10. The Palomas 1 right P₃ and P₄ in buccal (B), lingual (L), and occlusal (O) views. Scale: 1 cm

PLATE 6.11. The Palomas 1 right and left M₁s in buccal (B), lingual (L), and occlusal (O) views. Scale: 1 cm

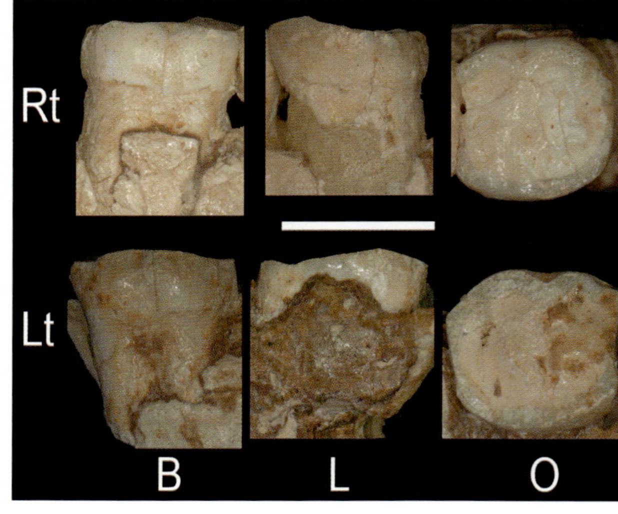

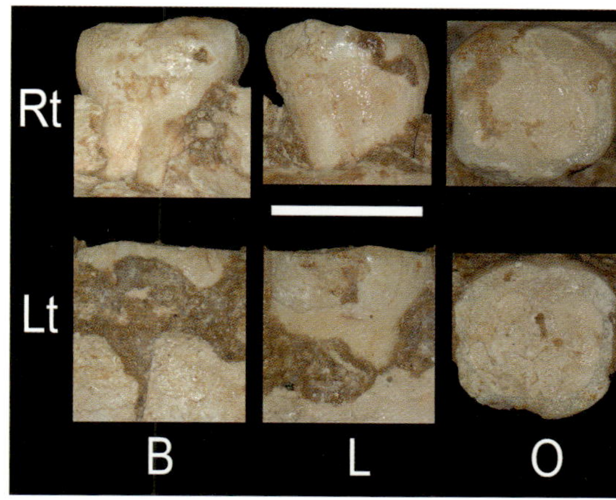

PLATE 6.12. The Palomas 1 right and left M$_2$s in buccal (B), lingual (L), and occlusal (O) views. Scale: 1 cm

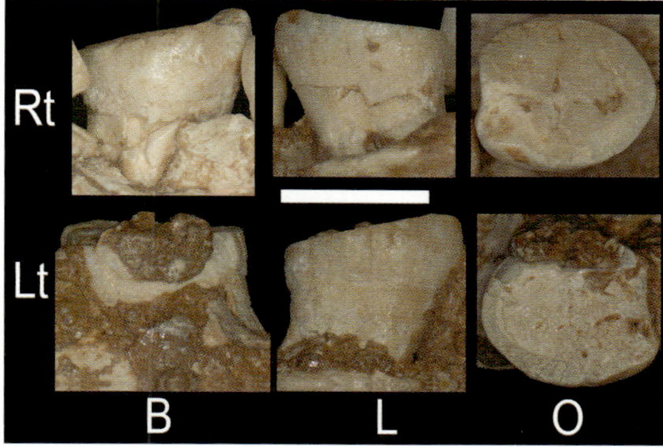

PLATE 6.13. The Palomas 1 right and left M$_3$s in buccal (B), lingual (L), and occlusal (O) views. Scale: 1 cm

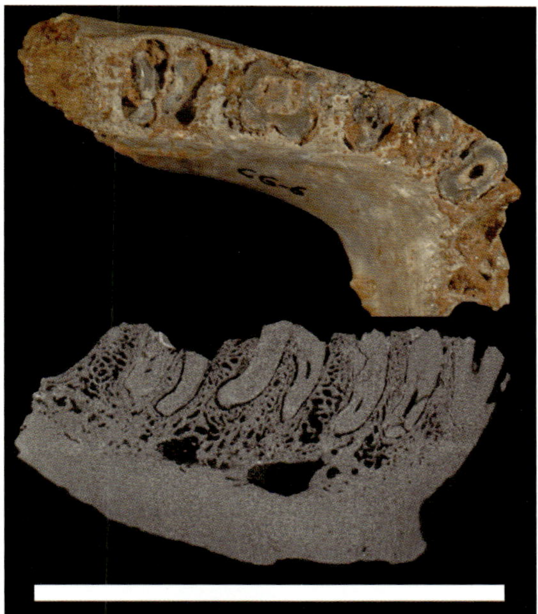

PLATE 6.14. The Palomas 6 mandible in superior (occlusal) view above and a micro-CT slice through the C$_1$ to M$_2$ and adjacent mandibular corpus. Photo scale: 5 cm

PLATE 6.15. Occlusal view of the Palomas 7 immature mandible (*left*) and lingual view of the C^1 largely formed crown in its crypt (*right*). Scales: 1 cm

PLATE 6.16. Mesial (M), distal (D), labial (La), and lingual (Li) views of the Palomas 18 left C$_1$. Scale: 1 cm

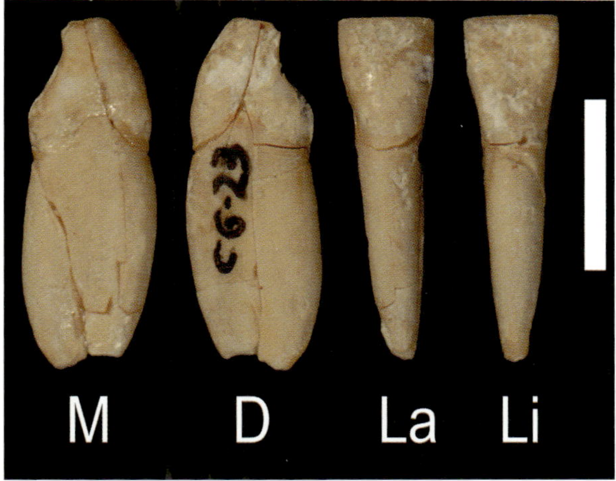

PLATE 6.17. Mesial (M), distal (D), labial (La), and lingual (Li) views of the Palomas 19 left I$_1$. Scale: 1 cm

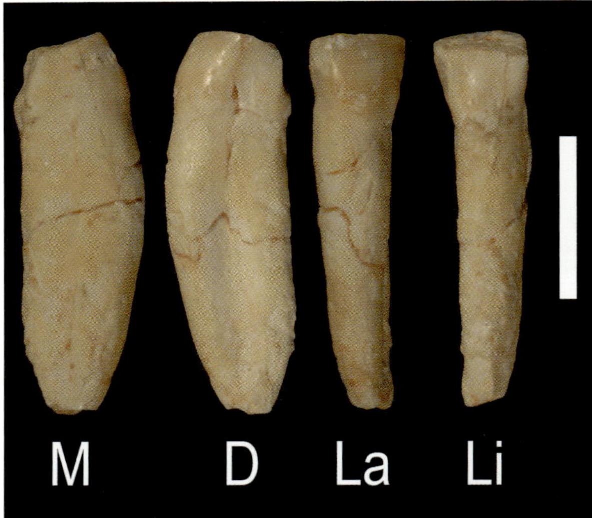

PLATE 6.18. Mesial (M), distal (D), labial (La), and lingual (Li) views of the Palomas 20 left I_2. Scale: 1 cm

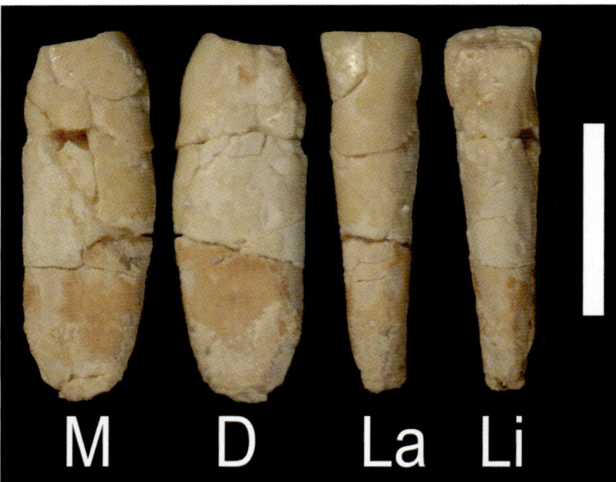

PLATE 6.19. Mesial (M), distal (D), labial (La), and lingual (Li) views of the Palomas 21 left I_1. Scale: 1 cm

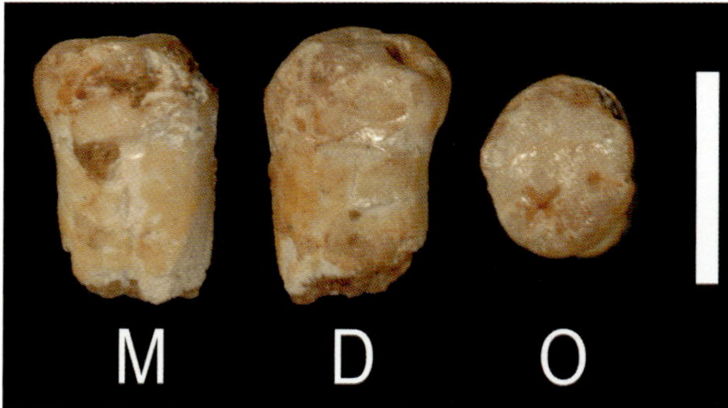

PLATE 6.20. The Palomas 22 left P_3 in mesial (M), distal (D), and occlusal (O) views. Scale: 1 cm

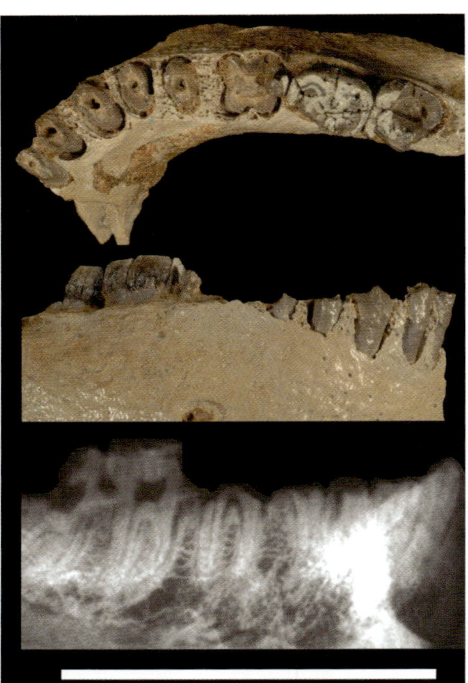

PLATE 6.21. The Palomas 23 right mandibular dentition in occlusal view (*top*), buccal view (*middle*), and buccolingual radiograph (*bottom*). The broken roots of the I_1 to P_4 are evident, along with the partial crowns of the M_1 to M_3. Scale: 5 cm

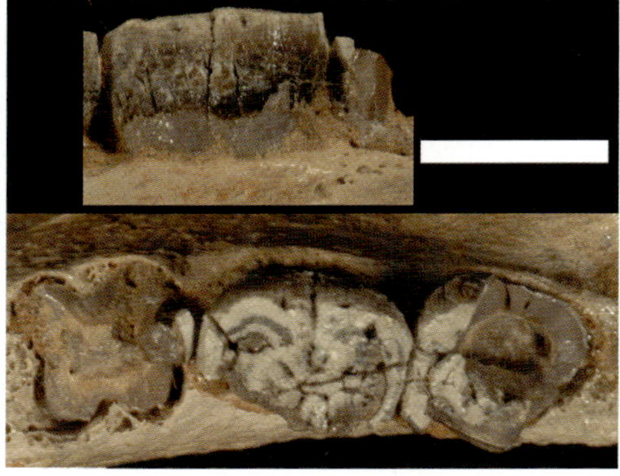

PLATE 6.22. Details of the Palomas 23 right molars. Above: buccal view of the M_2; below: occlusal view of the M_1 to M_3. Scale: 1 cm

PLATE 6.23. The Palomas 24 left I^1 in mesial (M), distal (D), labial (La), lingual (Li), and occlusal (O) views. Scale: 1 cm

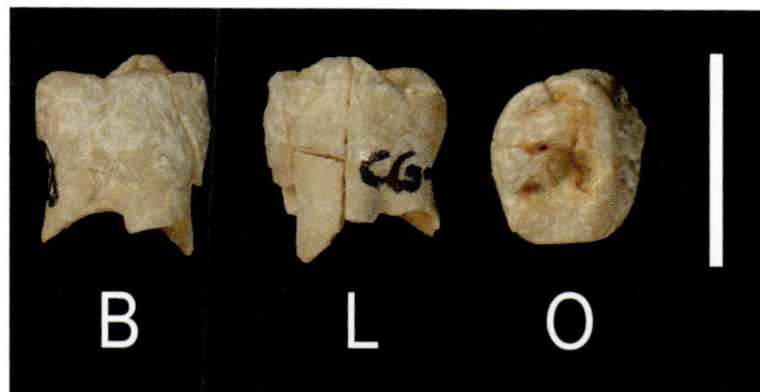

PLATE 6.24. Buccal (B), lingual (L), and occlusal (O) views of the Palomas 25 right dm_1. Scale: 1 cm

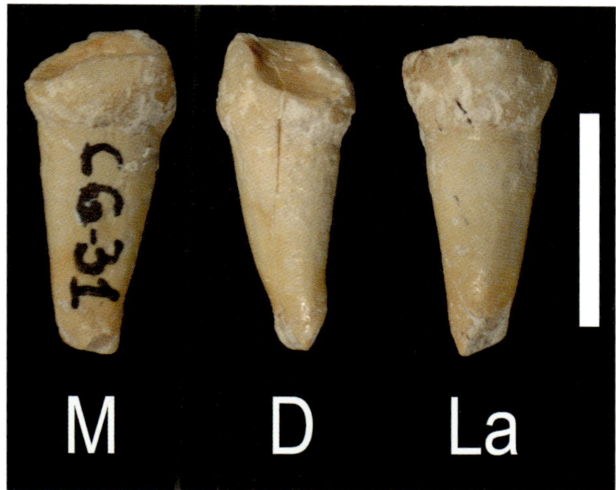

PLATE 6.27. Mesial (M), distal (D), and labial (La) views of the Palomas 31 left dc_1. Scale: 1 cm

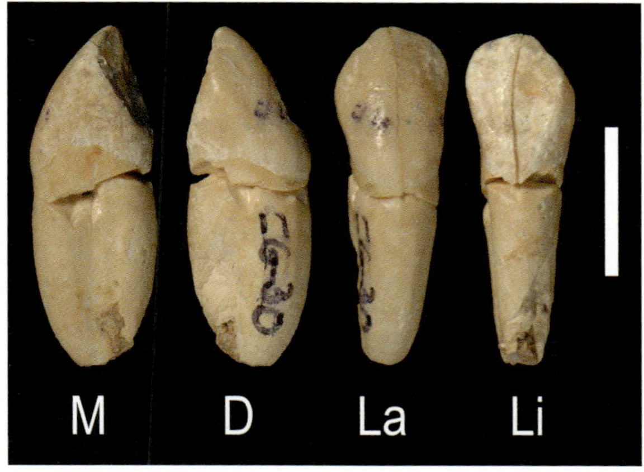

PLATE 6.25. Mesial (M), distal (D), labial (La), and lingual (Li) views of the Palomas 26 right C_1. Scale: 1 cm

PLATE 6.28. Partial roots of the Palomas 33, 46, and 55 teeth. The last is still partially embedded in matrix. Scale: 1 cm

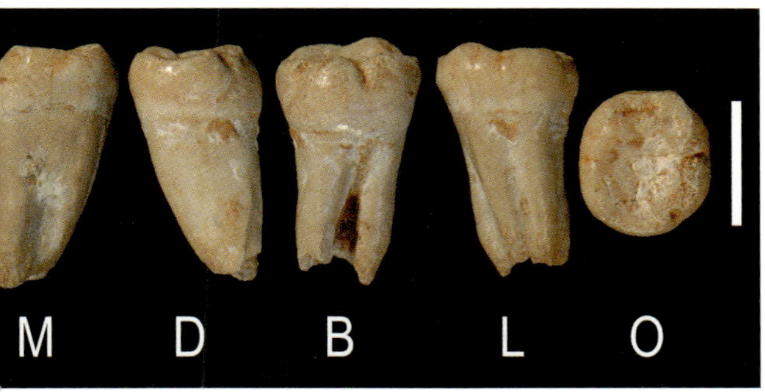

PLATE 6.26. The Palomas 29 right M_2 in mesial (M), distal (D), buccal (B), lingual (L), and occlusal (O) views. Scale: 1 cm

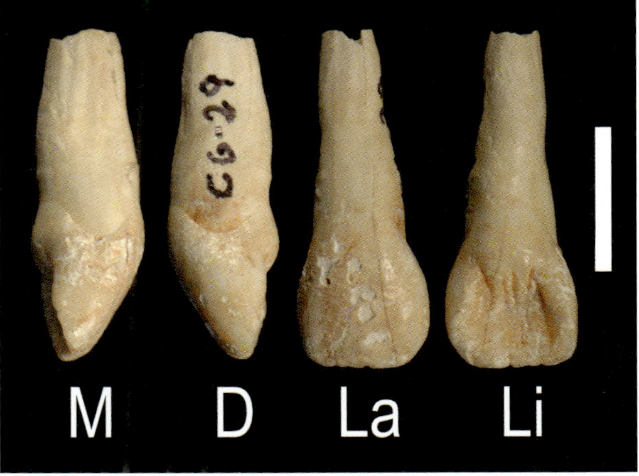

PLATE 6.29. The Palomas 34 left I^1 in mesial (M), distal (D), labial (La), and lingual (Li) views. Scale: 1 cm

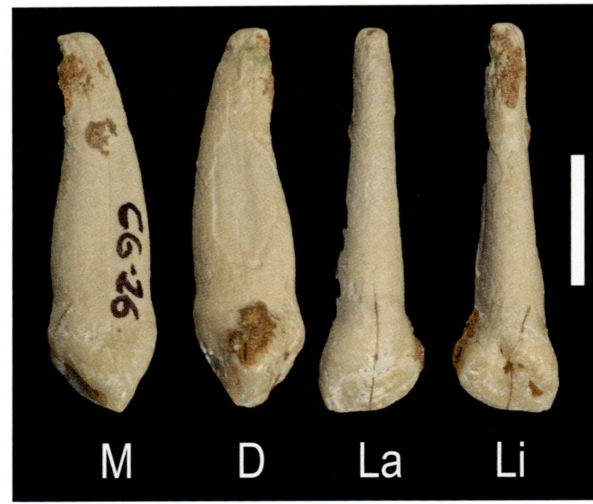

PLATE 6.30. The Palomas 35 left C¹ in mesial (M), distal (D), labial (La), and lingual (Li) views. Scale: 1 cm

PLATE 6.33. The Palomas 39 left di¹ in mesial (M), distal (D), labial (La), and lingual (Li) views. Scale: 1 cm

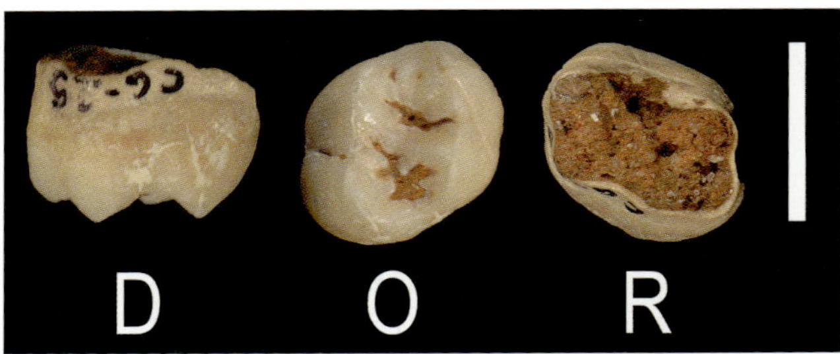

PLATE 6.31. The unerupted Palomas 36 left dm² in distal (D), occlusal (O), and radicular (R) views. Scale: 1 cm

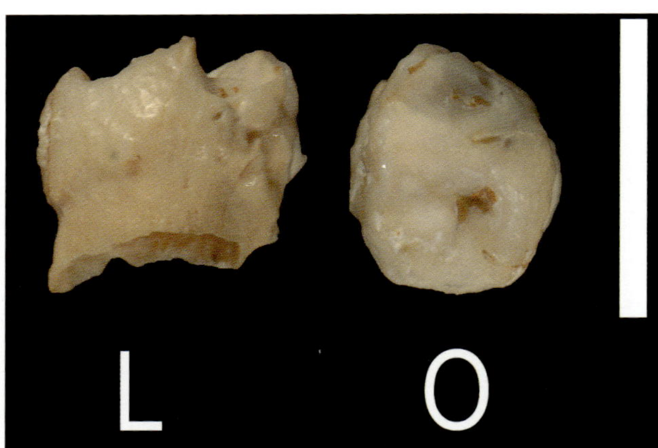

PLATE 6.34. The Palomas 40 left dm¹ in lingual (L) and occlusal (O) views. Scale: 1 cm

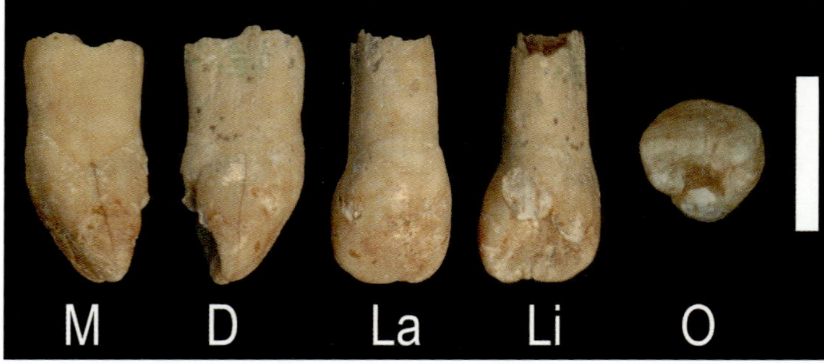

PLATE 6.32. The Palomas 37 right I² in mesial (M), distal (D), labial (La), lingual (Li), and occlusal (O) views. Scale: 1 cm

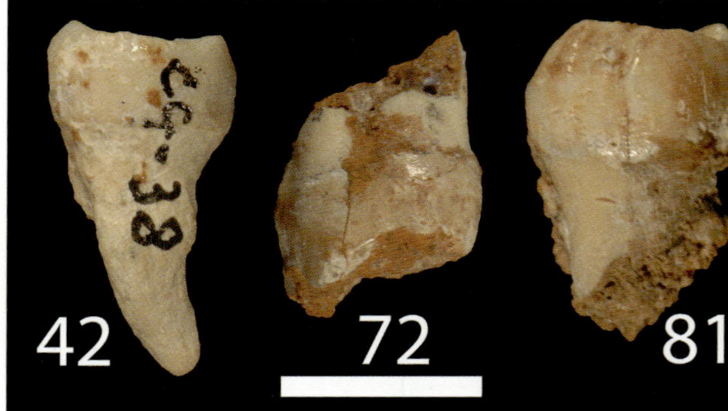

PLATE 6.35. The Palomas 42 mandibular molar, the Palomas 72 maxillary molar in mesial view, and the Palomas 81 M₁ or M₂ in distobuccal view. Scale: 1 cm

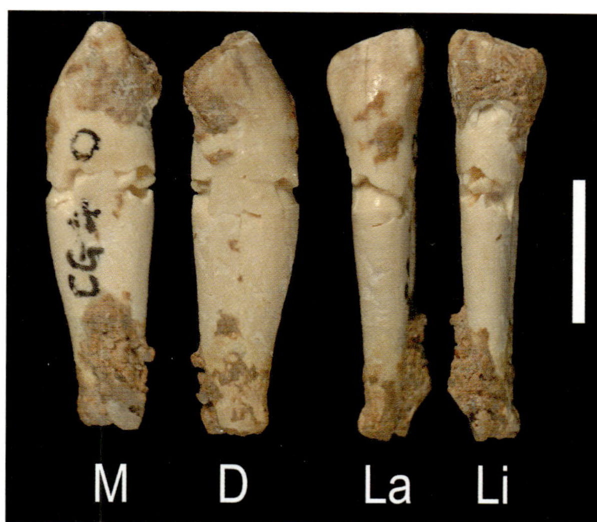

PLATE 6.36. The Palomas 44 right C_1 in mesial (M), distal (D), labial (La), and lingual (Li) views. Scale: 1 cm

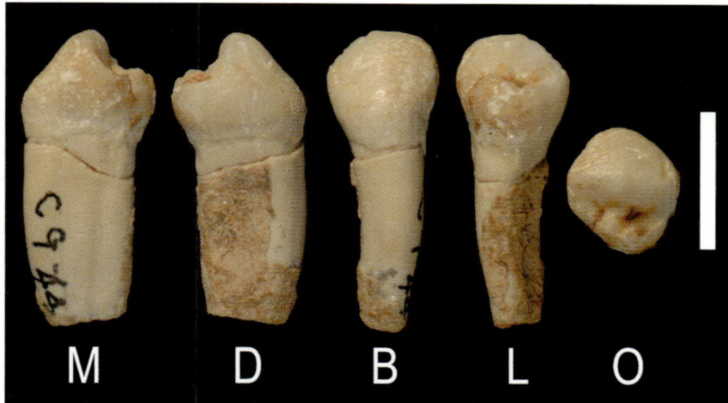

PLATE 6.37. Mesial (M), distal (D), buccal (B), lingual (L), and occlusal (O) views of the Palomas 45 right P_3. Scale: 1 cm

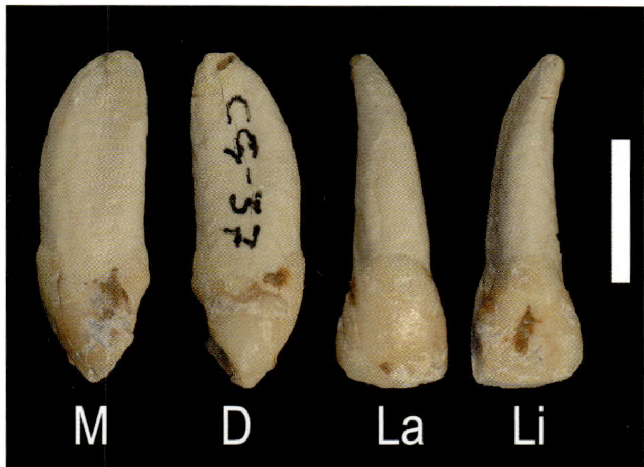

PLATE 6.38. The Palomas 48 right I^2 in mesial (M), distal (D), labial (La), and lingual (Li) views. Scale: 1 cm

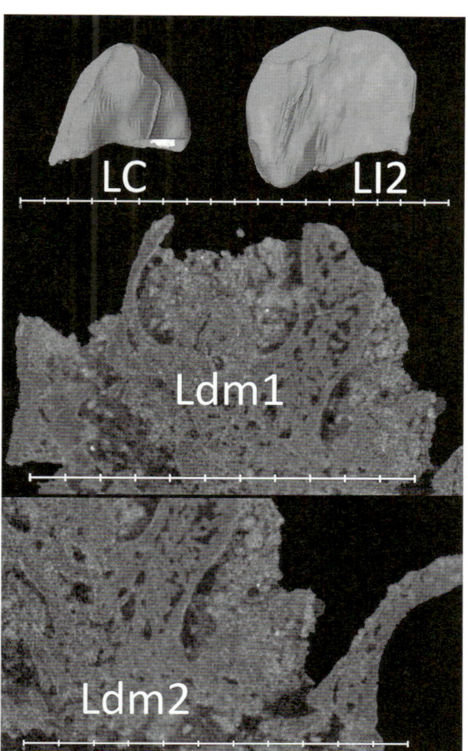

PLATE 6.39. Images of the Palomas 49 left I_2 and C_1 virtually extracted from the mandible and micro-CT slices through the crypts of the left dm_1 and dm_2. Scales: 100 μm for the C_1 and I_2 and 250 μm for the dm1 and dm_2 crypts

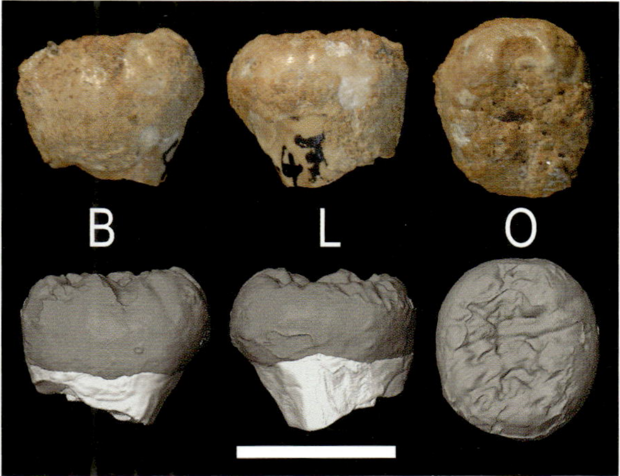

PLATE 6.40. Buccal (B), lingual (L), and occlusal (O) views of the Palomas 50 molar, with photos of the specimen *above* and micro-CT–derived views cleaned of the adhering matrix *below*. Scale: 1 cm

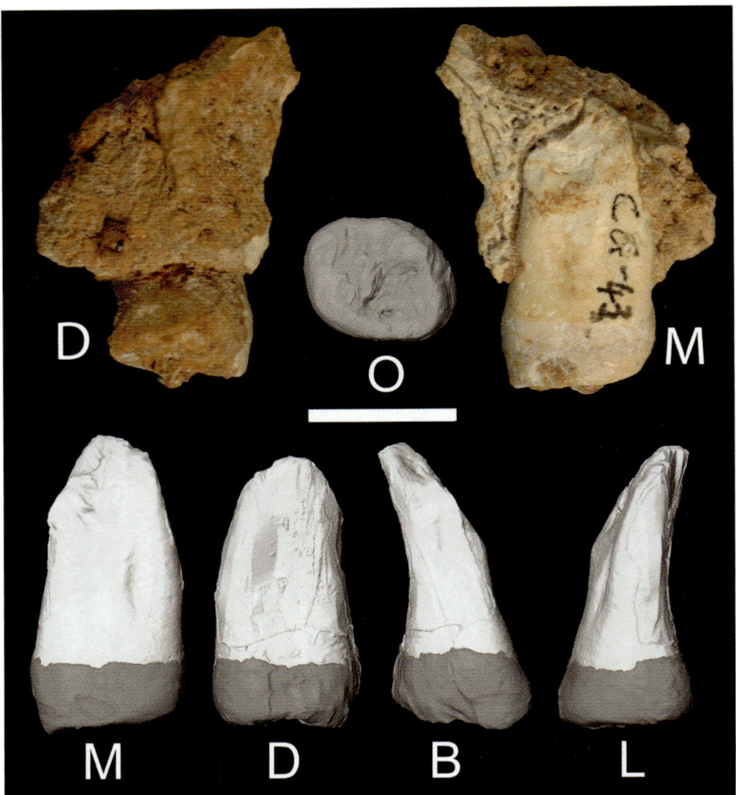

PLATE 6.41. Mesial (M) and distal (D) views of the Palomas 51 right M³ with its adhering maxilla and matrix, plus virtually exposed views of the tooth in occlusal (O), mesial (M), distal (D), buccal (B), and lingual (L) views. Scale: 1 cm

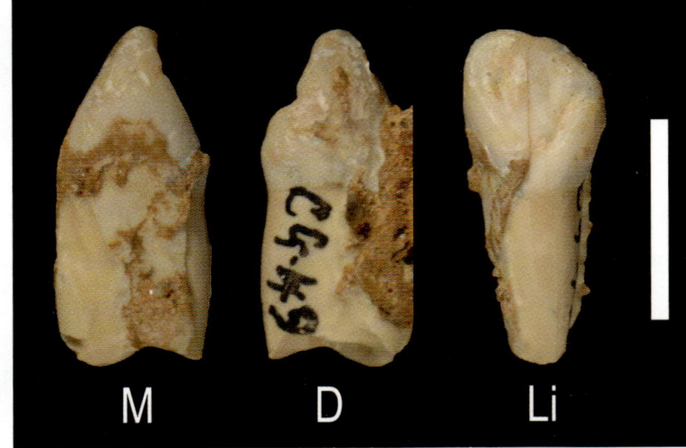

PLATE 6.43. Mesial (M), distal (D), and lingual (Li) views of the Palomas 54 right C₁. Scale: 1 cm

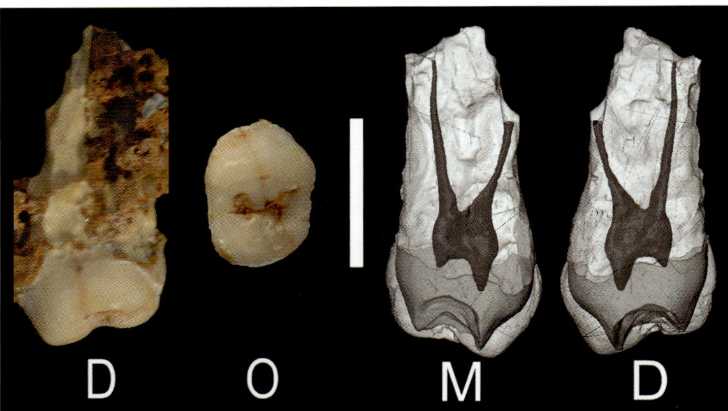

PLATE 6.42. The Palomas 53 P⁴ left in distal (D) and occlusal (O) views and in mesial (M) and distal (D) micro-CT views with the enamel cap, dentin, and pulp chamber evident. Scale: 1 cm

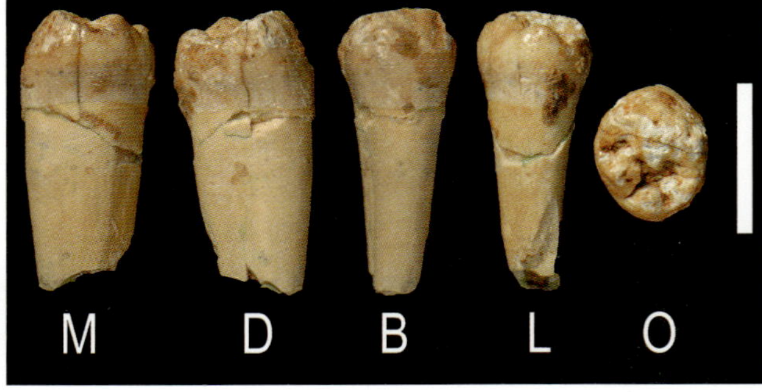

PLATE 6.44. The Palomas 57 right P₄ in mesial (M), distal (D), buccal (B), lingual (L), and occlusal (O) views. Scale: 1 cm

dentin (stage 4a–4b). The degree of the occlusal wear makes it not possible to evaluate the development of the essential crests and the marginal ridges. The shape of lingual contour indicates that the protocone was positioned mesially. The occlusal surface shows that the paracone and the protocone are separated by a well-defined, uninterrupted sagittal sulcus. This sulcus is confined to the central portion of the tooth and does not extend to either the mesial or distal margins, and hence there was an absence of a developmental groove.

Maxillary P⁴ Left

The root exhibits a transverse fracture on its upper third (Plate 6.4) . The crown has a large mesial interproximal facet (5.3 mm wide). The crown presents occlusal wear with the cusps worn off (stage 4a–4b) and exposed dentin. The degree of the occlusal wear makes it impossible to evaluate the development of the essential crests and the marginal ridges. The shape of lingual contour indicates that the protocone was positioned mesially. The occlusal surface shows that the paracone and the protocone are separated by a well-defined, uninterrupted sagittal sulcus. This sulcus is confined to the central portion of the tooth and does not extend to either the mesial or distal margins, and hence there was an absence of a developmental groove.

Maxillary M¹ Right

The right M¹ is largely complete (Plate 6.5). Lingually, the root lost a chip (4.2 × 2.6 mm) on the upper third of the distal interproximal margin, and it has been eroded postmortem. In lingual view, the tooth exhibits four transverse fractures on the root and a vertical microfracture on the crown running between the main cusps, not visible on the occlusal surface. The mesial interproximal facet is 2.9 × 6.1 mm, and the distal one is 2.1 × 5.4 mm. The occlusal wear is stage 4c, with dentin exposed more lingually than buccally. Although the degree of wear impedes the analysis of the crown morphology, the overall shape indicates that four

cusps were presents with a large hypocone (cusp 4 grade 4). There is no evidence of either a cusp 5 or a Carabelli's trait (grade 0 for each). There is a moderate buccal groove separating the paracone and metacone and another weaker lingual groove between protocone and hypocone. Based on exposure of the root bifurcation buccally, the tooth is nontaurodont.

Maxillary M¹ Left

Lingually, the tooth (Plate 6.5) exhibits a vertical microfracture on the crown running between the main cusps, but it is not visible on the occlusal surface. In addition, the root shows a transverse fracture in the middle, and the root apex was broken postmortem. The mesial interproximal facet is 6.8 mm wide. The crown presents stage 4c wear with large facets of exposed dentin on all the cusps. The occlusal wear is greater on this tooth than on the right M¹ and more lingually than buccally. Although the degree of wear makes it difficult to analyze the crown morphology, the overall shape indicates that four cusps were present with a large hypocone (cusp 4 grade 4). Neither a cusp 5 nor a Carabelli's trait was present (grade 0 for each). Based on exposure of the root bifurcation buccally, the tooth is nontaurodont.

Maxillary M² Right

Buccally, the root lost a chip 3.1 × 3.3 mm on the cementum-cervix junction and has a transverse fracture (Plate 6.6). The crown lost a chip of enamel 1.4 × 3.3 mm on the buccomesial margin. The mesial interproximal facet is 1.8 × 5.3 mm. The crown presents a worn occlusal surface with the cusps planed off (stage 4b) and dentin exposed more lingually than buccally. Although the degree of wear limits the analysis of the crown morphology, the overall shape indicates that four cusps were present with a moderate hypocone (cusp 4 grade 3). There is no evidence of a cusp 5 or a Carabelli's trait (grade 0 for each). Based on exposure of the root bifurcation buccally, the tooth is nontaurodont.

Maxillary M² Left

Buccally, the crown was eroded postmortem, it is missing fragments of the enamel, and the occlusal surface lost a chip 2.5 × 1.3 mm (Plate 6.6). Lingually, the root was eroded. There is a mesial interproximal facet 6.5 mm wide. The crown has occlusal wear with the cusps worn off (stage 4b) and the dentin exposed more lingually than buccally. The overall shape indicates that four cusps were present with a relatively large hypocone (cusp 4 grade 3). Neither a cusp 5 nor a Carabelli's trait appears to have been present (grade 0 for each).

Maxillary M³ Right

Lingually the crown exhibits a vertical microfracture that runs its length and a transverse fracture running across the cervix (Plate 6.7). The buccodistal root apex was lost postmortem. There is a mesial interproximal facet 1.9 × 5.4 mm. The crown has occlusal wear with the cusps worn off (stage 4a), especially buccolingually, and there is a facet with exposed dentin on the protocone. The overall shape indicates that four cusps were present with a moderate hypocone (cusp 4 grade 3). There is no evidence of a cusp 5 or a Carabelli's trait (grade 0 for each). Based on exposure of the root bifurcation lingually and buccally, the tooth is nontaurodont.

Maxillary M³ Left

Buccally the crown has lost two chips of enamel postmortem, a buccomesial one (2.8 × 3.3 mm) and a buccodistal one (1.7 × 2.3 mm) (Plate 6.7). The root was eroded lingually. There is a mesial interproximal facet 5.3 mm wide. The occlusal wear has removed the cusps (stage 5c) more distally than mesially and greater than on the right M³. The overall shape suggests that four cusps were present with a slight hypocone (cusp 4 grade 2). Cusp 5 and Carabelli's trait both appear to have been absent (grade 0 for each).

Mandibular I₁ Right

The crown of this tooth and the root were broken postmortem and were restored from several fragments (Plate 6.9). It was separated from its crushed alveolus during cleaning (Plate 5.1). The enamel is eroded with several pits, and a chip 1.7 × 1.0 mm is also missing, exposing the dentin, which must have been caused postmortem. The root apex is missing. The incisal edge of this tooth is heavily worn (stage 5), preserving approximately one-half of the crown. The lingual surface wear does not allow determination of whether mesial and distal marginal ridges or a cingulum was present.

Mandibular I₁ Left

Palomas 21 probably represents the left I₁ of Palomas 1. See description below.

Mandibular I₂ Left

Palomas 20 probably represents the left I₂ of Palomas 1. See description below.

Mandibular C₁ Right

The tooth is *in situ* in its alveolus; it has a sagittal microfracture on both the lingual and buccal sides, visible from the incisal edge and running lengthwise (Plate 6.9). There is another microfracture on the mesial marginal ridge, visible only on the lingual surface. The incisal edge of the crown is worn (stage 4) and inclined toward the P₃. There is a mesial interproximal facet 3.2 × 2.3 mm.

The lingual surface is moderately shovel-shaped, with well-developed mesial and distal marginal ridges (the latter larger than the former) that blend into a weak cingulum. The lower third of the distal marginal ridge exhibits two furrows on the distolingual surface. Even though the tooth has substantial occlusal wear, the remains of a moderate distal accessory ridge (grade 3) are discernible on the lingual surface.

Mandibular C₁ Left

The tooth was separated from its crushed alveolus during cleaning (Plates 5.1 and 6.9). It lost most of its enamel except for a 3.1 × 6.2 mm cervical fragment. Mesially, most of the root has been lost. The apical third of the root is absent. The incisal edge of the crown is worn (stage 4) and inclined toward the P₃. The lingual surface is moderately shovel-shaped, with a well-developed and worn distal marginal ridge that is bigger than the mesial marginal ridge. The mesial marginal ridge blends into a weak cingulum. Even though the tooth has substantial occlusal wear, the remains of a distal accessory ridge (grade 4) are discernible on the lingual surface.

Mandibular P₃ Right

The tooth (Plate 6.10) presents two microfractures, one running lengthwise on the mesial and distal margins and visible on the occlusal surface, and another one on the distolabial surface. The distal interproximal facet is 4.2 mm wide. The contour of the tooth crown is symmetrical. In occlusal view, the protoconid shows more wear than the metaconid (stage 4). The tooth possesses two lingual cusps; the metaconid, in mesiolingual position, is larger than a distolingual cusp (grade 3). The wear on the protoconid prevents assessment of whether a transversal crest was present

Mandibular P₃ Left

The Palomas 22 left P₃ likely represents the left P₃ of Palomas 1, given morphological, metrical, and attritional similarities. It is considered below.

Mandibular P₄ Right

The crown (Plate 6.10) presents several microfractures running lengthwise on all faces. In lingual view, the crown shows a vertical microfracture running between both cusps. In the same area, a chip of enamel 1.4 × 1.5 mm was lost postmortem. The protoconid and metaconid exhibit moderate

occlusal wear (stage 4), with exposed dentin. The tooth is symmetrical in occlusal view and exhibits two lingual cusps, one a distal accessory cusp and another a larger one, the metaconid, in a mesiolingual position (grade 3). The metaconid and protoconid are connected, forming a transverse crest (grade 2).

Mandibular M₁ Right

Vertically oriented microfractures can be observed along the enamel of the major cusps (Plate 6.11). The occlusal surface is worn to stage 5a, with exposed dentin on all of the cusps except the metaconid, more on the distal half than the mesial one and greater buccally than lingually. The occlusal surface is missing a chip of dentin 3.0 × 2.3 mm. The distal interproximal facet is 2.5 mm high by 5.3 mm wide. The degree of wear makes it difficult to assess the occlusal crown morphology, although the overall shape indicates that five cusps were present with a Y pattern. There is a moderate buccal groove separating the protoconid from the hypoconid. Due to the degree of wear, the presence/absence of a midtrigonid crest cannot be assessed, and the sulcal pattern is indistinct. There is no protostylid (grade 0).

Mandibular M₁ Left

In buccal view (Plate 6.11), two vertical microfractures can be observed along the enamel of the major cusps. The occlusal surface has the cusps worn off (stage 5a) and exposed dentin on all cusps, more in the distal half than in the mesial half and greater buccally than lingually. The crown lost a small chip of enamel 1.0 × 1.7 mm on the mesiobuccal surface. The mesial interproximal facet is 3.2 mm high by 6.8 mm wide. There is a distal interproximal facet 4.9 mm wide. There is a moderate buccal groove separating the protoconid from the hypoconid. The overall crown shape delimits five cusps. Due to the degree of wear, the presence/absence of a midtrigonid crest cannot be assessed. There is no protostylid (grade 0).

Mandibular M₂ Right

The enamel exhibits vertically oriented microfractures on all faces; mainly in lingual view, the crown presents a microfracture running lengthwise (Plate 6.12). The mesial interproximal facet is 2.4 mm high by 6.0 mm wide. The occlusal surface has the cusps worn off (stage 4b–4c), with some exposed dentin on each of the cusps; the largest dentin facet is on the protoconid. The occlusal wear is greater buccally than lingually. This tooth presents less occlusal wear than the M_1. The overall shape of the crown suggests that five cusps were present in a Y pattern. Due to the degree of wear, the presence/absence of a midtrigonid crest cannot be assessed. There is no protostylid (grade 0).

Mandibular M₂ Left

In lingual view (Plate 6.12), the crown presents the loss of a chip of enamel (1.0 × 2.5 mm) on the entoconid and a microfracture running lengthwise on the metaconid. The occlusal surface is strongly worn, and the cusps are worn off (stage 4c), with facets of exposed dentin on all the cusps. The largest dentin exposure is on the protoconid. The occlusal wear is greater buccally than lingually and more in the mesial than the distal half. This tooth presents less occlusal wear than the M_1 but more than the right M_2. The overall shape of the crown suggests that five cusps were present, although the degree of wear obscures the sulcal pattern. Due to the degree of wear, the presence/absence of a midtrigonid crest cannot be assessed. There is no protostylid (grade 0).

Mandibular M3 Right

The occlusal surface indicates that the cusps were worn (stage 4a) without dentin exposure on most of the cusps (Plate 6.13); there is a small dentin facet on the entoconid. The mesial interproximal facet is 3.1 mm high by 5.1 mm wide. This tooth shows less occlusal wear than the M_2 and M_1. The occlusal wear is greater buccally than lingually. In lingual view, the occlusal wear shows a pronounced downward slope in a mesiodistal direction. There is a slight buccal groove separating the protoconid from the hypoconid. The overall shape of the crown suggests that five cusps were present in a Y pattern. Due to the degree of wear, the presence/absence of a midtrigonid crest cannot be assessed. There is no protostylid (grade 0).

Mandibular M₃ Left

In lingual view (Plate 6.13), the crown lost a chip of enamel 2.2 × 2.2 mm. The occlusal surface is worn to stage 5b–5c with exposed dentin on the distal cusps. This tooth shows less occlusal wear than the M_2 and M_1 but more than the right M_3 (in general, the crowns of the molars show more attrition on the left side than on their antimeres). The occlusal wear is greater buccally than lingually and distally than mesially. In lingual view, the occlusal wear shows a pronounced downward slope in a mesiodistal direction. The overall shape of the crown suggests that five cusps were present, although the degree of wear obscures the limits between the cusps. Due to the degree of wear, the presence/absence of a midtrigonid crest cannot be assessed. There is no protostylid (grade 0).

Palomas 6 Mandible with C_1 to M_2 Roots Left

The Palomas 6 mandible retains most of the roots of the left C_1 to M_2 (Plates 5.5 and 6.14), but the crowns and the cervices of the teeth were lost through the partial burning of the mandible. Some general impressions of the roots can be obtained in superior view of the mandible and radiographically; for example, the M_1 and M_2 mesial and distal roots are separate close to their cervices and therefore not taurodont. The canine root appears to be relatively long. The roots are broken too far below the cervices to provide reliable cervix diameters or root lengths. There is a possible inflammation at the canine root apex. Little more can be ascertained from the preserved roots.

Palomas 7 Mandible with C_I and P_3 Crowns Left

Palomas 7 (Plates 5.9 and 6.15) consists of the partial left corpus of an immature mandible, from the distal deciduous canine socket to the mesial side of the M_I crypt, containing the C_I germ exposed medially, the germ for P_3 internally evident radiographically, and the alveoli for the dm_I and dm_2. Based on the degree of development of the C_I (between $Cr_{3/4}$ and Cr_c), the individual had an age at death between 3.5 and 4.5 years, and the modal ages of the P_3 (between $Cr_{1/2}$ and $Cr_{3/4}$) are 4.5 years (males) and 4.4 years (females). Together, these two teeth suggest an age at death of 3.5 to 4.5 years. The mesial M_I crypt is for a crown and not a root socket, indicating that the M_I had a complete crown but little root formation (Plate 5.9); it therefore supports the age assessment.

The only tooth region providing morphological data is the exposed lingual side of the C_I. The tooth has well-formed mesial and distal marginal ridges, a strong ridge through the vertical middle of the lingual surfaces, and a well-pronounced distal accessory ridge (grade 4). The interradicular septa for the dm_I and dm_2 indicate that these teeth were nontaurodont.

Palomas 18 Mandibular C_I Left

The crown and root of this tooth are completely formed (Plate 6.16). The crown has a vertical crack. The enamel was eroded *in situ*, showing several pits, such that the surface of the mesial margin was altered along its upper half. The crown is slightly worn (stage 2), with a small facet of exposed dentin, suggesting that it belonged to a subadult. The crown presents a small distal facet 1.7 × 1.7 mm. In addition, the crown has a linear hypoplasia.

The lingual surface is somewhat shovel-shaped, with a slight mesial marginal ridge and a strong distal marginal ridge. The crown has a well-formed distal accessory ridge (grade 4). Labially, the crown is asymmetric and convex in a mesial-distal direction. The distal crown margin is steeply angled. The root is well preserved,

showing a shallow longitudinal groove expressed mesially and distally with a rounded shape.

Palomas 19 Mandibular I_I Left

This tooth (Plate 6.17) presents a completely formed crown and root; the root apex was broken postmortem. The crown has a fracture visible on the incisal edge running mesiodistally and along the length of interproximal margins. The tooth exhibits several additional cracks running transversely and lengthwise. The root was restored from several fragments. The enamel was eroded *in situ* and shows several pits.

The incisal edge of the crown is worn, showing exposed dentin (stage 3). The crown presents an incomplete distal facet 2.3 × 1.9 mm. The completely developed root, the presence of incisal wear, and the distal facet indicate that the tooth was in functional occlusion. No mamelons are present. The lingual surface is not shovel-shaped (grade 0), and the cingular region is smooth. The labial surface is featureless. There are clear mesial and distal longitudinal root sulci, and the eroded root apex shows two radicular canals. The tooth also has a broad and shallow linear enamel hypoplasia on the cervical half of the crown.

It is possible that Palomas 19 derives from the same individual as Palomas 73 and 79 (see below), based on similar occlusal wear and dental enamel hypoplasia, although it appears to have been modestly younger in age based on the degree of occlusal wear.

Palomas 20 Mandibular I_2 Left

The crown of this tooth is completely formed, but the root apex was broken postmortem (Plate 6.18). The enamel and the cementum were eroded postmortem, creating several pits. At the middle of the tooth, the root exhibits a transverse fracture on the mesial surface that continues on the distal surface in a zigzag form. There is also a U-shaped fracture on the mesiobuccal surface of the root. Several cracks run in all directions on the tooth, some visible on the incisal edge of the crown. It may well represent the left I_2 of Palomas 1, and it shows morphological and morphometric similarities, including its wear pattern, with Palomas 21.

The incisal edge of this tooth is heavily worn (stage 5), preserving only half of the crown. The crown presents a small mesial interproximal facet 1.2×2.8 mm. There is a larger distal one, 4.2×3.4 mm. The lingual surface has slight and moderate development of the mesial and distal marginal ridges, but the extent of wear makes it not possible to provide a reliable ASUDAS grade. The moderately expressed cingulum blends into a slightly expressed median ridge. The root has a well-defined distal developmental groove. The root apex deflects distally.

Palomas 21 Mandibular I_1 Left

The crown of this tooth and the root are completely formed, but both were broken postmortem and restored from several fragments (Plate 6.19). If it represents the left I_1 of Palomas 1 (given comparisons with the cast of the unprepared Palomas 1 remains), the root apex has been lost since it was discovered. The incisal edge of this tooth is heavily worn (stage 5), preserving only one-third of the crown. The mesial and distal marginal ridges cannot be distinguished. There is no cingular development.

Palomas 22 Mandibular P_3 Left

The tooth consists of a completely formed crown but only a partially preserved root (≈ 7.4 mm preserved) (Plate 6.20). This tooth shows postmortem erosion of the enamel and the root, with several pits. The crown exhibits some microfractures vertically oriented on all the sides. It may represent the left P_3 of Palomas 1, given morphological and morphometric similarities to the *in situ* right P_3 of Palomas 1.

The crown (wear stage 4) has a heavily worn protoconid with exposed dentin. There is an interproximal mesial facet 3.1×3.8 mm and an

interproximal distal facet 2.6×3.7 mm. The tooth is asymmetrical in occlusal view and possesses two lingual cusps, a slightly worn metaconid is the largest and in a mesiolingual position, followed by the distolingual cusp (grade 3). The protoconid is connected by a transverse crest with the metaconid (grade 2). The essential crests define the distal border of a small anterior fovea. The distolingual cusp meets a large distal fossa.

Palomas 23 Mandible with Partial I_1 to M_3 Right

Palomas 23 (Plates 5.7 and 6.21) is a partially burned right mandibular corpus from the symphyseal midline to the base of the right ramus. The teeth were largely lost through the burning, having had all but the molar crowns broken off below the cervices (Plate 6.22). The I_1 to P_4 retain only partial roots, and none of their cervices, in the alveolar bone. The M_1 retains only a fragment of the crown against the M_2 interproximal surface. The M_2 crown is largely present, but the mesiolingual corner is gone. The M_3 preserves the mesial one-quarter of the crown and a small portion of the midlingual crown. The M_2 appears to have had 5 cusps, but damage and occlusal wear (stage 4c) make further observation difficult. The bifurcations of the roots for all three molars are close to the alveolar plane, and the teeth are therefore not taurodont.

Palomas 24 Maxillary I^1 Left

The crown and root are completely formed (Plate 6.23). The enamel was extensively eroded *in situ*, resulting in several pits, especially on the labial surface. Most of the incisal edge is chipped, with loss of a large chip distally. The cementum was also eroded, having several pits principally at the distal side. The preserved half of the incisal edge, toward the mesial surface of the crown, exhibits slight wear (stage 1–2).

Lingually, the crown exhibits strong shoveling (grade 6), with well-developed marginal ridges that circumscribe a deep sulcus, and both ridges converge on a strongly developed cingulum. The latter comprises a marked lingual tubercle with a free apex (tuberculum dentale grade 6). At the bases of both of the marginal ridges, there are marked interruption grooves dividing them from the cingulum. The labial crown surface is both mesiodistally and superoinferiorly convex (grade 4). The root deflects mesially.

Palomas 25 Mandibular dm₁ Right

The enamel of this tooth was eroded postmortem and has several pits (Plate 6.24). The crown shows cracks on all of the cusps running along its length, some visible on the occlusal surface. Three fragments were restored on the mesial surface and another one on the distal surface. A round and deep carious lesion is evident on the occlusal surface by the base of the hypoconid.

This tooth has a fully formed crown, but only 5.1 mm (in its larger length) of the root is preserved, due to an irregular postmortem break. The fragility of the root suggests that it had not completed its development, but the tooth is nonetheless worn occlusally. The crown has exposed dentin (stage 4c), mainly on the cusps of the buccal surface. The crown exhibits a mesial interproximal facet (2.1 × 3.1 mm) and a distal one (2.1 × 3.1 mm). The fragility of the root, the occlusal wear, and the distal interproximal facet indicating full occlusal eruption of the dm₂ together suggest a developmental age of three to six years.

The crown presents four cusps. The metaconid is the largest, followed by the hypoconid and the entoconid of subequal size, and then the worn protoconid. The protoconid and metaconid are connected by a strong midtrigonid crest that lost a distal fragment postmortem, exposing the dentin. The midtrigonid crest and a mesial marginal ridge lower than the crest enclose a moderately sized anterior fovea. The metaconid and hypoconid contact via a weak ridge. In mesiobuccal view, the protoconid exhibits a weak projection at the base, the tuberculum molare.

Palomas 26 Mandibular C₁ Right

This tooth (Plate 6.25) presents a completely formed crown and a nearly complete root, except for the apex. The crown presents a thin vertical crack running lengthwise, and there are several microfractures oriented vertically from the cervix. The enamel was eroded *in situ*, showing several pits. The root was broken postmortem, and even though it was reassembled, it has some loose fragments on the upper third of the mesial and distal surfaces. The crown is completely unworn (stage 1), with neither interproximal nor occlusal facets. The lack of wear and degree of root formation indicate a developmental age of 9 to 11 years.

The crown apex is distinct, being set off from the rest of incisal margin. The distal accessory ridge is absent (grade 0). The lingual surface presents trace shoveling with only the slightest development of the mesial and distal marginal ridges. The labial surface is featureless. The distal crown margin is slightly angled. The root shows a shallow groove expressed mesially with a rounded shape.

Palomas 27 Maxillary/Mandibular dc1

Whereabouts unknown. "A poorly preserved right maxillary or left mandibular deciduous canine, with characteristically Neanderthal dimensions" (Walker et al. 1998).

Palomas 29 Mandibular M₂ Right

The crown of this tooth is completely formed, and the root apices were broken postmortem (Plate 6.26). The enamel was eroded *in situ*, producing several pits, mainly on the occlusal surface. Vertically oriented microfractures run on all faces, but they are not detectable on the occlusal surface.

The crown exhibits a minor degree of occlusal wear (stage 3c), with slight dentin exposure on the protoconid. There is a mesial interproximal facet (1.8 × 3.5 mm). There is no distal interproximal facet. The crown possesses five cusps arranged in a Y pattern. The protoconid is the largest cusp, followed by the metaconid, the entoconid, the hypoconid, and a large hypoconulid (cusp 5 grade 4). The mesial accessory crests of the protoconid and metaconid are joined by a bridge of enamel forming a midtrigonid crest that is partially interrupted by a weak fissure (grade 1) but not a distal trigonid crest (grade 0). This crest forms the distal border of a distinct anterior fovea (grade 3). The mesial accessory ridge of the metaconid is well developed. There are two deep intercuspal grooves of separation on the buccal side, one between the protoconid and the hypoconid and another between the latter and the hypoconulid. The metaconid does not exhibit a deflecting wrinkle (grade 0). There is no protostylid (grade 0). There are two roots present, fused along their lengths on their lingual surfaces but separate buccally. The tooth is therefore nontaurodont.

Palomas 31 Mandibular dc$_1$ Left

This tooth has a heavily worn crown, and the root apex was eroded postmortem (Plate 6.27). The enamel was eroded and exhibits several pits. The crown is more worn lingually and distally than labially and mesially. The incisal edge of this tooth is heavily worn (stage 5), preserving only half of the crown. There is a distal interproximal facet 1.5 × 3.0 mm, but there is no mesial interproximal facet. The extent of the wear indicates that the tooth was functional for a number of years and suggests an age-at-death of six to eight years.

Palomas 33 Mandibular Incisor Root

The partial root (Plate 6.28) retains the lower one-half to two-thirds of a root with a length of 13.2 mm. It is probably from a mandibular central incisor. The root apex was broken postmortem.

Palomas 34 Maxillary I^1 Left

This tooth (Plate 6.29) presents a completely formed crown and an incompletely formed root; the latter is three-quarters (R$_{3/4}$) formed (≈12.8 mm). The enamel was eroded *in situ* and has several pits. Two microfractures run lengthwise toward the distal and mesial sides of the crown and the root. The crown is occlusally unworn (stage 1), with a mesial interproximal facet 2.8 × 2.0 mm. There is no distal interproximal facet. The lack of wear and degree of root formation suggest a developmental age of seven to eight years. The root presents several strongly marked linear cementum hypoplasias, and in addition exhibits a deformation on the distolabial surface at the middle of the root.

Three mamelons are present along the incisal edge, and their associated furrows are visible both on lingual and labial surfaces. Lingually, the crown exhibits strong shoveling (grade 5) with well-developed marginal ridges that meet a cingulum. The latter comprises three moderately sized lingual tubercles lacking a free apex (tuberculum dentale grade 3). At the base of both the distal and mesial marginal ridges are marked interruption grooves (more marked distally than mesially) dividing them from cingulum. The labial surface is both superoinferiorly and mesiodistally moderately convex (grade 4). The root shows a shallow longitudinal depression along the mesial surface.

Palomas 35 Maxillary C^1 Left

The crown and root of this tooth are completely formed (Plate 6.30). The crown presents a vertical fracture running lengthwise. The crown is moderately worn (stage 4), with exposed dentin. The enamel was eroded and has several small pits and a large one (2.3 × 2.5 mm) on the labial side. The root is eroded on the distal side. The crown shows a linear hypoplasia.

The lingual surface is moderately shovel-shaped (grade 4), with more pronounced development of the mesial marginal ridge (grade 3), and it is fused to the lingual tubercle. These ridges circumscribe a moderate sulcus. At the base of the well-developed distal marginal ridge is a marked interruption groove dividing it from the cingulum, and there is a thin and slightly developed distal accessory ridge (grade 2). The mesial marginal ridge and tuberculum dentale are not divided by a developmental groove. The cingulum comprises a moderately sized lingual tubercle lacking a free apex (tuberculum dentale grade 4). Labially, the crown is asymmetric in a mesial-to-distal direction and superoinferiorly convex.

Palomas 36 Maxillary dm² Left

The tooth is an unerupted maxillary left deciduous molar. The external crown morphology and size originally led to its identification as an M¹ or an M² (Walker et al. 2008), but its dental tissue proportions (see chapter 9) indicate that it is best considered as a dm². Its dimensions (although large; see chapter 7) and morphology are fully compatible with this revised identification. The crown is complete, although it has fine cracks. There is minimal (≈1.4 mm) root formation, and the tooth is therefore developmentally between R_i and $R_{1/4}$. These characters and the absence of any occlusal or interproximal wear (stage 1) suggest a developmental age about one year.

The tooth possesses four cusps in following configuration: paracone > protocone > hypocone > metacone. The hypocone shows a moderately large size (grade 5). There is no evidence of either a cusp 5 or a Carabelli's trait (grade 0). The crown exhibits well-developed essential crests, and a bridge of enamel (*crista obliqua*) connects the protocone and metacone. The paracone and protocone show a mesial accessory ridge between them that forms a small mesial fossa. The hypocone presents a mesial accessory cusp, and the metacone has a distal accessory cusp. In occlusal view, the cuspal outline resembles a rounded parallelogram (9.7 mm mesiodistally × 5.7 mm buccolingually). The buccal and lingual profiles are strongly angled toward the occlusal basin and in buccal profile the tooth is quite convex. In both

lingual and buccal views, there is a marked separation between the protocone and the hypocone (forming a marked groove visible lingually) and between the paracone and the metacone (with a tiny groove visible buccally).

Palomas 37 Maxillary I² Right

This tooth presents a completely formed crown and an incompletely formed root (Plate 6.32); the latter is one-half ($R_{1/2}$) formed (≈8.3 mm). The enamel was eroded and presents several pits. The crown exhibits four linear hypoplasias. There are no interproximal facets. The lack of wear (stage 1) and degree of root formation suggest a developmental age of five to seven years.

Three mamelons are present along the incisal edge; the central one is tiny, and their associated furrows are visible on the lingual surface. Lingually, the crown exhibits strong shoveling (grade 5), with well-developed marginal ridges that meet at the cingulum and form a deep lingual sulcus. The cingulum comprises a moderately sized lingual tubercle lacking a free apex because it was broken postmortem (grade 5). At the base of the distal marginal ridge there is a marked interruption groove dividing it from the cingulum. The crown is both mesiodistally and superoinferiorly convex (grade 4).

Palomas 38 Mandibular M₂ or M₃

Whereabouts are unknown. "A permanent mandibular second or third molar fragment, with severe crown attrition" (Walker et al. 1998).

Palomas 39 Maxillary di¹ Left

The crown is complete, and the root apex was broken postmortem (Plate 6.33). However, the size of the radicular canal indicates that the root was open but close to closure, except the apex (R_c). This tooth exhibits two cracks on the interproximal surfaces along its length, visible at the incisal distal edge, and several short microfractures on the crown and root. The cementum was lost postmortem on the labial surface.

The crown is unworn (stage 1); the lack of wear, the degree of root formation, and the lack of interproximal facets indicates a developmental age of 12 to 18 months. Lingually, it exhibits trace shoveling, with the distal marginal ridge more developed than the mesial marginal ridge. The extent of the mesial marginal ridge is partially obscured by matrix, but it was nonetheless small. In addition, lingually the tooth exhibits a weak cingulum not connected with the marginal ridges. The distal border of the crown is more angled than the mesial border. Labially, the tooth is featureless but has clear mesiodistal convexity.

Palomas 40 Mandibular dm$_1$ Left

This tooth has a fully formed and unworn crown, but only 2.0 mm of root ($\approx$R$_{1/4}$) is developed (Plate 6.34). The absence of occlusal wear and the degree of root development indicate an age of 6 to 10 months.

The crown presents four cusps. The protoconid is the largest cusp, followed by the metaconid and hypoconid, which are subequal in size, and finally the entoconid. The hypoconid presents a possible marginal tubercle; alternatively, there are five cusps, and this one is a small hypoconulid. The essential crests of the cusps are well developed. The protoconid and metaconid are connected by weak midtrigonid crest, with a large anterior fovea. From the mesial aspect, the marginal ridge is lower than the midtrigonid crest. There are no interproximal facets.

Palomas 41 Mandibular I$_1$ Right?

Whereabouts are unknown. "A probably right, mandibular medial permanent incisor, with an incompletely formed root like that in modern 6 year olds" (Walker et al. 1998).

Palomas 42 Mandibular M

The tooth (Plate 6.35) consists of one edge of a molar crown with the external side of a root, probably of a mandibular molar. The occlusal enamel surface has been planed off with dentin exposure (stage 3). There is no evidence of lesions.

Palomas 43 Maxillary I^2 Left

Whereabouts are unknown. "A probably left maxillary permanent lateral incisor—its long root might signal a canine—with a crown showing little attrition" (Walker et al. 1998).

Palomas 44 Mandibular C$_1$ Right

The crown and root of this tooth are completely formed (Plate 6.36). The crown has a vertical crack. The enamel was eroded *in situ* with several pits, and the lingual surface has lost a fragment at the cervix (approximately 1.0 × 3.3 mm). The root was broken postmortem, and it lacks some fragments in the cervical third. The incisal edge of the crown is moderately worn, showing exposed dentin (stage 4). Lingually there is matrix and loss of enamel at the cervix. The root apex deflects distally.

Palomas 45 Mandibular P$_3$ Right

This unworn tooth presents a completely formed crown and an incompletely formed root, with the root between half (R$_{1/2}$) and three-quarters (R$_{3/4}$) formed (13.7 mm) (Plate 6.37). The root exhibits a transverse crack around its the upper third. The occlusal wear is minimal, with a tiny facet on the protoconid with a point of exposed dentin (stage 2). There is a pit on the occlusal surface close to distolingual cusp. A distal interproximal facet (2.5 × 4.5 mm) is evident. The degree of root development indicates an immature individual, providing a developmental age of approximately 9 to 10 years.

The contour of the tooth crown is asymmetrical. The tooth possesses two lingual cusps. Of these, the metaconid in a mesiolingual position is the largest cusp, followed by the distolingual cusp (grade 3). Both the protoconid and the metaconid have well-developed essential crests connected by a weak transverse crest (grade 1). The protoconid exhibits a well-developed distal accessory ridge that meets a posterior fovea. The metaconid presents both mesial and distal marginal ridges, and a mesiolingual groove is present. The buccal aspect of the crown is superoinferiorly convex. The root is mesially convex when viewed from the lingual side and presents two shallow grooves mesially.

Palomas 46 Partial Root

Two-thirds of the root are preserved, with a maximum length of 15.6 mm (Plate 6.28). It is probably from a mandibular right premolar. The apex is closed.

Palomas 47 Mandibular I_1 Right

Whereabouts are unknown. "A right mandibular juvenile or adolescent mesial permanent incisor, with an incompletely formed root and no attrition of the crown" (Walker et al. 1998). The published measurements contrast with those of other Palomas (and Neandertal) I_1s, and they are therefore not included in Table 7.3.

Palomas 48 Maxillary I^2 Right

The crown and root are completely formed (Plate 6.38). The enamel was eroded *in situ* and has several pits. A microfracture runs lengthwise on the mesial surface of the crown. The root exhibits some microfractures running lengthwise, and the root apex was slightly broken postmortem. The crown is missing a small chip of enamel (1.2 × 1.5 mm) at the distobuccal surface.

The incisal surface is worn with exposed dentin (stage 3). There is a distal interproximal facet

2.9 × 0.9 mm, and the mesial side was eroded postmortem. The lingual surface is semishovel-shaped (grade 4), with well-developed mesial and distal marginal ridges that meet at the cingulum and form a deep lingual sulcus. The labial surface exhibits pronounced mesiodistal convexity (grade 4) and moderate superoinferior convexity. The root is convex mesially and labially, showing a shallow longitudinal depression along the mesial surface, and the apex deflects distally. It is mesiodistally compressed.

Palomas 49 Mandible with Deciduous Alveoli plus I_2 and C_1 Germs

The Palomas 49 mandible (Plate 5.10) retains portions of the alveoli from the right dm_2 to the left dm_2, largely complete for the dm_1s and left dm_2 and mostly lingual for the anterior teeth. The only deciduous teeth preserved consist of a mesial fragment of the left dm_1, an apical root fragment of the left dc_1, and the developing crowns of the left I_2 and C_1. All of the anterior deciduous teeth appear to have been in full occlusion prior to their postmortem loss, judging from their remaining alveoli. The degree of eruption of the deciduous molars is less clear.

The root sockets of the deciduous molars are largely present, but they are partially filled with hard matrix, especially those of the dm_2s (Plates 5.10 and 6.39). The dm_1s appear to have been close to full root formation, since there is the clear presence of the interradicular septum in the socket on the right side and a slight projection into the socket for the alveolar bone between the mesial and distal roots, especially lingually. The interradicular septum is also evident in the micro-CT image of the left dm_1 socket (Plate 6.39). The dm_1 roots should therefore have been at least at $R_{3/4}$, suggesting an age of about two years, closer to two and one-half years if the roots were close to R_c.

The almost complete left dm_2 socket has matrix around its internal edges and in its base, and its distal margin is absent, making details of its root developmental status obscure (Plate 6.39). The buccal and lingual superior margins of the space are slightly damaged, but the preserved edges suggest that there was little more than a thin lamina of bone on each side around the dm_2 crown. The mesial side of the right dm_2 socket is intact, and it rounds up to a sharp edge with the distal dm_1 socket (Plate 5.10). The full depth for its mesial root is evident, but it is unclear the extent to which the root was formed. These observations suggest that the dm_2s were close to alveolar eruption and possibly between alveolar and gingival eruption. Given that the ages for dm_2 emergence (gingival eruption) in recent humans are between one and one-half and two and one-half years postnatal, with full root formation closer to three years, an age of approximately two and one-half years seems to be also indicated by the dm_2.

Anterior symphyseal damage has resulted in loss of three of the permanent incisor crowns, but the developing crown of the left I_2 is preserved. Its degree of development approximates $Cr_{3/4}$ (Plate 6.39). Adjacent to the I_2 is the forming left C_1 crown, which is at approximately $Cr_{1/2}$. These degrees of crown formation indicate an age at death of two to three years, with a modal age of two and one-half years. These degrees of permanent tooth calcification and the inferences from the deciduous molar alveoli are consistent in providing an age at death for Palomas 49 of approximately two and one-half years.

The dm_1 and dm_2 alveoli suggest widely spaced mesial and distal roots. The virtual reconstructions of the left I_2 and C_1 indicate some formation of marginal ridges, but they are not sufficiently formed to permit scoring. Other dental morphological information is not evident.

Palomas 50 Mandibular M_3 Right

The tooth is a mandibular molar with the root broken off just below the cervix and substantial matrix on the occlusal and distal crown (Plate 6.40). The occlusal surface and most of the crown is covered in a thin layer of matrix, but it has been possible to clean the crown virtually. There is no apparent cuspal or interproximal wear, so it may have been unerupted or recently erupted. The crown exhibits five cusps, although the sulci between them are indistinct, given the complexity of the occlusal morphology. There is an anterior fovea and the midtrigonid crest is incomplete, because there is a sulcus between the two mesial cusps (grade 0). It lacks a protostylid.

The tooth was originally published as an M_1, M_2, or M_3 right (Walker et al. 2008), given the adhering matrix. The now evident occlusal morphology, however, suggests that it is an M_3, and it is so considered here. Note that the old designation (CG-47, visible in the lingual view) is duplicated with the missing Palomas 47.

Palomas 51 Maxillary M³ Right

Palomas 51 is a complete tooth with the distal and lingual alveolar bone; the whole distal surface and much of the occlusal surface are covered with hard, brecciated matrix (Plate 6.41). It has nonetheless been possible to extract the tooth surfaces virtually from the matrix. The occlusal surface possesses four cusps, with a relatively large hypocone (stage 4). The cusps were planed off, but there is no evidence of dentin exposure, so the occlusal wear should be at most stage 3. The mesial interproximal facet is 2.2×3.5 mm. There is no evidence of a Carabelli's trait (grade 0). The mesiobuccal and mesiolingual (lingual) roots are fused along their lengths.

Palomas 53 Maxillary P³ Left

This tooth presents the completely formed crown and the root (Plate 6.42), but the latter was eroded and squashed postmortem, resulting in a narrow lengthwise deformation. The crown is better preserved; the enamel is eroded with several pits mainly on the lingual face, and it exhibits vertically oriented microfractures on all of the surfaces. The lower half of the crown has matrix on it. There are no hypoplasias on the exposed occlusal half of the crown. Radiographic imaging of the tooth shows that there is a bifurcated pulp chamber into the damaged and joined roots, indicating the tooth is double-rooted (Plate 6.42). It is therefore considered to be a P^3 rather than a P^4 (*contra* Walker et al. 2008).

The crown shows occlusal wear that is limited to a small facet on mesiobuccal face and a facet on the paracone with exposed dentin (stage 3). The mesial and distal interproximal facets measure 2.2 × 4.9 mm and 2.1 × 5.5 mm, respectively; they lack subvertical grooves. The paracone and protocone are subequal in size and are separated by a well-defined and uninterrupted sagittal sulcus. This sulcus is confined to the central portion of the tooth and does not extend to either the mesial or distal margins. The protocone is mesially placed. The distal marginal ridge is thicker than the mesial marginal ridge. The essential crests of both the paracone and the protocone are well developed, and the paracone bifurcates near the sagittal sulcus (grade 2).

Palomas 54 Mandibular C_1 Right

This tooth retains a completely formed crown; the root was broken postmortem, preserving approximately two-thirds of its length (8.2 mm) (Plate 6.43). The crown has a vertical crack running lengthwise. The enamel was eroded *in situ* and has several pits. Labially, there is matrix down the inferior half of the crown and the root. Lingually, the surface of the root is clean. The crown is slightly worn (stage 2–3), with a tiny area of exposed dentin on its tip and a small mesial interproximal facet (2.9 × 1.0 mm); these features suggest that it belonged to a subadult between 11 and 13 years of age.

The lingual surface is moderately shovel-shaped, with pronounced development of a distal marginal ridge that blends into a weak cingulum (shoveling grade 3). In addition, there is a slightly developed distal accessory ridge (grade 2) and a moderate mesial accessory ridge.

Palomas 55 Maxillary Anterior Root

The apical two-thirds (length: 11.7 mm) of the root of an anterior maxillary tooth is preserved, adherent to matrix (Plate 6.28).

Palomas 57 Mandibular P_4 Right

The crown of this tooth is completely formed. The apical third of the root was broken postmortem (Plate 6.44); the diameter of the apical canal suggests incomplete root development. This tooth has a vertical crack in the crown that runs distomesially on the protoconid; another one runs obliquely from the cervix along the upper third of the root with some loss of fragments mesially and distally. The enamel was eroded *in situ* and has several pits. The crown exhibits minor occlusal wear (stage 1), and the lack of interproximal facets should indicate recent eruption at 10 to 12 years of age.

The tooth is largely symmetrical in occlusal view with a marked inclination of the distolingual border. There are three lingual cusps; the largest is the metaconid, in mesial position, followed by the distolingual and distal cusps (grade 9). The essential crest of the protoconid is bifurcated, and the protoconid and metaconid are connected, forming a well-developed transverse crest (grade 2). There is a moderate distal accessory ridge on the protoconid. The essential crests define the distal border of a large anterior fovea. The buccal aspect of the crown is superoinferiorly convex. The root presents a shallow groove distally in the middle; mesially, it is smooth.

Palomas 58 Mandibular M_3 Left

Palomas 58 is a largely complete mandibular molar without full closure of the root apices (Plate 6.45). It is largely covered in matrix and is adherent to a mass of hard breccia. However, virtual cleaning of the crown reveals well-preserved occlusal morphology. The best identification of the tooth is as a left mandibular third molar, given the clear arrangements of the buccal cusps. The occlusal surface seems unworn (stage 1) and consists of a rim near the crown margin with radiating tiny ridges coming out of a central basin. The absence of occlusal wear and the degree of development of the root indicate a developmental age of 18 to 20 years.

There are two small cusps mesially, a sagittal sulcus mesially, and then a rise to the mesiobuccal cusp. The buccal cusp is clearly delineated from the more mesial and distal ones, but the distinctions between the other cusps are hard to discern with the abundance of radiating fissures. There are at least five cusps, and possibly six. There is a minimal anterior fovea (grade 1) and no midtrigonid crest (grade 0). There is no protostylid (grade 0).

Palomas 59 Mandible with C_1 to M_2 Left

The Palomas 59 mandible retains the damaged distal I_1 alveolus, the damaged I_2 alveolus with a portion of the root in the mandible, the C_1 to M_2 in their alveoli, and then a damaged mesial M_3 alveolus (Plates 5.11 and 6.46). There is a distinct distal interproximal facet on the M_2, indicating that the M_3 was in full occlusion and hence the individual was mature at death. The five preserved teeth are presented here.

The mandible has an anomalous pattern of molar wear, with greater occlusal wear on the M_2 than on the M_1, associated with pathological changes at the M_2 radicular apex. There is also a dental carious lesion on the mesiobuccal M_2 and a probable one on the distal M_1. The mandible and teeth are generally well preserved, but there are traces of breccia remaining, especially in the interproximal spaces, and there has been minor chemical erosion to scattered areas of the surfaces.

Mandibular C_1 Left

The tooth (Plate 6.47) shows a vertical microfracture on both the lingual and buccal sides detectable

from the incisal edge and running lengthwise. The incisal edge of the crown is worn (stage 3) and inclined toward the P_3. The lingual surface is somewhat shovel-shaped, with moderately developed mesial and distal marginal ridges.

Mandibular P_3 Left

The tooth (Plate 6.48) presents vertical microfractures on both the lingual and buccal sides running lengthwise from the occlusal surface. In occlusal view, the crown shows wear stage 2, and the metaconid exhibits a small postmortem erosion facet. The contour of the tooth crown is asymmetrical, tending to a scalene triangular shape in outline. The tooth possesses two lingual cusps; the metaconid is in a central position and is larger than the distolingual cusp (grade 3). The metaconid has a well-developed essential crest, separated by a deep groove from the less-developed essential crest of the protoconid, and therefore does not form a transverse crest. The distal marginal ridge is thicker than the mesial marginal ridge. There is no mesiolingual sulcus.

Mandibular P_4 Left

The tooth (Plate 6.48) presents a vertical fracture on the lingual side of the metaconid; on the cervical third of the buccal side, a chip (2.7×3.3 mm) is missing, exposing the dentin. The enamel is chemically eroded with several pits. The crown has minor occlusal wear (stage 3), limited to small facets exposing portions of the dentin on the main cusps.

The tooth is asymmetrical in occlusal view and exhibits two lingual cusps: the metaconid, in a mesiolingual position (grade 2), and a distolingual accessory cusp. Both metaconid and protoconid have well-developed essential crests that meet to form a transverse crest (grade 2). This crest defines the distal border of a deep anterior fovea. In mesial view, the transverse crest is higher than the mesial marginal ridge. There is a well-developed distal accessory ridge on the protoconid. There is no mesiolingual sulcus.

Mandibular M₁ Left

On the buccal side, two vertically oriented micro-fractures can be observed along the enamel of the major cusps, but otherwise the tooth is well preserved, with only several erosion pits in the enamel (Plate 6.49). The occlusal surface is worn (stage 4b) more in the distal half than in the mesial half. The wear facets are small on the hypoconid and the metaconid, and they grow in size from protoconid to the entoconid, being largest on the hypoconulid.

The crown presents five cusps arranged in a Y pattern. The metaconid and the entoconid (which are equal in size) are the largest cusps, followed by the protoconid, the hypoconid, and a very large hypoconulid (cusp 5 grade 4). The lingual surface is oval in profile, while the buccal surface is tri-lobed, showing a strong buccal groove separating the protoconid from the hypoconid and a smaller groove separating the latter from the hypoconulid. The essential crests of the protoconid and metaconid meet to define the distal border of a large anterior fovea (grade 3), but they do not form a midtrigonid crest, because both crests are intersected by a weak sagittal sulcus (grade 0). The metaconid does not show a deflecting wrinkle (grade 0). The entoconid is bifurcated, and the hypoconid presents the essential cusp bifurcated and a mesial accessory ridge. There is no protostylid (grade 0).

Radiographic examination shows that the bifurcation of the roots is well below the alveolar plane. Hence the pulp chamber has a taurodontic form (Plate 6.46).

Mandibular M₂ Left

The wear pattern on this tooth is unusual (Plates 6.46 and 6.49). The occlusal surface is more worn (stage 4c) than the M₁, and the facets on the mesial cusps are larger than those on the distal cusps (see also chapters 10 and 12). The distal interproximal facet measures 2.7 × 3.6 mm, indicating the ante-mortem eruption of a now missing M₃.

The crown possesses five cusps arranged in a Y pattern. The protoconid and the metaconid are subequal in size followed by the entoconid, the hypoconid, and a large hypoconulid (cusp 5 grade 5). The buccal surface shows a strong buccal groove separating the protoconid from the hypoconid. The presence of an anterior fovea is unclear due to occlusal wear; a small one (grade 1) may be indicated by remnants of sulci branching mesially off of the central sagittal sulcus between the protoconid and the metaconid. The essential crests of protoconid and metaconid meet but do not form a midtrigonid crest, because both crests are intersected by a weak sagittal sulcus (grade 1). The metaconid does not present a deflecting wrinkle (grade 0). There is no protostylid (grade 0).

The radiograph of the dentition (Plate 6.46) shows that the roots remain as a single unit until to the apex of their combined root, with a pulp chamber that extends from the middle of the crown to the transverse apex. The tooth is there-fore fully taurodont.

Palomas 60 Maxillary P³ Right

The occlusal surface is largely exposed, but the crown is partially obscured buccally in hard matrix (Plate 6.50). The mesial cervix and root have a thin layer of matrix on them, but the distal side is buried in breccia. The tooth can nonetheless be largely exposed through virtual cleaning (Plate 6.50). The erosion on the mesial side makes it not possible to analyze the presence of a mesial inter-proximal facet. The crown is completely formed. The root has open root apices and is about three-quarters formed ($\approx R_{3/4}$). The unworn crown (stage 1) and the development of the root suggest a developmental age of 10 to 12 years.

The paracone is more developed than the protocone. Both exhibit well-developed essential crests (grade 2) that meet in a deep sagittal sul-cus; there is therefore no transverse crest (grade 0). The protocone is centrally placed. The distal marginal ridge is thicker than the mesial mar-ginal ridge. The buccal aspect of the crown is superoinferiorly convex. There is bifurcation of the roots on the distal side, but they are bridged on the mesial side, such that there is only a shallow longitudinal radicular groove mesially.

Palomas 61 Mandibular dm$_1$ Right

The lingual side of the crown is embedded in matrix and the distobuccal corner is broken away (the lost chip is 4.0 × 5.4 mm) (Plate 6.51). There is extensive dentin exposure (stage 6) and a clear mesial interproximal facet (2.4 × 1.3 mm). The root was broken off at the cervix, and one cannot tell if it was resorbed. Given the wear, it was fully in occlusion, but one cannot determine if it was shed antemortem.

Palomas 68 Maxillary P³ and P⁴ Right

Palomas 68 consists of the right P³ and P⁴ in position relative to each other in a lateral fragment of the right maxilla, but both of the teeth and the bone are covered in a layer of brown matrix (Plate 6.52). It has been possible to virtually extract them (Plate 6.52), exposing aspects of both the crowns and roots. The P⁴ crown has a linear hypoplasia around the upper third. The P⁴ shows little occlusal wear (stage 2a). The distal interproximal facet of the P⁴ (3.0 × 4.4 mm) is located entirely on the buccal cusp. Both crowns have the buccal and lingual cusps separated by deep sulci, such that transverse crests are absent. The protocone is the larger cusp, especially on the P⁴. The P⁴ is larger than the P³. The teeth have long roots (Table 7.5). The P³ root is double even though the two roots are fused except at the apices.

Palomas 69 Mandibular dm$_2$ Right

The tooth retains a fragment of the mesial and lingual crown with part of the mesial root (Plate 6.53). There was chemical erosion of the crown, but there is no evidence of either a mesial interproximal facet or occlusal wear (stage 1). There is

approximately two-thirds of the root formed. The tooth was therefore erupting or close to eruption at the time of death.

The essential crests of the protoconid and the metaconid meet and form the distal border, interrupted by a deep sagittal sulcus, of a marked anterior fovea, so there is an absence of a midtrigonid crest. The occlusal morphology is very close to that of Palomas 70, but because both are right, they represent different individuals.

Palomas 70 Mandibular dm$_2$ Right

The tooth is complete, even though the enamel was eroded *in situ* and has several pits (Plate 6.54). There is a large pit (1.4 mm wide) in the middle of the buccal shoulder that may well be a hypoplasia. The crown and root are completely formed; the crown is unworn (stage 1). The interproximal facets are not visible, because the mesial side has a thin layer of matrix on it and the distal side is eroded. The lack of wear and the degree of root formation indicate a developmental age of three to four years.

The crown presents five cusps arranged in a Y pattern with the metaconid being the largest cusp, followed by the protoconid and the hypoconid (they are of subequal size) and the entoconid and a large hypoconulid. The essential crests of the protoconid and the metaconid meet and form the distal border, interrupted by a deep sagittal sulcus, of a marked anterior fovea, so there is an absence of a midtrigonid crest. Mesially, a thick mesial marginal ridge defines the anterior fovea. There is a protostylid on the buccal surface of the protoconid. There is also a deep furrow between the hypoconid and hypoconulid. This tooth possesses two widely splayed roots. The mesial root has lingual and buccal components connected by a dentin plate of ≈2.7 mm; the buccal component is larger than the lingual one. The smaller (7.6 mm wide) distal root is comprised of a single component.

Palomas 71 Maxillary dm¹ Left

The tooth (Plate 6.55) retains the complete crown and roots, but the root apices are still open. The central sulcus of the crown and the interradicular space are filled with hard matrix, but it has been possible to extract the occlusal morphology virtually. The lingual crown is chemically etched, but the buccal crown is more intact.

Palomas 72 Maxillary M

The tooth is represented by a molar fragment, including both crown and root portions, that cannot be further identified (Plate 6.35). There is approximately one-third of the original root preserved. The occlusal morphology is not apparent, but the crown has a linear hypoplasia around the midcrown.

Palomas 73 Maxillary I¹ Right

This tooth presents only the labial half of a crown (Plate 6.56). The lingual surface of the crown and the root were broken and lost postmortem. The enamel was also eroded and has several pits. The incisal surface is worn with exposed dentin (stage 4). The crown exhibits two linear enamel hypoplasias. The labial surface is mesiodistally convex (grade 3).

Based on the presence and position of the enamel hypoplasias, degree of occlusal wear and similar labial microwear, Palomas 73 is likely to be paired with Palomas 79.

Palomas 74 Maxillary C¹ Left

The crown is completely formed, and the root is almost complete, with an open apex (Plate 6.57). The erosion of the enamel produced several small pits. The root presents a postmortem fracture on the lower third. Much of the root cementum was removed postmortem. The unworn crown (stage

1) presents a sagittal fracture running vertically. The lack of wear and degree of root formation suggest a developmental age of 10 to 12 years. The labial surface presents some transverse irregularities that may represent very minor dental enamel hypoplasias.

The lingual surface has only a slight shoveling (grade 2), with a slight development of the mesial marginal ridge and little development of the distal marginal ridge (grade 2). The mesial ridge is fused to the lingual tubercle, but a thin layer of matrix on lingual surface obscures the development of tuberculum dentale (grade ≥3). There is no distal accessory ridge (grade 0). Labially, the crown is asymmetric in a mesial-to-distal direction. The distal crown margin is steeply angled. The labial crown and the root are convex when viewed mesially.

Palomas 75 Mandibular P₃ Left

The tooth consists of the crown and adjacent root (Plate 6.58). The enamel was eroded *in situ*, showing several pits. The tooth exhibits a vertical crack running along its entire length. The root is partially preserved (2.1 mm). This tooth has a worn crown (stage 4) with exposed dentin on the protoconid. There is a distal interproximal facet 2.1 × 3.2 mm. The mesial interproximal side has matrix on it. The contour of the tooth crown is asymmetrical. The tooth seems to have two lingual cusps, but the degree of wear makes it not possible to verify the configuration. The buccal aspect of the crown is superoinferiorly convex.

Palomas 76 Mandibular P₃ Right

The tooth (Plate 6.59) consists of a completely formed crown but only a partially preserved root (≈7.4 mm). This tooth is unworn (stage 1), but the crown shows postmortem erosion of the enamel, with several pits and vertically oriented microfractures on all surfaces. The absence of both occlusal wear and interproximal facets suggests an age at death of 9 to 11 years.

The contour of the tooth crown is modestly asymmetrical. On the lingual side, a single cusp (grade 0), the metaconid, is in a central position. The protoconid and a distal accessory ridge form a moderate distal fossa. Both the protoconid and metaconid have moderately developed essentials crests, and there is a weak transverse crest between them (grade 1). There is no mesiolingual sulcus.

Palomas 78 Mandibular P_4 Left

The tooth (Plate 6.60) consists of a partially preserved crown with a root approximately 4.2 mm long. The mesial third of the crown has been broken off down to the cervix. The tooth exhibits several microfractures vertically oriented running along its length. The enamel was eroded *in situ* with several pits. The crown shows minor occlusal wear (stage 1); the lack of a distal facet could indicate recent eruption, which suggests a developmental age of 9 to 11 years.

The tooth possesses three lingual cusps; the metaconid in mesial position, followed by the distolingual and distal cusps; the three cusps are subequal in size (grade 8). There is a well-developed distal accessory ridge on the protoconid. The existence of a transverse crest between the protoconid and the metaconid is uncertain.

Palomas 79 Maxillary I^1 Left

The crown and root are completely formed (Plate 6.61). The upper half of the lingual side of the crown and the marginal ridges are chipped. The root apex was broken postmortem. A vertical fracture running into the upper third of the root is on the labial surface of the crown. The enamel was eroded *in situ* and has several pits. The incisal surface is worn with exposed dentin (stage 3–4).

The labial surface is both mesiodistally and superorinferiorly convex (grade 3). The tooth is otherwise featureless or relevant aspects (for example, marginal ridges) are too damaged to assess. The root exhibits a shallow longitudinal depression on the distal surface toward the apex. There is a hypoplasia on the midlabial surface.

Based on the presence and position of the enamel hypoplasias, degree of occlusal wear and similar labial microwear, Palomas 79 is likely to be paired with Palomas 73.

Palomas 80 Mandible with M_2 and M_3

Palomas 80 is a left hemimandible from the P_3/P_4 interdental septum to the middle of the ramus, retaining the erupting M_2 and the germ of the M_3 in its crypt (Plates 5.12 and 6.62). The radiograph shows that the M_2 crown formation stage is $R_{3/4}$, and that of the unerupted molar (M_3) is about $Cr_{3/4}$. These degrees of calcification provide a modal age at death of about 11 to 12 years postnatal. The M_2 is at alveolar eruption and close to gingival eruption; diverse recent human populations provide an average age for gingival eruption of approximately 11 years (range of means: 9.8 to 11.6 years). The premolars and M_1 are absent, and their alveoli are too damaged to indicate their stages of development. The M_2 and M_3 together indicate an age at death for Palomas 80 of 10 to 12 years.

Mandibular M_2 Left

The M_2 metaconid features a deep fracture, originating mesially in the anterior fovea, crossing the metaconid occlusally, and penetrating vertically between the metaconid and cusp 4 on the lingual side. Another fracture is visible between cusps 3 and 5 on the distolingual side. Chemical erosion is apparent at different points of the enamel surface.

The occlusal surface presents six cusps arranged in a Y pattern. The hypoconulid and the entoconulid are very small, the former a little larger than the latter (stages 3 and 2, respectively). The protoconid possesses a well-developed essential crest that meets the mesial accessory crest of the metaconid at the sagittal sulcus. These crests define the distal border of a large anterior fovea (grade 4). The distal accessory crest of the protoconid is very well developed and divided from the essential crest by a fissure, nearly forming a distinct cusp. This accessory crest meets the essential crest of the metaconid, which is bifurcated; the latter has the shape of a deflecting wrinkle (grade 2). A distal accessory crest of the metaconid is not present. A continuous midtrigonid crest is therefore absent (grade 0). The hypoconid possesses a well-developed essential crest and well-defined mesial and distal accessory crests. The entoconid has an essential crest that bifurcates midway to the occlusal basin. Cusp 6 is defined by both occlusal and distal fissures (grade 2). In occlusal view, both buccally and lingually, there is a marked separation between the protoconid and the hypoconid and between the metaconid and the entoconid. The lingual, buccal, and distal surfaces present furrows associated with the various cusps and cusplets.

Mandibular M$_3$ Left

The Palomas 80 M$_3$ is completely within its crypt, and it is visible only through an occlusal opening 4.5 mm anteroposterior and 2.5 mm mediolateral and virtually. It has been possible to extract it virtually (Plate 6.62), although the similar radiodensities of the crown and adherent matrix make clean separation of the two materials difficult. The resultant images of the crown nonetheless indicate a tooth with five cusps. The obscurity of most of the fine intercuspal sulci makes it not possible to determine whether it has a Y pattern or the presence of an anterior fovea. However, the distinct depression between the protoconid and the metaconid argues for the absence of the midtrigonal crest.

Palomas 81 Mandibular M$_1$ or M$_2$ Left

The tooth retains distobuccal and distal surfaces with minor erosion (Plate 6.35). There is no occlusal attrition (stage 1) and no distal interproximal facet, and the root may be incompletely formed (it is partly obscured in matrix). Presence of a midtrigonid crest is not recordable.

Palomas 82 Mandibular C$_1$ Left

The crown and root of this tooth are completely formed (Plate 6.63). The crown has a vertical fracture running lengthwise. The root was broken postmortem and was restored from several fragments. Even though it is covered with matrix, it is possible to see that the apex is open (just at R$_c$). Despite *in situ* erosion of the enamel, resulting in several pits, the incisal surface presents a minimal cusp tip, which indicates that the tooth was only slightly worn (stage 1); these features suggest that it belonged to a subadult between 10 and 12 years of age. The labial surface presents some transverse irregularities that may represent very minor dental enamel hypoplasias.

The crown has a small mesial facet (1.7 × 1.3 mm). There is a weakly developed distal accessory ridge (grade 2) that blends into a distal marginal ridge by a deflected wrinkle, creating a depression on the distal ridge 3.0 × 1.7 mm.

Labially, the crown is asymmetric in a mesial-to-distal direction. The crown apex is distinct, being set off from the rest of the incisal edge. The lingual surface is somewhat shovel-shaped, with a marked distal marginal ridge and a thin but moderately developed mesial marginal ridge. Both the crown and the root are labially superoinferiorly convex. The labial surface is featureless. The root presents a shallow groove distally.

Palomas 83 Mandibular dm$_2$ Right

The crown and the root of this tooth are completely formed, but the distal and mesiolingual roots were broken postmortem (Plate 6.64). There is sediment between the roots. The enamel was eroded postmortem and has several pits. The crown presents microfractures running lengthwise from the major cusps. The crown is moderately worn with dentin exposed on all cusps (stage 3c). A mesial interproximal facet 2.3 × 1.9 mm is present. There is no distal interproximal facet. The degree of occlusal wear, the development of the roots, and the presence of only a mesial interproximal facet suggest a developmental age between four and six years.

The crown presents seven cusps in a Y pattern. The worn protoconid is the largest cusp in extension, followed by the metaconid and the hypoconid (both are of subequal size), the entoconid, a median hypoconulid, an entoconulid of subequal size to the hypoconulid, and a median metaconulid. The essential crests of the protoconid and the metaconid are well developed and, although not connected, appear to form an anterior fovea; there is therefore no midtrigonid crest. Mesially, the fovea is defined by thick mesial marginal ridges. The buccal groove between the protoconid and hypoconid is wide and deep. Buccally, as well, the crown presents a marked furrow between the hypoconid and hypoconulid. This tooth also possesses two widely splayed roots, but partial destruction and sediment coating make their description difficult.

Palomas 84 Mandibular M$_1$ Left

The crown of this tooth is completely formed although probably unerupted; the root, broken postmortem lingually, is incompletely formed ($\approx$4.7 mm) (Plate 6.65). The root is between R$_{1/4}$ and R$_{1/2}$, which provides an age-at-death of three to five years. An enamel chip 6.3 × 6.0 mm is missing on the distolingual side, exposing the dentin. There are traces of matrix in the occlusal fissures. The crown is unworn (stage 1). There is no mesial facet, indicating that this tooth was unerupted, supporting the young age at death.

The crown presents five cusps arranged in a Y pattern. The protoconid is the largest cusp, followed by hypoconid and entoconid (which are equal in size) and by the metaconid and the hypoconulid. The latter is large (cusp 5 grade 5). The four major cusps present well-developed essential crests. There is no midtrigonid crest (grade 0). The essential crest of the metaconid deflects distally, forming a deflecting wrinkle, but it does not make contact with the entoconid (grade 2). The mesial accessory crest of the protoconid forms the distal border of a small anterior fovea (grade 1).

The entoconid presents a well-developed mesial accessory crest. Both lingual and buccal surfaces of the tooth are rounded in profile. There is no protostylid (grade 0).

Palomas 85 Maxillary di^2 Left

The crown of this tooth is completely formed, and the root was broken postmortem, preserving only 4.7 mm (Plate 6.66). The incisal edge of this tooth is heavily worn (stage 5) preserving only half of the crown or less. The enamel and the cementum were eroded postmortem, showing several pits. The broken root prevents knowing if resorption had occurred, but the substantial wear suggests an older childhood age. The lingual surface presents a slight and a moderate development of the distal and mesial marginal ridges, respectively.

Palomas 87 Mandibular P$_4$ Left

The crown of this tooth is completely formed, and the root is almost complete, with the apex still open (root between R$_{3/4}$ and R$_c$) (Plate 6.67). This provides an age at death of 10 to 12 years. The root was obliquely broken in the middle, but otherwise both crown and root are well preserved. There are five tiny hypoplastic pits on the buccal surface of the main cusp and cusplets near the occlusal surface. The crown exhibits minor occlusal wear, limited to a small facet on the mesiobuccal face (stage 1). The lack of interproximal facets should indicate recent eruption.

The tooth is asymmetrical in occlusal view with a marked inclination of the distolingual border. It possesses three lingual cusps; the largest is the metaconid, in mesial position, followed by the distolingual and distal cusps (grade 9). Both the protoconid and metaconid have well-developed essential crests that are separated by a deep fissure, and therefore they do not form a transverse crest. The essential crest of protoconid is bifurcated, and the essential crest of metaconid is trifurcated. There is a well-developed distal accessory ridge on the protoconid. The essential crests define the distal border of a large anterior fovea. The buccal aspect of the crown is supero-inferiorly convex. The root is distally convex when viewed from the lingual side.

Palomas 88 Mandibular dm$_2$ Left

The specimen consists of a left dm$_2$ with a piece of lateral mandibular corpus extending anteroinferiorly to the edge of the mental foramen (Plate 6.68). The crown and root are completely formed, but the enamel was eroded postmortem and exhibits several pits. The mesial root was broken postmortem. The occlusal surface and the overall crown are encrusted, which makes their analysis difficult.

The crown is unworn (stage 1–2), so the lack of occlusal wear and the degree of root formation suggest a developmental age of three to four years. The erosion of the enamel makes it not possible to detect any interproximal facets. The crown presents five cusps arranged in a Y pattern. The protoconid, the metaconid, and the hypoconid are of subequal size, followed by the entoconid and a large hypoconulid. The protoconid and metaconid are not connected; hence, there is no midtrigonid crest. Matrix prevents assessing the anterior fovea.

Palomas 89 Mandibular I$_2$ Right

The crown of this tooth is completely formed but unworn (stage 1) (Plate 6.69). The root was broken postmortem, preventing determination of whether it was complete. The lack of incisal wear suggests that it belonged to a subadult seven to nine years of age. The enamel was eroded postmortem, showing several pits. No mamelons are present. The lingual surface presents semishoveling (grade 3), with readily apparent distal and mesial marginal ridges and a moderately expressed cingulum. The labial surface is featureless, but it has a clear labial convexity.

Palomas 90 Maxillary I^1 Left

This tooth presents a completely formed crown and root (Plate 6.70). A chip of enamel 6.0 × 4.9 mm is missing on the labial side of the crown. The tooth was eroded *in situ*, producing several pits. Two microfractures run lengthwise on the distal and mesial surfaces of the crown. The root presents a surrounding fracture on the lower third, and the root apex was slightly broken postmortem. The incisal surface is worn with exposed dentin (stage 3).

The lingual surface is shovel-shaped (grade 3) and possesses a thick mesial marginal ridge. The cingulum forms a lingual tubercle lacking a free apex (tuberculum dentale grade 4). The labial surface is both mesiodistally and superorinferiorly convex (grade 4), but it is otherwise featureless. The root shows two longitudinal sulci on the distal surface and another one on the lingual side, both on the cervical half of the crown. In addition, there is a shallow longitudinal depression along the mesial surface of the root.

Palomas 91 Mandibular I$_2$ Left

The crown of this tooth is completely formed, but the root was broken postmortem (Plate 6.71). The incisal edge is eroded, but it is slightly worn (stage 3) with exposed dentin. The enamel was eroded *in situ* showing several pits. Lingually, there is a chip (1.7 × 2.0 mm) of lost crown and root at the cementoenamel junction. There is a mesial interproximal facet 4.4 × 1.5 mm.

In lingual view, the crown exhibits trace shovel shape (grade 1). The slight cingulum blends into a weakly expressed median ridge. Both the crown and the root are superoinferiorly convex. The labial surface is featureless with clear labial convexity. The root has a large distal sulcus and is mesiodistally compressed.

Palomas 93 Mandibular dc₁ Right

This tooth has a heavily worn crown, and the apex of the root was broken postmortem (Plate 6.72). This tooth was previously (Walker et al. 2008) identified as a maxillary deciduous canine.

The enamel was eroded and has several pits. The tooth exhibits a fracture running along the interproximal surfaces. The crown is more worn lingually and distally than labially and mesially. The incisal edge of this tooth is heavily worn (stage 5), preserving only half of the crown. There are no apparent interproximal facets, because of either the advanced wear or through postmortem erosion. The extent of wear indicates that the tooth was functional for many years and suggests that the individual died at six to eight years of age. There is no evidence of root resorption.

Palomas 94 Maxillary P⁴ Left

This tooth possesses a completely formed and unworn crown (stage 1) and a broken root (Plate 6.73). The crown has several erosion pits, and in occlusal view, a microfracture runs the mesiodistal length on the border of the buccal face. The lack of occlusal wear and interproximal facets suggests a developmental age of 10 to 12 years.

The paracone and protocone are subequal in size and are separated by a well-defined, uninterrupted sagittal sulcus. It therefore does not have a transverse crest (grade 0). The sagittal sulcus is confined to the central portion of the tooth and does not extend to either the mesial or distal margins. The distal marginal ridge is thicker than the mesial marginal ridge and presents a small accessory cusp. The essential crests of both the paracone and the protocone are well developed. The paracone shows a distal accessory ridge.

Palomas 95 Mandibular dc₁ Left

This tooth has a fully formed unworn crown (stage 1), but only 4.1 mm of the root is preserved (Plate 6.74). The root was broken postmortem, although the fragility of the root (especially lingually) suggests that it had not completed its development. The enamel was eroded showing several pits. The tooth exhibits a fracture running along its length on the interproximal surfaces, visible on the incisal edge. The lack of both occlusal wear and interproximal facets indicates that the crown was not in functional occlusion at the time of this individual's death, suggesting a developmental age of 12 to 18 months.

Lingually, the mesial marginal ridge of crown presents a trace shovel shape. The weakly developed median ridge blends into a slight cingulum with a free apex. The enamel is eroded on the distal margin of the tooth, which makes it not possible to determine the degree of development of its distal marginal ridge. The labial crown surface is both mesiodistally and buccolingually convex.

Palomas 98 Mandibular C₁ Right

The tooth retains the worn crown and a largely complete root, even though the apex was broken off (Plate 6.75). The crown is worn to stage 4–5, but it is encrusted with matrix and has exposure of the lingual and mesial enamel. The fusion of the root apex and the advanced occlusal wear indicate a mature individual.

The substantial occlusal wear does not allow assessment of the accessory ridges, although it is possible to detect the prominent marginal ridges. The labial surface is featureless.

Palomas 99 Mandibular I Left

Palomas 99 is a mandibular left incisor, probably an I₂ (Plate 6.76), but confirmation of it as an I₂ is not possible, given the damage to it. The tooth retains most of the root without the apex, plus a partial and matrix-encrusted crown. The mesial half of the crown is absent, and the remaining crown has been glued to the root across the cervix. It appears to have a prominent cingulum. Although broken, it is apparent that the root apex was closed. The occlusal surface is too damaged to assess wear, and no discrete traits can be observed.

Palomas 100 Mandibular $M_{1(2)}$ Left

Palomas 100 is a complete left mandibular molar, with mesial and distal interproximal facets. It therefore is an M_1 or an M_2 but more likely an M_1 (Plate 6.77). There is minor postmortem chipping of the enamel around the edges of the crown. The root was broken across in the middle and reassembled, and the apices of the two roots are in matrix, especially the distal one. The exposed mesial root has its apex closed. The occlusal enamel, although partly obscured by hard matrix, is planed off but with little or no dentin exposure. It therefore has stage 3 wear. The interproximal facets are both 5.4 mm in breadth; the mesial one is markedly concave transversely, whereas the distal one is flat.

The mesial root has clear buccal and lingual lobes, but the two sides of the distal root are less separated. The buccal bifurcation of the roots is 4.9 mm from the cervix, suggesting a modest degree of taurodontism.

Morphological Considerations and Comparisons

Given the limits of wear, preservation, and encrustation, various of the discrete morphological characteristics of the Palomas teeth have been detailed in the descriptions of the individual teeth. The majority of these features are summarized for the permanent teeth in Tables 6.2 to 6.5 and for the deciduous ones in Table 6.8, noting the grade score, categorizing them with respect to presence versus absence following the criteria of Bailey (2006), and listing those teeth for which the trait is not scoreable. The Palomas permanent tooth frequencies are then compared to sample summaries provided by Bailey (2006), Martinón-Torres et al. (2012), and Bailey and Zubov (2014) for western Eurasian Middle and Late Pleistocene humans (Tables 6.6 and 6.7). The deciduous ones are compared to data from Bailey (pers. comm.) in Table 6.9.

The Middle Pleistocene sample is predominantly from Atapuerca-SH. The Neandertal data from each source are provided separately, although both sources include data for teeth from the initial Late Pleistocene Krapina sample to the mid–MIS 3 Neandertals. The Upper Paleolithic (E/MUP) data are from Bailey (Bailey 2006; Bailey and Zubov 2014). Permanent tooth data are not included for the Middle Paleolithic modern humans (MPMH), because Martinón-Torres et al. (2012) pooled them together with remains spanning the Upper Paleolithic, and Bailey and Zubov included nonmodern African remains in their MPMH sample. Note, however, that although the two sources largely agree on most of their trait frequencies for the Neandertals (within 10% difference), they differ by more than 10% (in some cases substantially more than 10%) in 8 of 21 traits for which both sources provide Neandertal data (38% of the traits). It is unlikely that differences in sample size and composition are solely responsible for these differences, and hence issues of interobserver error and trait definition become relevant. Clearly, for those eight traits differences in scoring by different observers are significant, and one can query the extent to which such differences apply to other traits (as well as to our scoring of the Palomas teeth relative to the data of those authors). Consequently, little weight should be given to those traits for which the two sources differ markedly, and modest differences in the frequencies for other traits (even if statistically significant) should not be given biological significance.

Given these caveats, there are some traits for which the Neandertals, and to a similar extent the Middle Pleistocene archaic humans, differ from early modern humans (Tables 6.6 and 6.7). In the maxillary anterior dentition, the incisors of the Neandertal lineage are dominated by shoveling and labial convexity, and the Palomas I^1s follow the same pattern. However, although the Neandertals have a very high frequency of I^2 tuberculum dentale, only one of two Palomas I^2s shows similar development of this feature. The remainder of the maxillary dental features, canine and molar, provide little separation of the samples.

The Palomas P_4s are similar to the Neandertals in transverse crest development and multiple lingual cusps, but two-thirds of the E/MUP teeth also have the latter trait. The overwhelming majority of the Neandertal lineage mandibular molars exhibit midtrigonid crests, in contrast to the Upper Paleolithic sample. However, only one of eight Palomas M_1s, M_2s, and M_3s exhibits such a crest.

It is also possible (Tables 6.8 and 6.9) to compare some of the discrete occlusal traits of the Palomas deciduous teeth to a sample of Neandertals and small samples of early modern humans. The one Palomas di^1 (Palomas 39) conforms to the four Neandertals in being shoveled and labially convex, contrasting in the former trait with the one Upper Paleolithic specimen available. The Palomas 25 and 40 dm_1 traits mostly conform to the Neandertal pattern, which contrasts with the early modern humans mostly in presence of the midtrigonid crest and anterior fovea. The same two traits are more frequent in Neandertal dm_2s; the Palomas dm_2s follow the pattern of having large anterior foveae, but they lack continuous midtrigonid crests. Otherwise they are unexceptional for Late Pleistocene humans.

In these discrete (or ordinally ranked) morphological features, therefore, there is variation in the extent to which the Neandertals are distinct from at least Upper Paleolithic modern humans. In those traits in which the samples contrast, the Palomas teeth largely fall with the Neandertals, but not entirely. In particular, they have a dearth of midtrigonid crests on the mandibular permanent molars and the dm_2s, a trait that otherwise occurs in high frequencies among the Neandertals and their European Middle Pleistocene predecessors.

Summary

The Palomas sample preserves 110 teeth or portions thereof, as isolated teeth or in partial or largely complete alveoli. As such, the Palomas dental sample is one of the larger ones known for Late Pleistocene humans, albeit one that substantially consists of damaged teeth that provide limited morphological and paleobiological data. Teeth from immature individuals, especially juveniles, dominate the sample, which is of paleodemographic interest (see chapter 15), but also provide a substantial number of teeth with minimal occlusal and interproximal wear. Like other Late Pleistocene skeletal samples, this sample also has a dearth (or possible absence) of older adults, as would be indicated by heavily worn teeth.

Morphologically, the Palomas dental sample appears to be aligned principally with the contemporaneous western Eurasian Neandertals, but trait-by-trait considerations indicate that it is principally the maxillary central incisor configurations that place them close to the Neandertal sample distribution. In the other features in which Neandertal and early modern human teeth largely differ, the Palomas sample is variable and/or contrasts with the dominant Neandertal pattern. In these aspects, the occlusal morphology of the Palomas teeth echoes concerns (for example, Trinkaus 2006a; Liu et al. 2013; Wu et al. 2014) regarding within- and between-group variation in these (and other) features in the later Pleistocene and the identification of distinctive (autapomorphic) Neandertal features as opposed to ones that are ancestral and/or vary broadly across late archaic humans.

SP NO.	PREVIOUS OR FIELD NO.	IDENTIFICATION	MATURITY	DEVELOPMENT	PROVENIENCE	DISCOVERY DATE
1a	CG-1	Maxilla right with C^1 to M^3	mature	complete	Shaft upper breccia	1991
1b	CG-1	Maxilla left with C^1 to M^3	mature	complete	Shaft upper breccia	1991
1c	CG-1	Mandible right with I_1, C_1 to M_3	mature	complete	Shaft upper breccia	1991
1d	CG-1	Mandible left with C_1, M_1 to M_3 (+SP20 to 22)	mature	complete	Shaft upper breccia	1991
6	CG-6	Mandible corpus left with C_1 to M_2 roots	mature	complete	Hillside rubble	1993
7	CG-7	Mandible mesial corpus left with C_1 and P_3 germs	immature	incomplete	Mine level rubble	1993
18	CG-22	Canine mandibular left	immature	complete	Upper Cutting level 2d	27 July 1994
19	CG-23	Incisor 1 mandibular left	mature	complete	Upper Cutting level 2	27 July 1994
20	N1	Incisor 2 mandibular left (probably SP1)	mature	complete	Upper Cutting	1994/1996
21	N2	Incisor 1 mandibular left (probably SP1)	mature	complete	Upper Cutting	1994/1996
22	N3	Premolar 3 mandibular left (probably SP1)	mature	complete	Upper Cutting	1994/1996
23	CG-2	Mandible with left I_1 to right P_4 roots plus right M_1 to M_3	mature	complete	Hillside rubble	2 July 1995
24	CG-27	Incisor 1 maxillary left	immature	complete	Hillside rubble	4 July 1995
25	CG-28	Deciduous molar 1 mandibular right	immature	incomplete	Upper Cutting level 2f	4 July 1995
26	CG-30	Canine mandibular right	immature	incomplete	Upper Cutting level 2f	4 July 1995
27	CG-34 (1998)	Deciduous canine	immature	—	Hillside rubble	5 July 1995
29	CG-33 (CG-35)	Molar 2 mandibular right	immature	complete	Upper Cutting level 2h	6 July 1995
31	CG-31	Deciduous canine mandibular left	immature	complete	Upper Cutting level 2i	9 July 1995
33	CG-46	Root fragment (mandibular incisor?)	—	complete	Upper Cutting level 2i	10 July 1995
34	CG-29	Incisor 1 maxillary left	immature	incomplete	Upper Cutting level 2k	17 July 1995
35	CG-26	Canine maxillary left	immature	complete	Upper Cutting level IA	20 July 1995
36	CG-25	Deciduous molar 2 maxillary left	immature	incomplete	Upper Cutting level IA	21 July 1995
37	Code 8	Incisor 2 maxillary right	immature	incomplete	Upper Cutting level IA	27 July 1995
38	CG-36 (1998)	Molar 2-3 with mandibular fragment	—	—	Upper Cutting level IB	1995/1996
39	CG-35	Deciduous incisor 1 maxillary left	immature	incomplete	Upper Cutting level IA	27 July 1996
40	CG-42	Deciduous molar 1 mandibular left	immature	incomplete	Upper Cutting level IA	27 July 1996
41	CG-40 (1998)	Incisor 1 mandibular right	immature	incomplete	Upper Cutting level IA	27 July 1996
42	CG-38	Molar crown and root fragment	mature	complete	Upper Cutting level IB	28 July 1996
43	CG-39 (1998)	Incisor 2 maxillary left	immature?	—	Upper Cutting level IB	29 July 1996
44	CG-40	Canine mandibular right	mature	complete	Upper Cutting level IB	29 July 1996
45	CG-44	Premolar 3 mandibular right	immature	incomplete	Upper Cutting level IB	29 July 1996
46	CG-45	Root fragment	mature	complete	Upper Cutting level IB	29 July 1996
47	CG-47 (1998)	Incisor 1 mandibular right	immature	incomplete	Upper Cutting level IB	29 July 1996
48	CG-37	Incisor 2 maxillary right	mature	complete	Upper Cutting level IBa	31 July 1996

SP NO.	PREVIOUS OR FIELD NO.	IDENTIFICATION	MATURITY	DEVELOPMENT	PROVENIENCE	DISCOVERY DATE
49	CG-51 (CG-49)	Mandible with I_2 & C_1 germs	immature	incomplete	Upper Cutting level IA	2 August 1996
50	CG-47	Molar 3 mandibular right	mature ?	undetermined	Hillside rubble	6 August 1996
51	CG-43	Molar 3 maxillary right with alveolus	mature	complete	Upper Cutting level I	9 August 1996
53	CG-48	Premolar 3 maxillary left	mature	complete	Upper Cutting level 2c	13 August 1997
54	CG-49	Canine mandibular right	immature	undetermined	Upper Cutting level 2c	13 August 1997
55	CG-50	Root, anterior maxillary tooth	mature	complete	Upper Cutting level 2c	15 August 1997
57	Code 3, CG-65	Premolar 4 mandibular right	immature	incomplete	Upper Cutting level 2d	19 August 1997
58	Code 25, CG-71	Molar 3 mandibular left	mature	incomplete	Upper Cutting level 2d	29 July 1998
59	CG-70	Mandibular body left from I_2 to M_3, with C_1 to M_2	mature	complete	Upper Cutting level 2f	1 August 1998
60	Code 19, CG-72	Premolar 3 maxillary right	immature	incomplete	Upper Cutting level 2f	6 August 1998
61	Code 17, CG-73	Deciduous molar 1 right; mandibular fragment	immature	complete	Upper Cutting level 2f	9 August 1998
68	Code 21, CG-92	Premolars 3 & 4 right in maxillary fragment	immature	undetermined	Upper Cutting level IA	29 July 2001
69	Code 16, CG-94	Deciduous molar mandibular probably right fragment	immature	undetermined	Upper Cutting level IAc	6 August 2001
70	Code 13, CG-96	Deciduous molar 2 mandibular right	immature	complete	Upper Cutting level IAd	6 August 2001
71	Code 22, CG-93	Deciduous molar 1 maxillary left	immature	incomplete	Upper Cutting level IAd	6 August 2001
72	Code 26, CG-97	Molar maxillary left fragment	mature ?	undetermined	Upper Cutting level IAf	8 August 2001
73	Code 23	Incisor 1 maxillary right labial crown	mature ?	undetermined	Upper Cutting level IAh	25 July 2002
74	Code 12, CG-99	Canine maxillary left	immature	incomplete	Upper Cutting level IAi	29 July 2002
75	Code 20, CG-100	Premolar 3 mandibular left	mature	complete	Upper Cutting level IAe-h	9 August 2002
76	Code 4, CG-106	Premolar 3 mandibular right	immature	incomplete	Upper Cutting level 2a	25 July 2003
78	Code 2, CG-110	Premolar 4 mandibular left	immature	incomplete	Upper Cutting level 2c	30 July 2003
79	Code 11, CG-107	Incisor 1 maxillary left	mature ?	complete	Upper Cutting level 2a	31 July 2003
80	CG-105	Mandible left with M_2 and M_3 germ	immature	incomplete	Upper Cutting level 2d	5 August 2003
81	Code 24, CG-108	Molar 1 or 2 mandibular left	mature ?	undetermined	Upper Cutting level 2b	5 August 2003
82	Code 7, CG-109	Canine mandibular left	immature	incomplete	Upper Cutting level 2b	6 August 2003
83	Code 14, CG-111	Deciduous molar 2 mandibular right	immature	complete	Upper Cutting level 2c	7 August 2003

SP NO.	PREVIOUS OR FIELD NO.	IDENTIFICATION	MATURITY	DEVELOPMENT	PROVENIENCE	DISCOVERY DATE
84	Code 5	Molar 1 mandibular left	immature	incomplete	Upper Cutting level 2f	10 August 2003
85	Code 18, CG-116	Deciduous incisor 2 maxillary left	immature	complete	Upper Cutting loose dirt	10 August 2003
87	Code 1	Premolar 4 mandibular left	immature	incomplete	Upper Cutting level 2g	23 July 2004
88	Code 15	Deciduous molar 2 mandibular left with lateral corpus bone	immature	complete	Upper Cutting level 2g	23 July 2004
89	Code 6	Incisor 2 mandibular right	immature	undetermined	Upper Cutting level 2m	25 July 2004
90	Code 10	Incisor 1 maxillary left	mature	complete	Upper Cutting level 2e	27 July 2004
91	Code 9	Incisor 2 mandibular left	mature ?	undetermined	Upper Cutting level 2e	27 July 2004
93	Code 27	Deciduous canine mandibular right	immature	complete	Upper Cutting level 2n	27 July 2006
94	Code 29	Premolar 4 maxillary left	immature	undetermined	Upper Cutting level 2o	27 July 2006
95	Code 28	Deciduous canine mandibular left	immature	undetermined	Upper Cutting level 2o	28 July 2006
98	—	Canine mandibular right	mature	complete	Upper Cutting level 2c	31 July 2007
99	—	Incisor 2? mandibular left	mature	complete	Upper Cutting level 2c	31 July 2007
100	SP08002	Molar mandibular left	mature	complete	—	26 July 2008

TABLE 6.2. Summary of dental discrete traits for maxillary incisors, canines, and premolars

DISCRETE DENTAL TRAIT	PRESENT (ASUDAS SCORE)[a]	ABSENT (ASUDAS 0)[b]	NOT SCORABLE[c]
I^1			
Shoveling (SH)	SP24 (6), SP34 (5), SP90 (3)	—	SP73, SP79
Double shovel	—	SP24, SP34, SP73, SP79, SP90	—
Tuberculum dentale (TD)	SP24 (6), SP34 (3), SP90 (4),	—	SP73, SP79
Labial convexity (LC)	SP24 (4), SP34 (4), SP73 (3), SP79 (3), SP90 (4)	—	—
I^2			
Shoveling (SH)	SP37 (5), SP48 (4)	—	—
Tuberculum dentale (TD)	SP37 (5)	SP48	—
Labial convexity (LC)	SP37 (4), SP48 (4)	—	—
C^1			
Shoveling (SH)	SP35 (4), SP74 (2)	—	SP1R & L
Tuberculum dentale (TD)	SP1R (1), SP1L (≥2), SP35 (4), SP74 (≥3)	—	—
Distal accessory ridge (CDAR)	SP35 (2)	SP74	SP1R & L
Mesial marginal ridge (CMR)	SP35 (3), SP74 (2)	—	SP1R & L
P^3			
Essential crests	SP53 (2), SP60 (2)	—	SP1R & L, SP68
Transverse crest	—	(SP1R & L), SP53, SP60, SP68	—
P^4			
Essential crests	SP94 (2)	—	SP1R & L, SP68
PMDAC	SP94 (1)	—	—
Transverse crest	—	SP1R & L, SP68, SP94	—

[a] When applicable, the Sima de las Palomas (SP) specimen number is followed in parentheses by its ASUDAS score, the criteria for some of which are derived from Bailey (2002a). Right and left are indicated only for Palomas 1, since it is the only associated dentition with both sides largely present.
[b] Absence indicates either the absence of the trait for those that are scored as presence/absence or a score of zero (0) for those that are given ASUDAS rank order scores.
[c] Specimens for which most of the crown is present but the trait cannot be reliably scored given wear, damage, and/or obscuring matrix. Specimens for which the crown is largely absent or extensively damaged are not listed.

DISCRETE DENTAL TRAIT	PRESENT (ASUDAS SCORE)[a]	ABSENT (ASUDAS 0)[b]	NOT SCORABLE[c]
M^1			
Carabelli's cusp	—	SP1R & L	—
Hypocone C4	SP1R & L (4)	—	—
Metacone C2	SP1R (4) & L (5)	—	—
Cusp 5	—	SP1R & L	—
Number of cusps (NC)	SP1R & L (4)	—	—
PTYL	—	—	SP1R & L
Crista obliqua	—	—	SP1R & L
Taurodont	—	SP1R & L	—
M^2			
Carabelli's cusp	—	SP1R & L	—
Hypocone	SP1R & L (3)	—	—
Metacone	SP1L (4)	—	—
Number of cusps (NC)	SP1R & L (4)	—	—
Cusp 5	—	SP1R & L	—
Taurodont	—	SP1R	—
M^3			
Carabelli's cusp	—	SP1R & L, SP51	—
Hypocone	SP1R (3) & L (2), SP51 (4)	—	—
Number of cusps (NC)	SP1R & L (4)	—	—
Cusp 5	—	SP1R & L, SP51	—
Taurodont	—	SP1R	—

[a] When applicable, the Sima de las Palomas (SP) specimen number is followed in parentheses by its ASUDAS score, the criteria for some of which are derived from Bailey (2002a). Right and left are indicated only for Palomas 1, since it is the only associated dentition with both sides largely present.

[b] Absence indicates either the absence of the trait for those that are scored as presence/absence or a score of zero (0) for those that are given ASUDAS rank order scores.

[c] Specimens for which most of the crown is present but the trait cannot be reliably scored given wear, damage, and/or obscuring matrix. Specimens for which the crown is largely absent or extensively damaged are not listed.

TABLE 6.4. Summary of dental discrete traits for mandibular incisors, canines, and premolars

DISCRETE DENTAL TRAIT	PRESENT (ASUDAS SCORE)[a]	ABSENT (ASUDAS 0)[b]	NOT SCORABLE[c]
I_1			
Shoveling S(H)	—	SP19	SP1R, SP21, SP99
I_2			
Shoveling (SH)	SP89 (3), SP91 (1)	—	SP20
Labial convexity	SP89 (3), SP91 (4)	—	SP20
C_1			
Shoveling (SH)	SP1R & L (3), SP7 (2), SP18 (2), SP26 (1), SP54 (3), SP59 (1), SP82 (2)	—	SP44, SP98
Distal accessory ridge (CDAR)	SP1R (3) & L (4), SP18 (4), SP54 (2), SP82 (2), SP7 (4)	SP26	SP44, SP59, SP98
Tuberculum dentale (TD)	—	SP18, SP26, SP59	SP7, SP44
P_3			
Asymmetry	SP22, SP45, SP75, SP76, SP59	—	SP1R,
+1 lingual cusp (PLCV)	SP1R (3), SP22 (3), SP45 (3), SP59 (3)	SP76	SP75
Transverse crest	SP45 (1), SP22 (2), SP76 (1)	SP59	SP1R, SP75
Anterior fovea	SP22,	SP59	SP1R, SP75
Posterior fovea	SP22, SP45	SP59	SP1R, SP75
Distal accessory ridge	SP45, SP76	SP59	SP1R, SP75
P_4			
Asymmetry	SP59, SP87	SP1R, SP57	—
+1 lingual cusp (PLCV)	SP1R (3), SP57 (9), SP59 (2), SP78 (8), SP87 (9)	—	—
Transverse crest	SP1R (2), SP57 (2), SP59 (2)	SP87	SP78
Anterior fovea (AF)	SP57 (3), SP59 (3), SP87 (3)	—	—
Distal accessory ridge	SP57, SP59, SP78, SP87	—	SP1R

[a] When applicable, the Sima de las Palomas (SP) specimen number is followed in parentheses by its ASUDAS score, the criteria for some of which are derived from Bailey (2002a). Right and left are indicated only for Palomas 1, since it is the only associated dentition with both sides largely present.

[b] Absence indicates either the absence of the trait for those that are scored as presence/absence or a score of zero (0) for those that are given ASUDAS rank order scores.

[c] Specimens for which most of the crown is present but the trait cannot be reliably scored given wear, damage, and/or obscuring matrix. Specimens for which the crown is largely absent or extensively damaged are not listed.

DISCRETE DENTAL TRAIT	PRESENT (ASUDAS SCORE)[a]	ABSENT (ASUDAS 0)[b]	NOT SCORABLE[c]
M_1			
Y pattern MGP	SP1R (Y), SP59 (Y), SP84 (Y)	—	SP1L
Hypoconulid C5	SP59 (4), SP84 (5)	—	SP1R & L
Anterior fovea (AF)	SP59 (3), SP84 (1)	—	SP1R & L
Mid-trigonid crest (MTC)	—	SP59, SP84	SP1R & L
Mesial accessory ridge	SP59, SP84	—	SP1R & L, SP84
Deflecting wrinkle (DW)	SP84 (2)	SP59	SP1R & L
Protostylid (PTYL)	—	SP1R & L, SP59, SP84	—
Number of cusps (NC)	(SP1R & L (5)), SP59 (5), SP84 (5)	—	—
C6	—	SP59, SP84	SP1R & L
C7	—	SP59, SP84	SP1R & L
Taurodont	SP59	SP6, SP23	—
M_2			
Y pattern (MGP)	SP1R (Y), SP29 (Y), SP59 (Y), SP80 (Y)	—	SP1L, SP23
Hypoconulid C5	SP29 (4), SP59 (5), SP80 (3)	—	SP1R & L, SP23
C6	SP80 (2)	SP29, SP59	SP1R & L, SP23
C7	—	SP29, SP59, SP80	SP1R & L, SP23
Anterior fovea (AF)	SP29 (3), (SP59 (1)), SP80 (4)	—	SP1R & L, SP23
Mid-trigonid crest (MTC)	SP29 (1A)	SP59, SP80	SP1R & L, SP23
Mesial accessory ridge	SP29	—	SP1R & L, SP23, SP59
Deflecting wrinkle (DW)	SP80 (2)	SP29, SP59	SP1R & L, SP23
Protostylid (PTYL)	—	SP1R & L, SP29, SP59, SP80	SP23
Number of cusps (NC)	(SP1R & L (5)), (SP23 (5)), SP29 (5), SP59 (5), SP80 (6)	—	—
Taurodont	SP59	SP6, SP23, SP29	—
M_3			
Y pattern (MGP)	SP1R (Y), SP50, SP58	—	SP1L, SP23, SP80
Anterior fovea	SP50 (1), SP58 (1)	—	SP1R & L, SP23, SP80
Mid-trigonid crest (MTC)	—	SP50, SP58, SP80	SP1R & L, SP23
Protostylid (PTYL)	—	SP1R, SP50, SP58	SP1L, SP23, SP80
Number of cusps (NC)	(SP1R & L (5)), SP58 (≥5), (SP80 (5))	—	—
Taurodont	—	SP23	—

[a]When applicable, the Sima de las Palomas (SP) specimen number is followed in parentheses by its ASUDAS score, the criteria for some of which are derived from Bailey (2002a). Right and left are indicated only for Palomas 1, since it is the only associated dentition with both sides largely present.

[b]Absence indicates either the absence of the trait for those that are scored as presence/absence or a score of zero (0) for those that are given ASUDAS rank order scores.

[c]Specimens for which most of the crown is present but the trait cannot be reliably scored given wear, damage, and/or obscuring matrix. Specimens for which the crown is largely absent or extensively damaged are not listed.

TABLE 6.6. The Palomas frequencies for presence of select incisor, canine, and premolar dental nonmetric traits compared to those of Middle and Late Pleistocene western Eurasian humans

DENTAL TRAIT[a]	PALOMAS	MIDDLE PLEISTOCENE[b]	NEANDERTAL (M-T)[b]	NEANDERTAL (B&Z)[c]	E/MUP[c]
I^1 shoveling (2-6)	100% (3/3)	95.8% (23/24)	100% (18/18)	100% (33/33)	33.3% (3/9)
I^1 labial convexity (2-4)	100% (5/5)	100% (21/21)	100% (17/17)	93.9% (31/33)	16.7% (2/12)
I^1 double shoveling (2-6)	0.0% (0/5)	—	—	0.0% (0/31)	0.0% (0/12)
I^2 *tuberculum dentale* (2-6)[d]	50.0% (1/2)	71.4% (15/21)	96.7% (29/30)	92.9% (26/28)	50.0% (3/6)
C^1 distal accessory ridge (2-5)[e]	100% (1/1)	—	—	52.2% (12/23)	60.0% (3/5)
C^1 mesial ridge (1-3)[d]	100% (2/2)	90.5% (19/21)	73.7% (14/19)	35.7% (10/28)	20.0% (1/5)
P_4 asymmetry (>1)	50.0% (2/4)	—	—	71.7% (33/46)	16.7% (1/6)
P_4 transverse crest (>1)	75.0% (3/4)	64.0% (16/25)	63.6% (14/22)	69.6% (32/46)	0.0% (0/6)
P_4 lingual cusp number (>1)	100% (5/5)	100% (25/25)	100% (24/24)	100% (45/45)	66.7% (4/6)

Note: The number of individuals with the trait present and the total sample (present/n) are provided after the percent present.

[a] The ASUDAS and non-ASUDAS traits follow Scott and Turner (1997) and Bailey (2002a, 2006). The ASUDAS grades counted as indicating "presence" are provided after each trait. The Martinón-Torres et al. (2012) scores have been adjusted accordingly.

[b] The Middle Pleistocene and Neandertal (M-T) data are from Martinón-Torres et al. (2012), combining their Atapuerca and "*H. heidelbergensis*" samples into a Middle Pleistocene one. Since Martinón-Torres et al. (2012) included the Palomas data published in Walker et al. (2008), their numbers have been adjusted to remove the Palomas data.

[c] The Neandertal (B&Z) and Upper Paleolithic data are primarily from Bailey and Zubov (2014).

[d] Traits for which the Neandertal frequency differs by more than 10% between Martinón-Torres et al. (2012) and Bailey and Zubov (2014).

[e] The grade 1 of Martinón-Torres et al. (2012) subsumes grades 1 and 2 employed here, so they are not comparable and hence are not included.

TABLE 6.7. The Palomas frequencies for presence of select molar dental nonmetric traits compared to those of Middle and Late Pleistocene western Eurasian humans

DENTAL TRAIT[a]	PALOMAS	MIDDLE PLEISTOCENE[b]	NEANDERTAL (M-T)[b]	NEANDERTAL (B, B&Z)[c]	E/MUP[c]
M^1 Carabelli's (3+)[d]	0.0% (0/1)	5.0% (1/20)	40.0% (8/20)	75.7% (28/37)	46.7% (7/15)
M^1 cusp 5 (1-5)[d]	0.0% (0/1)	76.2% (16/21)	95.5% (21/22)	58.8% (20/34)	57.1% (8/14)
M^2 three-cusped (<2)	0.0% (0/1)	28.6% (6/21)	9.5% (2/21)	0.0% (0/46)	6.7% (1/15)
M_1 mid-trigonid crest (>0)	0.0% (0/2)	88.9% (24/27)	90.0% (27/30)	95.9% (47/49)	0.0% (0/16)
M_2 mid-trigonid crest (>0)	33.3% (1/3)	96.6% (28/29)	100% (23/23)	100% (39/39)	0.0% (0/16)
M_3 mid-trigonid crest (>0)	0.0% (0/3)	69.0% (20/29)	94.1% (16/17)	93.3% (14/15)	0.0% (0/16)[e]
M_1 cusp 6 (1-5)	0.0% (0/2)	7.7% (2/26)	48.3% (14/29)	40.0% (14/35)	17.6% (3/17)
M_1 cusp 7 (1-4)[d]	0.0% (0/2)	53.8% (14/26)	69.0% (20/29)	25.9% (14/54)	10.0% (2/20)
M_1 deflecting wrinkle (3)[d]	0.0% (0/2)	12.0% (3/25)	3.4% (1/29)	20.0% (8/40)	7.1% (1/14)
M_2 Y pattern	100% (4/4)	41.9% (13/31)	70.8% (17/24)	78.8% (41/52)	52.9% (9/17)
M_2 four-cusped (4)	0.0% (0/5)	20.0% (6/30)	8.3% (2/24)	0.0% (0/56)	46.2% (6/13)
M_3 four-cusped (4)	0.0% (0/3)	20.0% (6/30)	12.5% (2/16)	—	—
M_1 anterior fovea (>1)	50.0% (1/2)	83.8% (31/37)	81.3% (26/32)	88.6% (31/35)	52.6% (10/19)[e]
M_2 anterior fovea (>1)[d]	66.7% (2/3)	79.3% (23/29)	72.7% (16/22)	88.5% (23/26)	50.0% (10/20)[e]
M_3 anterior fovea (>1)[d]	0.0% (0/2)	65.4% (17/26)	83.3% (15/18)	93.9% (13/14)	46.7% (7/15)[e]

Note: The number of individuals with the trait present and the total sample (present/n) are provided after the percent present.
[a] The ASUDAS and non-ASUDAS traits follow Scott and Turner (1997) and Bailey (2002a, 2006). The ASUDAS grades counted as indicating presence are provided after each trait. The Martinón-Torres et al. (2012) scores have been adjusted accordingly.
[b] The Middle Pleistocene and Neandertal (M-T) data are from Martinón-Torres et al. (2012), combining their Atapuerca-SH and "*H. heidelbergensis*" samples into a Middle Pleistocene one.
[c] The Neandertal (B, B&Z) and Upper Paleolithic data are primarily from Bailey and Zubov (2014), with additional data from Bailey (2006) for M_3s and anterior foveae.
[d] Traits for which the Neandertal frequency differs by more than 10% between Martinón-Torres et al. (2012) and Bailey and Zubov (2014).
[e] The Upper Paleolithic data, from Bailey (2006), pooled together with Early/Mid (E/MUP) and Late (LUP) European Upper Paleolithic data. For the M_3 midtrigonid crest, the E/MUP frequency would remain at 0.0%, but with a smaller sample size.

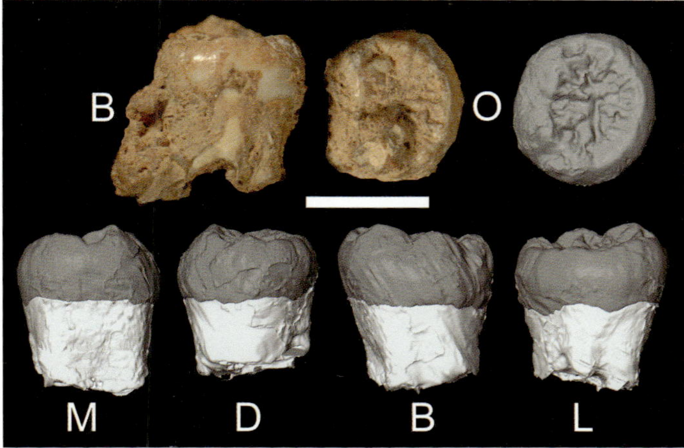

PLATE 6.45. Buccal (B) and occlusal (O) photographs of the Palomas 58 left M$_3$, plus mesial (M), distal (D), buccal (B), lingual (L), and occlusal (O) views of the virtually cleaned Palomas 58 molar. Scale: 1 cm

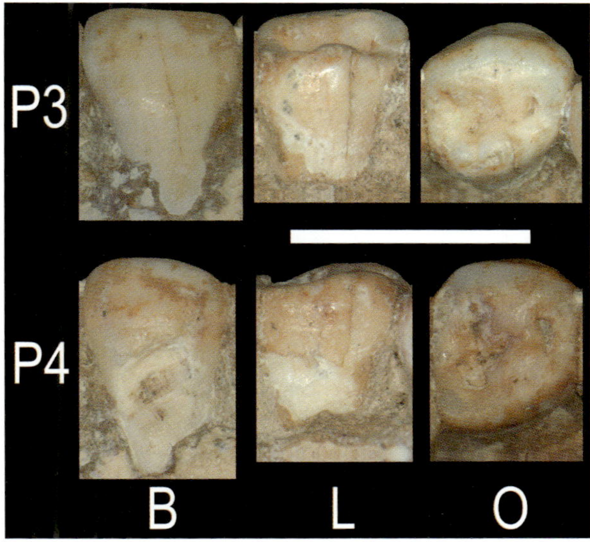

PLATE 6.48. The Palomas 59 left P$_3$ and P$_4$ in buccal (B), lingual (L), and occlusal (O) views. Scale: 1 cm

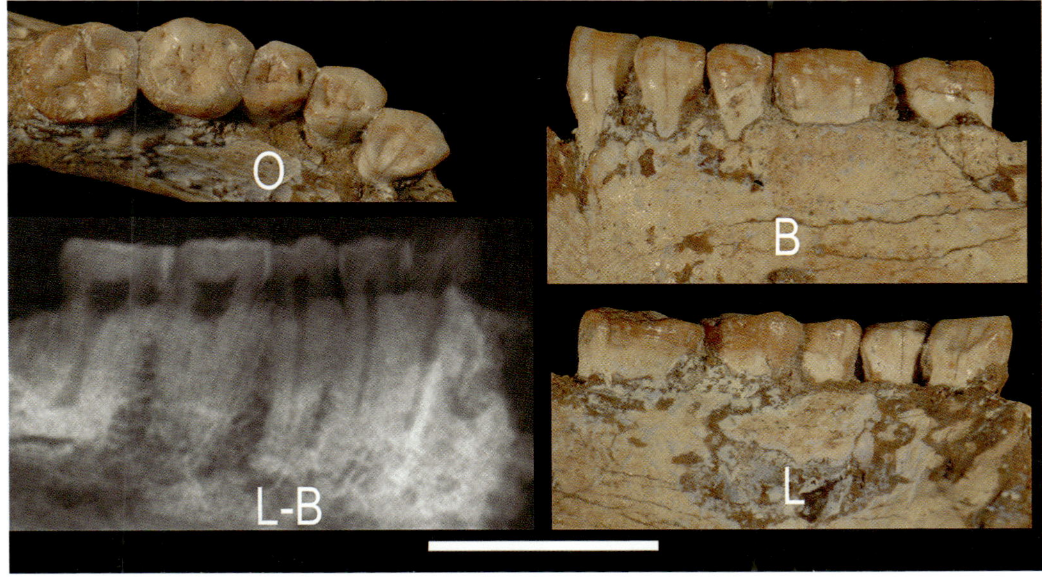

PLATE 6.46. Views of the Palomas 59 left mandibular dentition, all of which is in its alveoli. Views are occlusal (O), buccal (B), and lingual (L). The lingual-buccal (L-B) radiograph is provided. Scale: 2 cm

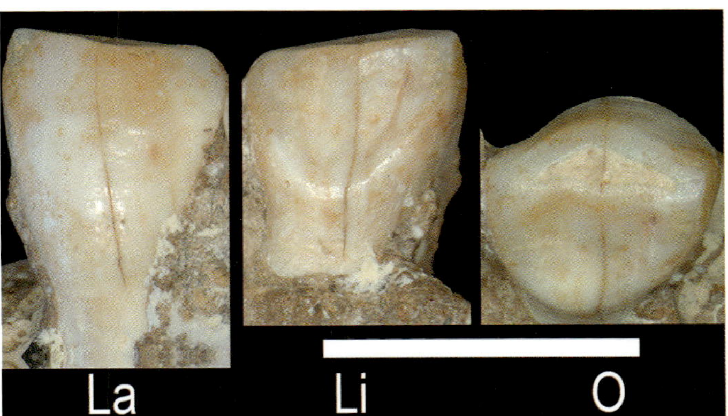

PLATE 6.47. The Palomas 59 left C$_1$ in labial (La), lingual (Li), and occlusal (O) views. Scale: 1 cm

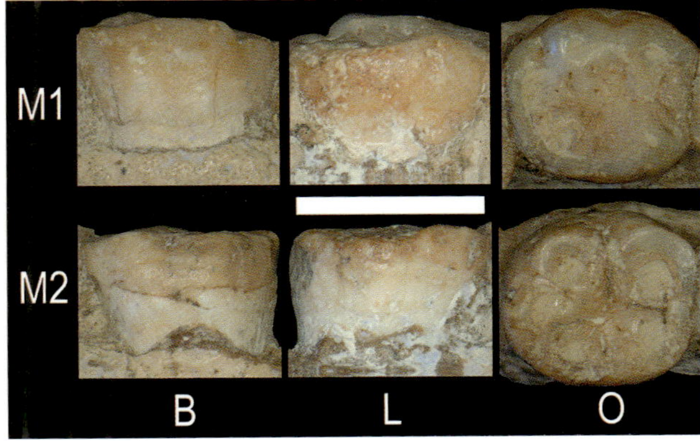

PLATE 6.49. The Palomas 59 left M$_1$ and M$_2$ in buccal (B), lingual (L), and occlusal (O) views. Scale: 1 cm

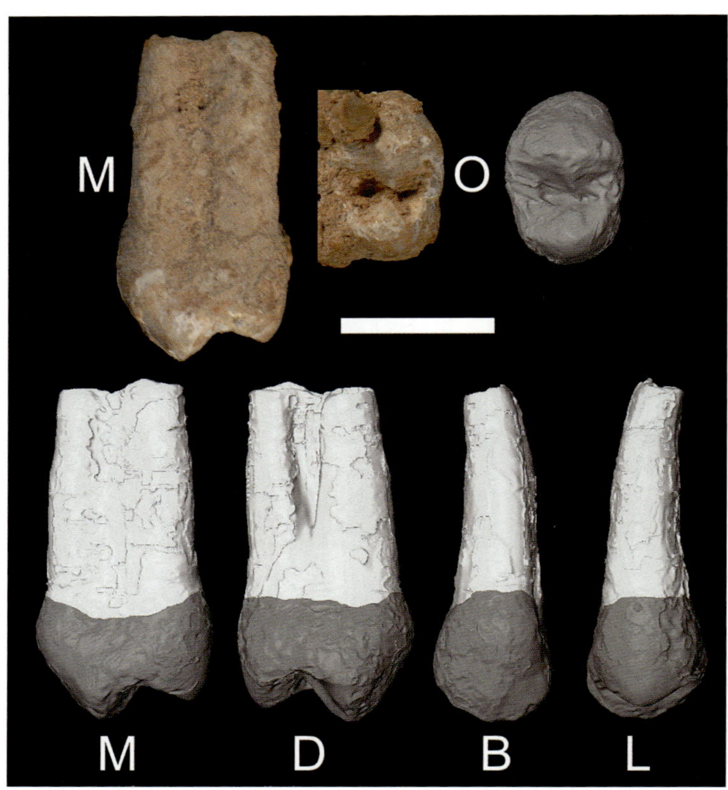

PLATE 6.50. Mesial (M) and occlusal (O) photographs of the Palomas 60 right P³, plus mesial (M), distal (D), buccal (B), lingual (L), and occlusal (O) views of the virtually cleaned tooth. Scale: 1 cm

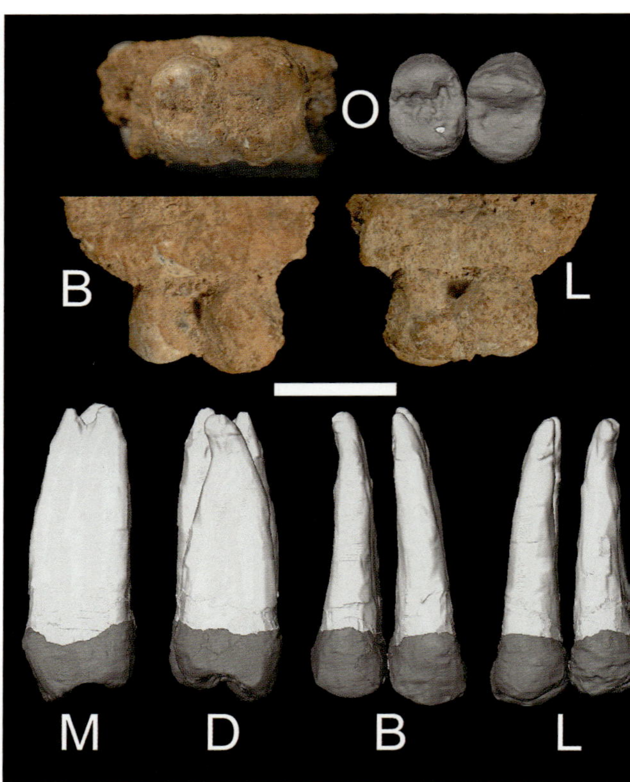

PLATE 6.52. The Palomas 68 right P³ and P⁴, with lateral maxillary alveolar bone in buccal (B), lingual (L), and occlusal (O) photographs, plus occlusal (O), mesial (M), distal (D), buccal (B), and lingual (L) virtually cleaned views of the paired teeth. Scale: 1 cm

PLATE 6.51. Buccal (B) and occlusal (O) views of the Palomas 61 right dm_1. The lingual matrix has been removed photographically. Scale: 1 cm

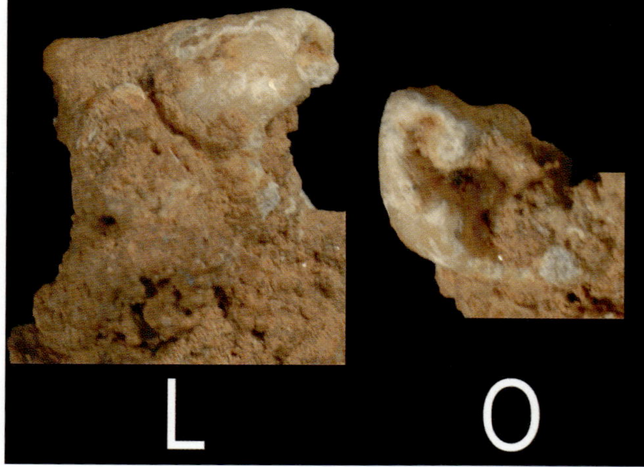

PLATE 6.53. Lingual (L) and occlusal (O) views of the partial right dm_2 of Palomas 69. Scale: 1 cm

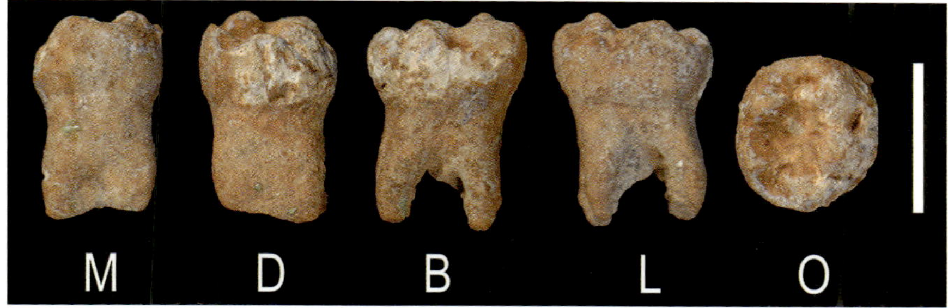

PLATE 6.54. Mesial (M), distal (D), buccal (B), lingual (L), and occlusal (O) views of the Palomas 70 right dm_2. Scale: 1 cm

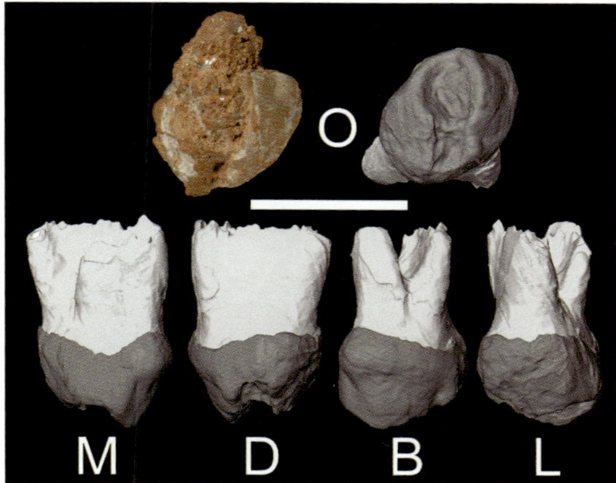

PLATE 6.55. The Palomas 71 left dm^1 in occlusal view (O) and the virtually cleaned tooth in mesial (M), distal (D), buccal (B), lingual (L), and occlusal (O) views. Scale: 1 cm

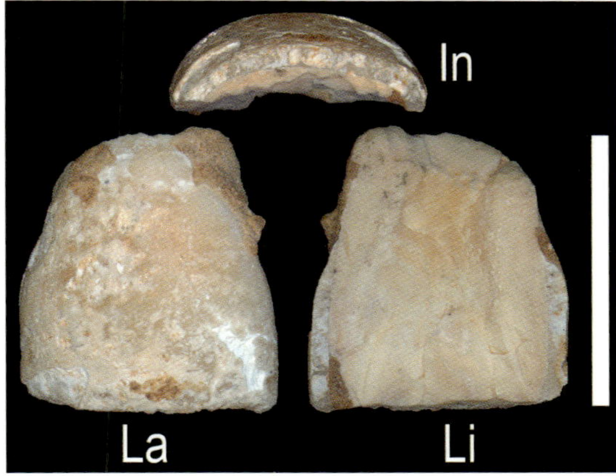

PLATE 6.56. Labial (La), lingual (Li), and incisal (In) views of the labial crown portion of the Palomas 73 right I^1. Scale: 1 cm

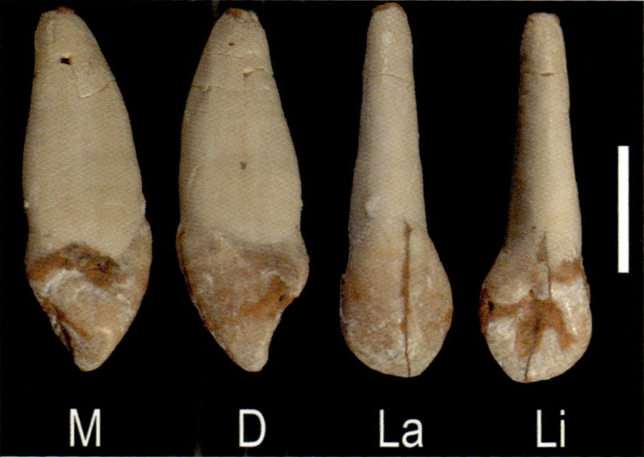

PLATE 6.57. Mesial (M), distal (D), labial (La), and lingual (Li) views of the Palomas 74 left C^1. Scale: 1 cm

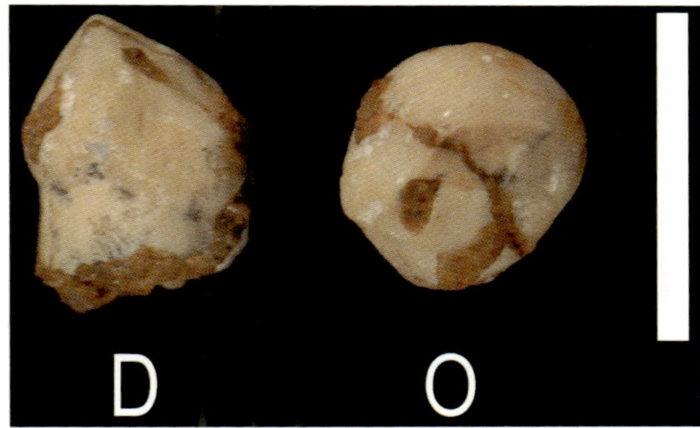

PLATE 6.58. The Palomas 75 left P_3 in distal (D) and occlusal (O) views. Scale: 1 cm

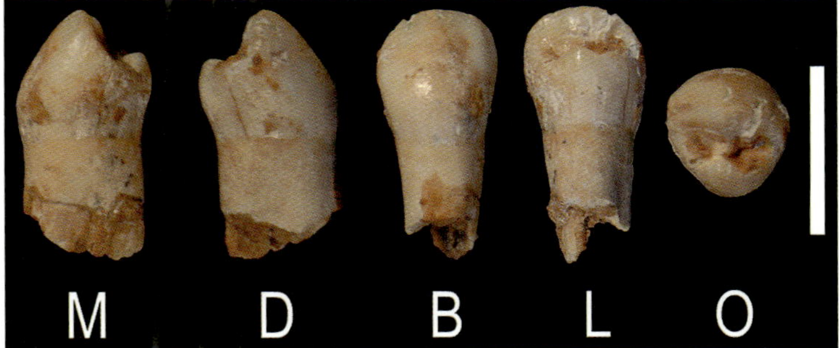

PLATE 6.59. The Palomas 76 right P_3 in mesial (M), distal (D), buccal (B), lingual (L), and occlusal (O) views. Scale: 1 cm

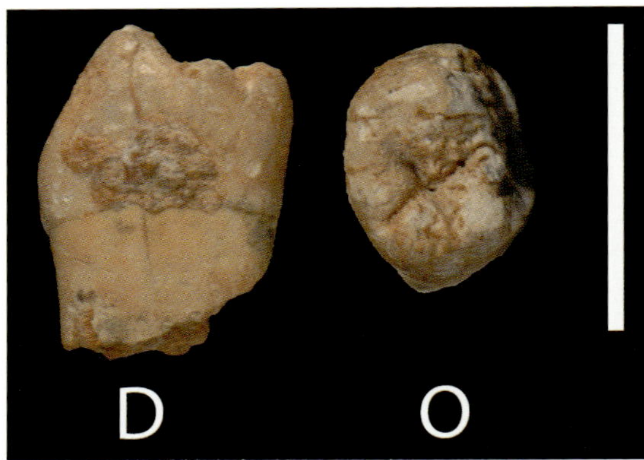

PLATE 6.60. Distal (D) and occlusal (O) views of the Palomas 78 left P_4. The mesial crown was removed postmortem. Scale: 1 cm

PLATE 6.63. The Palomas 82 left C^1 in mesial (M), distal (D), labial (La), and lingual (Li) views. Scale: 1 cm

PLATE 6.61. The Palomas 79 left I^1 in mesial (M), distal (D), labial (La), and lingual (Li) views. Scale: 1 cm

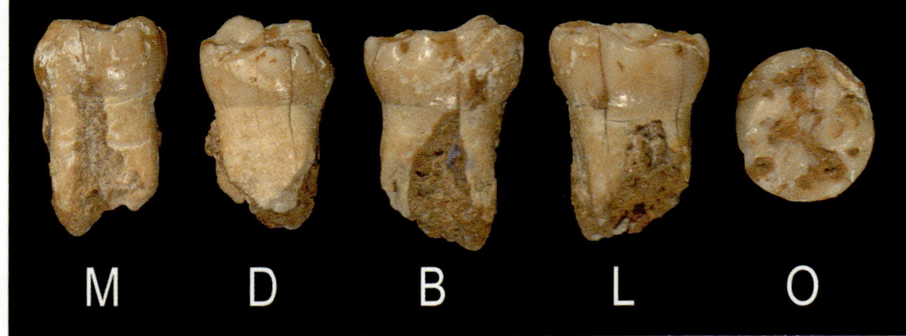

PLATE 6.64. Mesial (M), distal (D), buccal (B), lingual (L), and occlusal (O) views of the Palomas 83 right dm_2. Scale: 1 cm

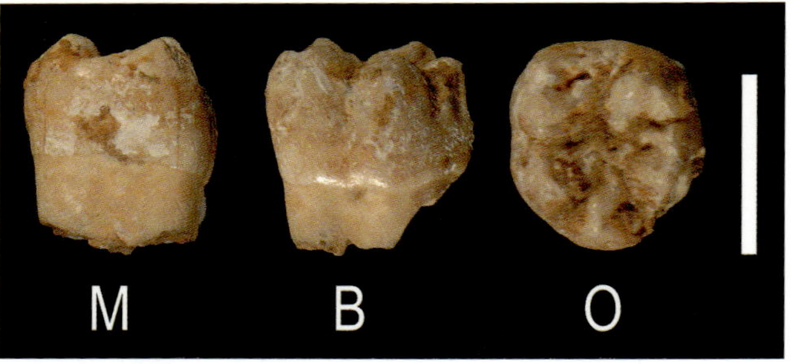

PLATE 6.65. The Palomas 84 left M_1 in mesial (M), buccal (B), and occlusal (O) views. Scale: 1 cm

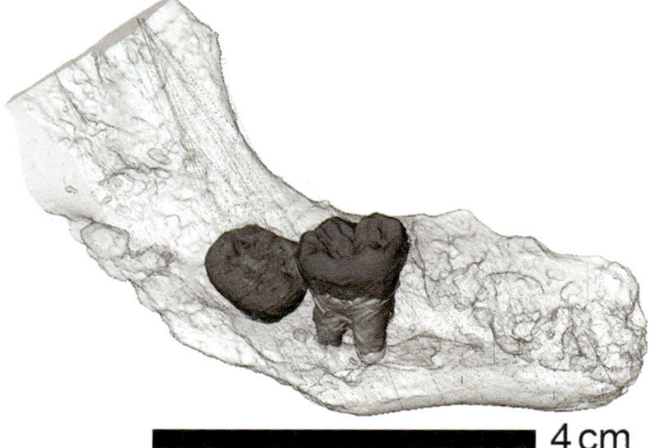

PLATE 6.62. The Palomas 80 left M_2 in occlusal view (above left), a virtual extraction of the left M_3 from its crypt, and a virtual reconstruction of the mandible with the M_2 and the M_3 in their crypts. The irregularities of the M_3 surface are due to difficulties in segmenting the M_3 from its surrounding matrix. Upper scale for the M_2: 1 cm; lower scale for the mandible: 4 cm. See also Plate 5.12.

PLATE 6.66. The Palomas 85 left di^2 in labial (La), lingual (Li), mesial (M), distal (D), and occlusal (O) views. Scale: 1 cm

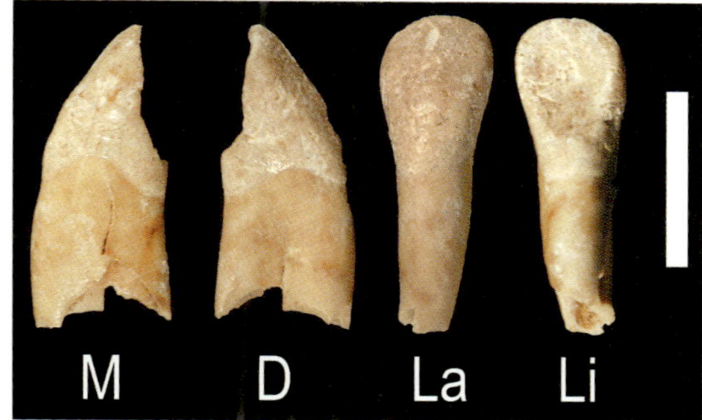

PLATE 6.69. The Palomas 89 I$_2$ right in mesial (M), distal (D), labial (La), and lingual (Li) views. Scale: 1 cm

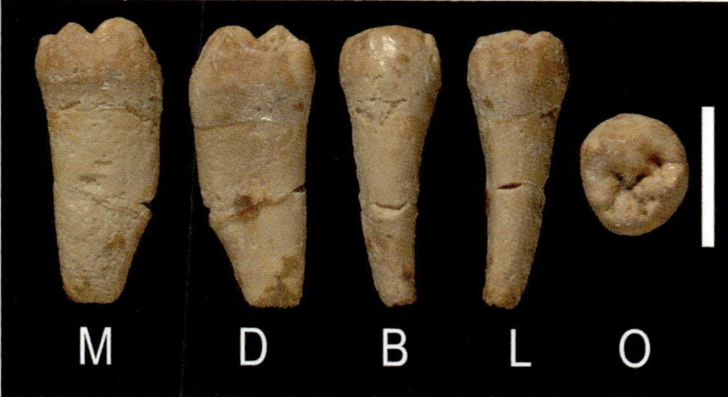

PLATE 6.67. Mesial (M), distal (D), buccal (B), lingual (L), and occlusal (O) views of the Palomas 87 left P$_4$. Scale: 1 cm

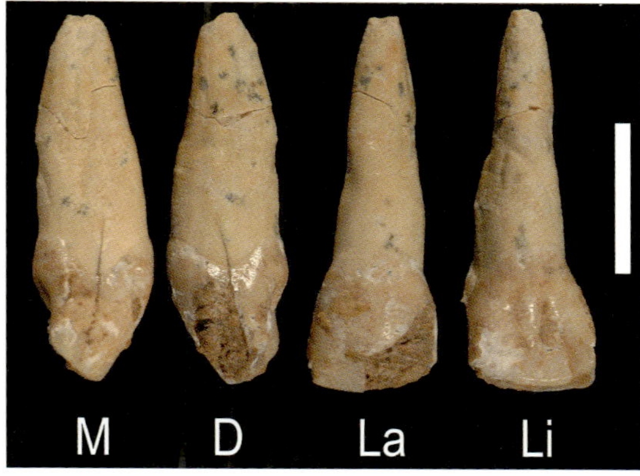

PLATE 6.70. Mesial (M), distal (D), labial (La), and lingual (Li) views of the Palomas 90 left I^1. Scale: 1 cm

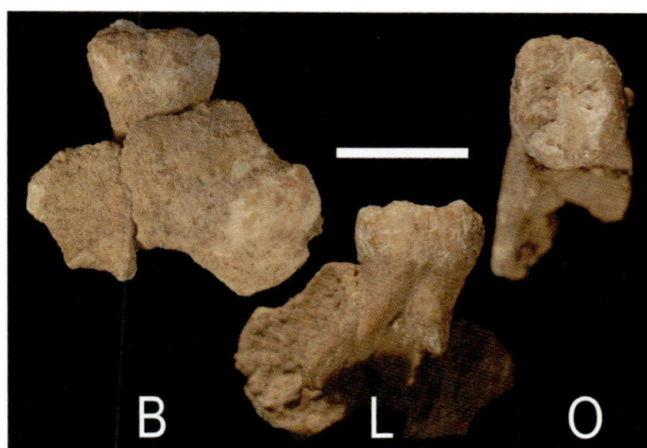

PLATE 6.68. The Palomas 88 left dm$_2$ in buccal (B), lingual (L), and occlusal (O) views. The tooth is joined to its lateral alveolus and that of the distal dm$_1$, and the superoposterior margin of the mental foramen is present below the dm$_1$. Scale: 1 cm

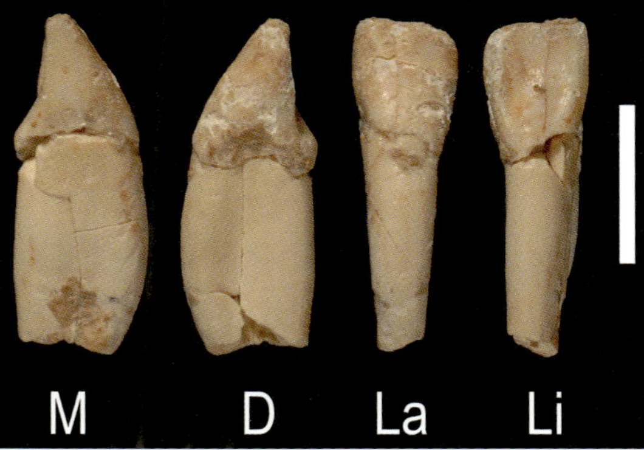

PLATE 6.71. The Palomas 91 left I$_2$ in mesial (M), distal (D), labial (La), and lingual (Li) views. Scale: 1 cm

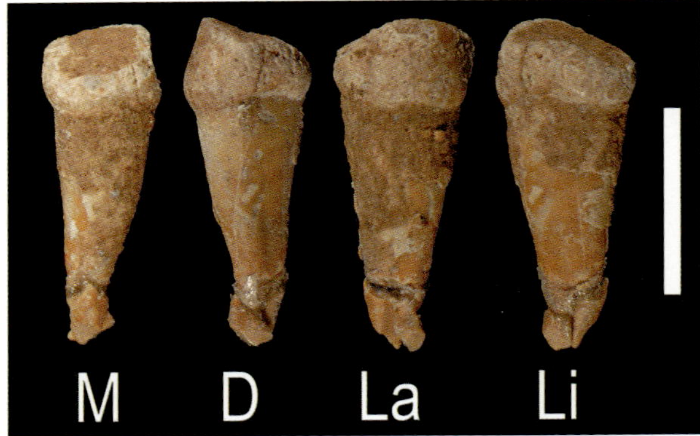

PLATE 6.72. The Palomas 93 dc_1 right in mesial (M), distal (D), labial (La), and lingual (Li) views. Scale: 1 cm

PLATE 6.75. The Palomas 98 right C_1 in mesial (M), distal (D), labial (La), and lingual (Li) views. Scale: 1 cm

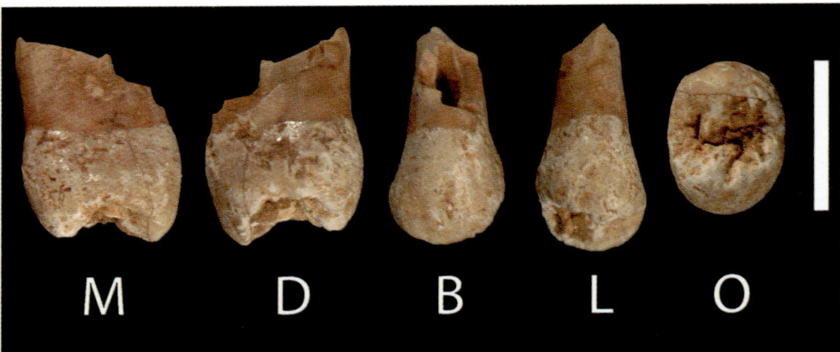

PLATE 6.73. Mesial (M), distal (D), buccal (B), lingual (L), and occlusal (O) views of the Palomas 94 left P_4. Scale: 1 cm

PLATE 6.76. The Palomas 99 I_1 or I_2 left in mesial (M), distal (D), and labial (La) views. Scale: 1 cm

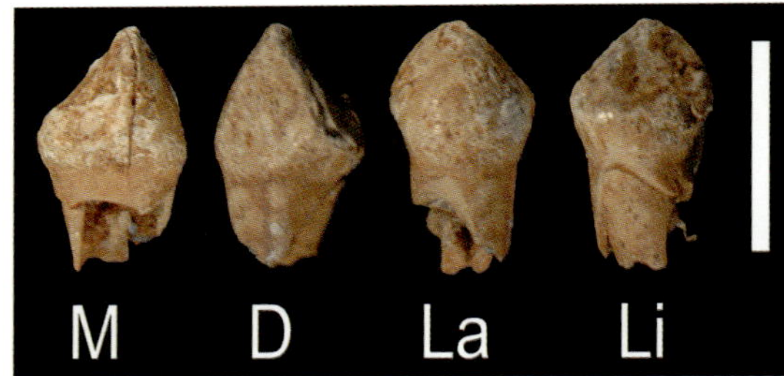

PLATE 6.74. The Palomas 95 left dc_1 in mesial (M), distal (D), labial (La), and lingual (Li) views. Scale: 1 cm

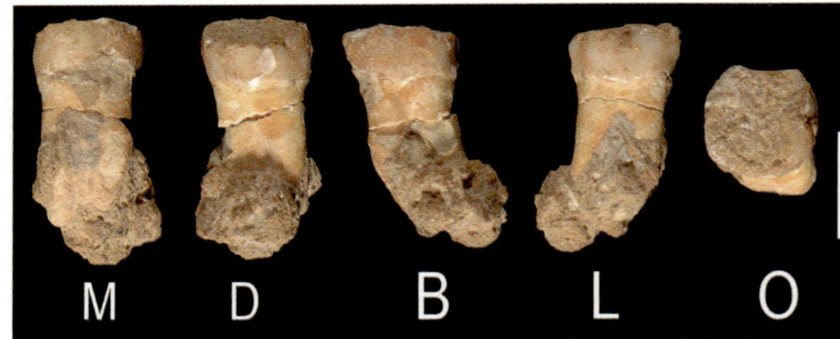

PLATE 6.77. Mesial (M), distal (D), buccal (B), lingual (L), and occlusal (O) views of the Palomas 100 left M_1 (or M_2). Scale: 1 cm

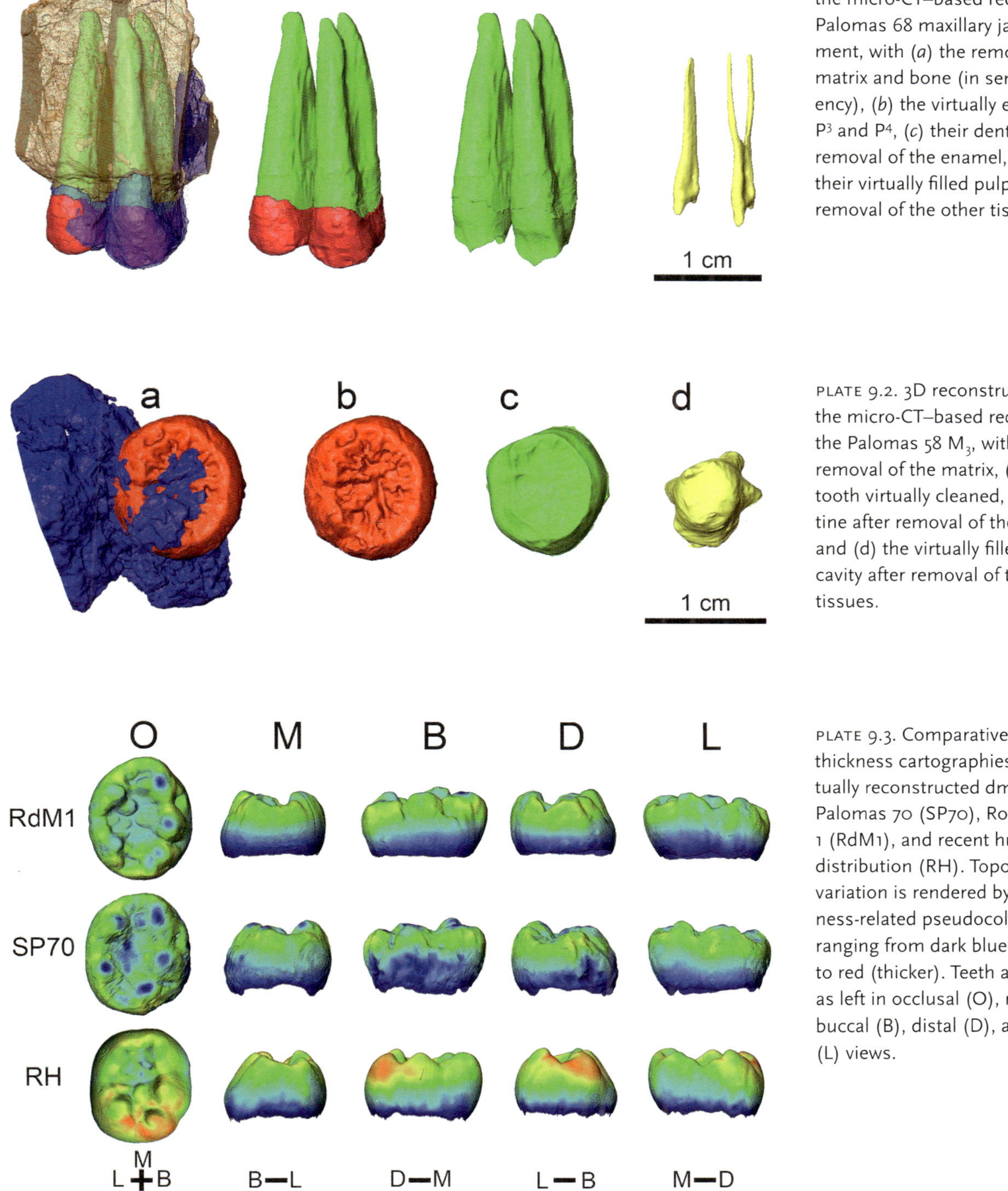

PLATE 9.1. 3D reconstruction of the micro-CT–based record of the Palomas 68 maxillary jaw fragment, with (*a*) the removal of the matrix and bone (in semitransparency), (*b*) the virtually extracted P³ and P⁴, (*c*) their dentine after removal of the enamel, and (*d*) their virtually filled pulp cavity after removal of the other tissues.

PLATE 9.2. 3D reconstruction of the micro-CT–based record of the Palomas 58 M_3, with (*a*) the removal of the matrix, (*b*) the tooth virtually cleaned, (*c*) the dentine after removal of the enamel, and (d) the virtually filled pulp cavity after removal of the other tissues.

PLATE 9.3. Comparative enamel thickness cartographies of the virtually reconstructed dm_2 crowns of Palomas 70 (SP70), Roc de Marsal 1 (RdM1), and recent human mean distribution (RH). Topographic variation is rendered by a thickness-related pseudocolor scale ranging from dark blue (thinner) to red (thicker). Teeth are shown as left in occlusal (O), mesial (M), buccal (B), distal (D), and lingual (L) views.

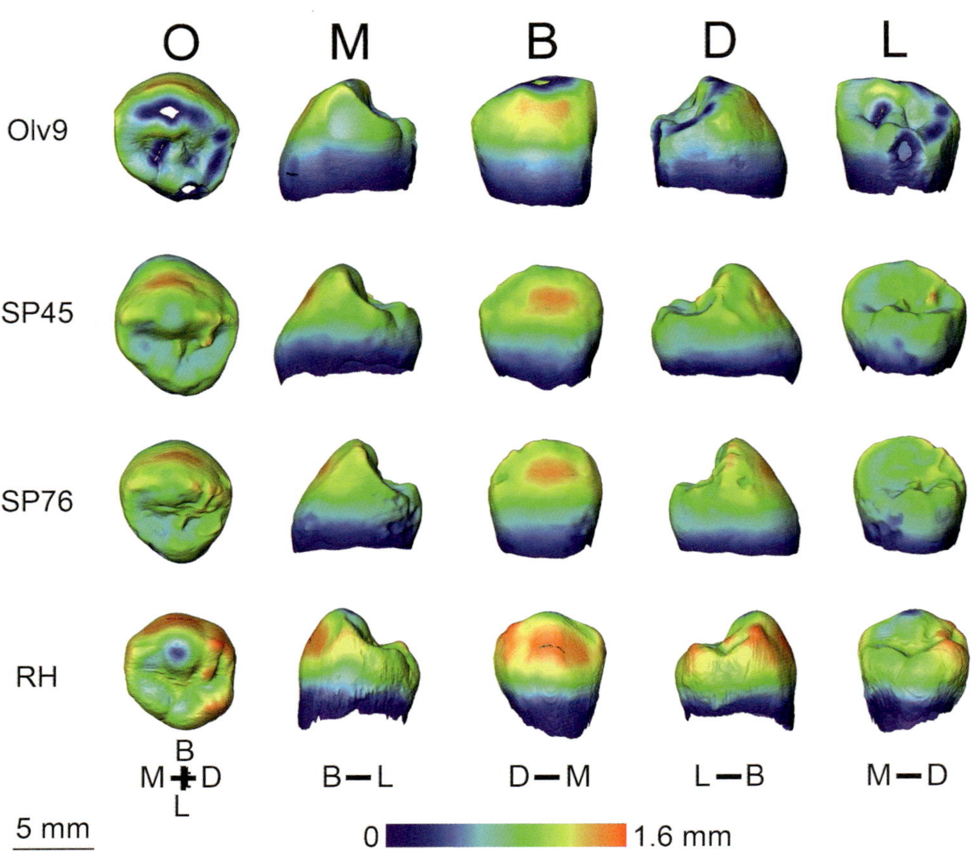

PLATE 9.4. Comparative enamel thickness cartographies of the virtually reconstructed P_3 crowns of Palomas 45 (SP45) and 76 (SP76), Oliveira 9 (Olv9), and recent human mean distribution (RH). Topographic variation is rendered by a thickness-related pseudocolor scale ranging from dark blue (thinner) to red (thicker). Teeth are shown as left in occlusal (O), mesial (M), buccal (B), distal (D), and lingual (L) views.

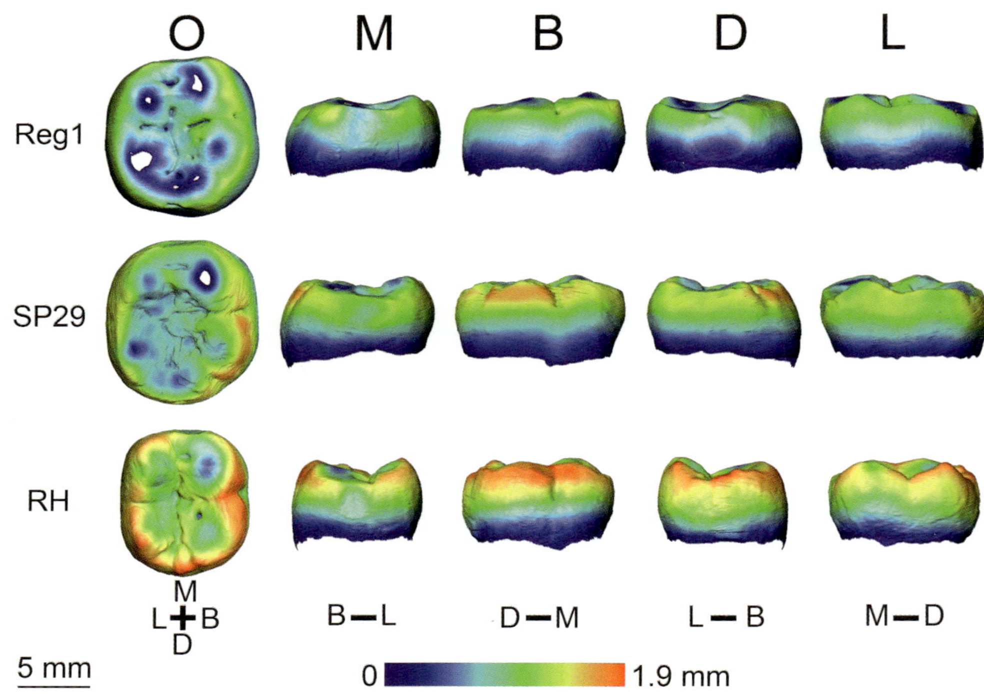

PLATE 9.5. Comparative enamel thickness cartographies of the virtually reconstructed M_2 crowns of Palomas 29 (SP29), Regourdou 1 (Reg1), and recent human mean distribution (RH). Topographic variation is rendered by a thickness-related pseudocolor scale ranging from dark blue (thinner) to red (thicker). Teeth are shown as left in occlusal (O), mesial (M), buccal (B), distal (D), and lingual (L) views.

DISCRETE DENTAL TRAIT	PRESENT (ASUDAS SCORE)[a]	ABSENT (ASUDAS 0)[b]	NOT SCORABLE[c]
di¹			
Shoveling	SP39 (2)	—	—
Labial convexity	SP39 (3)	—	—
dm²			
Carabelli's cusp	—	SP36	—
Hypocone C4	SP36 (5)	—	—
Metacone C2	SP36 (4)	—	—
Cusp 5	—	SP36	—
Number of cusps (NC)	SP36 (4)	—	—
Crista obliqua	SP36	—	—
Taurodont	—	—	SP36
dm₁			
4 cusp	SP25, SP40	—	SP61
Mid-trigonid crest	SP25, SP40	—	SP61
Anterior fovea	SP25, SP40	—	SP61
Tuberculum molare	SP25 (small)		SP61
Taurodont	—	SP7	SP25, SP40
dm₂			
Y pattern	SP70 (Y), SP83 (Y), SP88 (Y)	—	—
Hypoconulid (cusp 5)	SP70 (large), SP88 (large)	—	—
Cusp 6	SP83	(SP70)	—
Cusp 7	SP83	(SP70)	—
Anterior fovea	SP69, SP70, SP83	—	SP88
Mid-trigonid crest	—	SP69, SP70, SP83, SP88	—
Protostylid (PTYL)	SP70	—	—
Taurodont	—	SP7, SP83	—

[a] When applicable, the Sima de las Palomas (SP) specimen number is followed in parentheses by its ASUDAS score, the criteria for some of which are derived from Bailey (2002a).

[b] Absence indicates either the absence of the trait for those that are scored as presence/absence or a score of zero (0) for those that are given ASUDAS rank order scores.

[c] Specimens for which most of the crown is present but the trait cannot be reliably scored given wear, damage, and/or obscuring matrix. Specimens for which the crown is largely absent or extensively damaged are not listed.

TABLE 6.9. Comparative dental occlusal discrete traits for the Palomas deciduous maxillary di^1 and the mandibular dm$_1$ and dm$_2$

DENTAL TRAIT[a]	PALOMAS	NEANDERTALS	MPMH[b]	UPPER PALEOLITHIC[c]
di^1 shoveling (>1)	100% (1/1)	100% (4/4)	—	0.0% (0/1)
di^1 labial convexity (>1)	100% (1/1)	100% (4/4)	100% (1/1)	100% (1/1)
dm$_1$ 4 cusp (≥4 cusps)[d]	100% (2/2)	91.7% (11/12)	100% (3/3)	90.9% (10/11)
dm$_1$ mid-trigonid crest (>0)	100% (2/2)	93.3% (14/15)	33.3% (1/3)	0.0% (0/11)
dm$_1$ anerior fovea (>1)	100% (2/2)	92.9% (13/14)	50.0% (1/2)	41.7% (5/12)
dm$_1$ *tuberculum molare* (>1)	0.0% (0/1)	37.5% (3/8)	0.0% (0/2)	0.0% (0/8)
dm$_2$ Y pattern	100% (3/3)	93.3% (14/15)	100% (3/3)	92.3% (12/13)
dm2 hypoconulid (cusp 5) (>0)	100% (2/2)	100% (21/21)	100% (3/3)	100% (17/17)
dm$_2$ cusp 6 (>0)	50.0% (1/2)	0.0% (0/11)	33.3% (1/3)	0.0% (0/12)
dm$_2$ cusp 7 (>1)	50.0% (1/2)	5.6% (1/18)	0.0% (0/3)	0.0% (0/12)
dm$_2$ anterior fovea (>1)	100% (3/3)	90.0% (18/20)	33.3% (1/3)	33.3% (4/12)
dm$_2$ mid-trigonid crest (>1)	0.0% (0/4)	78.9% (15/19)	33.3% (1/3)	8.3% (1/12)
dm$_2$ protostylid (>pit)	100% (1/1)	40.0% (4/10)	0.0% (0/3)	0.0% (0/10)

Source: Comparative data courtesy of S. E. Bailey.

[a]The ASUDAS and non-ASUDAS traits follow Scott and Turner (1997) and Bailey (2002a, 2006). The ASUDAS grades counted as indicating presence are provided after each trait.

[b]MPMH = Middle Paleolithic modern humans, from Qafzeh and Skhul.

[c]The Upper Paleolithic sample includes both Early/Mid and Late Upper Paleolithic specimens, given the small sample sizes available for Upper Paleolithic deciduous teeth.

[d]One Neandertal dm$_1$ has three cusps; one Upper Paleolithic dm$_1$ has three cusps, and two of them have five cusps.

The Palomas Dental Remains

Size and Proportions

7

BEATRIZ PINILLA AND ERIK TRINKAUS

THE ASSESSMENT OF Middle and Late Pleistocene human dental remains has focused extensively on measures of dental size across the dental arcade as well as with respect to individual teeth (see, for example, Weidenreich 1937; Wolpoff 1971; Twiesselmann 1973; Smith 1977a; Frayer 1978; Bermúdez de Castro 1986; Trinkaus 2004). Most of these considerations have assumed that tooth size is largely under genetic control, although some degree of environmental influence might be involved (Goose 1967; Alvesalo and Tigerstedt 1974; Townsend and Brown 1978; Bernal et al. 2010). Dental dimensions are therefore of interest with respect to the affinities of the Palomas Neandertals, and they are possibly related to aspects of their masticatory function.

The comparisons have been primarily with respect to crown dimensions, although trends in root length have also been assessed (for example, Bailey 2005; Trinkaus et al. 2013). The ultimate concerns have been primarily with respect to the well-established reduction in overall dental dimensions through the evolution of genus *Homo*, but there are also concerns regarding the extent to which individual tooth dimensions experienced reduction or stasis. The latter issue is reflected in part in individual tooth size, but it is better viewed through assessments of relative size through the dental arcade, independent of the absolute size of the teeth (for example, Stefan

and Trinkaus 1998). Relative proportions are also relevant to assessing the affinities of individual specimens or samples. From these analyses there has been a general inference that average postcanine dental dimensions changed little across samples through the western Eurasian Middle and Late Pleistocene, at least until the last glacial maximum, and that any average differences have been largely outweighed by within-sample variation. Yet there appear to have been substantial reductions in anterior dental dimensions, especially with the transition to early modern humans.

A subset of the Palomas dental remains has been noted for their small crowns yet long roots relative to the larger Neandertal sample (Walker et al. 2008). The total Palomas dental sample, however, has not been previously assessed, relative to broader Middle and Late Pleistocene dental samples, for overall dental size or for proportions along the arcades.

Dimensional Methods

In order to assess the postcanine dental crown dimensions, we used two complementary techniques: crown linear diameters and measures of total occlusal area (TOA). The labiolingual breadths of the anterior dental remains are also compared.

Crown diameters (Tables 7.2, 7.3, 7.5 to 7.7, 7.15, and 7.16) were obtained with calipers following standard techniques, with the plane of the buccal surface providing the crown orientation for molar crowns and the bucco(labio)lingual midline providing the orientation for the anterior teeth and premolars. In addition to crown diameters, measurements are provided, as available and accessible, given the preservation of the teeth, for cervical diameters, root maximum diameters, crown heights, and root lengths. The figures represent values obtained by M. J. Walker, A. V. Lombardi, J. Zapata, and E. Trinkaus, and they were checked for consistency against the original specimens by Trinkaus.

A number of the Palomas postcanine crowns have been reduced mesiodistally through interproximal wear, as have many of the Pleistocene comparative specimens. The mesiodistal diameters of the anterior teeth have been similarly reduced through both interproximal and occlusal wear. In the tables, the mesiodistal crown dimensions of those reduced teeth are indicated with square brackets. There is little difference in sphericity values across the samples (Table 7.1), and consequently the crown diameter comparisons employ solely bucco(labio)lingual diameters. Given the standard publication of crown diameters, these comparisons maximize samples sizes for the fossil groups. The data derive from personal measurement of the original specimens and from the primary published descriptions of the fossil teeth. Available antimeric values were averaged prior to the comparisons to avoid duplication, even though averaging reduces the variances slightly.

For the premolars and molars, the total occlusal area was measured for the sufficiently complete Palomas dental crowns and available teeth in the comparative samples. Dental occlusal crowns were photographed perpendicular to the camera focal plane, treating each tooth as if it were isolated (also when it was in anatomical position) following the methodology described by Wood and Abbott (1983), Suwa et al. (1994, 1996), Pérez et al. (2006), and Grine et al. (2009). The camera used was a Nikon D1H, and most of the sample was photographed at the Unidad de Tratamiento de Imágenes y Soporte Informático del Servicio Científico Técnico of the Universitat de Barcelona (PCB-UB). The camera was placed on a tripod and was leveled horizontally. All images included a scale to allow the size calibration. Images obtained presented a resolution of 3000 × 2008 pixels.

Data were obtained using IMAT or Photoshop CS (Adobe Inc.) and ImageJ (NIH). IMAT is an automatic software developed by the PCB-UB that allows the measurement of the area once the perimeter has been delimited (Estebaranz et al. 2004; Galbany and Pérez-Pérez 2006). Other researchers (Deter 2009; Alrousan 2009) have obtained similar results using Sigma Scan Pro5 (SPSS Inc.). The method requires the delimitation of the occlusal perimeter, which can be performed with either IMAT directly or with Photoshop CS and ImageJ. Once delimited, the total occlusal area (TOA) is computed. As with crown diameters, values for antimeres were averaged prior to inclusion in sample distributions.

The intraobserver error associated with this method is low (2.0% in the image acquisition and 1.8% in the area measurement). Those errors are similar to ones described elsewhere for dental measurements (for example, Frayer 1977; Shah et al. 2005). Maximum errors on the TOA analyses were found on premolars (3.4%) and teeth with a high level of interproximal wear (1.6%). This is in agreement with the idea that small surfaces are more affected by errors, as is the interproximal wear, because it represents a reconstructed area.

In addition to occlusal areas, IMAT calculates a sphericity value of the dental crown contour as a ratio of the minimum to maximum diameters. Therefore, a sphericity value of 1.0 describes a circle, and subcircular contours provide values <1.0.

In addition, the incisor and canine labial root lengths were compared across the samples. Data are available principally for Neandertals and some Early/Mid Upper Paleolithic (E/MUP) humans; limited data are available for Middle Pleistocene and Middle Paleolithic modern humans (Table 7.14).

The Body Size Issue

The assessments here are primarily of absolute dental (crown and root) dimensions. However, the three Palomas specimens providing body mass estimates (Palomas 77, 92, and 96) are among the smallest of the known adult Neandertals, with mean body mass estimates of 64.9, 61.7, and 59.9 kg, respectively (chapter 14). The latter two are the smallest body mass estimates currently available for the Neandertals, most closely approximated by the Krapina 209 (62.6 kg) and Tabun 1 (63.3 kg) females. Most of the remainder of the Palomas postcrania are also among the smallest of the Late Pleistocene humans. Because some of the Palomas teeth have been noted as having small crowns, it was assessed whether there is a relationship between dental dimensions and estimated body mass. Using a pooled Late Pleistocene, pre–last-glacial-maximum sample, least squares regressions of summed mandibular bucco(labio)lingual diameters versus estimated body mass provided r^2 values of 0.018, 0.034, and 0.091 for the summed M_1 to M_3 (n = 22), M_1 plus M_2 (n = 29), and the summed I_1 to C_1 (n = 24), respectively. None of them provide a slope significantly different from zero at the p < 0.05 level. Issues of body mass within the range of variation of the Palomas sample should therefore not be of concern in the assessments of dental size.

Anterior Dentition Crown Comparisons

Human anterior teeth (incisors and canines) through the later Pleistocene show substantial reduction in their average labiolingual crown diameters, especially with respect to the incisors and less so with the canines (Table 7.4; Fig. 7.1). On average, the Neandertals have the largest crown diameters, although they are generally similar to those of the Middle Pleistocene sample. In this context, the Palomas incisors and canines as a sample are unexceptional. One of the

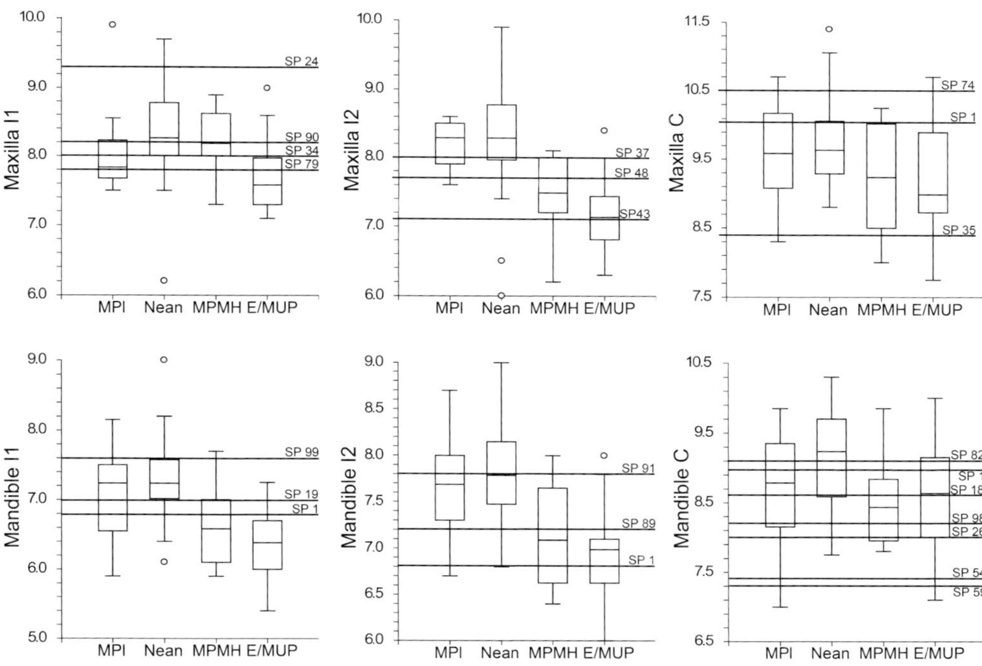

FIG. 7.1. Comparative labiolingual crown diameters of the Palomas (*horizontal lines*) and Middle and Late Pleistocene (*box plots*) human anterior teeth. For the individual Palomas values, see Tables 7.2 and 7.3. MPl: western Eurasian Middle Pleistocene humans; Nean: Neandertals; MPMH: Middle Paleolithic modern humans; E/MUP: Early and Mid Upper Paleolithic modern humans.

I's, Palomas 24, is among the largest Neandertal I's and beyond the early modern human ranges of variation, but the three other I's fall comfortably within the ranges of the comparative samples (Table 7.2). The other relatively large anterior tooth is the Palomas 74 C¹, which is toward the top of the overall ranges yet within those of the Middle Pleistocene, Neandertal, and Upper Paleolithic samples. There are also a few relatively small Palomas anterior teeth. The Palomas 43 I² is smaller than all but two very small Neandertal I²s, and the Palomas 35 C¹ is among the smaller of the later Pleistocene specimens, as are the Palomas 54 and 59 C$_I$s (Tables 7.2 and 7.3; Fig. 7.1).

These anterior dental comparisons therefore largely confirm (at least for the incisors) previous assessments of a substantial reduction in anterior teeth in the Middle and Late Pleistocene. In this context, the Palomas teeth, although some of them are moderately large or small, largely remain with the ranges of variation of the Neandertal sample and completely within the combined range of variation of the Middle Pleistocene plus Neandertal samples.

Postcanine Crown Comparisons

The postcanine teeth have been assessed using both buccolingual crown diameters (Tables 7.4 to 7.7) and total occlusal areas (Table 7.8). Across the later Pleistocene samples, there is some reduction in the premolar crown dimensions (Tables 7.9 to 7.13; Figs. 7.2 and 7.3).

In the maxillary teeth, both the diameters and the occlusal areas show a decrease in average size through the Middle and Late Pleistocene. In the mandibular premolar diameters, the modest decrease is evident only through the three more recent samples, whereas it is evident in the occlusal areas from the Middle Pleistocene onwards. It is likely that the more modest Middle Pleistocene mandibular premolar diameters are due to the relatively small teeth of the large Atapuerca-SH sample, which provides crown diameters but not occlusal areas.

The Palomas premolars are almost all quite

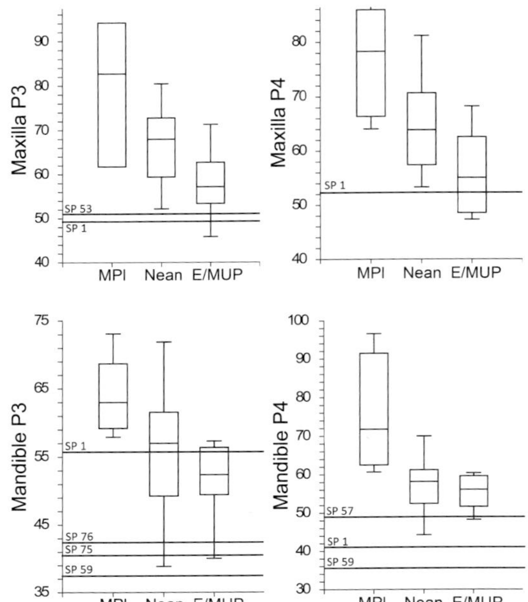

FIG. 7.2. Comparative total occlusal areas of the Palomas (*horizontal lines*) and Middle and Late Pleistocene (*box plots*) human premolars. For the individual Palomas values, see Table 7.8. For sample abbreviations, see Fig. 7.1.

small relative to the comparative samples in their occlusal areas, the one exception being the Palomas 1 P$_3$ (Fig. 7.2). In the crown diameters, the Palomas sample appears to be quite variable, spanning much of the ranges of variation of the comparative samples (Fig. 7.3). The consistently small mandibular premolars are those of Palomas 59, although they are joined by Palomas 75 in the P$_3$ comparison and, to a lesser extent, Palomas 78 in the P$_4$ comparison (Table 7.6).

In the molar comparisons (Tables 7.9 to 7.12; Figs. 7.4 and 7.5), there is little consistent directional change through the Middle and Late Pleistocene. The Middle Pleistocene occlusal areas are mostly large compared to the more recent samples, but that sample lacks data on the large Atapuerca-SH sample with its more modest dental dimensions. In the mandibular molar crown diameters, the Middle Pleistocene sample has among the smaller of the teeth. In these comparisons, the Palomas 59 M$_I$ and M$_2$ have among the smallest dimensions, and they are variably joined

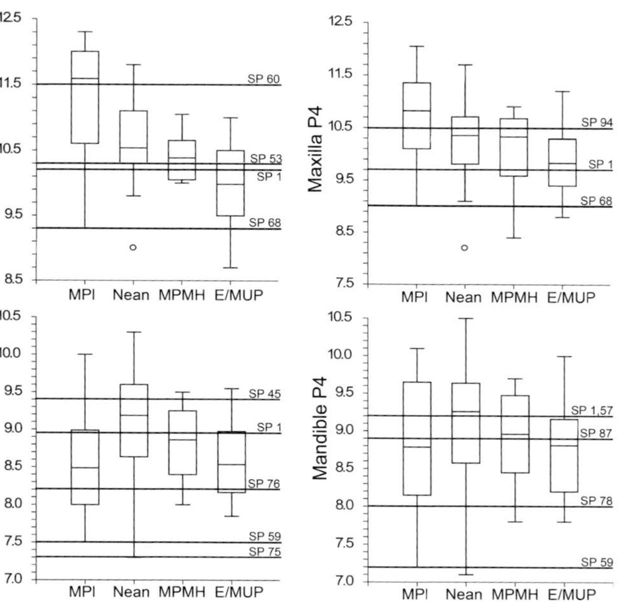

FIG. 7.3. Comparative buccolingual crown diameters of the Palomas (*horizontal lines*) and Middle and Late Pleistocene (*box plots*) human premolars. For the individual Palomas values, see Tables 7.5 and 7.6. For sample abbreviations, see Fig. 7.1. For the P_3 and P_4, the Arago 13 high outliers (11.5 mm each) are not shown.

by Palomas 1 (Tables 7.7 and 7.8). The remainder of the Palomas teeth are generally well within the Middle and Late Pleistocene variation, although none of them are above the median of the Neandertal sample.

The two methods for assessing the postcanine crown dimensions of the Palomas Neandertals place them either well within the larger Neandertal variation or at the lower limits of its size distributions. The Palomas 59 mandibular dentition, and to a lesser extent the Palomas 1 teeth, stand out as rather small for the Late Pleistocene.

Anterior versus Posterior Dental Crown Proportions

Given the decreases in anterior dental dimensions but little change in posterior dental dimensions, it is appropriate to assess the relative sizes of these two portions of the mandibular dental arcade, especially because it has been shown (Stefan and Trinkaus 1998) that such comparisons

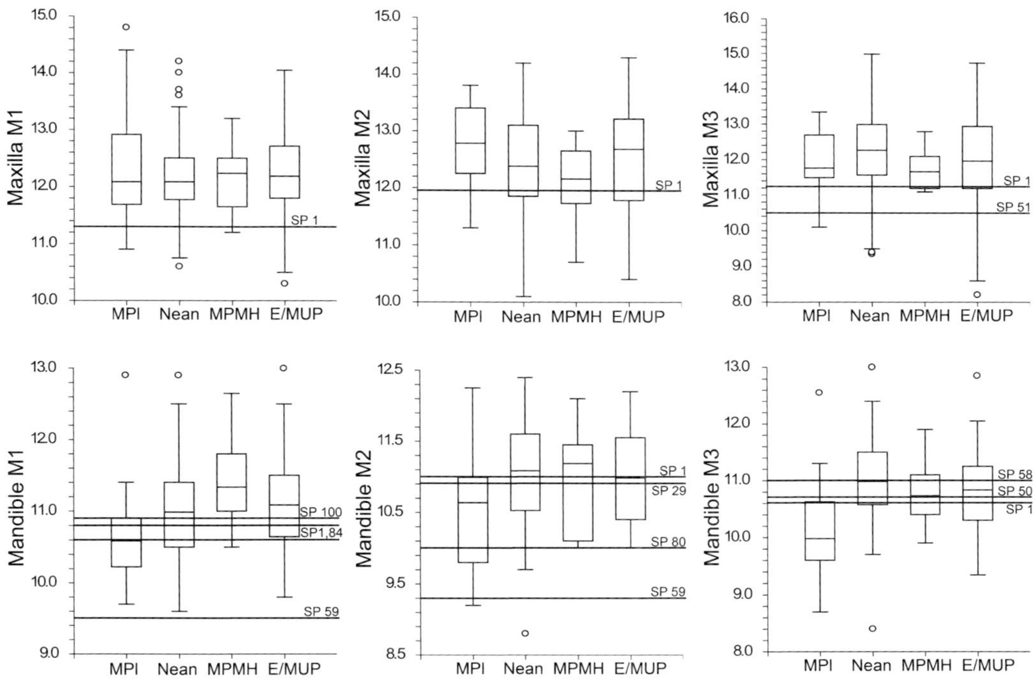

FIG. 7.4. Comparative buccolingual crown diameters of the Palomas (*horizontal lines*) and Middle and Late Pleistocene (*box plots*) human molars. For the individual Palomas values, see Tables 7.5 and 7.7. For sample abbreviations, see Fig. 7.1. For the M^2 and M_2, respectively, the Arago 1 (15.9 mm) and 13 (13.8 mm) high outliers are not shown.

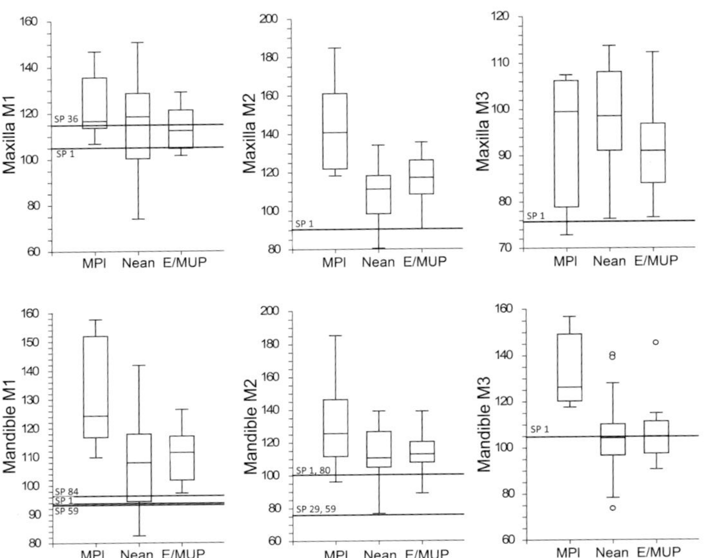

FIG. 7.5. Comparative total occlusal areas of the Palomas (*horizontal lines*) and Middle and Late Pleistocene (*box plots*) human molars. For the individual Palomas values, see Table 7.8. For sample abbreviations, see Fig. 7.1.

serve to separate the Neandertals from early modern humans. Only one Palomas specimen, Palomas 1, permits assessment along a full mandibular dental arcade, given the inclusion of the probably associated Palomas 20 and 21 incisors. Palomas 59 provides proportions for a more limited arcade, from C_1 to M_2.

To assess the proportions between anterior and posterior dental dimensions, the summed anterior labiolingual dimensions are compared to summed posterior buccolingual dimensions. For the anterior teeth, the sum of the I_1 to P_3 crown diameters is employed for Palomas 1 (in this case including P_3 with the incisors and canines), and C_1 plus P_3 is used for both Palomas 1 and 59. In the posterior dentition, the sum of the P_4 to M_3 crown diameters is used, but P_4 to M_2 is also employed, given the absence of the M_3s of Palomas 59 and a number of comparative specimens. In the I_1 to P_3 summed values (Table 7.13), there is a clear average reduction in anterior dental dimensions between the two archaic human samples and the two early modern human ones; there is less of a reduction in the more limited C_1 plus P_3 comparison. There is little change across the comparative samples in the P_4 to M_2 and the P_4 to M_3 values. Palomas 1 falls near the middles of the comparative samples in this measure, but

Palomas 59 is rather small, providing the smallest P_4 to M_2 value.

When the summed I_1 to P_3 anterior values are plotted against summed posterior tooth values (Fig. 7.6), there is minimal overlap between the archaic and early modern human samples when M_3 is included; there is a modest overlap when M_3 is not included. For a given posterior dental value, the archaic humans, Middle and Late Pleistocene, have generally larger anterior teeth than the early modern humans. Palomas 1 is among the Neandertals, although it is close to the boundary between the two groups.

Reducing the number of anterior and posterior teeth to be able to include Palomas 59 in the comparisons (Fig. 7.7) increases the overlap between the later archaic and early modern human samples. Palomas 1 remains principally with the Neandertals in its dental proportions. It is harder to assess the proportions of Palomas 59, because it is distinctly smaller than any of the other specimens, being approached only by Châteauneuf 2 (Tillier 1979). Continuing the regression lines of the larger Neandertal and Upper Paleolithic samples to the size of the Palomas 59 teeth places it in the overlap zone of their distributions.

The two Palomas mandibular dental arcades permitting anterior to posterior dental proportions therefore are consistent with their placement among the Neandertals. Yet Palomas 1, with the full set of eight measurements, has among the relatively smaller anterior dentitions, and Palomas 59, with only five teeth to assess, is ambiguous as to its affinities.

Anterior Dental Root Lengths

It has been noted (Weidenreich 1937; Vlček 1969; Bailey 2005; Trinkaus et al. 2014a) that archaic *Homo* anterior teeth tend to have relatively long roots, whereas those of early modern humans are generally shorter. Comparisons of available anterior dental root lengths (Table 7.14) largely support this contrast, with a substantial reduction between later archaic and early modern humans in their average root lengths. There are limited data available for the Middle Paleolithic modern humans, but their few measurements are close to the Upper Paleolithic values. The Middle Pleistocene sample is also small, but the means available for the canines are close to those provided by Bermúdez de Castro and Rosas (2001) for a larger number of the Atapuerca-SH

canines (C^1: 21.6 ± 1.3 mm, n = 13; C_1: 19.1 ± 1.4 mm, n = 15).

In this context the Palomas anterior dental root lengths are modest for Neandertals (Table 7.14). A few of their root lengths are large, including those for the Palomas 43 I^2, the Palomas 35 C^1, the Palomas 20 I_2, the Palomas 44 C_1, and the Palomas 98 C_1 (Tables 7.2 and 7.3). But other lengths are quite small, even excluding the very small I_1 length for Palomas 96. Nonetheless, a couple of the Palomas canine values are below the Upper Paleolithic means. Therefore, the Palomas anterior dental root lengths generally overlap the Neandertal and early modern human distributions.

Mandibular Deciduous Crown Comparisons

Previous assessments of deciduous dental dimensions in later Pleistocene humans have documented little consistent change between samples (P. Smith 1978; Tillier 1979; Hillson and Trinkaus 2002). Comparisons of the buccolingual diameters of the dc_1 to dm_2 show similar dimensions for the deciduous molars, although the E/MUP deciduous canines are somewhat smaller than

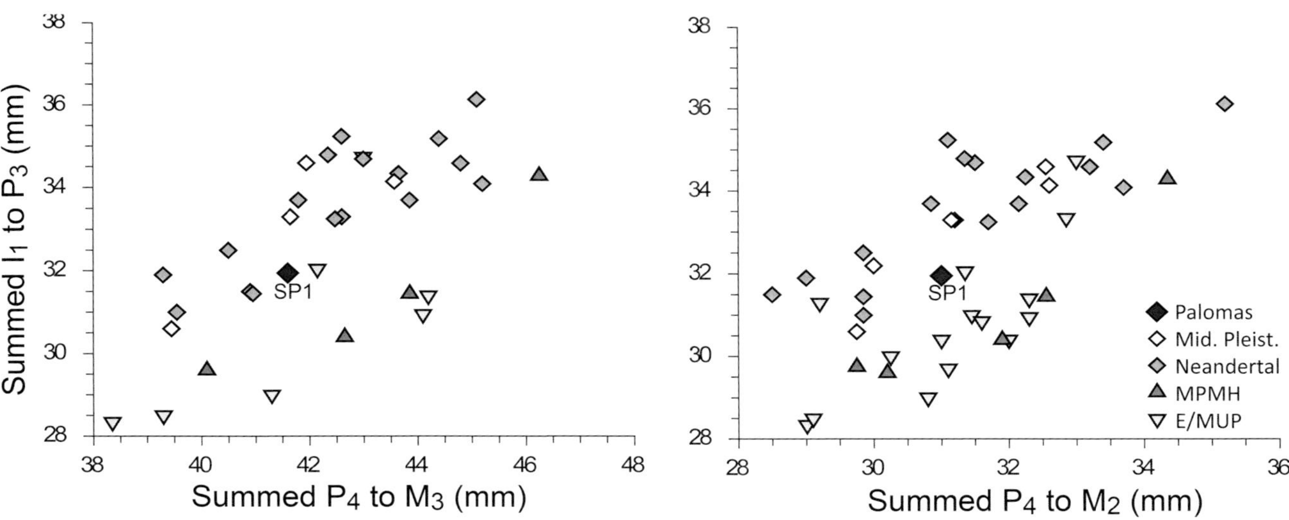

FIG. 7.6. Comparison of the Palomas 1 (SP1) summed mandibular anterior (I_1 to P_3) versus posterior (P_4 to M_3 and P_4 to M_2) buccolingual crown diameters (Table 7.13) versus Middle and Late Pleistocene samples. Palomas 1 includes the probably associated Palomas 20 and 21 incisors. Sample abbreviations as in Fig. 7.1.

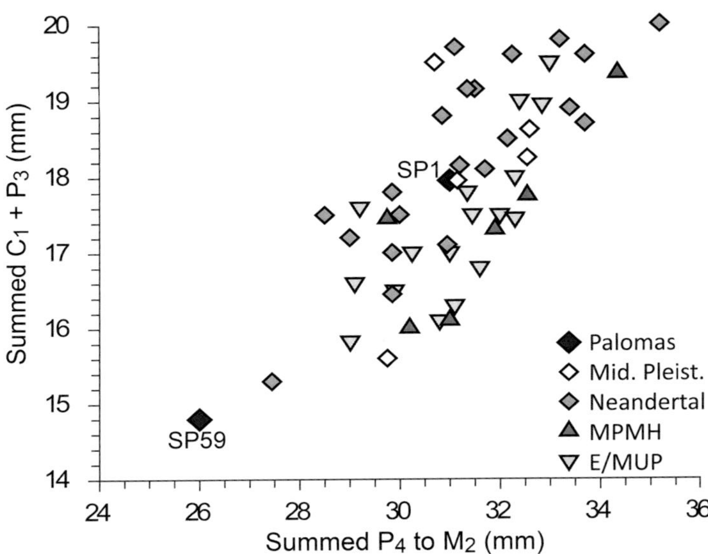

FIG. 7.7. Comparison of the Palomas 1 (SP1) and 59 (SP59) summed mandibular C_1 to P_3 versus posterior (P_4 to M_2) buccolingual crown diameters (Table 7.13) versus Middle and Late Pleistocene samples. Sample abbreviations as in Fig. 7.1.

the Neandertal interquartile range (Table 7.15; Fig. 7.8). Yet the Palomas dm_2 values are below all of the medians, although within most of the sample ranges. Moreover, the dm_1 crown breadths are below all of the interquartile ranges, and one (Palomas 40) is matched by only one low Neandertal value (Châteauneuf 2). Similar comparisons of the few maxillary deciduous teeth to the limited later Pleistocene samples place the Palomas 39 di^1 and the Palomas 36 dm^2 among the larger teeth, the Palomas 85 di^2 among the smaller ones, and the Palomas 71 dm^1 in the middle of the comparative samples. The comparative samples remain insignificantly different (K-W p-values: di^1, 0.303; di^2, 0.104; dm^1, 0.688; dm^2, 0.224).

Summary

The Palomas Neandertal dentitions therefore provide a substantial number of measures of their crown (and root) dimensions. Given the isolated nature of all of them except those of Palomas 1 and 59, however, assessments beyond absolute size are limited. In general, the Palomas teeth follow the size patterns for later Pleistocene humans, despite considerable variability within tooth categories. This conclusion applies especially if one looks across the dental arcades and does not focus on individual tooth positions.

A few of the teeth are notable for their large

those of the Middle Pleistocene and Middle Paleolithic samples (Fig. 7.8). Yet those three teeth do not have significantly different crown diameters across the four samples (K-W p-values: dc_1, 0.234; dm_1, 0.222; dm_2, 0.398).

The Palomas dc_1 crown breadths are among the higher values, with Palomas 93 and 95 exceeding the early modern human ranges and

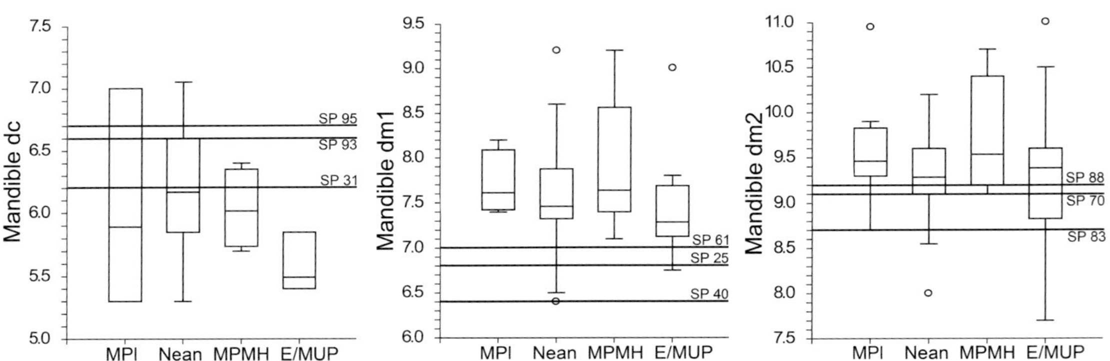

FIG. 7.8. Comparative buccolingual crown diameters of the Palomas (*horizontal lines*) and Middle and Late Pleistocene (*box plots*) human mandibular deciduous canines and molars. For the individual Palomas values, see Table 7.15. For sample abbreviations, see Fig. 7.1.

crown dimensions, especially the Palomas 24 I[1] and the Palomas 74 C[1]. More of the Palomas teeth fall in the smaller portions of the later Pleistocene ranges of variation. In the anterior dentition, these teeth include the Palomas 43 I[2], the Palomas 35 C[1], the Palomas 20 I$_2$, and the Palomas 54 and 59 C$_1$s. In the postcanine dentition, these smaller teeth include the Palomas 1 M[1]s, the Palomas 51 M[3], the Palomas 57 P$_4$, and the Palomas 59 P$_3$ to M$_2$. Yet most of the Palomas dental dimensions fall comfortably within Neandertal (and later Pleistocene human) ranges of variation.

This tendency for the Palomas dental sample to be smaller on average than the overall Neandertal sample could be taken as foreshadowing the long-term trends in human dental dimensions. Yet there is little difference in postcanine tooth size from the Middle Pleistocene until the end of the Pleistocene, and the primary reduction is in the anterior teeth with the emergence of early modern humans. The Palomas sample has small teeth throughout the arcades, but it also has some rather large anterior ones.

More important, the overall contrast in anterior tooth dimensions between late archaic and early modern humans is reflected in anterior versus posterior dental crown proportions, independent of the overall size of the dentition. In this, when a full set of mandibular crown breadths is available, there is only minimal overlap of the samples. The Palomas 1 proportions are close to the overlap between the two groups of humans but are comfortably included within the Neandertal sample in these proportions. An attempt was made to include the less complete Palomas 59 dentition in anterior to posterior dental proportions, but its more limited number of teeth, especially the absence of the incisors, and its very small size make its affinities in this aspect unclear.

TABLE 7.1. Premolar and molar mean sphericity values for the Palomas and comparative sample teeth

TOOTH	PALOMAS	MID. PLEIST.	NEANDERTALS	E/MUP
Maxilla				
P[3]	0.912 (2)	0.937 (3)	0.923 (22)	0.919 (7)
P[4]	0.894 (1)	0.939 (4)	0.930 (27)	0.921 (8)
M[1]	0.952 (1)	0.951 (6)	0.943 (26)	0.936 (12)
M[2]	0.911 (1)	0.952 (6)	0.939 (24)	0.935 (13)
M[3]	0.921 (1)	0.936 (4)	0.928 (19)	0.913 (7)
Mandible				
P$_3$	0.962 (4)	0.960 (6)	0.937 (26)	0.962 (8)
P$_4$	0.941 (2)	0.959 (4)	0.949 (24)	0.957 (5)
M$_1$	0.950 (3)	0.954 (6)	0.941 (43)	0.955 (15)
M$_2$	0.957 (4)	0.956 (10)	0.961 (32)	0.964 (13)
M$_3$	0.952 (1)	0.963 (6)	0.962 (30)	0.961 (10)

Note: The individual Palomas values are in Table 7.8. Sample sizes (by individual) are provided in parentheses. Mid. Pleist. = Middle Pleistocene archaic humans; E/MUP = Early and Mid Upper Paleolithic modern humans.

TABLE 7.2. Dimensions of the Palomas anterior maxillary teeth, in millimeters

SP NO.	SIDE	CRN MD	CRN BL	CRV MD	CRV BL	ROOT MD	ROOT BL	CRN HT	RT HT B	RT HT L	TOTAL HT
Maxilla I¹											
24	Left	(9.1)	9.3	6.3	(7.8)	5.6	7.8	(13.0)	16.4	16.4	(27.4)
34	Left	9.2	8.0	6.3	7.0	4.5	5.4	12.4	>12.8	>12.3	>23.8
73	Right	[9.9]	—	—	—	—	—	—	—	—	—
79	Left	>8.9	7.8	6.9	6.8	6.1	6.5	[10.5]	(14.8)	(15.5)	[24.8]
90	Left	[9.0]	8.2	5.9	7.3	5.2	6.8	[11.0]	(16.3)	(16.0)	[25.7]
Maxilla I²											
37	Right	7.9	8.0	5.5	7.3	5.1	7.8	11.3	—	—	>16.7
43	Left	6.9	7.1	5.0	7.3	3.6	6.5	8.4	20.4	—	28.8
48	Right	>7.8	7.7	5.2	7.2	4.3	7.1	[10.2]	15.1	14.3	[23.5]
Maxilla C¹											
1	Right	[8.9]	9.8	6.6	9.6	—	(9.2)	8.2	—	(15.6)	—
1	Left	[8.4]	10.3	6.5	9.6	—	—	—	—	—	—
35	Left	7.9	8.4	5.1	7.5	4.5	7.0	[9.7]	21.1	21.1	[29.5]
74	Left	9.0	10.5	5.8	9.6	5.2	9.5	13.3	17.9	18.1	29.6

Note: Numbers after > indicate broken or incompletely developed root, and the measurement of the preserved or developed portion is provided; numbers in parentheses indicate minor damage, and the original measurement is estimated with little probable error; numbers in square brackets indicate a worn tooth, occlusally and/or interproximally, and the measurement is of the remaining portion. Crn = crown; Crv = cervix; Rt = root; MD = mesiodistal; BL = buccolingual or labiolingual; B = buccal; L = lingual; Ht = height.

TABLE 7.3. Dimensions of the Palomas anterior mandibular teeth, in millimeters

SP NO.	SIDE	CRN MD	CRN BL	CRV MD	CRV BL	ROOT MD	ROOT BL	CRN HT	RT HT B	RT HT L	TOTAL HT
Mandibular I₁											
1	Right	[4.7]	6.8	4.2	6.7	4.0	7.2	4.2	—	—	—
19	Left	[5.5]	7.0	3.6	6.7	3.7	7.7	[8.5]	(15.0)	(15.0)	>20.6
21 (1)	Left	[5.0]	6.4	4.3	6.3	3.5	6.8	[4.4]	14.8	13.9	[19.6]
99	Left	—	7.6	4.5	6.9	3.8	7.1	—	(15.8)	(15.0)	20.2
Mandibular I₂											
20 (1)	Left	[5.6]	6.8	4.4	6.7	4.6	7.3	[5.4]	18.2	16.8	[23.6]
89	Right	6.3	7.2	3.9	7.0	3.6	8.0	10.8	—	—	>17.9
91	Left	[6.8]	7.8	4.5	7.6	4.0	8.3	[10.5]	—	—	>21.7
Mandibular C₁											
1	Right	[7.9]	8.9	5.9	8.7	—	—	[7.6]	—	—	—
1	Left	[7.7]	9.0	5.9	8.9	—	—	—	—	—	—
18	Left	7.8	8.6	5.6	8.7	4.4	9.1	9.7	18.2	18.2	26.6
26	Right	7.6	8.2	5.6	8.1	5.0	8.0	12.3	(14.5)	(14.5)	>23.3
44	Right	[6.9]	—	4.8	7.3	4.1	7.5	[9.2]	20.9	21.4	[28.5]
54	Right	7.2	(7.4)	—	—	—	—	—	—	—	>17.2
59	Left	[7.0]	7.3	(4.9)	6.8	—	—	8.9	—	—	—
82	Left	(7.9)	9.1	6.1	—	4.7	9.6	12.1	(17.0)	16.6	26.3
98	Right	—	8.0	5.3	7.3	4.0	7.9	6.5	19.8	18.6	25.3

Note: Numbers after > indicate a broken or incompletely developed root, and the measurement of the preserved or developed portion is provided; numbers in parentheses indicate minor damage, and the original measurement is estimated with little probable error; numbers in square brackets indicate a worn tooth, occlusally and/or interproximally, and the measurement is of the remaining portion. Crn = crown; Crv = cervix; Rt = root; MD = mesiodistal; BL = buccolingual or labiolingual; B = buccal; L = lingual; Ht = height.

TABLE 7.4. Comparison of labiolingual crown diameters of the anterior dentition for the Palomas teeth and the Middle and Late Pleistocene comparative samples, in millimeters [mean ± SD (n)]

GROUP	MAXILLA I^1	MAXILLA I^2	MAXILLA C^1	MANDIBLE I_1	MANDIBLE I_2	MANDIBLE C_1
Palomas	8.1 ± 1.0 (4)	7.1, 7.7, 8.0	8.4, 10.1, 10.5	6.5, 6.8, 7.0	7.3 ± 0.4 (4)	8.3 ± 0.8 (6)
Middle Pleist.	8.0 ± 0.6 (18)	8.2 ± 0.4 (11)	9.6 ± 0.7 (18)	7.1 ± 0.6 (13)	7.6 ± 0.5 (19)	8.7 ± 0.8 (21)
Neandertals	8.3 ± 0.7 (42)	8.3 ± 0.7 (48)	9.8 ± 0.6 (42)	7.3 ± 0.6 (28)	7.8 ± 0.5 (42)	9.1 ± 0.7 (46)
MPMH	8.1 ± 0.5 (12)	7.5 ± 0.6 (11)	9.2 ± 0.8 (10)	6.6 ± 0.6 (11)	7.1 ± 0.6 (10)	8.5 ± 0.6 (10)
E/MUP	7.7 ± 0.5 (29)	7.2 ± 0.5 (24)	9.2 ± 0.8 (29)	6.4 ± 0.5 (30)	6.9 ± 0.4 (32)	8.6 ± 0.7 (29)
K-W p-value	<0.001	<0.001	0.065	<0.001	<0.001	0.017

Note: Middle Pleist. = Middle Pleistocene archaic humans; MPMH = Middle Paleolithic modern humans; E/MUP = Early and Mid Upper Paleolithic modern humans; K-W p-value = Kruskal-Wallis p-values across the four comparative samples.

TABLE 7.5. Dimensions of the Palomas postcanine maxillary teeth, in millimeters

SP NO.	SIDE	CRN MD	CRN BL	CRV MD	CRV BL	ROOT BL	CRN HT	RT HT B	RT HT L	TOTAL HT
Maxillary P³										
1	Right	[7.0]	10.3	—	—	—	[7.3]	—	—	—
1	Left	[7.3]	10.1	5.7	(8.5)	—	[7.5]	—	—	—
53	Left	[7.2]	(10.3)	—	—	—	5.9	—	—	—
60	Right	—	11.5	—	10.1	9.9	—	>13.4	—	>22.8
68	Right	—	(9.4)	—	—	—	—	(19.1)	(18.4)	—
Maxillary P⁴										
1	Right	[6.0]	9.9	—	—	—	[6.2]	—	—	—
1	Left	[5.3]	9.5	—	—	—	6.1	—	—	—
68	Right	—	(9.0)	—	—	—	—	(18.8)	(17.7)	—
94	Left	7.7	10.5	5.1	9.1	—	7.9	—	—	>13.8
Maxillary M¹										
1	Right	[10.2]	11.2	8.6	10.6	—	[6.9]	—	13.4	—
1	Left	[10.1]	11.4	8.7	10.3	—	[5.0]	—	—	—
Maxillary M²										
1	Right	[10.2]	12.0	—	—	—	—	—	—	—
1	Left	[9.4]	11.9	—	10.8	—	[5.8]	12.3	—	[18.0]
Maxillary M³										
1	Right	(10.0)	11.5	—	10.8	—	[6.6]	—	—	—
1	Left	[8.6]	(11.0)	—	—	—	—	—	—	—
51	Right	(8.5)	10.5	—	10.0	—	[4.2]	15.7	—	[20.4]

Note: Numbers after > indicate broken or incompletely developed root, and the measurement of the preserved or developed portion is provided; numbers in parentheses indicate minor damage, and the original measurement is estimated with little probable error; numbers in square brackets indicate a worn tooth, occlusally and/or interproximally, and the measurement is of the remaining portion. Crn = crown; Crv = cervix; Rt = root; MD = mesiodistal; BL = buccolingual or labiolingual; B = buccal; L = lingual; Ht = height.

TABLE 7.6. Dimensions of the Palomas mandibular premolars, in millimeters

SP NO.	SIDE	CRN MD	CRN BL	CRV MD	CRV BL	ROOT MD	ROOT BL	CRN HT	RT HT B	RT HT L	TOTAL HT
Mandibular P₃											
1	Right	[7.5]	9.0	5.6	7.7	—	—	[5.6]	—	—	—
22 (1)	Left	[7.2]	8.9	5.3	7.7	—	—	[5.8]	—	—	>13.2
45	Right	7.9	9.4	5.1	7.9	4.2	8.0	9.1	—	—	>21.5
59	Left	[6.3]	7.5	—	6.8	—	—	6.9	—	—	—
75	Left	[7.7]	7.3	—	—	—	—	[5.6]	—	—	>8.2
76	Right	7.2	8.2	5.2	7.1	4.7	7.3	8.9	—	—	>15.4
Mandibular P₄											
1	Right	[6.9]	9.2	5.3	8.6	—	—	[5.0]	—	—	—
57	Right	7.5	9.2	5.6	8.1	4.5	7.9	7.3	>12.2	—	>19.2
59	Left	[5.9]	7.2	—	—	—	—	—	—	—	—
78	Left	—	8.0	5.2	7.3	—	—	7.4	—	—	>11.0
87	Left	7.5	8.9	5.3	7.8	4.7	7.8	7.4	(14.5)	(15.0)	>20.7

Note: Numbers after > indicate broken or incompletely developed root, and the measurement of the preserved or developed portion is provided; numbers in parentheses indicate minor damage, and the original measurement is estimated with little probable error; numbers in square brackets indicate a worn tooth, occlusally and/or interproximally, and the measurement is of the remaining portion. Crn = crown; Crv = cervix; Rt = root; MD = mesiodistal; BL = buccolingual or labiolingual; B = buccal; L = lingual; Ht = height.

TABLE 7.7. Dimensions of the Palomas mandibular molars, in millimeters

SP NO.	SIDE	CRN MD	CRN BL	CRV MD	CRV BL	ROOT MD	ROOT BL	CRN HT	RT HT B	RT HT L	TOTAL HT
Mandibular M₁											
1	Right	[10.6]	10.7	9.5	8.5	—	—	[4.8]	—	—	—
1	Left	[10.9]	10.9	9.7	—	—	—	[5.0]	—	—	—
23	Right	—	—	(9.0)	(9.0)	—	—	—	—	—	—
59	Left	[10.3]	9.5	—	8.1	—	—	5.7	—	—	—
84	Left	11.8	10.6	9.4	9.3	—	—	8.2	—	—	>11.8
100	Left	—	10.9	9.2	8.5	—	—	6.0	13.5	—	19.5
Mandibular M₂											
1	Right	[10.8]	11.0	10.2	9.7	—	—	[4.3]	—	—	—
1	Left	[11.1]	(11.0)	—	—	—	—	—	—	—	—
23	Right	11.3	—	9.9	8.3	—	—	5.8	—	—	—
29	Right	11.9	10.8	9.5	9.5	8.2	8.0	7.1	(14.2)	(14.4)	20.9
59	Left	(10.5)	9.3	—	8.2	—	—	4.9	—	—	—
80	Left	12.0	(10.0)	—	—	—	—	—	—	—	—
Mandibular M₃											
1	Right	[11.7]	10.6	9.8	—	—	—	[6.1]	—	—	—
1	Left	[12.0]	—	9.9	—	—	—	—	—	—	—
23	Right	—	—	(9.2)	8.8	—	—	5.8	—	—	—
50	Right	11.7	10.7	—	—	—	—	7.8	—	—	—
58	Left	12.4	(11.0)	—	—	—	—	—	—	—	>13.2

Note: Numbers after > indicate broken or incompletely developed root, and the measurement of the preserved or developed portion is provided; numbers in parentheses indicate minor damage, and the original measurement is estimated with little probable error; numbers in square brackets indicate a worn tooth, occlusally and/or interproximally, and the measurement is of the remaining portion. Crn = crown; Crv = cervix; Rt = root; MD = mesiodistal; BL = buccolingual or labiolingual; B = buccal; L = lingual; Ht = height.

TABLE 7.8. Total occlusal areas (TOAs) and sphericity values for the Palomas premolars and molars, in square millimeters

SP NO.	TOOTH	TOA-RIGHT	TOA-LEFT	SPHERICITY-RIGHT	SPHERICITY-LEFT
Maxilla					
1	P^3	48.9	49.6	0.897	0.896
53	P^3	—	51.1	—	0.935
1	P^4	52.8	51.9	0.884	0.894
1	M^1	106.2	103.8	0.929	0.938
1	M^2	—	90.4	—	0.911
1	M^3	—	75.7	—	0.921
Mandible					
1	P_3	55.7	—	0.957	—
59	P_3	—	37.4	—	0.969
75	P_3	40.7	—	0.968	—
76	P_3	42.7	—	0.952	—
1	P_4	41.0	—	0.911	—
57	P_4	48.7	—	0.963	—
59	P_4	—	35.6	—	0.950
1	M_1	101.5	86.3	0.937	0.949
59	M_1	—	93.5	—	0.942
84	M_1	—	96.1	—	0.964
1	M_2	103.1	97.7	0.963	0.944
29	M_2	75.3	—	0.970	—
59	M_2	—	75.9	—	0.942
80	M_2	—	100.4	—	0.964
1	M_3	104.7	—	—	0.952
81	M_{1-2}	—	98.1	—	0.964

TABLE 7.9. Comparison of buccolingual crown diameters of the maxillary postcanine teeth for the Palomas teeth and the Middle and Late Pleistocene comparative samples, in millimeters [mean ± SD (n)]

GROUP	MAXILLA P^3	MAXILLA P^4	MAXILLA M^1	MAXILLA M^2	MAXILLA M^3
Palomas	10.4 ± 0.9 (4)	9.0, 9.7, 10.5	11.3	11.9	10.5, 11.3
Middle Pleist.	11.2 ± 0.9 (11)	10.7 ± 0.8 (12)	12.3 ± 1.0 (20)	13.0 ± 1.1 (16)	11.9 ± 0.9 (15)
Neandertals	10.6 ± 0.6 (40)	10.3 ± 0.7 (38)	12.2 ± 0.8 (50)	12.4 ± 0.9 (43)	12.1 ± 1.2 (38)
MPMH	10.4 ± 0.4 (9)	10.1 ± 0.7 (12)	12.1 ± 0.6 (19)	12.1 ± 0.7 (10)	11.7 ± 0.6 (7)
E/MUP	10.0 ± 0.6 (31)	9.9 ± 0.6 (32)	12.2 ± 0.8 (50)	12.5 ± 0.9 (46)	11.8 ± 1.3 (35)
K-W p-value	<0.001	0.009	0.933	0.121	0.477

Note: Middle Pleist. = Middle Pleistocene archaic humans; MPMH = Middle Paleolithic modern humans; E/MUP = Early and Mid Upper Paleolithic modern humans; K-W p-value = Kruskal-Wallis p-values across the four comparative samples.

TABLE 7.10. Comparison of buccolingual crown diameters of the mandibular postcanine teeth for the Palomas teeth and the Middle and Late Pleistocene comparative samples, in millimeters [mean ± SD (n)]

GROUP	MANDIBLE P_3	MANDIBLE P_4	MANDIBLE M_1	MANDIBLE M_2	MANDIBLE M_3
Palomas	8.6 ± 0.8 (5)	8.3 ± 0.9 (6)	10.4 ± 0.6 (4)	10.5 ± 0.8 (4)	10.6, 10.7, 11.0
Middle Pleist.	8.7 ± 0.9 (17)	9.0 ± 1.0 (22)	10.7 ± 0.6 (28)	10.7 ± 1.0 (22)	10.2 ± 0.9 (17)
Neandertals	9.1 ± 0.7 (46)	9.2 ± 0.7 (52)	11.0 ± 0.6 (67)	11.1 ± 0.8 (57)	11.0 ± 0.8 (57)
MPMH	8.8 ± 0.5 (8)	8.9 ± 0.6 (8)	11.4 ± 0.6 (15)	10.9 ± 0.7 (11)	10.7 ± 0.6 (9)
E/MUP	8.5 ± 0.5 (28)	8.8 ± 0.5 (30)	11.1 ± 0.7 (50)	11.0 ± 0.7 (51)	10.9 ± 0.8 (25)
K-W p-value	0.002	0.062	<0.001	0.216	0.004

Note: Middle Pleist. = Middle Pleistocene archaic humans; MPMH = Middle Paleolithic modern humans; E/MUP = Early and Mid Upper Paleolithic modern humans; K-W p-value = Kruskal-Wallis p-values across the four comparative samples.

TABLE 7.11. Comparative maxillary total occlusal areas (TOAs) for the Palomas teeth and the Middle and Late Pleistocene comparative samples, in square millimeters [mean ± SD (n)]

GROUP	MAXILLA P^3	MAXILLA P^4	MAXILLA M^1	MAXILLA M^2	MAXILLA M^3
Palomas	49.3, 51.1	52.3	105.0	90.4	75.7
Middle Pleist.	61.8, 83.0, 94.2	76.9 ± 10.3 (4)	122.5 ± 14.3 (6)	143.7 ± 24.2 (6)	94.9 ± 15.3 (4)
Neandertals	67.1 ± 8.2 (22)	65.1 ± 8.2 (27)	116.1 ± 19.1 (26)	109.7 ± 13.9 (24)	98.2 ± 11.9 (19)
E/MUP	58.1 ± 8.2 (7)	56.2 ± 7.7 (8)	114.0 ± 9.5 (12)	116.6 ± 12.2 (13)	91.8 ± 11.1 (7)
K-W p-value	0.034	0.006	0.626	0.003	0.425

Note: Middle Pleist. = Middle Pleistocene archaic humans; MPMH = Middle Paleolithic modern humans; E/MUP = Early and Mid Upper Paleolithic modern humans; K-W p-value = Kruskal-Wallis p-values across the four comparative samples.

TABLE 7.12. Comparative mandibular total occlusal areas (TOAs) for the Palomas dentitions and the Middle and Late Pleistocene comparative samples, in square millimeters [mean ± SD (n)]

GROUP	MANDIBLE P_3	MANDIBLE P_4	MANDIBLE M_1	MANDIBLE M_2	MANDIBLE M_3
Palomas	37.4, 40.7, 42.7, 55.7	35.6, 41.0, 48.7	93.5, 93.9, 96.1	75.9, 75.9, 100.4, 100.4	104.7
Middle Pleist.	64.0 ± 5.7 (6)	75.4 ± 15.5 (4)	131.1 ± 18.9 (6)	130.9 ± 26.5 (10)	132.5 ± 15.8 (6)
Neandertals	55.9 ± 8.3 (26)	58.2 ± 6.7 (24)	107.4 ± 14.8 (43)	112.1 ± 16.2 (32)	104.2 ± 15.3 (30)
E/MUP	51.8 ± 5.6 (8)	55.8 ± 4.7 (5)	110.2 ± 9.0 (15)	113.9 ± 12.1 (13)	106.9 ± 15.4 (10)
K-W p-value	0.013	0.016	0.014	0.098	0.003

Note: Middle Pleist. = Middle Pleistocene archaic humans; E/MUP = Early and Mid Upper Paleolithic modern humans; K-W p-value = Kruskal-Wallis p-values across the four comparative samples.

TABLE 7.13. Comparison of summed mandibular buccolingual diameters for anterior (I_1 to P_3) versus posterior (P_4 to M_2 and P_4 to M_3) for Palomas 1 and the Middle and Late Pleistocene comparative samples, in millimeters [mean ± SD (n)]

INDIVIDUAL OR GROUP	$C_1 + P_3$	I_1 TO P_3	P_4 TO M_2	P_4 TO M_3
Palomas 1	18.0	32.0	31.0	41.6
Palomas 59	14.8	—	26.0	—
Middle Pleist.	17.7 ± 1.3 (8)	33.0 ± 1.6 (5)	31.7 ± 2.8 (9)	42.8 ± 3.7 (7)
Neandertals	18.3 ± 1.3 (29)	33.8 ± 1.7 (18)	31.3 ± 1.9 (29)	42.6 ± 2.2 (26)
MPMH	17.3 ± 1.1 (7)	30.9 ± 1.8 (6)	31.4 ± 1.7 (7)	43.0 ± 2.3 (5)
E/MUP	17.4 ± 1.0 (20)	30.8 ± 1.7 (16)	30.9 ± 1.4 (22)	41.5 ± 2.2 (9)
K-W p-value	0.017	<0.001	0.836	0.528

Note: Middle Pleist. = Middle Pleistocene archaic humans; MPMH = Middle Paleolithic modern humans; E/MUP = Early and Mid Upper Paleolithic modern humans; K-W p-value = Kruskal-Wallis p-values across the four comparative samples.

TABLE 7.14. Comparison of labial root lengths (cervix to apex) of the anterior dentition for the Palomas teeth and the Middle and Late Pleistocene comparative samples, in millimeters [mean ± SD (n)]

GROUP	MAXILLA I^1	MAXILLA I^2	MAXILLA C^1	MANDIBLE I_1	MANDIBLE I_2	MANDIBLE C_1
Palomas[a]	14.8, 16.3, 16.4	15.1, 16.0, 20.4	15.6, 17.9, 21.1	10.7, 14.9, 15.0, 15.8	16.8	14.5, 17.0, 18.2, 19.8, 20.9
Middle Pleist.	18.6 ± 0.7 (5)	17.8 ± 1.3 (4)	20.6 ± 2.4 (4)	18.0, 20.0	—	20.3 ± 2.1 (7)
Neandertals	17.3 ± 1.1 (16)	17.1 ± 1.6 (15)	22.1 ± 2.1 (18)	16.8 ± 1.6 (10)	16.0 ± 1.4 (13)	20.6 ± 2.5 (17)
MPMH	—	—	—	13.7, 13.9	—	14.8, 16.1
E/MUP	13.6 ± 1.7 (12)	14.8 ± 1.6 (6)	17.2 ± 3.6 (7)	13.4 ± 1.8 (11)	14.4 ± 1.4 (13)	17.0 ± 2.1 (12)
K-W p-value	<0.001	0.022	0.010	0.002	0.010	0.001

Source: Data from Matiegka (1934), McCown and Keith (1939), Patte (1960), Brabant and Sahly (1964), Vlček (1969, 1993), Sakura (1970), Bermúdez de Castro (1988), Trinkaus et al. (2003, 2010, 2013, 2014a), Bailey (2005), Bailey and Hublin (2006), Doboş et al. (2010), Hershkovitz et al. (2011), and Le Cabec et al. (2013).

Note: Middle Pleist. = Middle Pleistocene archaic humans; MPMH = Middle Paleolithic modern humans; E/MUP = Early and Mid Upper Paleolithic modern humans; K-W p-value = Kruskal-Wallis p-values across the four comparative samples.

[a]The values of 16.0 mm for the I^2 and 10.7 mm for the I_1 are from Palomas 96.

TABLE 7.15. Dimensions of the Palomas deciduous teeth, in millimeters

SP NO.	SIDE	CRN MD	CRN BL	CRV MD	CRV BL	ROOT MD	ROOT BL	CRN HT	RT HT B	RT HT L	TOTAL HT
Maxillary di¹											
39	Left	7.8	6.1	5.7	4.7	5.2	3.8	7.7	(11.5)	(11.0)	>17.3
Maxillary di²											
85	Left	[4.7]	4.7	3.5	4.3	3.1	4.3	[3.5]	—	—	>8.2
Maxillary dm¹											
71	Left	—	9.1	—	8.0	—	—	6.8	>6.2	—	>13.2
Maxillary dm²											
36	Left	10.5	11.3	7.9	10.8	—	—	5.2	—	—	>9.6
Mandibular dc₁											
31	Left	[7.1]	6.2	5.5	4.8	4.2	4.5	[5.1]	(12.1)	(11.5)	[15.5]
93	Right	[7.4]	6.6	5.7	5.7	5.5	4.3	[5.1]	13.2	13.3	[17.4]
95	Left	7.0	6.7	5.0	5.6	—	—	[8.0]	—	—	>12.1
Mandibular dm₁											
25	Right	8.1	6.8	7.1	5.4	—	—	[5.4]	—	—	>9.6
40	Left	8.3	6.4	—	—	—	—	—	—	—	>7.9
61	Right	9.0	7.0	—	—	—	—	[5.0]	—	—	—
Mandibular dm₂											
70	Right	10.5	9.1	7.9	7.5	8.3	7.7	6.1	8.3	8.0	14.0
83	Right	10.2	8.7	7.4	7.0	7.3	7.1	5.4	—	6.6	>13.3
88	Left	10.4	9.2	8.2	8.1	7.6	—	6.4	—	—	>15.3

Note: Numbers after > indicate broken or incompletely developed root, and the measurement of the preserved or developed portion is provided; numbers in parentheses indicate minor damage, and the original measurement is estimated with little probable error; numbers in square brackets indicate a worn tooth, occlusally and/or interproximally, and the measurement is of the remaining portion. Crn = crown; Crv = cervix; Rt = root; MD = mesiodistal; BL = buccolingual or labiolingual; B = buccal; L = lingual; Ht = height.

TABLE 7.16. Dimensions of Palomas teeth of uncertain or imprecise identification, in millimeters

SP NO.	TOOTH	SIDE	CRN MD	CRN BL	CRV MD	CRV BL	ROOT MD	ROOT BL	CRN HT	RT HT B	TOTAL HT
27	dc	—	8.5	6.5	6.0	6.0	5.0	4.5	10.0	13.0	22.0
38	M₂ or M₃	—	—	—	—	—	—	—	(5.7)	—	(18.3)
72	M	—	—	10.4	—	—	—	—	5.2	—	—
81	M₁ or M₂	Left	—	—	—	—	—	—	6.8	—	—

Note: Numbers in parentheses indicate minor damage, and the original measurement is estimated with little probable error. Crn = crown; Crv = cervix; Rt = root; MD = mesiodistal; BL = buccolingual or labiolingual; B = buccal; L = lingual; Ht = height.

The Palomas Dental Remains
Shape Morphometrics

8

BEATRIZ PINILLA

THE AIM OF THIS CHAPTER is to use geometric morphometrics (GM) to assess the affinities of the Sima de las Palomas postcanine dental crown shapes with samples of Neandertals and early modern humans from Europe. Recent analyses using GM on different human fossil samples (for example, Martinón-Torres 2006; Gómez-Robles et al. 2007, 2008, 2011b; Gómez-Robles 2010) have noticed the general similarity of Late Pleistocene samples (Neandertals and early modern humans) with respect to their postcanine teeth, thereby making it often difficult to assess whether a specific premolar or molar has affinities to one sample or another on the basis of overall size and shape. However, other analyses (for example, Bailey and Lynch 2005; Benazzi et al. 2011a) have been able to achieve some level of distinction between these two groups using GM. The Palomas postcanine teeth were therefore assessed using GM to complement the information derived from their coronal and radicular discrete morphologies and dimensions (chapters 6 and 7).

Materials

Despite the large sample of postcanine teeth from the Sima de las Palomas (chapter 6), only ten of them were used in the GM analyses, given variable preservation and wear. The teeth analyzed include one P^4 (Palomas 94), three P_3s (Palomas 45, 59, and 76), three P_4s (Palomas 57, 59, and 87), one M_1 (Palomas 84), one M_2 (Palomas 29), and one dm_2 (Palomas 70). The teeth therefore represent the maxillary P^4 and the mandibular P_3, P_4, M_1, M_2, and dm_2.

The comparative samples consist of high-resolution replicas of the original teeth curated in the Anthropology Section, Departamento Biología Animal, at the Universitat de Barcelona and the Department of Anthropology, Washington University; most of the material was collected in 2003 by A. Pérez-Pérez (Galbany et al. 2004a) and by E. Trinkaus from 1978 to 1996. The fossil humans studied include Neandertals ranging in time from Marine Isotope Stage (MIS) 7/6 to mid–MIS 3 and early modern humans from MIS 5c and from mid–MIS 3 to late MIS 2; the Qafzeh 5, 9, and 11 Middle Paleolithic modern humans (MPMH) are variably included with the Upper Paleolithic sample, and the Upper Paleolithic sample includes both Early/Mid Upper Paleolithic (E/MUP) and Late Upper Paleolithic (LUP) specimens. The analyzed sample includes only those teeth for which the occlusal relief can be clearly identified for landmark definition. The Neandertal sample (100 teeth) is separated into those from Krapina (56 teeth) and those from various other European sites (44 teeth). There are 35 early modern human teeth in the analysis, for a total sample of 135 teeth.

Methods

Geometric morphometrics can be defined as a quantitative description of shape that assesses the form of the displacement of landmarks, in the plane (2D) and the space (3D) of the landmarks, preserving the spatial order of the data by super-imposing homologous landmarks of the analyzed specimens. The landmarks provide a means of sampling form (Bookstein 1997; O'Higgins 2000) and, by definition, need to be homologous points among the analyzed specimens, in terms of both location and function (Bookstein 1991; O'Higgins 2000; Robinson et al. 2002; Zelditch et al. 2004; Catalano et al. 2010). In this project, landmarks were located at maximum (cusp tip) or minimum (valley) points of teeth (Landmark type II of Bookstein 1991), following the methodological approach developed by Martinón-Torres (2006) and Gómez-Robles et al. (2007); their locations were easily replicable, and they could be found on all of the specimens.

On the molars, the landmarks are the apices of the four main cusps (cusp 1: protocone/protoconid; cusp 2: paracone/metaconid; cusp 3: metacone/hypoconid; and cusp 4: hypocone/entoconid). On the M_1 the hypoconulid was also included. On the premolars the landmarks are the apices of the cusps (type II) and foveas (the deepest point of the fovea or the point in which the central groove meets the lateral grooves; types I and II).

Semilandmarks are usually defined as a set of points that draw a curve and function as type III landmarks (Bookstein 1991). These landmarks are defined as extreme points with at least one deficient coordinate. They are not completely homologous among individuals and are the least reliable points (Gómez-Robles 2010). However, semilandmarks, as opposed to type III landmarks, have fewer degrees of freedom than the number of coordinates that define their position. They are dependent on two other points (contiguous semilandmarks) so that each position is considered relative to other coordinates. They are normally analyzed as "sliding" landmarks along the curve (Pérez et al. 2006). Even though semilandmarks

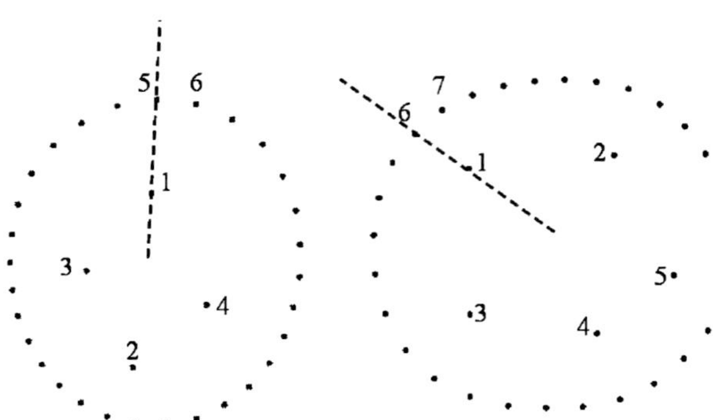

FIG. 8.1. Landmarks on each tooth type (*left*: premolars; *right*: mandibular molars and dm$_2$). The first semilandmark was always defined as the crown contour intercept of the line from the centroid to the first landmark; the remainder of the semilandmarks were marked from this first semilandmark to the right in a clockwise direction.

are not homologous between specimens, the area (contour/curve) they define should be (Slice 2005; Pérez et al. 2006).

The semilandmarks were defined from 30 lines automatically drawn on each tooth's digital image from the centroid to the edges. The centroid was derived from the four most extreme cusps in molars and from the two main cusps and the two foveas (mesial and distal) on the premolars. The presence of more cusps (Gómez-Robles et al. 2011a) was not taken into account, but the absence of a necessary cusp meant the deletion of the specimen from the sample. The centroid was generated with MakeFan 6.0 (IMB software; Sheets 2001). From this point, 30 radial equidistant lines were created for the full 360° (Fig. 8.1). The semilandmarks are the points of intersection of these radial lines and the crown contour. If the four cusps/points were not centered (and this might be more common on premolars than molars), the outline closer to the centroid would be overrepresented, while the opposite outline would appear underrepresented.

The Procrustes distance method was employed following Bernal (2007) and Gómez-Robles (2010). The available software minimizes the distances between the landmarks, employing a minimum of rotation, translation, and scaling to achieve this alignment. Therefore, the values obtained represent the amount of difference between homologous landmark positions, with reference to the mean value (or centroid) of the available sample.

The left antimere was employed for the analyses. In order to maximize sample sizes, when the left tooth was absent or when the landmarks were less clear, the right tooth was mirror-imaged with Adobe Photoshop CS3.

Since the cusp tips used are type II landmarks, wear is a clear handicap in sample selection. Only teeth with degrees of wear of 3 or less of Johansson et al. (2001), 2+/3 of Molnar (1971), and 3b of Smith (1984) were considered for the GM analysis. Semilandmarks, on the other hand, may be affected by interproximal wear more intensively than by occlusal wear, and thus the original contours were reconstructed across the interproximal facets (following Wood and Abbott 1983) to avoid shape changes due to interproximal wear.

The TPS package was used for digitizing (TpsUtil and TpsDig) and to analyze most of the data (TpsRelw) (Rohlf 1998, 2005, 2007). Linear discriminant analysis (LDA) was performed using SPSS v. 18 (2009).

Results

Maxillary P⁴

The principal components analysis (PCA) shows that Palomas 94 presents values on the first two principal components (PCs) (53. 5% of the variation) close to the overall consensus values (Fig. 8.2). The morphology of the maxillary P⁴ is not significantly different between the Neandertals and early modern humans. There is a great variability of the Neandertal P⁴s and especially within the Krapina sample. There is no clear chronological pattern in the P⁴ shape, although significant differences occur between the later modern humans (the sample limited to the MIS 2 Cisterna 1 and the MIS 3/2 Pataud 1) and the MIS 6/5 (Krapina and Montmaurin) Neandertals on principal component 2 (PC2) (21.1% of the variability; p = 0.017; F = 3.868). This distinction can be observed in the PCA, with the Pataud 1 and Cisterna 1 individuals showing lower PC2 values. The morphological variation represented on this PC might be related to a mesiodistal constriction of the premolar (or a buccolingual elongation).

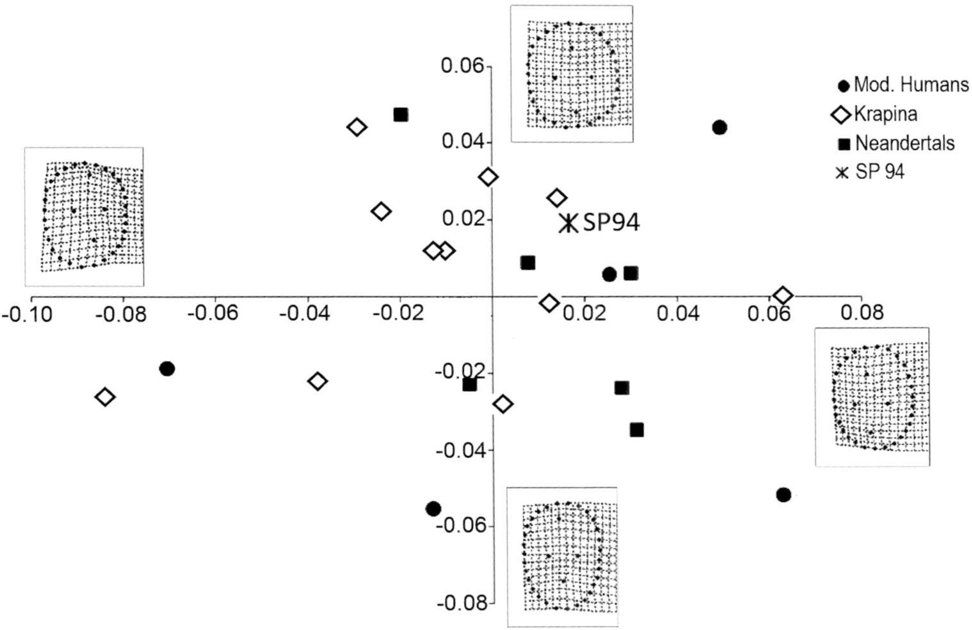

FIG. 8.2. The first two components of the PCA of the P⁴ (54.5% of the variability). The tooth configurations diagrammed represent the extremes of the variables and not any one tooth.

The LDA between early modern humans and Neandertals (including five PCs: 82.4% of the variability) shows correct classifications of 81.8% (68.2% after cross-validation), even though the differences are not significant on the derived function (p = 0.562; r = 0.448). Palomas 94 is classified as Neandertal instead of early modern human (Palomas 94: 89.6% post hoc probability).

Mandibular P_3

The first two PCs (Fig. 8.3; 55.5% of the variability) show a P_3 shape distribution in which the early modern humans are consistently similar in shape, whereas the Neandertals and the Krapina P_3s are more dispersed. The Palomas P_3s also show a great heterogeneity. Most of the Krapina (80%) and half of the Neandertal P_3s present a more rounded shape, which might be related to the presence of and space occupied by the lingual groove (less marked in the Krapina teeth) and the mesial marginal ridge. Both of these aspects are important in the positive

values of PC1 (38.0% of the variability) and in the negative values on PC2 (17.5% of the variability), whereas the opposite pattern is related to a simpler shape configuration with varying buccalization of the protoconid. Similarly, negative values on both axes represent an increase in the relative distance between the metaconid and the posterior fovea, which might be explained by the presence of accessory lingual cusps. The Palomas positions in the plot are heterogeneous, and the shapes of the three teeth are therefore markedly variable.

One-way ANOVA shows significant differences only on PC8 (only 2.2% of the variability) and only when considering the Krapina, Neandertal, Upper Paleolithic, and MPMH teeth separately (p = 0.004; F = 6.734). Differences occur between MPMH (limited to Qafzeh 11 and Qafzeh 9) and Upper Paleolithic humans (p = 0.028) and Neandertals (p = 0.031). PC8 is mainly explaining a diminution of the relative space occupied by the grooves (anterior and posterior) as well as the mesial marginal ridge and the lingual groove. The Qafzeh teeth tend

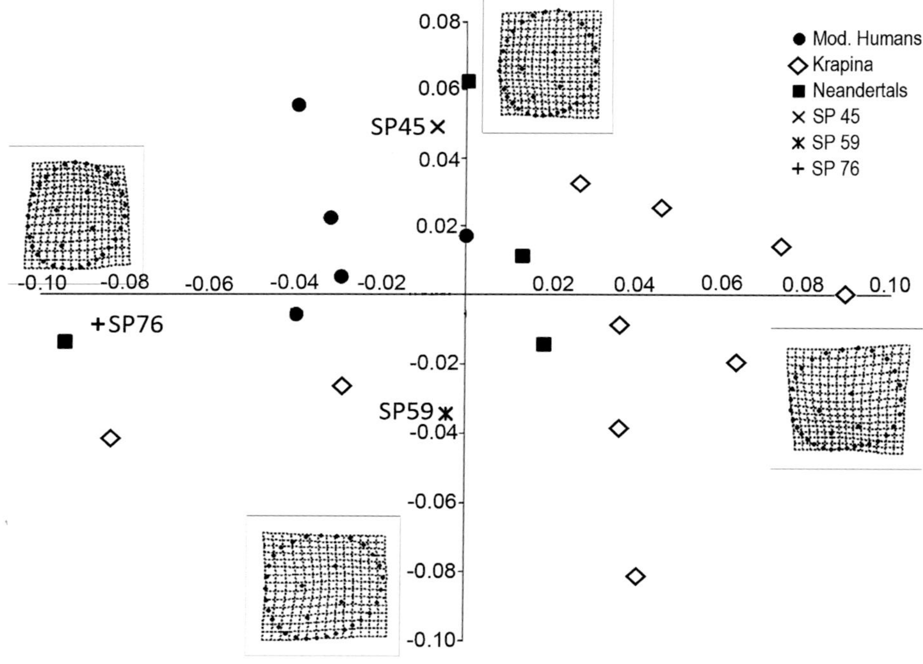

FIG. 8.3. The first two components of the PCA of the P_3 (55.5% of the variability). The tooth configurations diagrammed represent the extremes of the variables and not any one tooth.

to show a smaller distance between the grooves and a more rounded shape due to the smaller influence of the lingual groove on the outline. On the other hand, the Upper Paleolithic and Neandertal teeth show a greater distance between the grooves and a strong mesial marginal ridge, which influences the outline shape. However, those differences are small, and PC8 represents only ≈2% of the variation.

LDA functions (derived from the four first PCs: 80.0% of the variability) are not significant (p = 0.250; r = 0.547), and the post hoc probability of classification is smaller than that observed on the maxillary teeth (68.4% and 57.9% after cross-validation). The heterogeneity observed on the PCA is also represented in the post hoc classification of the Palomas teeth. Palomas 45 and 59 are classified as Neandertals (even though the classification post hoc probabilities are small: 60.1% and 68.5%, respectively), whereas Palomas 76 is classified as an early modern human with a 51.6% of post hoc probability (marginally different from an ambiguous 50%).

Mandibular P$_4$

The PCA configuration of the first two PCs (Fig. 8.4; 57.3% of the variability) shows a heterogeneous distribution, especially considering the sites with more than one tooth: Krapina (with individuals in all the quadrants) and Sima de las Palomas. It is, however, not the case for Qafzeh, for which both individuals (Qafzeh 9 and 11) are close to each other with negative values on both axes but close to the consensus configuration. The position of the only Upper Paleolithic specimen, Dolní Věstonice 13, is also close to the consensus configuration. The three Palomas P$_4$s are consistently to the right of the plot, although Palomas 57 is the only one with positive values on both axes. The main distinction of the extremes of PC1 (33.4% of the variability) is related to the lingual groove area, as well as the protoconid and metaconid positions: mesially oriented for the negative values or distally oriented for the positive values. In both cases, the movement affects both cusps at the same time. The distinction along the PC2 (23.9% of the variability) extreme

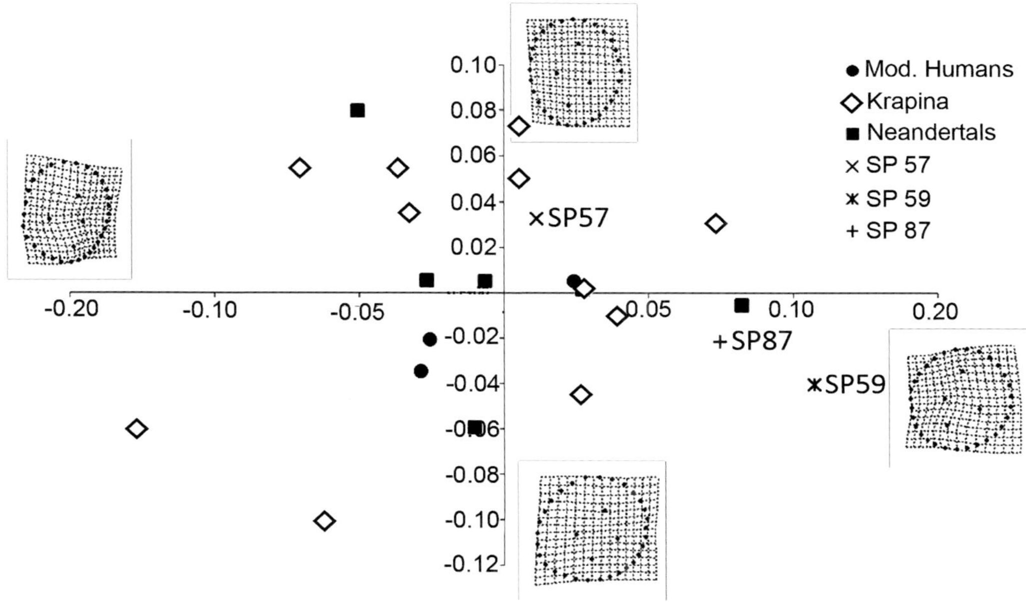

FIG. 8.4. The first two components of the PCA of the P$_4$ (57.3% of the variability). The tooth configurations diagrammed represent the extremes of the variables and not any one tooth.

configurations is related to accessory cusps and the marginal ridge areas, which affect the outline (more rounded for negative values and more elongated buccolingually with positive values). No significant differences occur between the samples, even when separating the Krapina ones from the sample of Neandertals, nor is there a chronological tendency observed.

The LDA of the first five PCs (85.9% of the variability; p = 0.830, r = 0.367) shows a post hoc probability of classification of 84.2% (73.7% after cross-validation). Note that when only as many PCs as the minimum number of specimens in the least represented group, in this case early modern humans (n = 3), are used, they account for less than 75% of the variability, and therefore too much information would have been lost. Therefore, in this case five PCs were used. The three Palomas P_4s, despite their heterogeneous distribution on the PCA configuration, are classified as Neandertals (post hoc probabilities of 79.1%, 55.3%, and 74.5%, respectively). Similarly incomplete separation of Late Pleistocene P_4 samples based on crown outlines was found by Bailey and Lynch (2005).

Mandibular M$_1$

The first two PCs (Fig. 8.5; 46.6% of the variability) show a substantial overlap among the samples. The early modern humans (87.5% of the individuals) tend to show positive values on PC2 (21.2% of the variability), except for Les Rois 50-40. Although their position is close to the consensus, the main distinction along the PC2 is related to a distal position of the protoconid and the hypoconid, together with a more parallel (buccolingual) arrangement of the entoconid-hypoconulid for the negative values. Moreover, for positive values the entoconid area is constrained. Neandertals are heterogeneously distributed, although almost half of the sample (46%) is restricted to the lower right quadrant (positive values on PC1 and negative values on PC2). Similarly, Krapina teeth (80.0% of the teeth) are restricted to positive values on PC1. PC1 (25.8% of the variability) shape variation is related to mesiodistal outline constriction: a more oval shape for positive values and a more rounded shape for negative values. These changes might be related to the presence or absence and relative importance of structures such as the anterior

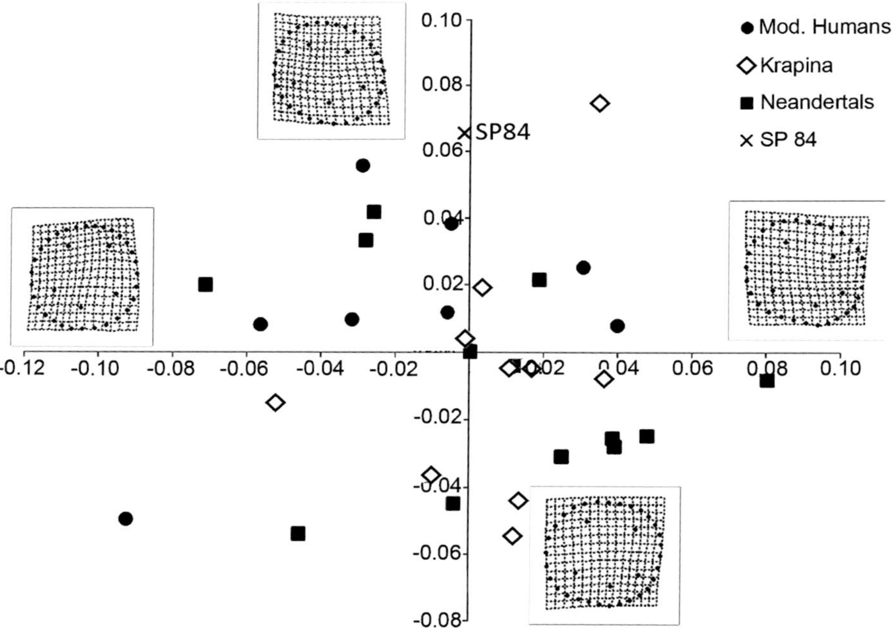

FIG. 8.5. The first two components of the PCA of the M$_1$ (46.6% of the variation). The tooth configurations diagrammed represent the extremes of the variables and not any one tooth.

fovea, the midtrigonid crest, the mesial marginal accessory tubercles, and the entoconulid. The metaconulid, on the other hand, might not be very important for the first two PCs, although it might be responsible for the shape observed in individuals with positive values on PCI and negative ones on PC2, in which the distance between the metaconid and the entoconid is higher.

A one-way ANOVA shows significant differences in M_1 shape between Neandertals and early modern humans on the PC4 (p = 0.025, F = 5.608) that nonetheless account for only 10.3% of the variability. The main difference between the extreme shape configurations is due to the already mentioned mesiodistal elongation, an increase in the distance between the metaconid and the entoconid (which might be related to cusp 7 presence) and between the protoconid and the hypoconulid (although the hypoconid remains more or less stable).

The LDA function derived from the first eight PCs (88.9% of the variability) is significant (p = 0.016, r = 0.726), with post hoc correct classifications of 90.3% (80.6% after cross-validation). In this analysis, Palomas 84 is classified as an early modern human with an 84.4% post hoc probability. Considering the PC4 configuration, in which Palomas 84 is in the middle of the early modern human variation but towards the limits of the Neandertal variation, the classification of Palomas 84 makes sense. It presents an oval shape, which is elongated mesiodistally, with a greater distance between the hypoconid and the hypoconulid than that of the consensus shape for the Neandertals.

Mandibular M_2

The PCA distribution (PCI and PC2, 53.9% of the variability) shows a heterogeneous distribution of all the groups considered (Fig. 8.6). Palomas 29 presents positive values on both components. PCI (33.9% of the variability) is characterized mainly by the distal position of all of the cusps in the positive values. This characteristic might be explained by the presence of the cusp 5 in the distal area. Actually, the negative value configurations are not only characterized by this "cusp mesialization" but also by a relatively higher distance between the hypoconid and the entoconid.

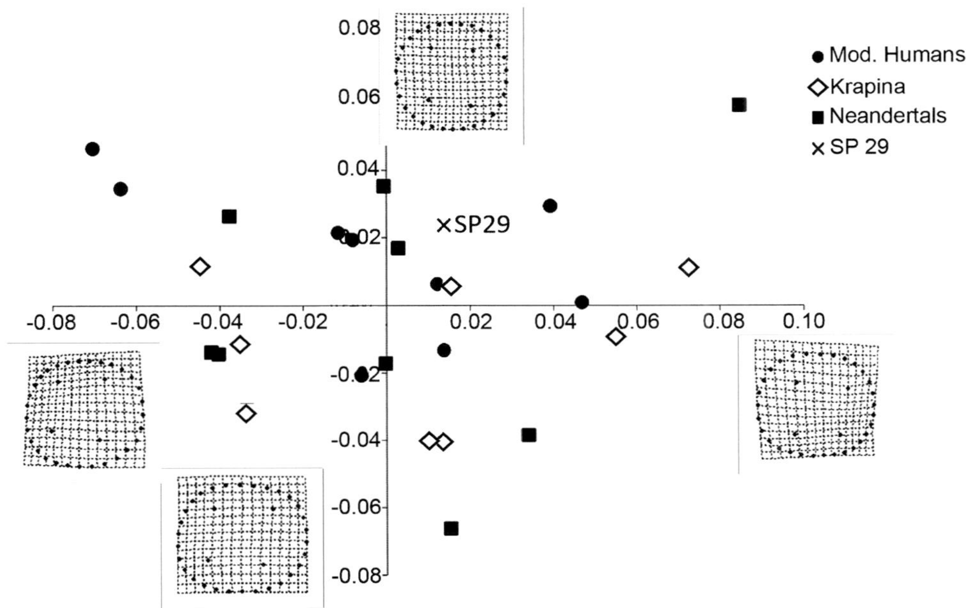

FIG. 8.6. The first two components of the PCA of the M_2 (53.9% of the variability). The tooth configurations diagrammed represent the extremes of the variables and not any one tooth.

PC2 (20.0% of the variability) might also represent the presence or absence of the hypoconulid, because the negative value configurations show an increase in the buccodistal area. However, contrary to the distinction on PC1, cusp movement is not as marked.

There is a significant difference between the Neandertals and the early modern humans on PC7 (p = 0.040; F = 4.712), but its importance is uncertain because it accounts for only 3.1% of the variability. The variation in this case is related to the early modern human "centralization" of cusps, especially the movement of the entoconid.

The LDA-derived function (including the first nine PCs; 94.3% of the variability) is not significant (p = 0.392; r = 0.609) in distinguishing Neandertal versus early modern human M_2 shape. Palomas 29 is classified as Neandertal with a post hoc probability of 83.2%. Yet the percentage of correct classification of the LDA function is only 81.5% (decreasing to 59.3% after cross-validation).

Mandibular dm₂

The first two PCs for the dm_2 (54.0% of the variation) show (Fig. 8.7) that the early modern humans are a very dispersed group. The Neandertals have considerable variation on PC1, but they are mainly restricted to the negative values on PC2 (83.3% of the individuals; only the late Middle Pleistocene La Chaise–Suard 13 provides a positive value). In the Krapina sample, most of them provide positive values on PC1 (83.3% of the individuals); only Krapina 52 (mandible B) furnishes a negative PC1 value. The Palomas 70 shape is represented by a slightly positive value on PC1 and a negative one on PC2, similar to most of the Neandertals.

PC1 shape variation (31.0% of the variability) is mainly the result of an increase in the entoconid area, producing an outline of a more elongated tooth with positive values. This change in shape seems to be unrelated to the presence of a cusp 6 or 7 in the teeth. PC2 (23.1% of the variation) is related to the buccalization of the metaconid,

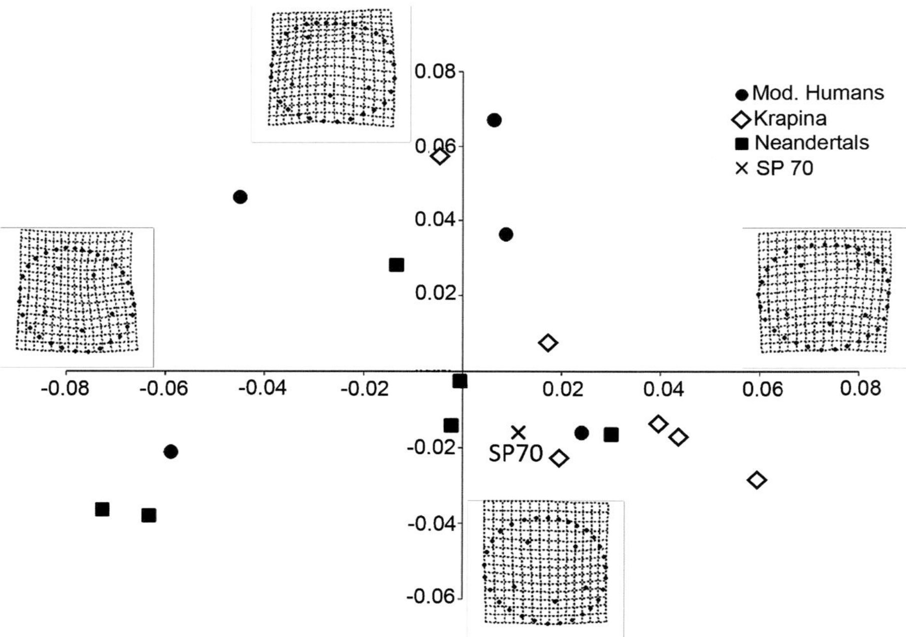

FIG. 8.7. The first two components of the PCA of the dm_2 (54.0% of the variability). The tooth configurations diagrammed represent the extremes of the variables and not any one tooth.

entoconid, and hypoconid primarily. Small variation in the mesial and distal outlines is also involved.

A one-way ANOVA shows no significant differences between the Neandertals and the early modern humans, nor does it when considering Krapina separately. Yet if the samples are divided according to chronological (MIS) periods, significant differences occur with respect to PC1 (p = 0.034; F = 3.885). However, Bonferroni post hoc analysis shows that the differences are constrained to MIS 3 versus MIS 6 (p = 0.031) and disappear when MIS 3 is segmented into Neandertals versus early modern humans—especially since the MIS 3 early modern human is limited to Fontéchevade 2, whose stratigraphic provenience is unclear (Chase and Teilhol 2009).

The LDA function performed with the first five PCs (88.1% of the variation) is not significant (p = 0.268; r = 0.630), and post hoc correct classification is only 76.5% (64.7% after cross-validation). Palomas 70 is classified as a Neandertal with a 74.8% post hoc probability.

Discussion

The teeth analyzed from Sima de las Palomas show geometric proportions similar to the overall mean values, not differing from the overall sample. The samples considered show differences in some shape configuration PCs but also a substantial amount of overlap. This result holds whether the Neandertals and early modern humans are compared or Krapina on the one hand and Qafzeh on the other hand are separated out. Only with respect to the M_1 are the differences from the ANOVA linked to a PC with a percentage of variation higher than 3.2%. Related to this aspect, only the LDA functions from that tooth are significant, but even a mild multiple comparison (Bonferroni) correction would make them nonsignificant at p < 0.05.

Despite this substantial overlap across the samples, the discriminant analyses after cross-validation show correct classification percentages high enough to provide some confidence in the morphometric shape analyses. Yet two of the teeth (the P_3 and the M_2) have cross-validation correct classification percentages of only 57.9% and 59.3%, respectively, and the dm_2 is not much higher (64.7%). Only the M_1 is classified at >80%—at 80.6%. These low levels of correct classification probability reinforce the substantial overlap in these measures across the later Pleistocene samples, evident in the other measures. Similar results have been evident in other analyses (for example, Bailey 2002a, 2002b, 2004 a, 2004b, 2006; Gambier et al. 2004; Gómez-Robles et al. 2007, 2008; Bailey et al. 2009; Gómez-Robles 2010; see chapter 6), especially if only a single tooth is considered.

The Sima de las Palomas teeth are consistently classified within the Neandertal group with post hoc probabilities of 51.6% to 91.9%. Only one case truncates this pattern: the mandibular M_1. In this case, Palomas 84 is classified as a modern human with an 84.4% post hoc probability due to its oval shape, its mesiodistal elongation, and the increase in the distance between the hypoconid and the hypoconulid. This result is even more important when considering that it is in the M_1 where the differences between the samples are most evident (10.3% of the variation in the PC is different) and the LDA function is also significant (p = 0.016). Other analyses of the mandibular M_1s have described them as largely similar between Neandertals and early modern humans, varying only in the higher percentages of the midtrigonid crest on the Neandertal M_1s (Zubov 1992; Bailey 2002a, 2002b; chapter 6) and, to a lesser extent, their higher percentages of cusps 6 and 7 and anterior foveae plus the lack of deflecting wrinkles. In most of the teeth analyzed, the differences observed in the extreme PC configurations might be related to the presence or absence of nonprimary cusps (for example, accessory cusps or tubercles, cusps 5, 6, or 7, etc.), which have been shown to have a substantial influence on geometric morphometric analyses (Gómez-Robles et al. 2011a).

The P^4 has been considered to be not very informative, due to the similarities of the maxillary premolars through the Middle and Late Pleistocene

(Bermúdez de Castro et al. 1999; Gómez-Robles 2010); however, on the mandibular premolars some researchers have found a central compression of the cusps in Neandertals, a tendency of this group to present a mesiolingual groove and a more asymmetrical lingual contour, and a high percentage of the presence of the transverse crest among early modern humans (Bailey 2002a, 2006; Bailey and Lynch 2005; Martinón-Torres 2006; Gómez-Robles et al. 2008). None of these features seem to have significantly affected the distributions on the analyses (positive values on PC1 and PC2 on P_3 might be influenced by this pattern, however, which would explain why the early modern humans tend to present negative values on PC1).

The mandibular molar analyses and shapes are more complex (Bailey 2002a). As mentioned, M_1 differences between the samples are small, and the ranges of variation of the structures observed, although different in percentages, do overlap widely. The PC1 and PC2 plot (Fig. 8.5) shows a substantial overlap of the groups, and the above-mentioned features of each group cannot be easily determined from the extreme shape configurations of the first two PCs nor the PC4 (which showed significant differences between the two groups).

The M_2s tend to show a similar occlusal surface to that of the M_1. Among the Neandertals, cusps 5, 6, and 7 are common, as well as the midtrigonid crest and the anterior fovea. In this case, even though it is not sufficiently clear, the negative values on PC1 (Fig. 8.6) might be representing the absence or reduction of the midtrigonid crest. However, this is not the case for PC7 (which is significantly different between populations, yet represents only 3.1% of the variation).

Finally, deciduous teeth have been little analyzed in terms of geometric morphometrics (for example, Benazzi et al. 2011a). Occlusal complexity may decrease with the emergence of Upper Paleolithic modern humans, although P. Smith (1978) considered that deciduous dental shape (and size) differences between the Neandertals and early modern humans were mainly restricted to the incisors and first molars, and little attention was paid to the dm_2. What tendencies exist (Fig. 8.7) occur with respect to the first PC (31.0% of the variation). The difference is mainly related to a more mesiodistally ovally elongated crown on the geologically older Neandertal dm_2s. Yet, the association of this configuration with more complex surfaces involving accessory cusps is not straightforward. In the case of the Palomas 70 dm_2, its position differs from that of most of the MIS 3 specimens. It is nonetheless classified as a Neandertal with a 74.8% post hoc probability.

Conclusion

Even though GM cannot respond to all the questions about the tooth shape, it allows us to analyze the form of a given structure quantitatively (Adams et al. 2004). The use of individual teeth, as opposed to dental arcades, limits the analysis of group similarities and differences and the assignment of individual specimens to *a priori* samples. Yet these considerations provide an additional perspective on human dental variation and affinities in the later Pleistocene.

The postcanine dental shapes of Neandertals and early modern humans are quite similar, and significant differences on the employed landmarks and semilandmarks are limited. However, percentages of post hoc classification with cross-validation are moderately high, and the Palomas isolated teeth are classified as Neandertals in 90.0% of the teeth studied (9 out of 10).

The Palomas Dental Remains

Enamel Thickness and Tissue Proportions

9

PRISCILLA BAYLE, MONA LE LUYER, AND KATHARINE A. ROBSON BROWN

RECENT ADVANCES IN DEVELOPMENTAL biology, quantitative genetics, and comparative anatomy at meso- and microstructural levels have shown that a wealth of unique information crucial in paleoanthropology is stored within tooth tissues (see, for example, Dean 2006; Macchiarelli et al. 2006, 2008; Schwartz and Dean 2008; Skinner et al. 2008; Smith 2008, 2013; Smith and Tafforeau 2008; Guatelli-Steinberg 2009; Bondioli et al. 2010; Braga et al. 2010; Kupczik and Hublin 2010; Bailey et al. 2011; Dean and Cole 2013; Pampush et al. 2013; Crevecoeur et al. 2014; Horvath et al. 2014; Martinón-Torres et al. 2014; Hlusko 2015; Smith et al. 2015). Notably, enamel thickness and dental tissue proportions have been largely studied to assess taxonomy, phylogeny, and masticatory biomechanical constraints in fossil and extant hominoids (for example, see Molnar and Gantt 1977; Martin 1985; Beynon and Wood 1986; Schwartz 2000; Kono 2004; Grine 2005; Suwa et al. 2009; Bayle et al. 2010; Mahoney 2010, 2013; Zanolli et al. 2010, 2014; Benazzi et al. 2011a; Smith et al. 2012; Le Luyer et al. 2014; Skinner et al. 2015). Noninvasive access to this information is possible solely by using advanced virtual imaging techniques. Besides studies on methodological aspects of the use of micro-CT–based techniques to explore and measure the internal structure of teeth in two and three dimensions (for example, Avishai et

al. 2004; Suwa and Kono 2005; Olejniczak and Grine 2006; Olejniczak et al. 2007; Bondioli et al. 2010; Benazzi et al. 2014), a growing number of researchers use these approaches to characterize their variation and evolution in fossil and extant hominins (reviewed in Macchiarelli et al. 2013).

Many of these studies have focused on Neandertals and their comparison with modern humans (Zilberman and Smith 1992; Molnar et al. 1993; Macchiarelli et al. 2006; Bayle 2008a, 2008b; Olejniczak et al. 2008a; Bayle et al. 2009a, 2009b; Crevecoeur et al. 2010; Toussaint et al. 2010; Benazzi et al. 2011b; Fornai et al. 2014). As a whole, Neandertal deciduous and permanent teeth—mainly molars—have been described as having volumes of enamel similar to those of modern humans, but with that enamel deposited over a topographically more complex enamel-dentine junction (EDJ) surface and with larger dentine volumes, thus showing thinner average and relative enamel thicknesses. However, few complete dentitions have been studied, and limited quantitative information is available on anterior teeth and premolars. On two-dimensional (2D) slices, Smith et al. (2012) measured similar average enamel thicknesses in the incisors of a few Neandertals and modern humans. More important, our knowledge of Neandertal variation in enamel thickness and dental tissue proportions is limited and does not permit the evaluation of the existence and nature of potential chrono-geographical trends in these fossil humans. Even if current sample sizes do not allow a proper evaluation, Olejniczak et al. (2008a) noticed that the variation encompassed by Neandertal molars spanning a substantial geographical and temporal interval was lower than in the recent humans sampled.

Here we report the results of the micro-CT–based analysis of enamel thickness and dental tissue proportions in the teeth from Sima de las Palomas and compare them to the data available for Middle, Late Pleistocene, and Holocene archaic and modern humans. Besides providing the unique opportunity to study the homogeneity of these parameters in a constrained chrono-spatial context, this report substantially increases the Neandertal variation known so far, notably for the deciduous and anterior permanent dentition.

Materials and Methods

From the complete data set of teeth from Palomas (see chapter 6), 31 teeth have been studied for enamel thickness and tissue proportions (Table 9.1). All of the teeth selected were well preserved, free of damage, and unworn to slightly worn on their occlusal aspect.

Teeth were scanned on Skyscan 1172 x-ray equipment of the imaging laboratory from the Department of Archaeology and Anthropology of the University of Bristol that had been specially transported to the Área de Antropologia Física of the Universidad de Murcia. Acquisitions were realized according to the following parameters: 100 kV voltage, 100 µA current, an Al/Cu filter, a rotation step between 0.3° and 0.6°, and a frame averaging of 4. We used Nrecon v. 1.6.6 (Skyscan) to reconstruct the final volumes with an isotropic voxel size ranging from 21 µm for isolated teeth to 36 µm for jaw fragments.

Following the half-maximum height method (Spoor et al. 1993; Coleman and Colbert 2007), a semiautomatic threshold-based segmentation with manual corrections was conducted using Avizo v. 7 (VSG). Because of their good preservation and modest degree of mineralization and diagenetic change, no particular technical difficulties were encountered in segmenting the Palomas teeth, all of which are characterized by a rather distinct contrast between enamel and dentine. However, the virtual extraction of the teeth embedded within jaws was complicated by the expected low contrast between bone and dentine, and the removal of the hard matrix covering some teeth and jaws also made the segmentation process harder (see examples of virtual "cleaning" and dental tissue extraction in Plates 9.1 and 9.2). Crowns were then digitally isolated from the roots (Olejniczak et al. 2008a), and three-dimensional (3D) surface models were generated using a constrained smoothing algorithm.

For describing 3D tissue proportions and enamel thickness, six linear, surface, and volumetric variables were digitally measured from the crowns using Avizo v. 7 or calculated: the volume of the enamel cap (Evol, mm^3), the volume of the dentine and pulp (Dvol, mm^3), the percentage of the crown volume that is dentine and pulp (%Dvol), the surface area of the EDJ (SEDJ, mm^2), the average enamel thickness (3D AET, mm), and the scale-free relative enamel thickness (3D RET). 3D AET is the average straight-line distance between the EDJ and the outer enamel surface, calculated as the quotient of the enamel volume and EDJ surface area, while 3D RET is the 3D AET multiplied by 100 and scaled by the cube root of the coronal dentine and pulp volume, which yields a scale-free value that allows comparisons between specimens of different proportions of tooth size and/or body size (Martin 1985; Olejniczak et al. 2008a).

Two-dimensional dental tissue proportions and enamel thickness were measured on virtual buccolingual cross-sections through the two mesial dentine horn tips for the molars (Martin 1985; Olejniczak et al. 2008a), through the main dentine horn tip for the premolars (Feeney et al. 2010), and through the dentine horn tip and both maximum buccal and lingual cervical enamel extension for the anterior teeth (Feeney et al. 2010). Seven linear and surface variables were measured from the crowns using MPSAK

v. 2.9 (available in Dean and Wood 2003) or calculated from the bicervical diameter (BCD, mm), the enamel area (c, mm²), the dentine and pulp area (b, mm²), the percentage of the crown volume that is dentine and pulp (%b), the EDJ length (e, mm), the 2D AET (mm), and the 2D RET. For slightly worn crowns, corrections of outer enamel surface were made prior to measurements of 2D variables. Reconstructions of the removed enamel were performed based on the morphology observed for unworn teeth in virtual buccolingual cross-sections (Smith et al. 2012).

Intra- and interobserver rates assessed by Priscilla Bayle and Mona Le Luyer were lower than 5% for all the measured variables. For the teeth providing the most numerous data—that is, di¹, dm², M³, and M₂—standard box and whisker plots revealing the interquartile range (25th–75th percentiles: boxes), 1.5 interquartile ranges (whiskers), and median values (black line) were represented for 3D AET and 3D RET. For the same tooth types, adjusted Z-scores (Maureille et al. 2001; Scolan et al. 2012) were computed to compare the 3D dental tissue proportions and enamel thickness values of the Palomas specimens to the means and standard deviations of the comparative groups. This statistical method allows the comparison of unbalanced samples, which is often the case when dealing with the fossil record, by using the Student's inverse t distribution. Note that the +1.0 to –1.0 interval in these Z-scores comprises 95% of the variation in the reference sample. In the cases of the di¹ and the dm₂, plots of the log of 2D AET against the log of coronal dentine and pulp area were used to illustrate the relationship between AET and tooth size (see Skinner et al. 2015) and to show the placement of the Palomas specimens relative to other Neandertals and modern humans.

3D maps of topographic enamel thickness distribution were created with the segmented enamel and crown dentine components (Macchiarelli et al. 2008), where differences in enamel distribution are rendered by a thickness-related pseudocolor scale ranging from dark blue (thinner) to red (thicker) in color. This method allows one to map topographic thickness variation at the outer enamel surface and facilitates the site-specific and synthetic comparative assessments of this variable.

The comparative sample used for 3D and/or 2D enamel thickness and tissue proportions analyses consists of unworn or lightly worn teeth (wear stage ≤4 following Smith 1984) belonging to Middle Pleistocene archaic humans (MPl), Neandertals (Neand), Middle Paleolithic modern humans (MPMH), Aterian humans, Early/Mid Upper Paleolithic modern humans (E/MUP), a Late Upper Paleolithic specimen (LUP), and Neolithic (NH) and recent (RH) modern humans, detailed as follows:

The Middle Pleistocene (MPl) specimens are from Mauer (M₃), Tighenif (dm¹, I₂, P₃, P₄, M₂, M₃), Thomas Quarry (P³, P⁴, M³), and Steinheim (P⁴, M³) (Zanolli et al. 2010, 2014; Smith et al. 2012; Zanolli and Mazurier 2013). The Neandertal ones are from Combe-Grenal (M₃), Devil's Tower (dm², dm₁, dm₂), Engis (di¹), Ehringsdorf (I²), El Sidrón (M³), Krapina (di¹, dm¹, dm², dm₂, M₂, M₃), La Chaise Abri Bourgeois-Delaunay (P₄, M₂, M₃), La Chaise Abri Suard (dc₁, dm₁, dm₂, M₂, M₃), La Quina (M₃), Le Moustier (M³, M₂, M₃), Oliveira (P₃), Pech de l'Azé (di¹, dm¹, dm²), Regourdou (P₃, P₄, M₂, M₃), Roc de Marsal (di¹, dm¹, dm², dc₁, dm₁, dm₂), Spy (di¹, dc₁), Scladina (M₂), and Subalyuk (dm²) (Bayle 2008a, unpublished data; Olejniczak et al. 2008a; Bayle et al. 2009a, 2010, 2013; Crevecoeur et al. 2010; Kupczik and Hublin 2010; Smith et al. 2012; Willman et al. 2012; Fornai et al. 2014; NESPOS Database 2015). Note that the origins by specimen for the data provided by Smith et al. (2012) are not available; therefore, the list of sites for which the 2D mean data are provided by them (for I¹ to P⁴ and I₂ to P₄) may not include some Neandertal sites. The Aterian data derive from Dar Es Soltane (M₂), El Haroura (M₂, M₃), and Témara (M₂, M₃) (Kupczik and Hublin 2010).

The MPMH specimens are from Qafzeh (P³, P⁴). The E/MUP data are from Dolní Věstonice (dm¹, dm²), Lagar Velho (di¹, dc₁, dm₂), and La Rochette (dm¹, dm²) (Bayle et al. 2010; Fornai et al. 2014), and the LUP values are all from La Madeleine (dc₁, dm₁, dm₂) (Bayle et al. 2009b).

For Holocene humans, the Neolithic data are from Gurgy (di^1, dm^1, dm^2, dc$_i$ to dm$_2$, I^1 to P^4, I$_2$ to P$_4$, and M$_2$) (Le Luyer 2016). The recent human (RH) data are of worldwide origin (all tooth types) (Kono 2004; Grine 2005; Macchiarelli et al. 2006; Bayle 2008a, unpublished data; Olejniczak et al. 2008b; Smith et al. 2008; Bayle et al. 2010; Feeney et al. 2010; Mahoney 2010, 2013; Zanolli and Mazurier 2013; Fornai et al. 2014). Those recent human samples, for which the original by-specimen data are not available, have summary data provided in the tables and are referenced there.

Results and Discussion

Deciduous Teeth

Results of the 3D dental tissue proportions and enamel thickness comparative analysis in the selected deciduous maxillary and mandibular teeth are shown in Tables 9.2 and 9.3, respectively. Those of the 2D analysis are shown in Tables 9.4 and 9.5.

As a sample, the Palomas teeth largely fit within the Neandertal range of variation, which exhibits similar enamel volume/surface deposited over larger dentine volume/surface and EDJ area/length relative to that of modern humans, including the Upper Paleolithic and Neolithic individuals from Dolní Věstonice, Lagar Velho, La Rochette, La Madeleine, and Gurgy. As illustrated in the box plots of Figs. 9.1 and 9.2 for di^1 and dm$_2$, this pattern in Palomas results in low 3D AET and 3D RET values—ones that are closer to the Neandertal than to the modern human range of variation. In this context the Neolithic sample is characterized by relatively large volume-to-surface ratio of dentine and pulp and low enamel thickness values, often lower than those in the Upper Paleolithic specimens (Figs. 9.1 and 9.2; Tables 9.2 to 9.5; Le Luyer 2016). Even if this is probably highly linked with a higher wear degree in the Neolithic humans than in the rest of the sample (see Tables 9.4 and 9.5 showing higher AET and RET values reported for 2D measurements corrected for occlusal wear), these results enhance our poor knowledge of the variation of

past and present modern human populations for these traits. Also, the only Middle Pleistocene deciduous tooth available for these parameters, the Tighenif dm^1, shows particularly high coronal dentine and pulp volume and low 3D AET and 3D RET values (Zanolli et al. 2010).

When considering the intra-arcade variation (Tables 9.2 to 9.5), our data confirm the distal increase in 3D and 2D AET and RET along the human deciduous tooth row observed in previous studies (for example, Zilberman et al. 1992; Bayle et al. 2009a, 2010; Crevecoeur et al. 2010; Mahoney 2010; Zanolli et al. 2010) and linked to morphological and/or functional factors (reviewed in Mahoney 2013). Furthermore, the differences between Neandertals and modern

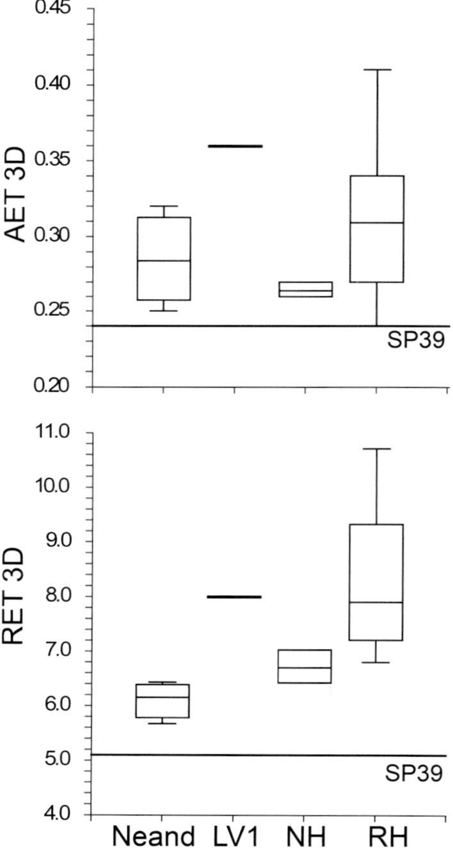

FIG. 9.1. Box plots of 3D AET and RET in the Palomas 39 di^1 compared to Neandertals (Neand), the Mid Upper Paleolithic individual from Lagar Velho (LV1), Neolithic humans (NH), and recent humans (RH).

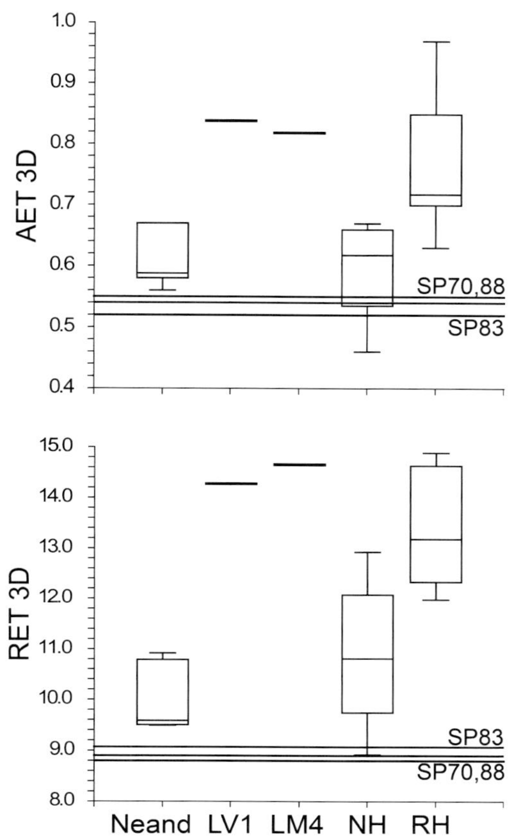

FIG. 9.2. Box plots of 3D AET and RET in the Palomas 70, 83, and 88 dm$_2$ compared to Neandertals (Neand), the Upper Paleolithic individuals from Lagar Velho (LV1) and La Madeleine (LM4), Neolithic humans (NH), and recent humans (RH). For the recent human sample, a high outlier of 18.93 is not shown.

humans is lower in the deciduous incisors (≈1% difference for 2D AET and ≈9% for 2D RET) than in the canines and molars (≈6%–10% for 2D AET and ≈11%–15% for 2D RET), as observed for 2D AET on the permanent dentition by Smith et al. (2012).

Adjusted Z-scores for di^1 and dm$_2$ (Figs. 9.3 and 9.4) confirm the proximity of Palomas deciduous teeth with Neandertals in all 3D and 2D variables. However, the data from Palomas substantially extend the previously known range of variation in Neandertal deciduous teeth. Indeed, the Palomas 29 di^1 has particularly low average and relative enamel thicknesses (Fig. 9.1), even if all but two of the measures fit within the Neandertal range (Fig. 9.3). As is shown by the bivariate plot of the log of 2D AET against the log of coronal dentine and pulp area (Fig. 9.5), being in essence a visual representation of 2D RET (Skinner et al. 2015), big teeth (with large dentine and pulp area) tend to have thin enamel, even if the bulky Krapina d21 tooth has thicker enamel than Palomas 39. Along the same lines, the smallest dm$_2$ from the Palomas sample, Palomas 83, shows higher 3D and 2D RET values than Palomas 70 and Palomas 88 (Fig. 9.2; Tables 9.3 and 9.5). Generally, this tooth aligns better with the recent human range than do its counterparts (Fig. 9.4), even if the three teeth clearly fit within the Neandertal variation, as is shown by the bivariate plot of 2D

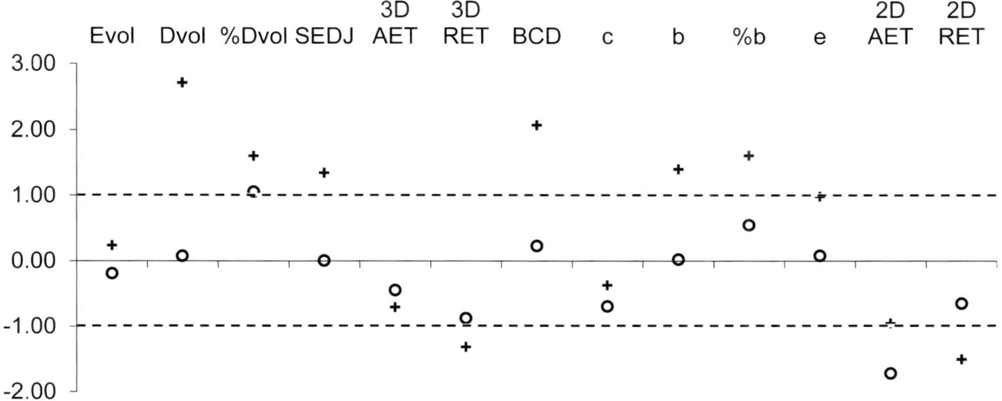

FIG. 9.3. Adjusted Z-scores of 3D and 2D variables assessed in the Palomas 39 di^1 crown and compared to the variation expressed by Neandertals (*circles*) and recent humans (*plus signs*). The solid line passing through zero represents the mean, and the dotted lines correspond to the estimated 95% limit of variation expressed for the two comparative samples. See "Materials and Methods," in chapter 9 for the definitions of the variables.

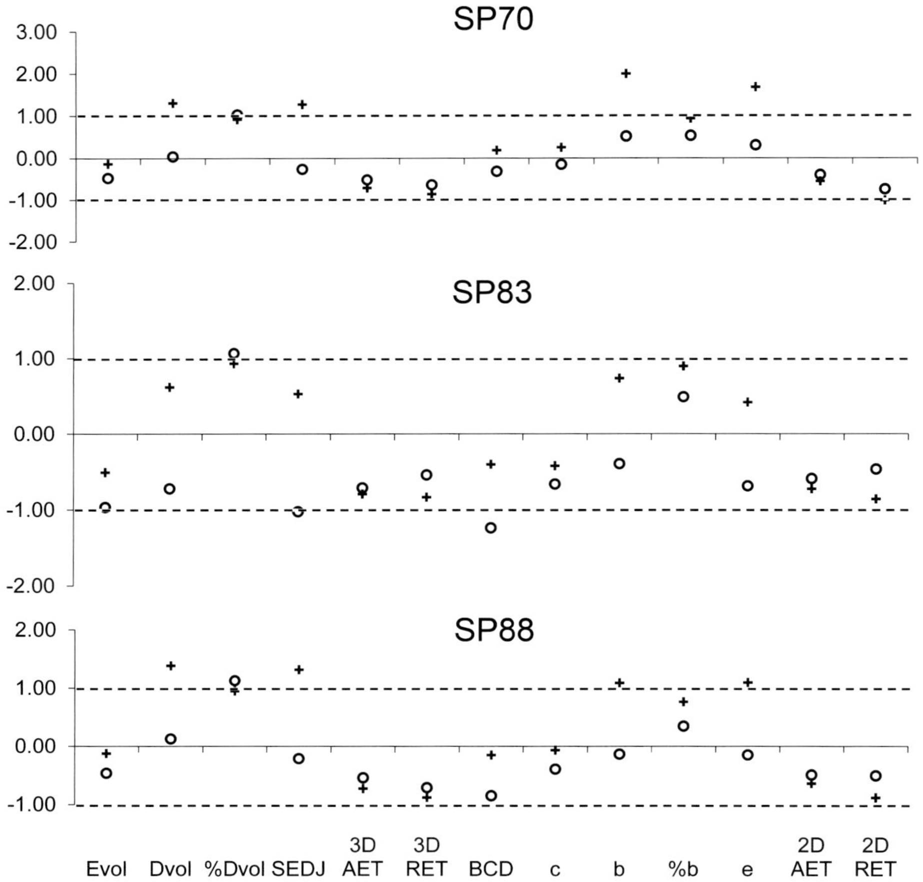

FIG. 9.4. Adjusted Z-scores of 3D and 2D variables assessed in the Palomas 70, 83, and 88 dm$_2$ crowns and compared to the variation expressed by Neandertals (*circles*) and recent humans (*plus signs*). The full line passing through zero represents the mean, and the dotted lines correspond to the estimated 95% limit of variation expressed for the two comparative samples. See "Materials and Methods" in chapter 9 for the definitions of the variables.

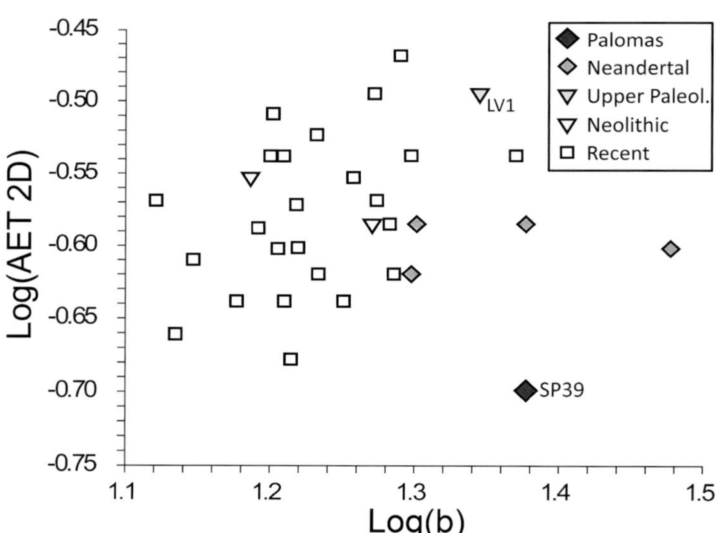

FIG. 9.5. Plot of 2D AET (log) against coronal dentine and pulp area (log) for the Palomas 39 di^1 compared to Neandertals from Engis (Engis 2), Krapina (Kr-d21), Roc de Marsal (RdM1), Spy (Spy VI), the Upper Paleolithic individual from Lagar Velho (LV1), Neolithic humans from Gurgy (NH), and recent humans (RH).

AET against the log of coronal dentine and pulp area (Fig. 9.6). The very low enamel thicknesses reported for the Palomas 25 dm$_1$, particularly in 3D, is partially linked to its bad preservation state and the presence of a carious lesion (Plate 6.24; chapter 12).

On intratooth comparisons, Palomas deciduous teeth are thinly enameled on the whole crown, as are other Neandertal teeth (see Plate 9.3 providing maps of topographic thickness variation at the outer enamel surface in the Palomas 70 dm$_2$ compared to Roc de Marsal 1 and the recent human mean distribution). Maximum differences between the Neandertal specimens and the recent human condition are shown on the buccodistal aspect of the hypoconulid, which is particularly thick in the modern specimen. These distinct patterns between Neandertals and modern humans may reflect differences in bite force magnitudes exerted among cusps during chewing (see, for example, Kay and Hiiemae 1974; Kay 1975; Molnar and Ward 1977; Macho and Spears 1999; Le Luyer et al. 2014 for functional significance of intratooth variation in enamel thickness).

Permanent Teeth

Results of the 3D dental tissue proportions and enamel thickness comparative analysis in the selected permanent maxillary and mandibular teeth are shown in Tables 9.6 to 9.10, respectively. Those of the 2D analysis are shown in Tables 9.11 to 9.15.

As with the deciduous teeth, the permanent teeth from Palomas fit the Neandertal range of variation. All teeth but the upper incisors have enamel volume/surface area ratios similar to those of modern humans, in conjunction with larger dentine volume/surface values and EDJ area/length ratios than in modern humans, including the Middle Paleolithic and Neolithic modern humans from Qafzeh and Gurgy as well as the Aterian humans. Thus, the Palomas permanent teeth are also characterized by low AET and RET values, as shown in the box plots of 3D data for M³ and M$_2$ (Figs. 9.7 and 9.8), even if the recent human variation largely encompasses the variation of the other groups for some tooth types, as was previously noted by Olejniczak et al. (2008a). Also, the Middle Pleistocene data available in 3D (from Tighenif) tend to have large dentine volumes and low AET and RET values, partially linked with the presence of a moderate degree of occlusal wear (Zanolli and Mazurier 2013), whereas the 2D data (from Mauer, Tighenif, Thomas Quarry, and Steinheim) show large enamel and dentine areas and relatively high AET and RET values (Smith et al. 2012; Zanolli et al. 2014).

In this context, as for the di¹, the Palomas upper permanent incisors are distinguished from the rest of the Palomas sample by having a higher proportion of enamel volume to surface area than in the modern samples, resulting in comparable or even slightly higher 2D and 3D AET (Tables 9.6 and 9.11). Even if functional interpretations have been suggested to explain this pattern previously observed on 2D slices (Smith et al. 2012),

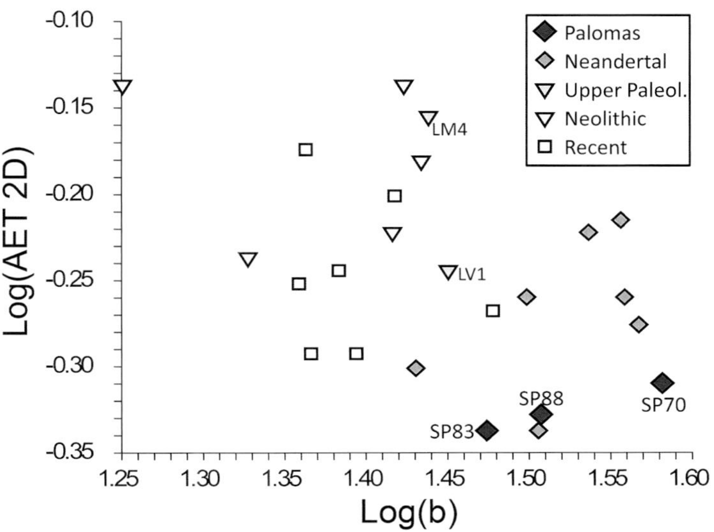

FIG. 9.6. Plot of 2D AET (log) against coronal dentine and pulp area (log) for the Palomas 70, 83, and 88 dm$_2$ compared to Neandertals (Neand) from Devil's Tower (DT1), Krapina (Kr-d62, and Kr-d68), La Chaise Abri Suard (S14–5, S37, and S42), Roc de Marsal (RdM1), the Upper Paleolithic individuals from Lagar Velho (LV1) and La Madeleine (LM4), Neolithic humans from Gurgy, and recent humans (RH).

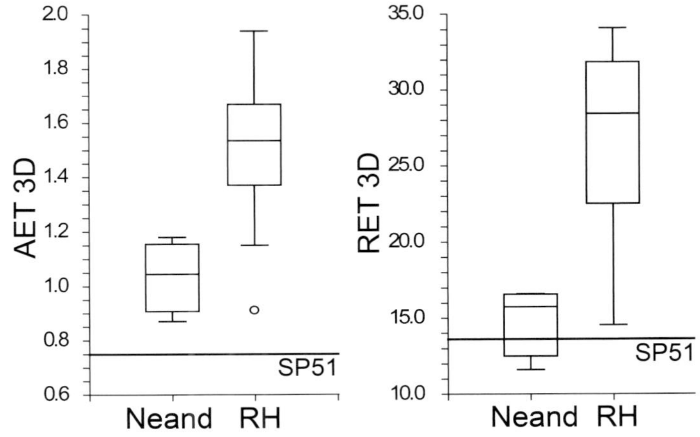

FIG. 9.7. Box plots of 3D AET and RET in the Palomas 51 M³ compared to Neandertals (Neand) and recent humans (RH).

future investigations are needed to unlock the genetically and/or functionally related factors sustaining these observations.

Adjusted Z-scores for M³ and M$_2$ (Figs. 9.9 and 9.10) confirm the proximity of Palomas permanent teeth to those of Neandertals in all 3D

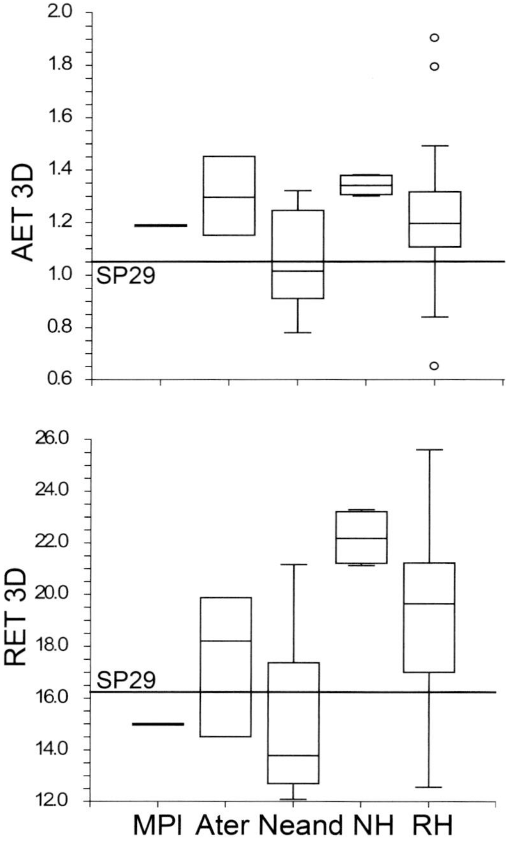

FIG. 9.8. Box plots of 3D AET and RET in the Palomas 29 M₂ compared to Neandertals (Neand), the Middle Pleistocene Tighenif 2 (MPI), Aterian (Ater), Neolithic humans (NH), and recent humans (RH). Among the recent humans, a high outlier for AET (2.30) and two high outliers for RET (30.68 and 40.71) are not shown.

and 2D variables. However, the data from Palomas again substantially extend the previously known ranges of variation in Neandertal teeth, as illustrated by the Palomas 51 M³, a tooth showing low ratios of enamel, dentine, and pulp volumes to surface areas, low proportions of EDJ surface area to length, and low 2D and 3D AET and RET values (Figs. 9.7 and 9.9). Furthermore, substantial variation is observed within the Palomas sample for some tooth positions, as is illustrated by the variation among the four lower canines or among the P_3s, with the bulkier teeth (Palomas 82 and 45, respectively, for the C_1 and P_3) showing the lower 2D and 3D enamel thickness values (Tables 9.8 and 9.13 for the C_1; Tables 9.9 and 9.14 for the P_3).

As with the deciduous teeth, Palomas permanent teeth are thin-enameled on the whole crown, as are other Neandertal teeth (see Plates 9.4 and 9.5, providing maps of topographic thickness variation at the outer enamel surface in the Palomas 45 and 76 P_3s compared to Oliveira 9 and a recent human mean distribution and in the Palomas 29 M_2 compared to Regourdou 1 and a recent human mean distribution, respectively). Again, maximum differences between the Neandertal specimens and the recent human condition are shown on the buccal face of the crowns, which are particularly thick in the modern specimen towards the distal

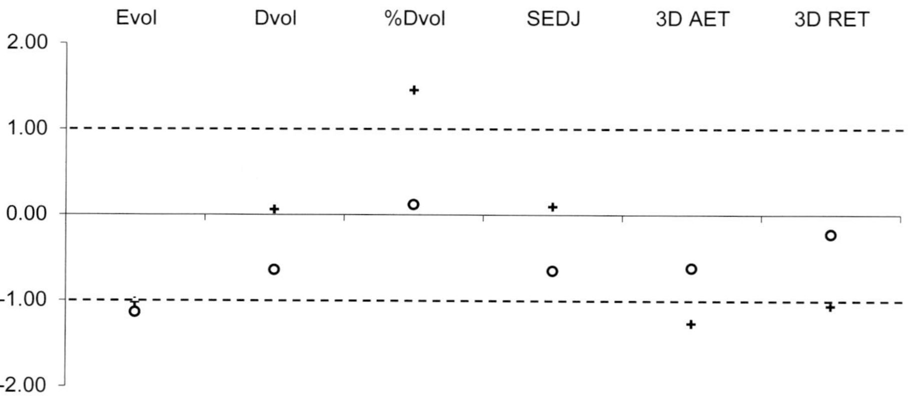

FIG. 9.9. Adjusted Z-scores of 3D variables assessed in the Palomas 51 M³ crown and compared to the variation expressed by Neandertals (*circles*) and recent humans (*plus signs*). The solid line passing through zero represents the mean, and the dotted lines correspond to the estimated 95% limit of variation expressed for the two comparative samples. See "Materials and Methods" in chapter 9 for the definitions of the variables.

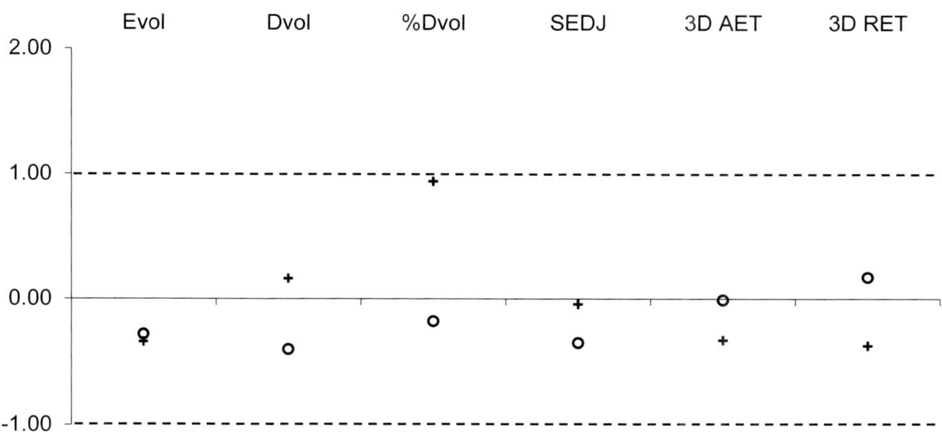

FIG. 9.10. Adjusted Z-scores of 3D variables assessed in the Palomas 29 M_2 crown and compared to the variation expressed by Neandertals (*circles*) and recent humans (*plus signs*). The solid line passing through zero represents the mean, and the dotted lines correspond to the estimated 95% limit of variation expressed for the two comparative samples. See "Materials and Methods," in chapter 9 for the definitions of the variables.

region, reflecting the higher solicitation of the buccal cusps of mandibular teeth during chewing (Molnar and Ward 1977; see Le Luyer et al. 2014 for a review).

Summary

The micro-CT–based survey of the Palomas teeth increases by two times or sometimes three times the number of Neandertal specimens known so far for 2D dental tissue proportions and enamel thickness. Most important, it provides previously unreported evidence on 3D dental tissue proportions and enamel thickness in Neandertal incisors and canines. It also offers the unique opportunity to study the variation of these traits in a constrained chronospatial context.

As a whole, the Palomas internal tooth structure aligns with the Neandertal range of variation, for both the deciduous and the permanent dentition and for all tooth positions. Notably, as shown on permanent teeth of other Neandertals, the deciduous and permanent upper incisors from Palomas also show a different signal from the rest of the dentition. However, the Palomas data extend the Neandertal variation known so far for several teeth. They also provide substantial variation within tooth types, suggesting that the intra- and interpopulation variation in Neandertal internal dental structure is far from being documented.

TABLE 9.1. Inventory of the human dental remains from the Sima de las Palomas that have been studied for enamel thickness and tissue proportions

SP NO.	TOOTH	SIDE	MATURATIONAL STAGE[a]	WEAR DEGREE[b]	ENAMEL PRESERVATION QUALITY
39	di^1	left	G	1/2	poor; loss at the cervix on the labial and lingual faces
71	dm^1	left	G	1	intermediate; covered with hard matrix
36	dm^2	left	D	1	good
95	dc_1	left	N/A	2	intermediate; eroded, several pits
25	dm_1	right	N/A	3	bad; carious lesion and several pits
70	dm_2	right	H	1	good; some pits
83	dm_2	right	H	3/4	good; eroded
88	dm_2	left	H	2/3	good; eroded
34	I^1	left	G	2	intermediate; several pits
37	I^2	right	E	1	very good
48	I^2	right	H	3	poor; eroded, and loss on the distal face
35	C^1	left	H	3	poor; several pits and a marked hypoplasia
74	C^1	left	G	1	intermediate; eroded, several pits
53	P^3	left	H	2	intermediate; covered with hard matrix
60	P^3	right	G	1	intermediate; buccal face covered with hard matrix
68	P^3	right	H	1	intermediate; covered with hard matrix
68	P^4	right	H	2/3	intermediate; covered with hard matrix
94	P^4	left	N/A	1	good; several pits
51	M^3	right	H	1	good; occlusal and distal faces covered with matrix
89	I_2	right	N/A	1	very good
18	C_1	left	H	2/3	good; several pits
26	C_1	right	G	1	good; several pits, and vertical fracture
54	C_1	right	N/A	2/3	bad; eroded, several pits, labial face covered with matrix
82	C_1	left	H	1/2	bad; several important pits
45	P_3	right	G	1	very good
76	P_3	right	N/A	1	good; several pits, and vertical fracture
57	P_4	right	N/A	1	good
87	P_4	left	G	1	intermediate; eroded, several pits, hypoplasia
29	M_2	right	H	2	intermediate; eroded, several pits
50	M_3	right	N/A	1	intermediate; occlusal face covered with matrix
58	M_3	left	N/A	1	intermediate; occlusal face covered with matrix

[a]Scored following Demirjian et al. (1973); see also Moorrees et al. (1963a, 1963b) calcification scores in chapter 6.
[b]Scored following Smith (1984).

TABLE 9.2. 3D dental tissue proportions and enamel thickness values for Palomas maxillary deciduous teeth compared to those for Middle Pleistocene, Neandertal, Early and Mid Upper Paleolithic, Neolithic, and recent humans

TOOTH	SAMPLE			EVOL (MM^3)	DVOL (MM^3)	%DVOL	SEDJ (MM^2)	3D AET (MM)	3D RET
di^1	Palomas 39			24.86	107.91	81.28	102.47	0.24	5.10
	Neand, n=4								
		Mean (SD)		29.34 (6.58)	102.12 (21.22)	77.71 (0.95)	102.19 (13.44)	0.28 (0.03)	6.11 (0.33)
			Range	24.24–38.98	87.27–133.34	76.65–78.90	93.21–122.11	0.25–0.32	5.67–6.43
	E/MUP: LV1 right			28.78	87.74	75.30	80.83	0.36	8.01
	Neolithic: GLN05-258-16			14.94	67.98	81.99	56.99	0.26	6.42
	Neolithic: GLN04-244-30			16.20	58.42	78.29	59.41	0.27	7.03
	RH, n=24								
		Mean (SD)		22.27 (5.18)	53.44 (9.52)	70.72 (3.12)	72.09 (10.69)	0.31 (0.04)	8.19 (1.11)
			Range	13.93–35.07	38.43–78.70	64.54–76.50	57.53–93.69	0.24–0.41	6.80–10.71
dm^1	Palomas 71			51.17	147.93	74.30	133.35	0.38	7.26
	MPl: Tighenif right			42.79	178.25	81.00	152.41	0.28	4.99
	Neand: RdM1 left			50.60	126.76	71.47	120.25	0.42	8.38
	Neolithic, n=3								
		Mean (SD)		48.73 (13.14)	108.64 (13.94)	69.38 (4.12)	102.17 (20.69)	0.48 (0.08)	9.98 (1.68)
			Range	34.40–60.21	93.45–120.87	64.96–73.10	80.05–121.05	0.43–0.57	8.62–11.86
	RH, n=3								
		Mean (SD)		67.90 (6.84)	99.45 (9.87)	59.40 (3.22)	103.67 (6.33)	0.65 (0.05)	14.17 (1.48)
			Range	60.63–74.20	88.58–107.87	56.26–62.70	98.44–110.71	0.60–0.70	12.74–15.69
dm^2	Palomas 36			180.22	303.27	62.73	212.96	0.85	12.60
	Neand, n=3								
		Mean (SD)		131.71 (40.96)	244.39 (45.42)	65.37 (2.66)	193.84 (16.77)	0.67 (0.15)	10.70 (1.66)
			Range	107.69–179.01	214.11–296.62	62.36–67.38	182.14–213.05	0.58–0.84	9.54–12.60
	Neolithic, n=3								
		Mean (SD)		124.50 (12.22)	187.43 (6.68)	60.13 (3.15)	160.97 (7.67)	0.78 (0.11)	13.58 (2.06)
			Range	114.02–137.92	181.15–194.45	56.77–63.04	152.91–168.19	0.70–0.90	12.16–15.94
	RH, n=9								
		Mean (SD)		116.81 (23.11)	174.39 (26.29)	60.03 (1.78)	149.87 (17.17)	0.77 (0.07)	13.87 (0.81)
			Range	94.75–165.65	138.40–231.90	57.92–63.55	132.45–189.85	0.71–0.88	12.78–15.29

Note: Evol = volume of the enamel cap (mm^3); Dvol = volume of the dentine and pulp (mm^3); %Dvol = percentage of crown volume that is dentine and pulp; SEDJ = surface area of the enamel-dentine junction (mm^2); 3D AET = average enamel thickness (mm); 3D RET = scale-free relative enamel thickness; MPl = Middle Pleistocene; Neand = Neandertal; E/MUP = Early and Mid Upper Paleolithic; RH = recent humans; LV1 = Lagar Velho 1; RdM1 = Roc de Marsal 1.

TABLE 9.3. 3D dental tissue proportions and enamel thickness values for Palomas mandibular deciduous teeth compared to those for Neandertal, Early and Mid Upper Paleolithic, Late Upper Paleolithic, Neolithic, and recent humans

TOOTH	SAMPLE	EVOL (MM³)	DVOL (MM³)	%DVOL	SEDJ (MM²)	3D AET (MM)	3D RET
dc₁	Palomas 95	28.85	82.03	73.98	92.03	0.31	7.21
	Neand, n=4						
	Mean (SD)	32.68 (5.34)	89.69 (15.66)	73.23 (2.60)	92.93 (9.16)	0.35 (0.04)	7.87 (0.97)
	Range	25.52–37.86	76.87–110.91	70.86–75.84	85.51–104.34	0.30–0.39	7.02–8.73
	E/MUP: LV1 right	36.73	68.66	65.15	73.49	0.50	12.21
	LUP: LM4 right	32.87	59.76	64.51	70.76	0.46	11.88
	Neolithic, n=3						
	Mean (SD)	31.99 (4.12)	58.90 (7.39)	64.76 (3.02)	68.77 (5.64)	0.46 (0.03)	11.96 (0.86)
	Range	29.04–36.69	50.62–64.84	62.54–68.20	63.21–74.49	0.44–0.49	10.97–12.50
	RH, n=5						
	Mean (SD)	37.61 (12.70)	53.02 (12.91)	58.86 (3.52)	67.04 (10.26)	0.56 (0.14)	14.85 (3.01)
	Range	24.97–58.08	35.03–71.05	55.02–64.38	49.78–75.42	0.41–0.78	10.71–18.89
dm₁	Palomas 25	(29.67)	(121.31)	(80.35)	(94.64)	(0.31)	(6.33)
	Neand, n=6						
	Mean (SD)	60.59 (8.24)	160.67 (17.24)	72.61 (2.29)	139.13 (12.06)	0.44 (0.05)	8.04 (0.94)
	Range	53.01–73.51	130.71–179.59	69.83–74.98	119.04–153.24	0.39–0.50	7.06–9.14
	LUP: LM4 left	54.96	94.23	63.16	110.53	0.50	10.93
	Neolithic, n=3						
	Mean (SD)	51.53 (6.85)	110.81 (7.12)	68.30 (2.99)	108.65 (3.96)	0.47 (0.06)	9.89 (1.37)
	Range	47.21–59.43	103.58–117.82	65.13–71.07	104.72–112.64	0.43–0.55	8.69–11.39
	RH, n=8						
	Mean (SD)	54.22 (9.70)	87.92 (14.54)	61.85 (2.85)	95.01 (12.02)	0.57 (0.07)	12.91 (1.72)
	Range	43.97–73.73	68.91–114.61	56.98–65.22	73.99–111.55	0.49–0.66	10.78–15.25
dm₂	Palomas 70	98.69	230.25	70.00	180.93	0.55	8.90
	Palomas 83	81.38	191.06	70.13	155.87	0.52	9.07
	Palomas 88	98.94	234.72	70.35	182.54	0.54	8.79
	Neand, n=7						
	Mean (SD)	115.39 (13.47)	228.12 (19.72)	66.46 (1.32)	189.58 (12.59)	0.61 (0.05)	9.95 (0.63)
	Range	94.95–131.16	195.49–244.45	63.96–67.76	170.16–206.93	0.56–0.67	9.49–10.92
	E/MUP: LV1 right	143.74	199.73	58.15	172.11	0.84	14.29
	LUP: LM4 left	125.00	173.20	58.08	152.85	0.82	14.67
	Neolithic, n=5						
	Mean (SD)	88.55 (23.40)	172.55 (38.38)	66.24 (3.43)	145.22 (24.51)	0.60 (0.08)	10.89 (1.44)
	Range	55.19–117.71	126.86–215.92	61.98–71.36	119.79–174.84	0.46–0.67	8.92–12.93
	RH, n=7						
	Mean (SD)	104.79 (17.68)	155.82 (21.79)	59.83 (4.24)	138.07 (12.90)	0.76 (0.11)	14.17 (2.34)
	Range	81.76–127.98	130.14–183.56	52.04–64.89	117.84–151.26	0.63–0.97	11.98–18.93

Note: Evol = volume of the enamel cap (mm³); Dvol = volume of the dentine and pulp (mm³); %Dvol = percentage of crown volume that is dentine and pulp; SEDJ = surface area of the enamel-dentine junction (mm²); 3D AET = average enamel thickness (mm); 3D RET = scale-free relative enamel thickness; MPl = Middle Pleistocene; Neand = Neandertal; E/MUP = Early and Mid Upper Paleolithic; RH = recent humans; LV1 = Lagar Velho 1; LM4 = La Madeleine 4.

TABLE 9.4. 2D dental tissue proportions and enamel thickness values for Palomas maxillary deciduous teeth compared to those for Neandertal, Early and Mid Upper Paleolithic, Neolithic, and recent humans

TOOTH	SAMPLE		BCD (MM)	C (MM²)	B (MM²)	%B	E (MM)	2D AET (MM)	2D RET
di¹	**Palomas 39**		**5.55**	**3.04**	**21.25**	**87.48**	**15.24**	**0.20**	**4.33**
	Neand, n=5								
		Mean (SD)	5.15 (0.40)	3.94 (0.41)	21.08 (3.70)	84.10 (1.52)	14.95 (0.98)	0.26 (0.02)	5.78 (0.59)
		Range	4.84–5.84	3.44–4.50	17.70–26.76	82.94–86.63	13.86–16.27	0.24–0.30	4.91–6.37
	E/MUP: LV1 right		4.88	4.71	19.72	80.72	14.88	0.32	7.13
	Neolithic: GLN05-258-16		4.48	3.32	16.62	83.36	12.59	0.26	6.46
	Neolithic: GLN04-244-30		4.23	3.45	13.69	79.87	12.31	0.28	7.58
	RH, n=24								
		Mean (SD)	4.23 (0.30)	3.50 (0.59)	15.23 (2.03)	81.31 (1.81)	13.18 (0.99)	0.26 (0.03)	6.81 (0.78)
		Range	3.61–4.70	2.54–4.91	11.78–20.89	77.88–84.23	11.64–15.09	0.21–0.34	5.60–8.14
	RH, n=10								
		Mean (SD)	—	3.25 (0.48)	15.63 (1.39)	82.79 (1.21)	13.33 (0.60)	0.24 (0.02)	[6.16]
dm¹	**Palomas 71**		**7.36**	**5.92**	**31.63**	**84.24**	**16.27**	**0.36**	**6.47**
	Neand, n=4								
		Mean (SD)	—	6.88 (0.99)	33.15 (2.97)	82.81 (1.88)	17.49 (0.57)	0.39 (0.05)	6.82 (0.71)
		Range	—	5.74–7.74	29.34–36.57	81.02–85.44	16.77–18.08	0.34–0.44	5.90–7.45
	E/MUP: Dolní Věstonice 36-2		—	10.12	32.87	76.46	18.04	0.56	9.78
	E/MUP: La Rochette		—	8.25	32.9	79.95	16.65	0.50	8.64
	Neolithic, n=3								
		Mean (SD)	6.69 (1.06)	8.76 (0.72)	22.33 (5.21)	71.15 (6.93)	14.47 (2.09)	0.62 (0.14)	13.54 (4.80)
		Range	5.68–7.80	8.25–9.59	16.51–26.54	63.27–76.30	12.49–16.65	0.50–0.77	9.61–18.90
	RH, n=24								
		Mean (SD)	—	7.60 (1.18)	28.32 (2.69)	78.85 (2.35)	16.45 (1.37)	0.46 (0.06)	8.71 (1.18)
		Range	—	6.00–10.11	23.73–35.39	74.52–83.20	14.70–21.30	0.32–0.60	6.08–11.25
	RHᵃ, n=24								
		Mean (SD)	—	8.12 (1.40)	28.01 (3.07)	77.53 (2.10)	16.44 (1.21)	0.49 (0.06)	[9.33]
dm²	**Palomas 36**		**10.56**	**14.78**	**33.97**	**69.68**	**19.81**	**0.75**	**12.80**
	Neand, n=9								
		Mean (SD)	—	11.39 (1.98)	36.73 (5.32)	76.36 (1.96)	18.94 (1.43)	0.60 (0.07)	9.89 (0.82)
		Range	—	8.33–14.42	29.64–44.64	74.17–80.16	16.49–21.36	0.48–0.70	8.74–11.33
	E/MUP: Dolní Věstonice 36-2		—	17.45	36.36	67.57	19.23	0.91	15.05
	E/MUP: La Rochette		—	19.41	38.03	66.21	20.17	0.96	15.60
	Neolithic, n=3								
		Mean (SD)	7.68 (1.59)	13.78 (0.40)	27.17 (0.66)	66.35 (1.01)	16.81 (0.40)	0.82 (0.04)	15.74 (0.91)
		Range	6.58–9.51	13.35–14.15	26.43–17.72	65.63–67.50	16.46–17.24	0.77–0.85	14.70–16.36
	RH, n=40								
		Mean (SD)	—	12.64 (1.69)	29.74 (2.93)	71.41 (1.40)	18.10 (1.01)	0.66 (0.02)	12.06 (0.82)
		Range	—	9.60–16.67	26.74–37.25	66.53–77.08	16.49–20.51	0.52–0.86	9.16–15.36
	RHᵃ, n=22								
		Mean (SD)	—	12.21 (2.29)	33.01 (4.00)	72.99 (1.78)	17.95 (1.86)	0.68 (0.13)	[11.84]
	RHᵇ, n=10								
		Mean (SD)	8.9 (0.5)	16.10 (3.00)	36.10 (4.00)	[69.16]	[19.14]	[0.84]	14.00 (2.00)
		Range	8.30–9.90	10.80–21.10	32.20–43.00	—	—	—	10.00–16.00

Note: BCD = bicervical diameter (mm); c = enamel area (mm²); b = dentine and pulp area (mm²); %b = percentage of the crown that is dentine and pulp; e = length of the enamel-dentine junction (EDJ, mm); 2D AET = average enamel thickness; 2D RET = scale-free relative enamel thickness; MPI = Middle Pleistocene; Neand = Neandertal; E/MUP = Early and Mid Upper Paleolithic; RH = recent humans; LV1 = Lagar Velho 1.

ᵃFrom Mahoney (2013). Only mean and SD values are provided. Mean values calculated from published measurements are in square brackets.

ᵇFrom Grine (2005). Mean values calculated from published measurements are in square brackets.

TABLE 9.5. 2D dental tissue proportions and enamel thickness values for Palomas maxillary deciduous teeth compared to those for Neandertal, Early and Mid Upper Paleolithic, Late Upper Paleolithic, Neolithic, and recent humans

TOOTH	SAMPLE		BCD (MM)	C (MM^2)	B (MM^2)	%B	E (MM)	2D AET (MM)	2D RET
d$_c$	Palomas 95		(5.96)	(5.11)	(27.51)	(84.33)	(16.04)	(0.32)	(6.07)
	Neand, n=4								
		Mean (SD)	5.63 (0.69)	5.45 (1.36)	24.62 (3.46)	82.08 (1.99)	16.35 (0.92)	0.33 (0.07)	6.63 (0.94)
		Range	4.77–6.45	3.57–6.82	20.22–28.37	80.62–84.99	15.04–17.20	0.24–0.40	5.28–7.44
	E/MUP: LV1 right		5.13	5.98	18.32	75.40	14.27	0.42	9.78
	LUP: LM4 right		5.19	7.07	21.18	74.98	15.31	0.46	10.03
	Neolithic, n=3								
		Mean (SD)	4.93 (0.29)	5.73 (0.75)	17.62 (1.85)	75.50 (0.71)	13.15 (0.50)	0.43 (0.04)	10.36 (0.61)
		Range	4.64–5.21	5.03–6.51	15.62–19.25	74.72–76.12	12.57–13.45	0.40–0.49	9.90–11.06
	RH, n=4								
		Mean (SD)	4.68 (0.28)	5.27 (0.55)	16.58 (2.00)	75.84 (1.32)	13.13 (1.18)	0.40 (0.02)	9.90 (0.89)
		Range	4.42–5.07	4.65–5.95	13.75–18.20	74.72–77.74	11.54–14.20	0.37–0.42	8.74–10.87
	RH^a, n=20								
		Mean (SD)	—	4.92 (1.02)	18.52 (2.53)	79.01 (1.51)	14.05 (1.20)	0.35 (0.05)	[8.13]
dm$_1$	Palomas 25		(5.89)	(4.42)	(26.38)	(85.64)	(14.42)	(0.31)	(5.97)
	Neand, n=6								
		Mean (SD)	6.50 (0.37)	7.15 (0.42)	32.74 (1.79)	82.08 (0.63)	17.30 (0.74)	0.41 (0.01)	7.22 (0.24)
		Range	5.96–7.01	6.45–7.64	30.25–34.75	80.94–82.69	15.86–17.86	0.40–0.43	6.98–7.53
	LUP: LM4 left		6.28	8.37	28.04	77.02	16.76	0.50	9.43
	Neolithic, n=3								
		Mean (SD)	6.31 (0.50)	7.29 (1.05)	25.61 (0.26)	77.90 (2.31)	15.82 (0.42)	0.46 (0.06)	9.09 (1.06)
		Range	5.87–6.85	6.46–8.47	25.31–25.80	75.28–79.67	15.45–16.27	0.41–0.52	8.16–10.25
	RH, n=8								
		Mean (SD)	5.66 (0.48)	6.57 (1.27)	21.87 (2.86)	76.93 (2.91)	14.19 (1.37)	0.46 (0.07)	9.94 (1.63)
		Range	4.89–6.22	4.81–8.38	17.06–26.59	73.11–80.77	11.72–15.93	0.36–0.57	8.07–12.42
	RH^a, n=21								
		Mean (SD)	—	6.22 (0.97)	23.48 (3.01)	79.05 (1.86)	14.82 (0.81)	0.42 (0.05)	[8.66]
dm$_2$	Palomas 70		6.99	9.77	38.16	79.62	20.09	0.49	7.87
	Palomas 83		6.36	7.75	29.78	79.35	16.82	0.46	8.44
	Palomas 88		6.63	8.81	32.15	78.49	18.59	0.47	8.36
	Neand, n=7								
		Mean (SD)	7.21 (0.26)	10.37 (1.52)	33.42 (3.53)	76.33 (2.38)	19.09 (1.26)	0.54 (0.05)	9.38 (0.77)
		Range	6.89–7.60	7.61–11.90	26.94–36.92	74.32–80.80	16.67–20.81	0.46–0.61	8.07–10.24
	E/MUP: LV1 right		6.82	13.89	28.20	66.99	24.21	0.57	10.81
	LUP: LM4 left		6.47	12.16	27.43	69.29	17.48	0.70	13.28
	Neolithic, n=5								
		Mean (SD)	6.55 (0.40)	10.07 (1.51)	23.76 (4.06)	70.08 (3.15)	15.22 (1.72)	0.66 (0.07)	13.75 (2.19)
		Range	6.04–7.14	7.99–12.05	17.81–27.15	65.17–72.68	13.00–16.58	0.58–0.73	11.83–17.35
	RH, n=7								
		Mean (SD)	6.79 (0.41)	9.01 (1.15)	24.89 (2.55)	73.41 (2.55)	15.74 (0.99)	0.57 (0.06)	11.51 (1.37)
		Range	6.07–7.23	7.66–10.49	22.84–30.04	68.88–76.07	14.38–17.36	0.51–0.67	9.93–13.90

TABLE 9.5. **Continued.**

TOOTH	SAMPLE		BCD (MM)	C (MM^2)	B (MM^2)	%B	E (MM)	2D AET (MM)	2D RET
	RH[a], n=22								
		Mean	—	10.50 (1.95)	28.29 (3.74)	72.93 (1.83)	16.88 (1.33)	0.62 (0.11)	[11.69]
	RH[b], n=10								
		Mean (SD)	6.7 (0.70)	12.50 (1.70)	28.50 (3.90)	[69.51]	[16.72]	[0.75]	14.00 (2.00)
		Range	5.40–7.50	10.40–16.50	24.40–35.20	—	—	—	11.00–17.00

Note: BCD = bicervical diameter (mm); c = enamel area (mm²); b = dentine and pulp area (mm²); %b = percentage of the crown that is dentine and pulp; e = length of the enamel-dentine junction (EDJ, mm); 2D AET = average enamel thickness; 2D RET = scale-free relative enamel thickness; MPl = Middle Pleistocene; Neand = Neandertal; E/MUP = Early and Mid Upper Paleolithic; LUP = Late Upper Paleolithic; RH = recent humans; LV1 = Lagar Velho 1; LM4 = La Madeleine 4.
[a]From Mahoney (2010, 2013). Only mean and SD values are provided. Mean values calculated from published measurements are in square brackets.
[b]From Grine (2005). Mean values calculated from published measurements are in square brackets.

TABLE 9.6. 3D dental tissue proportions and enamel thickness values for Palomas anterior permanent maxillary teeth compared to those for Neandertal, Neolithic, and recent humans

TOOTH	SAMPLE		EVOL (MM³)	DVOL (MM³)	%DVOL	SEDJ (MM²)	3D AET (MM)	3D RET
I¹	Palomas 34		127.61	186.45	59.37	197.47	0.65	11.31
	Neolithic, n=7							
		Mean (SD)	96.59 (23.41)	143.80 (23.39)	60.13 (5.49)	150.46 (23.15)	0.64 (0.09)	12.19 (1.83)
		Range	53.68–128.85	113.87–178.86	54.39–71.08	113.59–172.52	0.47–0.75	9.28–14.89
I²	Palomas 37		117.53	150.34	56.12	155.37	0.76	14.23
	Palomas 48		92.18	148.26	61.66	144.15	0.64	12.08
	Neand, n=3							
		Mean (SD)	153.00 (0.59)	184.23 (16.82)	54.56 (2.27)	188.54 (5.97)	0.81 (0.03)	14.29 (0.78)
		Range	152.65–153.68	166.76–200.31	52.21–56.75	183.00–194.86	0.78–0.84	13.39–14.77
	Neolithic, n=7							
		Mean (SD)	67.58 (16.64)	75.50 (14.22)	53.11 (4.64)	97.39 (16.13)	0.69 (0.08)	16.31 (1.98)
		Range	39.39–90.15	59.71–98.09	46.50–62.04	73.91–124.10	0.53–0.81	13.30–19.87
	RH, n=5							
		Mean (SD)	60.57 (7.17)	70.61 (13.77)	53.62 (1.96)	95.51 (9.77)	0.63 (0.03)	15.42 (0.97)
		Range	53.44–72.28	59.18–94.28	51.54–56.61	85.81–111.10	0.60–0.67	14.29–16.52
C¹	Palomas 35		77.67	134.96	63.47	135.97	0.57	11.14
	Palomas 74		168.36	230.79	57.82	187.97	0.90	14.60
	Neolithic, n=4							
		Mean (SD)	119.69 (11.23)	136.32 (15.26)	53.20 (2.39)	127.86 (3.87)	0.94 (0.08)	18.20 (1.43)
		Range	105.46–132.73	115.65–151.02	50.42–56.00	122.56–131.61	0.86–1.04	16.93–20.25

Note: Evol = volume of the enamel cap (mm³); Dvol = volume of the dentine and pulp (mm³); %Dvol = percentage of crown volume that is dentine and pulp; SEDJ = surface area of the enamel-dentine junction (mm²); 3D AET = average enamel thickness (mm); 3D RET = scale-free relative enamel thickness; Neand = Neandertal; RH = recent humans.

TABLE 9.7. 3D dental tissue proportions and enamel thickness values for Palomas posterior permanent maxillary teeth compared to those for Neandertal, Neolithic, and recent humans

TOOTH	SAMPLE		EVOL (MM³)	DVOL (MM³)	%DVOL	SEDJ (MM²)	3D AET (MM)	3D RET
P³	Palomas 53		111.89	152.78	57.72	133.86	0.84	15.64
	Palomas 60		164.16	197.09	54.56	168.28	0.98	16.76
	Palomas 68		85.15	111.97	56.80	112.80	0.75	15.66
	Neolithic, n=5							
		Mean (SD)	128.90 (8.38)	110.97 (19.10)	46.04 (3.41)	113.34 (12.48)	1.14 (0.10)	24.01 (3.35)
		Range	119.10–141.78	88.37–140.95	41.00–49.85	98.68–132.46	1.06–1.29	20.57–28.93
P⁴	Palomas 68		83.94	96.16	53.39	102.55	0.82	17.87
	Palomas 94		149.58	158.12	51.39	142.51	1.05	19.41
	Neolithic, n=3							
		Mean (SD)	135.40 (17.33)	110.44 (16.46)	44.88 (1.13)	111.70 (9.86)	1.21 (0.10)	25.30 (1.89)
		Range	118.26–152.91	97.84–129.07	43.61–45.77	104.52–122.94	1.10–1.29	23.84–27.43
M³	Palomas 51		115.51	186.76	61.79	153.57	0.75	13.16
	Neand, n=4							
		Mean (SD)	223.44 (26.70)	339.87 (67.03)	60.05 (4.07)	217.28 (27.50)	1.04 (0.13)	14.97 (2.33)
		Range	194.84–259.02	271.89–421.54	57.32–66.11	181.20–248.17	0.87–1.18	11.61–16.60
	RH, n=14							
		Mean (SD)	206.41 (39.85)	180.00 (50.80)	46.10 (4.83)	143.57 (46.41)	1.50 (0.27)	27.20 (5.94)
		Range	142.22–317.72	103.53–288.06	37.83–52.10	82.48–246.78	0.91–1.94	14.54–34.10

Note: Evol = volume of the enamel cap (mm³); Dvol = volume of the dentine and pulp (mm³); %Dvol = percentage of crown volume that is dentine and pulp; SEDJ = surface area of the enamel-dentine junction (mm²); 3D AET = average enamel thickness (mm); 3D RET = scale-free relative enamel thickness; Neand = Neandertal; RH = recent humans.

TABLE 9.8. 3D dental tissue proportions and enamel thickness values for Palomas anterior permanent mandibular teeth compared to those for Middle Pleistocene, Neolithic, and recent humans

TOOTH	SAMPLE		EVOL (MM³)	DVOL (MM³)	%DVOL	SEDJ (MM²)	3D AET (MM)	3D RET
I₂	Palomas 89		62.85	81.43	56.44	106.82	0.59	13.57
	MPI: Tighenif isolated tooth		45.2	90.77	66.76	106.42	0.42	9.45
	Neolithic, n=6							
		Mean (SD)	54.78 (9.33)	68.65 (7.39)	55.77 (3.86)	91.27 (9.35)	0.60 (0.05)	14.62 (1.35)
		Range	42.67–65.16	55.16–74.33	52.74–62.71	77.11–104.17	0.51–0.65	12.23–16.20
	RH: UdP2		47.58	48.11	50.28	78.03	0.61	16.76
C₁	Palomas 18		88.74	134.71	60.29	135.97	0.65	12.73
	Palomas 26		95.94	132.05	57.92	133.62	0.72	14.10
	Palomas 54		68.76	115.66	62.71	112.78	0.61	12.51
	Palomas 82		99.39	168.43	62.89	154.67	0.64	11.64
	Neolithic, n=5							
		Mean (SD)	92.84 (15.77)	116.46 (24.32)	55.49 (1.28)	121.78 (11.34)	0.76 (0.06)	15.58 (0.45)
		Range	72.73–108.73	86.19–141.37	54.24–57.16	108.00–131.99	0.67–0.82	15.10–16.21
	RH, n=6							
		Mean (SD)	70.72 (15.59)	89.69 (23.77)	55.70 (3.35)	106.22 (22.55)	0.67 (0.09)	15.15 (2.42)
		Range	54.19–93.75	69.00–130.58	51.66–60.34	82.58–145.51	0.56–0.82	11.63–17.69

Note: Evol = volume of the enamel cap (mm³); Dvol = volume of the dentine and pulp (mm³); %Dvol = percentage of crown volume that is dentine and pulp; SEDJ = surface area of the enamel-dentine junction (mm²); 3D AET = average enamel thickness (mm); 3D RET = scale-free relative enamel thickness; MPl = Middle Pleistocene; RH = recent humans.

TABLE 9.9. 3D dental tissue proportions and enamel thickness values for Palomas permanent mandibular premolars compared to those for Neandertal, Neolithic, and recent humans

TOOTH	SAMPLE	EVOL (MM³)	DVOL (MM³)	%DVOL	SEDJ (MM²)	3D AET (MM)	3D RET
P$_3$	**Palomas 45**	**105.01**	**130.44**	**55.40**	**118.89**	**0.88**	**17.42**
	Palomas 76	**93.00**	**99.18**	**51.61**	**98.45**	**0.94**	**20.41**
	MPl: Tighenif 2	101.96	202.75	66.54	165.53	0.62	10.48
	Neand: Oliveira 9	102.26	155.19	60.28	140.80	0.73	13.52
	Neand: Regourdou 1 right	86.01	135.15	61.11	117.85	0.73	14.22
	Neolithic: GLN04-201	108.80	109.53	50.17	108.29	1.00	21.00
	Neolithic: GLN06-215B	83.82	85.32	50.44	92.13	0.91	20.67
	RH: UdP2	104.63	89.52	46.11	98.11	1.07	23.84
P$_4$	**Palomas 57**	**117.88**	**135.65**	**53.50**	**121.16**	**0.97**	**18.94**
	Palomas 87	**120.02**	**121.58**	**50.32**	**112.20**	**1.07**	**21.59**
	MPl: Tighenif 2	169.43	224.49	56.99	190.88	0.89	14.60
	Neand: La Chaise BD9	82.81	135.43	62.06	128.10	0.65	12.59
	Neand: Regourdou 1 left	87.59	112.41	56.21	111.31	0.79	16.30
	Neolithic: GLN04-201	129.62	121.86	48.46	112.98	1.15	23.14
	Neolithic: GLN06-215B	105.37	94.26	47.22	99.34	1.06	23.31
	RH: UdP2	110.52	78.20	41.44	86.64	1.28	29.83

Note: Evol = volume of the enamel cap (mm³); Dvol = volume of the dentine and pulp (mm³); %Dvol = percentage of crown volume that is dentine and pulp; SEDJ = surface area of the enamel-dentine junction (mm²); 3D AET = average enamel thickness (mm); RET = scale-free relative enamel thickness; MPl = Middle Pleistocene; Neand = Neandertal; RH = recent humans.

TABLE 9.10. 3D dental tissue proportions and enamel thickness values for Palomas permanent mandibular molars compared to those for Middle Pleistocene, Neandertal, Aterian, Neolithic, and recent humans

TOOTH	SAMPLE	EVOL (MM³)	DVOL (MM³)	%DVOL	SEDJ (MM²)	3D AET (MM)	3D RET
M$_2$	Palomas 29	199.89	271.29	57.58	190.16	1.05	16.24
	MPl: Tighenif 2	373.06	502.42	57.39	312.54	1.19	15.01
	Neand, n=9						
	Mean (SD)	244.45 (64.89)	370.72 (101.72)	59.90 (5.32)	230.68 (47.41)	1.06 (0.19)	14.97 (3.03)
	Range	146.30–308.44	250.03–519.34	50.17–67.53	182.48–302.91	0.78–1.32	12.08–21.15
	Aterian, n=3						
	Mean (SD)	342.97 (32.95)	415.75 (77.03)	54.58 (4.02)	265.85 (34.58)	1.30 (0.15)	17.54 (2.74)
	Range	306.74–371.16	358.26–503.27	50.96–58.91	236.69–304.06	1.15–1.45	14.51–19.86
	Neolithic, n=4						
	Mean (SD)	234.10 (31.29)	224.08 (42.19)	48.76 (1.84)	174.13 (19.08)	1.34 (0.04)	22.20 (1.07)
	Range	198.38–272.96	181.59–282.10	46.73–50.82	150.57–197.16	1.30–1.38	21.11–23.27
	RH, n=30						
	Mean (SD)	237.50 (52.91)	248.98 (66.85)	50.93 (3.41)	194.50 (47.96)	1.26 (0.31)	20.31 (5.23)
	Range	126.23–361.91	135.34–426.01	42.66–57.32	95.40–285.35	0.65–2.30	12.56–40.71
	RH[a], n=13						
	Mean (SD)	252.40 (36.10)	272.10 (59.00)	[52.00]	194.80 (27.10)	1.30 (0.13)	[20.00]
M$_3$	Palomas 50	220.36	228.03	50.86	172.01	1.28	20.97
	Palomas 58	235.47	284.69	54.73	191.40	1.23	18.70
	MPl: Tighenif 2	372.24	371.53	49.95	255.48	1.46	20.27
	Neand, n=10						
	Mean (SD)	207.95 (45.40)	303.72 (73.19)	59.26 (6.11)	190.20 (22.51)	1.10 (0.22)	16.45 (3.51)
	Range	140.28–262.18	244.62–485.90	49.11–68.63	142.85–228.03	0.78–1.41	12.32–22.28
	Aterian: El Haroura	353.65	437.15	55.28	280.31	1.26	16.62
	Aterian: Témara	267.13	262.03	49.52	196.14	1.36	21.28
	RH, n=14						
	Mean (SD)	240.63 (67.12)	242.28 (90.29)	49.43 (3.78)	182.85 (45.89)	1.32 (0.20)	21.63 (3.24)
	Range	144.59–362.14	122.75–405.49	44.71–55.44	121.09–273.48	1.04–1.85	17.78–27.77

Note: Evol = volume of the enamel cap (mm³); Dvol = volume of the dentine and pulp (mm³); %Dvol = percentage of crown volume that is dentine and pulp; SEDJ = surface area of the enamel-dentine junction (mm²); 3D AET = average enamel thickness (mm); RET = scale-free relative enamel thickness; MPl = Middle Pleistocene; Neand = Neandertal; RH = recent humans.

[a]From Kono (2004). Only mean and SD values are provided. Mean values calculated from published measurements are in brackets.

TABLE 9.11. 2D dental tissue proportions and enamel thickness values for Palomas permanent anterior maxillary teeth compared to those for Neandertal, Neolithic, and recent humans

TOOTH	SAMPLE		BCD (MM)	C (MM²)	B (MM²)	%B	E (MM)	2D AET (MM)	2D RET
I¹	**Palomas 34**		**7.03**	**14.54**	**42.70**	**74.59**	**24.19**	**0.60**	**9.20**
	Neand[a], n=5								
		Mean	—	15.61	47.20	75.15	24.71	0.63	9.19
	Neolithic, n=7								
		Mean (SD)	6.14 (0.45)	13.59 (1.52)	31.91 (7.25)	69.65 (3.69)	20.79 (2.33)	0.65 (0.04)	11.81 (1.78)
		Range	5.63–6.89	11.78–15.91	22.75–40.94	64.31–72.94	17.94–23.55	0.60–0.69	10.50–14.55
	RH[a], n=32								
		Mean	—	13.46	32.73	70.86	21.55	0.62	10.91
I²	**Palomas 37**		**7.29**	**15.39**	**35.93**	**70.02**	**22.95**	**0.67**	**11.19**
	Palomas 48		**7.23**	**12.30**	**36.91**	**75.01**	**21.93**	**0.56**	**9.23**
	Neand[a], n=5								
		Mean	—	17.01	41.60	70.98	25.4	0.67	10.45
	Neolithic, n=7								
		Mean (SD)	5.70 (0.38)	12.59 (2.44)	25.36 (2.70)	67.02 (2.60)	18.72 (2.02)	0.67 (0.06)	13.29 (1.05)
		Range	5.22–6.27	9.56–17.36	21.88–28.72	62.33–70.02	16.36–22.42	0.56–0.77	11.87–14.73
	RH[a], n=31								
		Mean	—	12.16	26.78	68.77	19.04	0.64	12.51
C¹	**Palomas 35**		**7.40**	**11.48**	**40.62**	**77.97**	**20.00**	**0.57**	**9.01**
	Palomas 74		**8.93**	**22.52**	**57.03**	**71.69**	**25.07**	**0.90**	**11.90**
	Neand[a], n=4								
		Mean	—	20.25	55.52	73.27	24.28	0.83	11.20
	Neolithic, n=4								
		Mean (SD)	7.56 (0.33)	18.41 (2.13)	37.88 (4.77)	67.24 (1.96)	20.76 (1.36)	0.89 (0.06)	14.44 (1.13)
		Range	7.25–8.03	16.11–21.08	30.74–40.58	65.61–69.83	18.78–21.66	0.83–0.97	13.09–15.47
	RH Asia[b], n=7								
		Mean (SD)	7.82 (0.44)	19.07 (2.78)	43.68 (6.59)	[69.61]	22.42 (2.14)	0.85 (0.11)	[12.87]
	RH Northern Europe[b], n=21								
		Mean (SD)	7.52 (0.56)	18.79 (2.20)	40.85 (6.53)	[68.49]	21.27 (1.88)	0.89 (0.09)	[13.92]
	RH Southern Africa[b], n=11								
		Mean (SD)	7.41 (0.72)	19.06 (3.04)	41.05 (7.53)	[68.29]	20.68 (2.42)	0.92 (0.07)	[14.39]

Note: BCD = bicervical diameter (mm); c = enamel area (mm²); b = dentine and pulp area (mm²); %b = percentage of the crown that is dentine and pulp; e = length of the enamel-dentine junction (EDJ, mm); 2D AET = average enamel thickness; 2D RET = scale-free relative enamel thickness; Neand = Neandertal; RH = recent humans.

[a]From Smith et al. (2008). Only mean values are provided.

[b]From Feeney et al. (2010). Only mean and SD values are provided. Mean values calculated from published measurements are in brackets.

Table 9.12. 2D dental tissue proportions and enamel thickness values for Palomas permanent posterior maxillary teeth compared to those for Middle Pleistocene, Neandertal, Middle Paleolithic modern, Neolithic, and recent humans

TOOTH	SAMPLE		BCD (MM)	C (MM^2)	B (MM^2)	%B	E (MM)	2D AET (MM)	2D RET
P[3]	Palomas 53		8.99	16.94	47.38	73.66	21.67	0.78	11.36
	Palomas 60		10.08	24.87	54.49	68.66	24.49	1.02	13.76
	Palomas 68		8.52	13.73	40.08	74.49	20.71	0.66	10.47
	MPI[a], n=2 (Thomas Quarry 1 and 3)								
		Mean	—	28.05	55.94	66.60	22.59	1.24	16.50
	Neand[a], n=3								
		Mean	—	22.95	52.31	69.51	23.49	0.98	13.57
	MPMH: Qafzeh 15		—	26.98	45.01	62.52	21.46	1.26	18.74
	Neolithic, n=5								
		Mean (SD)	7.65 (0.32)	20.90 (0.71)	35.99 (4.29)	63.11 (2.90)	20.42 (1.17)	1.03 (0.07)	17.22 (2.05)
		Range	7.30–8.11	19.87–21.55	31.43–43.02	59.44–67.04	19.54–22.35	0.95–1.10	14.43–19.58
	RH Asia[b], n=10								
		Mean (SD)	8.61 (0.81)	26.65 (2.37)	43.92 (5.30)	[62.24]	21.78 (1.49)	1.23 (0.09)	[18.56]
	RH Northern Europe[b], n=17								
		Mean (SD)	8.25 (0.80)	23.58 (2.56)	41.47 (5.27)	[63.75]	20.87 (1.52)	1.13 (0.12)	[17.55]
	RH Southern Africa[b], n=10								
		Mean (SD)	8.32 (0.51)	22.06 (4.58)	37.97 (6.07)	[63.25]	19.67 (1.50)	1.11 (0.16)	[18.01]
P[4]	Palomas 68		8.04	13.37	36.43	73.16	19.56	0.68	11.32
	Palomas 94		9.56	22.00	44.12	66.73	21.19	1.04	15.63
	MPI: Thomas Quarry 3		—	29.69	54.25	64.63	22.51	1.32	17.91
	MPI: Steinheim 1		—	23.46	40.33	63.22	20.66	1.14	17.88
	Neand[a], n=4								
		Mean	—	23.13	48.09	67.52	22.28	1.04	15.03
	MPMH: Qafzeh 15		—	26.81	42.61	61.38	20.42	1.31	20.11
	Neolithic, n=3								
		Mean (SD)	7.98 (0.43)	21.53 (0.66)	35.29 (5.05)	61.93 (2.71)	19.83 (0.94)	1.09 (0.06)	18.40 (1.69)
		Range	7.55–8.40	20.79–22.02	31.30–40.97	60.09–65.04	18.86–20.74	1.04–1.16	16.58–19.93
	RH[a], n=26								
		Mean	—	22.90	39.18	63.11	19.99	1.14	18.55
M[3]	Palomas 51		9.79	14.09	33.00	70.07	17.75	0.79	13.82
	MPI: Thomas Quarry 3			31.44	46.17	59.49	20.15	1.56	22.96
	MPI: Steinheim 1			22.02	33.35	60.23	17.66	1.25	21.59
	Neand, n=5								
		Mean (SD)	—	25.15 (4.29)	46.44 (7.24)	64.84 (3.44)	21.38 (1.31)	1.17 (0.14)	17.26 (1.83)
		Range	—	20.74–32.00	38.03–57.68	60.49–69.84	20.25–23.62	0.99–1.35	14.30–18.87
	RH[a], n=52								
		Mean		27.13	40.96	60.16	19.70	1.38	21.76
	RH[c], n=10								
		Mean (SD)	10.20 (0.80)	26.00 (4.40)	36.20 (5.40)	[58.20]	—	—	24.00 (4.00)
		Range	8.70–11.50	18.80–34.80	26.30–44.40	—	—	—	18.00–29.00

Note: BCD = bicervical diameter (mm); c = enamel area (mm^2); b = dentine and pulp area (mm^2); %b = percentage of the crown that is dentine and pulp; e = length of the enamel-dentine junction (EDJ, mm); 2D AET = average enamel thickness; 2D RET = scale-free relative enamel thickness; MPI = Middle Pleistocene; Neand = Neandertal; MPMH = Middle Paleolithic modern humands; RH = recent humans.

[a]From Smith et al. (2008). Only mean values are provided.

[b]From Feeney et al. (2010). Only mean and SD values are provided. Mean values calculated from published measurements are in brackets.

[c]From Grine (2005). Mean values calculated from published measurements are in brackets.

TABLE 9.13. 2D dental tissue proportions and enamel thickness values for Palomas permanent anterior mandibular teeth compared to those for Neandertal, Neolithic, and recent humans

Tooth	Sample		BCD (mm)	c (mm^2)	b (mm^2)	%b	e (mm)	2D AET (mm)	2D RET
I_2	Palomas 89		6.97	10.28	29.54	74.19	20.03	0.51	9.44
	Neand[a], *n=3*								
		Mean	—	12.51	36.59	74.52	21.69	0.58	9.54
	Neolithic, n=7								
		Mean (SD)	5.86 (0.47)	10.61 (1.61)	22.70 (5.16)	67.69 (3.27)	17.57 (1.94)	0.60 (0.04)	12.88 (1.61)
		Range	4.89–6.18	8.04–12.55	12.39–27.22	60.66–71.01	14.12–19.61	0.55–0.66	11.12–16.18
	RH[a], *n=12*								
		Mean	—	11.40	27.49	70.69	19.23	0.59	11.28
C_1	Palomas 18		8.2	12.42	38.07	75.40	20.21	0.61	9.96
	Palomas 26		8.21	14.44	34.68	70.60	18.66	0.77	13.14
	Palomas 54		6.83	10.22	31.45	75.47	18.70	0.55	9.75
	Palomas 82		8.51	12.07	43.87	78.43	22.32	0.54	8.17
	Neand[a], *n=2*								
		Mean	—	16.13	45.37	73.77	22.50	0.72	10.57
	Neolithic, n=6								
		Mean (SD)	6.99 (0.94)	13.70 (2.59)	32.49 (9.73)	69.39 (5.00)	18.97 (3.57)	0.73 (0.08)	13.44 (4.03)
		Range	5.19–7.83	9.58–16.30	13.91–39.87	59.20–71.93	11.93–22.13	0.61–0.80	10.93–21.53
	RH Asia[b], *n=8*								
		Mean (SD)	7.44 (0.37)	17.55 (1.46)	36.27 (2.50)	[67.39]	19.59 (1.29)	0.90 (0.09)	[14.88]
	RH Northern Europe[b], *n=19*								
		Mean (SD)	7.97 (0.65)	18.90 (3.17)	45.08 (9.05)	[70.46]	22.48 (2.73)	0.84 (0.09)	[12.52]
	RH Southern Africa[b], *n=7*								
		Mean (SD)	7.33 (0.68)	16.78 (2.04)	41.49 (6.80)	[71.20]	21.12 (1.62)	0.80 (0.10)	[12.42]

Note: BCD = bicervical diameter (mm); c = enamel area (mm²); b = dentine and pulp area (mm²); %b = percentage of the crown that is dentine and pulp; e = length of the enamel-dentine junction (EDJ, mm); 2D AET = average enamel thickness; 2D RET = scale-free relative enamel thickness; Neand = Neandertal; RH = recent humans.

[a]From Smith et al. (2008). Only mean values are provided.

[b]From Feeney et al. (2010). Only mean and SD values are provided. Mean values calculated from published measurements are in brackets.

TOOTH	SAMPLE		BCD (MM)	C (MM²)	B (MM²)	%B	E (MM)	2D AET (MM)	2D RET
P₃	Palomas 45		7.77	15.16	37.16	71.02	18.84	0.80	13.20
	Palomas 76		7.23	14.63	30.42	67.52	17.25	0.85	15.38
	Neand: Oliveira 9		7.66	15.27	35.76	70.08	18.76	0.81	13.61
	Neand[a], n=2								
		Mean	—	17.61	38.87	68.82	19.68	0.89	14.38
	Neolithic: GLN04-201		6.41	16.01	31.38	66.22	17.77	0.90	16.08
	Neolithic: GLN06-215B		6.26	13.06	26.69	67	15.68	0.83	16.12
	RH Asia[b], n=7								
		Mean (SD)	7.13 (0.36)	18.49 (2.51)	35.11 (5.07)	[65.50]	18.62 (1.41)	0.99 (0.09)	[16.76]
	RH Northern Europe[b], n=13								
		Mean (SD)	7.03 (0.40)	18.93 (2.34)	35.87 (2.63)	[65.46]	18.63 (0.87)	1.02 (0.13)	[16.97]
	RH Southern Africa[b], n=9								
		Mean (SD)	6.84 (0.66)	18.30 (3.05)	30.93 (5.73)	[62.83]	17.16 (1.53)	1.06 (0.09)	[19.06]
P₄	Palomas 57		8.38	16.92	34.02	66.78	19.00	0.89	15.27
	Palomas 87		7.91	16.56	33.67	67.03	18.35	0.90	15.56
	Neand[a], n=2								
		Mean		20.59	39.05	65.48	20.69	1.00	16.06
	Neolithic: GLN04-201		7.00	18.77	32.92	63.69	17.69	1.06	18.49
	Neolithic: GLN06-215B		6.73	17.60	29.35	62.52	17.51	1.00	18.55
	RH[a], n=17								
		Mean	—	21.79	32.88	60.14	18.12	1.20	21.19

Note: BCD = bicervical diameter (mm); c = enamel area (mm²); b = dentine and pulp area (mm²); %b = percentage of the crown that is dentine and pulp; e = length of the enamel-dentine junction (EDJ, mm); 2D AET = average enamel thickness; 2D RET = scale-free relative enamel thickness; Neand = Neandertal; RH = recent humans.

[a]From Smith et al. (2008). Only mean values are provided.

[b]From Feeney et al. (2010). Only mean and SD values are provided. Mean values calculated from published measurements are in brackets.

TABLE 9.15. 2D dental tissue proportions and enamel thickness values for Palomas permanent mandibular molars compared to those for Middle Pleistocene, Neandertal, Neolithic, and recent humans

TOOTH	SAMPLE	BCD (MM)	C (MM^2)	B (MM^2)	%B	E (MM)	2D AET (MM)	2D RET
M$_2$	Palomas 29	8.54	16.98	36.46	68.23	18.77	0.90	14.98
	MPl: Tighenif 2	—	25.2	47.7	65.43	21.1	1.19	17.29
	Neand, n=6							
	Mean (SD)	—	21.02 (3.11)	42.80 (5.16)	67.11 (1.52)	20.39 (1.97)	1.03 (0.08)	15.75 (0.99)
	Range	—	17.10–24.62	33.85–47.56	65.28–69.69	17.78–22.89	0.94–1.14	14.21–16.80
	Neolithic, n=4							
	Mean (SD)	8.40 (0.20)	21.57 (1.84)	33.52 (5.13)	60.65 (2.63)	18.24 (1.34)	1.18 (0.08)	20.61 (2.21)
	Range	8.14–8.60	19.25–23.17	25.99–36.84	57.44–63.63	16.34–19.47	1.08–1.24	17.78–23.12
	RH[a], n=47							
	Mean	—	22.15	34.44	60.86	18.54	1.20	20.54
	RH[b], n=13							
	Mean (SD)	—	21.62 (2.46)	34.74 (6.29)	61.64	18.21 (1.57)	1.19 (0.14)	[20.22]
	RH[c], n=10							
	Mean (SD)	8.80 (0.70)	23.50 (4.60)	34.80 (6.60)	[59.69]	—	—	22.00 (3.00)
	Range	7.30–9.60	13.90–28.60	18.00–42.90	—	—	—	18.00–27.00
M$_3$	Palomas 50	8.03	18.78	29.19	60.84	16.54	1.14	21.02
	Palomas 58	9.33	18.71	30.66	62.11	17.36	1.08	19.46
	MPl: Mauer	—	24.05	34.58	58.98	18.98	1.27	21.55
	Neand, n=6							
	Mean (SD)	—	18.44 (2.09)	35.26 (5.15)	65.57 (1.85)	18.51 (0.99)	0.99 (0.07)	16.81 (1.06)
	Range	—	16.29–22.00	30.58–42.92	63.34–68.69	17.56–19.91	0.92–1.10	15.26–18.33
	RH[a], n=45							
	Mean		22.86	33.37	59.35	18.32	1.25	21.72
	RH[c], n=10							
	Mean (SD)	8.80 (0.70)	24.50 (4.30)	34.80 (6.20)	[58.68]	—	—	22.00 (3.00)
	Range	7.50–9.00	18.60–32.50	26.90–44.30	—	—	—	19.00–27.00

Note: BCD = bicervical diameter (mm); c = enamel area (mm²); b = dentine and pulp area (mm²); %b = percentage of the crown that is dentine and pulp; e = length of the enamel-dentine junction (EDJ, mm); 2D AET = average enamel thickness; 2D RET = scale-free relative enamel thickness; MPl = Middle Pleistocene; Neand = Neandertal; RH = recent humans.

[a]From Smith et al. (2008). Only mean values are provided.

[b]From Kono (2004). Only mean and SD values are provided. Mean values calculated from published measurements are in brackets.

[c]From Grine (2005). Mean values calculated from published measurements are in brackets.

10 The Palomas Dental Remains

Postcanine Wear

BEATRIZ PINILLA, ALEJANDRO ROMERO, AND ALEJANDRO PÉREZ-PÉREZ

CONSIDERATIONS OF THE NEANDERTALS have frequently focused on reflections of their diets and subsistence. These assessments have involved primarily faunal analyses (for example, Adler et al. 2006; Grayson and Delpech 2003; Stringer et al. 2008; Brown et al. 2011), but they have also employed stable isotopes (for example, Bocherens et al. 2005; Richards and Trinkaus 2009), molar wear (for example, El-Zaatari et al. 2011; Fiorenza et al. 2011), organic residues (for example, Hardy et al. 2001; Hardy and Moncel 2011; Henry et al. 2014), and dietary residues in dental calculus (for example, Henry 2010; Henry et al. 2011, 2014), the last also at the Sima de las Palomas (Salazar-García et al. 2013). Although a number of these analyses have been concerned with whether there were any meaningful differences between Neandertal and early modern human diets or their modes of nutritional acquisition, there have also been considerations of geographical and chronological variation and whether there are correlations between environmental parameters and diet (beyond food species availability).

There have also been applications of postcanine buccal microwear analysis to both comparisons of Neandertal versus modern human dietary profiles (Lalueza et al. 1996) and assessments of whether that food consumption was constrained by environmental parameters (Pérez-Pérez et al. 2003; Pinilla 2012). These studies have built on

buccal microwear analyses among extant and extinct primates (for example, Puech et al. 1981, 1986; Pérez-Pérez et al. 1994; Galbany et al. 2011) with applications to past populations (for example, Puech 1979; Puech et al. 1982; Pérez-Pérez et al. 1999; Pinilla et al. 2011; Pinilla 2012; Estebaranz et al. 2009, 2012).

At the same time, there have been assessments of the degree and pattern of macroscopic occlusal wear on the dentition to assess both the overall attritional and abrasive environment of the individuals and aspects of their diets (for example, Molnar 1971; Smith 1984; Kaifu et al. 2003; Deter 2009; Fiorenza et al. 2011). The comparisons have been frequently in terms of foraging versus food-producing societies, with some indications of more subtle patterning in wear through the dental arcade. There have been attempts to assess differential dental wear rates (for example, Smith 1977a, 1977b) as indications of dietary abrasiveness. Yet an independent measure of the age at death is necessary for such assessments (Skinner 1997). These assessments are therefore largely limited to immature individuals who would have had relatively little dental wear, although the few mature associated skeletons with postcranial age-at-death indicators (for example, Heim 1976; Trinkaus 1983; Hillson 2006; Trinkaus et al. 2014a) can be considered.

It is in this context that the postcanine dental

wear of the Sima de las Palomas dental remains is evaluated here. It involves assessments of buccal microwear on those teeth that are sufficiently preserved and of dentin exposure on those with sufficiently advanced occlusal wear. As detailed in chapter 6, the Palomas sample contains an abundance of dental remains, mostly isolated teeth but also ones in the Palomas 1 arcades and the Palomas 59 mandibular corpus. Yet the sample also consists of a number of immature teeth, and crown erosion, enamel chipping, and adherent matrix are common. The number of individuals appropriate for postcanine wear analysis is therefore limited, but it has the potential to provide insights into Neandertal dental attrition from southern Europe.

Materials and Methods

Buccal Microwear Materials

The human dental sample from the Sima de las Palomas includes 33 postcanine teeth deriving from 15 individuals with the potential to provide data on buccal microwear (Table 10.1). Anterior teeth (incisors and canines) are not included, given that their microwear is a combination of dietary and nonmasticatory factors (chapter 11; see also Trinkaus 1983; Bermúdez de Castro et al. 1988; Lalueza 1992; Lalueza and Frayer 1997; Lozano et al. 2008; Krueger 2011; Estalrrich and Rosas 2013). Thirteen of the postcanine teeth are isolated specimens, 4 come from Palomas 59, and 16 derive from Palomas 1. However, given the poor surface preservation of the Palomas remains, only 9 teeth from four of these individuals provide reliable data.

The comparative samples include Neandertals (n = 23) from MIS 4 and 3 and Upper Paleolithic early modern humans (n = 31) from MIS 3 and 2. The Neandertals are associated with the Middle Paleolithic except for the initial Upper Paleolithic Saint Césaire 1. The Upper Paleolithic sample includes 25 E/MUP specimens and 10 LUP ones (not all of which are included in all of the comparisons); the former are from MIS 3, and the latter are from MIS 2. All are from European sites;

only well-preserved teeth were selected. Younger immature specimens (especially deciduous teeth) are not included, as their buccal microwear pattern is very heterogeneous (Pérez-Pérez et al. 1994; Pinilla et al. 2011).

In order to assess the possible effects of climate and latitude on buccal microwear patterns among the Neandertals, they were divided into "cold" versus "mild" subsamples (bearing in the mind the Late Pleistocene climatic fluctuations), and into "north" versus "south" ones (Table 10.2). They were also separated into MIS 4 versus MIS 3 samples, given the currently available, and sometimes imprecise, dating of the specimens.

Dentin Exposure Materials

Thirty-seven teeth deriving from 17 individuals from the Sima de las Palomas were assessed for occlusal dentin exposure (DE) (Table 10.1). Of those teeth, 18 are from Palomas 1 and 4 are from Palomas 59. Moreover, the Palomas 1 M_3s present broken occlusal surfaces that make it impossible to quantify their dentin exposure areas. Of the remaining 15 teeth, only 3 (Palomas 29, 75, and 83) exhibit dentin exposure, whereas the remaining 12 were either unerupted (and hence unworn) or exhibit only faceting of the occlusal enamel.

The comparative sample is composed of all of the individuals in which the occlusal area was clearly observable, ideally from which microwear signatures were obtainable. These characteristics limit the sample to 57 individuals with 183 teeth. Eighty teeth derive from early modern humans, including specimens from the MIS 5 site of Skhul; the MIS 3 sites of Barma Grande, Brassempouy, Caldeirão, Dolní Věstonice, Mladeč, Pataud, Pavlov, Předmostí, and Les Rois; and the MIS 2 sites of Cisterna, Farincourt, Lachaud, Le Mourin, and Saint-Germain-la-Rivière. Seventy-one teeth come from Late Pleistocene Neandertals, from the sites of Banyoles, Breuil, Devil's Tower, Guattari, Kůlna, Malarnaud, Montgaudier, Le Moustier, La Quina, Švédův stůl (Ochoz), Subalyuk, Tabun, and Vindija. In addition, 32 teeth were measured from the European Middle Pleistocene sites Arago, Mauer, Montmaurin, Pontnewydd, and Steinheim.

Buccal Microwear Methods

Microwear on teeth is the physical and microscopic alteration of the tooth surface due to the interaction of the particles present in ingested food and the enamel. The enamel will be scratched when the particles ingested are harder than the enamel (≈4–5 Gpa in the Mohs scale, although see Sanson et al. 2007 for exceptions). Those particles can be intrinsic, present within the food (Baker et al. 1959; Piperno 1988; Lalueza and Pérez-Pérez 1994; Gügel et al. 2001; Piperno et al. 2004; Xia et al. 2015), or contaminants that are ingested with the food but not part of it, such as sand, dust, and ash (Peters 1982; Teaford and Lytle 1996; Mahoney 2006).

The analytical procedure follows the methodology of Pérez-Pérez et al. (1994, 1999, 2003), Lalueza et al. (1996), Galbany et al. (2004a, 2005), and Estebaranz et al. (2009, 2012). To produce high-resolution casts, Coltène President regular body polyvinylsiloxane impression material was used (Galbany et al. 2004a). Casts were obtained using Feroca polyurethane or epoxy resins and centrifuged to remove air bubbles. Casts were micrographed on the buccal surface with a Leica 360 Scanning Electron Microscope (PCB-UB) and a Hitachi S3000N (SSTT-IUA), following Pérez-Pérez et al. (1994) and Galbany et al. (2004a). Images were taken at 10–15 Kv and 100x magnification, with a 35–40 mm work distance. Each image was cropped to a 0.56 mm² square, and the scratches were recorded with Sigma Scan V (SPSS) in a semiautomatic procedure, registering the quantity, the length, and the slope of the manually identified scratches.

Fifteen variables were determined, including the number (N), length (X), and standard deviation of the length (S) of the horizontal (0–22° and 158–180°, H), vertical (67–112°, V), mesiodistal (112–157°, MD), and distomesial (22–67°, DM) scratches. The final three variables correspond to the sum of the densities (NT), the mean of the lengths (XT), and the mean of the standard deviations (ST). A compendium of the 15 variables defines the buccal microwear pattern. The proportion of horizontal (NH/NT) to vertical (NV/NT)

scratches has been demonstrated to be useful in distinguishing mainly carnivorous versus mostly vegetarian populations among modern hunter-gatherers (Lalueza et al. 1996), and therefore that method has been used for the fossils.

To avoid interobserver error, all images were analyzed by Beatriz Pinilla. The intraobserver error obtained is ≈6–8%, which is similar to the errors previously described in the literature (Grine et al. 2002; Galbany et al. 2005). Statistical analyses were performed using SPSS v. 18. All of the variables considered followed normal distributions (p ≥ 0.05, Kolmogorov-Smirnov test), and therefore parametric tests were applied. A Levene test was performed, and in most cases variables passed the homogeneity of variances test. For the Neandertal analyses, in three out of six analyses (50%), one variable did not pass the tests. The variables involved are NH, SMD, and ST (6.7% of the variables). In all the cases, if a variable did not pass the test, it was deleted from further parametric analyses. In the interpopulation variation, all of the microwear variables passed the test, and parametric comparisons were made including the 15 microwear variables. The analyses include single-classification ANOVA (p ≥ 0.05 and Bonferroni post hoc), Pearson correlation, and linear discriminant analyses (all of them performed with 1,000 permutations).

Dentin Exposure Methods

Dentin exposure (DE) was assessed using digital images of the occlusal surface (see also chapter 8). Dental occlusal crowns were photographed perpendicular to the camera focal plane, treating each tooth as if it were isolated following the methodology of Suwa et al. (1994, 1996), Pérez et al. (2006), and Grine et al. (2009). A Nikon D1H camera was used, and the sample was photographed at the Unidad de Tratamiento de Imágenes y Soporte Informático del Servicio Científico Técnico of the Universitat de Barcelona (PCB-UB). The camera was placed on a tripod and leveled horizontally. All images include a scale to allow size calibration. The images have a resolution of 3000 × 2008 pixels. Data were

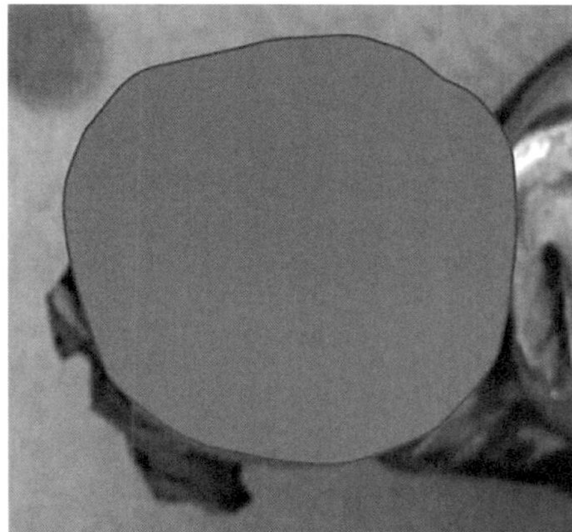

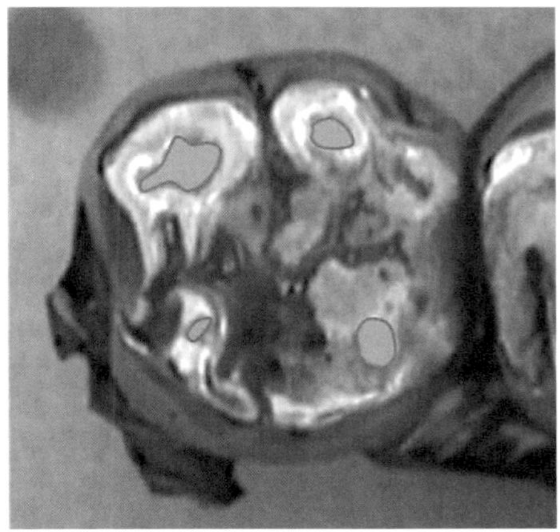

FIG. 10.1. Diagram of the method used in the dentin exposure (DE) results. *Top*: the delimitation of the total occlusal area (TOA, the solid area); *bottom*: the delimitation of the dentin exposure (the solid ameboid areas on the cusps). Using the scale in each of the steps allows the measurement of the areas (TOA and DE) and the calculation of dentin exposure as a percentage of the TOA. Individual: Vindija right M_1 VI-11.39 (206). Scale: 10 mm

obtained using ImageJ 1.43u (National Institutes of Health, USA) and IMAT, a software developed by the PCB-UB that allows the measurement of the area once the perimeter has been delimited (Estebaranz et al. 2004; Galbany and Pérez-Pérez 2006). Other researchers have used Sigma Scan Pro5 (SPSS Inc.) (Deter 2009; Alrousan 2009) or ImageJ, obtaining similar results. The method requires the delimitation of the occlusal perimeter (Fig. 10.1), which can be performed with IMAT directly or with Photoshop CS and ImageJ. The dentin was measured in the same image, together with the total occlusal area (TOA; chapter 7),

following Skinner (1997) and Deter (2009). DE was recorded by dividing the dentine area by TOA and multiplying this value by 100 to get a DE percentage: $DE = (\sigma (DE\ area_1 + DE\ area_2 + . . .+ DE\ area_n) / TOA) \times 100$, in which each DE area is the area of one dentin patch.

The average intraobserver error associated with this methodology is 4.7%, higher than that observed in the TOA measurement (1.8%; see chapter 7). The increase in the error might be related to the size of the structure (being a higher percentage of the total in individuals with smaller DEs). However, when the DE is scaled to the TOA and a percentage is obtained, the average error decreases to 2.8%. Statistical analyses were performed with SPSS 18.

Buccal Microwear Results

Palomas Preservation

All of the teeth were checked with a 30x magnifying glass to exclude specimens with postmortem damage (such as erosion and abrasion) or noneruption (King et al. 1999; Martínez and Pérez-Pérez 2004). Of the Palomas postcanine sample, 80.4% were excluded due to poor preservation (Table 10.1; Fig. 10.2). Only four individuals (19.6%) were included in the results (Fig. 10.3). Two of those individuals are Palomas 1 and 59, and both of them present more than one well-preserved tooth. In the case of Palomas 1, 22.2% (4 out of 18) of the teeth were analyzed; in the case of Palomas 59, masticatory scratches could be studied on 3 of 4 postcanine teeth. The main cause of exclusion was erosion (Table 10.1). In most of the teeth, the enamel is eroded, and the resulting micrographs show several pits and microfractures. Some of the teeth also exhibit vertical cracks, although their presence is not directly related to the exclusion of the tooth. In some teeth, sediment has not been completely removed, which makes the study of the complete enamel surface impossible. However, in none of the analyzed cases does the sediment cover the total buccal crown, and therefore this has not been a cause of exclusion. In 14 cases out of

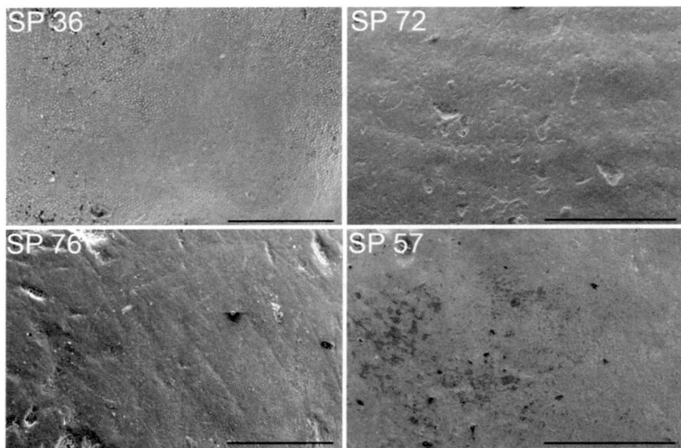

FIG. 10.2. SEM micrographs showing poorly pre-served buccal surfaces of the Palomas 36 left dm^2, the Palomas 72 left M^1, the Palomas 76 right P$_3$, and the Palomas 57 right P$_4$. The bar scales are 200 μm (images obtained at 200x) for Palomas 36 and 76 and 500 μm (images obtained at 100x) for Palomas 72 and 57. Palomas 36 shows a clear prism exposure pattern on the buccal surface, and Palomas 76 shows dietary scratches visible and identifiable at 200x (but the individual is not analyzable at 100x). Palomas 72 and 57 show an eroded surface with no-diet related microwear pattern. On the Palomas 72 image, micro-cracks and pits are easily distinguished.

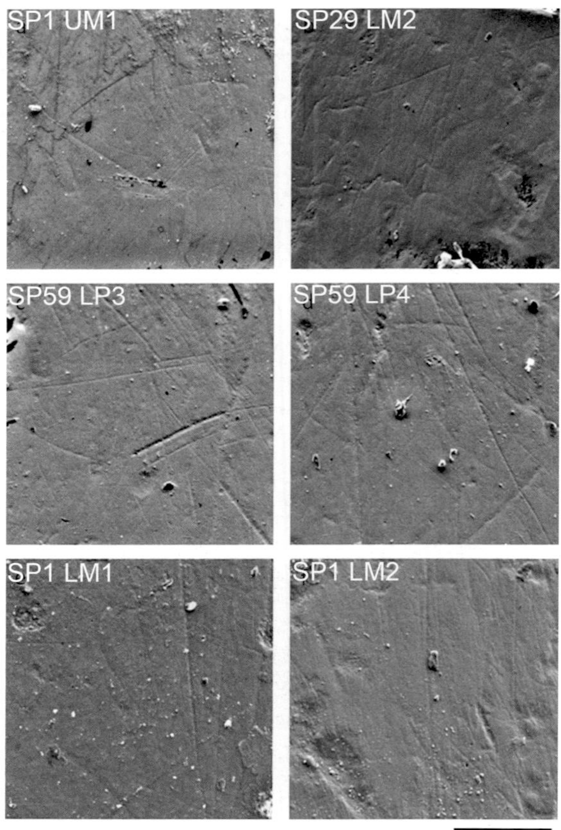

FIG. 10.3. SEM micrographs at 100x showing well-pre-served buccal enamel tooth surfaces of the Palomas 1 right M^1, M$_1$, and M$_2$; the Palomas 29 right M$_2$; and the Palomas 59 left P$_3$ and P$_4$. Even though some of these teeth have erosion, the standardized 0.56 mm^2 surface has been possible to observe, and therefore the scratches have been analyzed.

the 32 teeth excluded due to erosion (43.8%), the cause of exclusion also included other problems (such as broken teeth). Palomas 36 was excluded because the crown had not erupted (Fig. 10.2), since clear prisms are exposed.

It should be noted that, in the analysis of the microwear signatures in a smaller area, increas-ing the magnification would have allowed a wider sample size (Palomas 72 in Fig. 10.2). However, the standardized methodology of micrographing buccal enamel surfaces at 100x was followed.

Palomas Buccal Microwear

Only four individuals from Sima de las Palo-mas have been effectively analyzed. Their varia-tion in scratch densities is high (more than 100 scratches), although the variation of the maximum scratch lengths does not surpass 70 μm. Density is heterogeneous with respect to the orientations, without a clear tendency in the four individuals.

The main density of Palomas 1 occurs with verti-cal scratches, whereas on Palomas 29 and 75 the mesiodistal scratches are dominant. On Palomas 59 more than 35% of the scratches are distomesial (Table 10.3).

There is a tendency for the stratigraphically younger Palomas specimens, those from the Upper Cutting infilling, to have an increase in the density of the scratches (NT) and the scratch lengths (XT) (Fig. 10.4). In the proportions of ver-tical and horizontal scratches, two of the strati-graphically younger specimens (Palomas 59 and 75) show a higher proportion of horizontal scratches (NH/NT) and a lower one of vertical scratches (NV/NT) than do Palomas 1 and 29, so there is not a clear separation in the ratios according to the stratigraphic units. In this comparison, the youngest individual (Palomas 59) has the longest scratches and the highest percentage of horizontal scratches; it should nonetheless be mentioned that Palomas 59 presents caries (Walker et al., 2011c; chapter 12), which may have affected that individual's diet (Hillson 2008).

Palomas and Neandertal Microwear Signatures

The Palomas individuals remain more or less stable in density when compared to the Neandertal sample, although at the lower density range of Neandertal variation (Fig. 10.4). When analyzing the proportions derived from Lalueza et al. (1996) (Fig. 10.5), the results indicate that the Palomas dentitions are well within the range of the Neandertals. Palomas 1 and 29 have average values on both variables, and Palomas 59 and 75 have average values for NV/NT but slightly lower than average ones for NH/NT. Their positions in the plot are close to the mixed-diet recent hunter-gatherer range in these measures, and they are most similar to those from arid-tropical environments (Andamanese, Veddahs, Khoisan, Tasmanians, and Australians).

The main differences within the Neandertal sample are related to the geographical area (Table 10.4). Significant differences between northern and southern samples occur only on the distomesial scratch density (NDM p = 0.011; F = 7.834), although there is a tendency for the northern individuals to present lower scratch densities (NT), lower NH/NT values, and slightly higher scratch lengths (XT) and NV/NT values. These measures indicate that southern individuals had a more abrasive diet and likely incorporated a higher percentage of vegetables in their diet than did the northern groups. This tendency is confirmed when the geographic ranges are segmented into four categories: northwestern, southwestern, central Mediterranean, and central European. Significant differences occur in all of the length

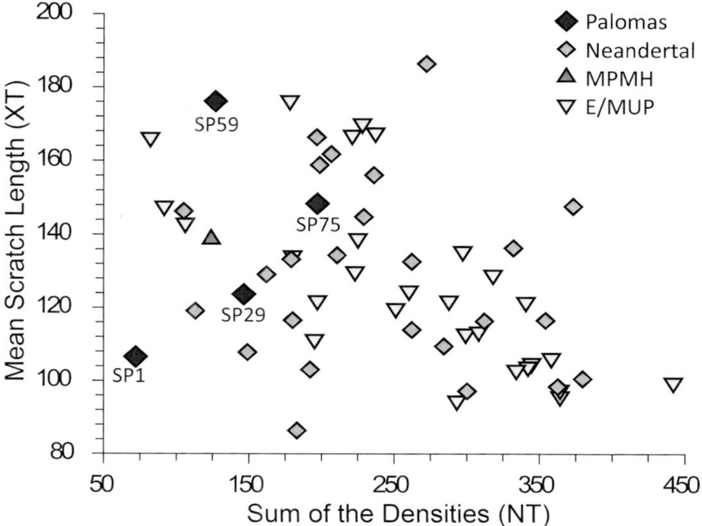

FIG. 10.4. The mean lengths of the scratches (XT) versus the sums of the scratch densities (NT) for the Palomas specimens, Neandertals, a Middle Paleolithic modern human (MPMH), and Early/Mid Upper Paleolithic modern humans (E/MUP). Comparative data from Pinilla (2012) and Pinilla et al. (2014).

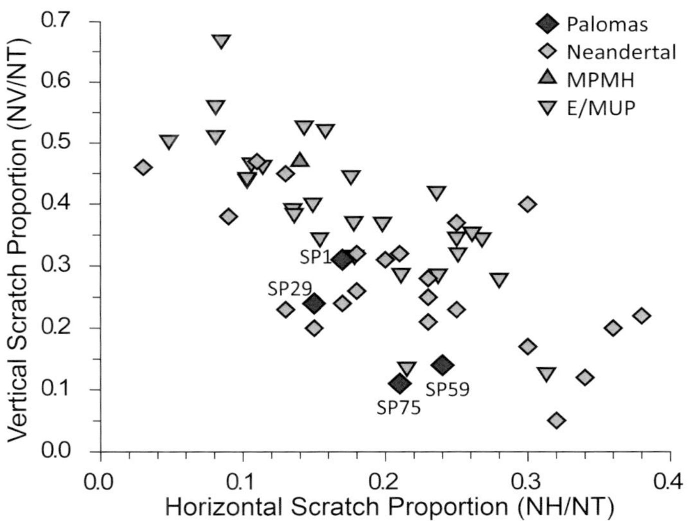

FIG. 10.5. The vertical scratch proportion (NV/NT) versus the horizontal scratch proportion (NH/NT) for the Palomas specimens, Neandertals, a Middle Paleolithic modern human (MPMH), and Early/Mid Upper Paleolithic modern humans (E/MUP). Note that the sum for each specimen does not reach 100%, because the oblique scratches (mesiodistal and distomesial) are not included. Comparative data from Pinilla (2012) and Pinilla et al. (2014).

variables (XH p = 0.003, F = 6.717; XV p = 0.030, F = 3.700; XMD p = 0.044, F = 3.255; XDM p = 0.002, F = 7.035; and XT p = 0.001, F = 7.732). The main differences are related to the distinction between the southern groups (XH: NW-CM p = 0.023; SW-CM p = 0.005; XMD: SW-CM p = 0.038; XDM: SW-CM p = 0.006; SW-CE p = 0.005; and XT: SW-CE p = 0.014; SW-CM p = 0.002). However, the southern individuals (CM and SW) do not present values higher than the 0.25 for the vertical scratches proportion, whereas both surpass 0.2 for the horizontal scratch index, supporting the inference that the southern individuals consumed a higher quantity of abrasive vegetables than did the northern populations. The variation in diet in the contrasting climatic regimes was minimal, and no microwear variable shows significant differences even though the milder environment inhabitants tend to show slightly more and longer scratches than do colder climate individuals (Table 10.4).

None of the LDA shows any significance for the derived functions, and the maximum post hoc correct classifications with cross-validation do not surpass ≈60% (although without the cross-validation, correct classifications reach ≈95%).

The density and length of the scratches in teeth from Palomas present a pattern closer to that found in individuals from northern and cold environments (Table 10.4). As already noted, the density of their scratches is low, and the scratches tend to be long, which fits better with the dietary pattern observed in northern latitudes and colder climates. However, the indices bring Palomas dentitions close to the southern remains, especially on the horizontal scratch proportion, suggesting a mixed diet with a greater importance of plant food than in the north.

Neandertal and Modern Human Buccal Microwear

Differences between the buccal microwear patterns of Neandertals and those of the Upper Paleolithic modern humans occur mainly in the density of the scratches (abrasiveness of the diet) (Table 10.5). One-way ANOVA indicates that significant differences occur for the vertical scratch density (NV p < 0.001; F = 14.485) and the total density of scratches (NT p = 0.030; F = 4.987), Upper Paleolithic humans having more scratches in both cases. Moreover, the correlation between density and length of scratches (NT to XT) is only significant for the early modern humans (p < 0.001; r = −0.719) but not for the Neandertals (p = 0.255; r = −0.227) (Fig. 10.4). The pattern is maintained if only the MIS 3 specimens are considered, although only the vertical scratch density is significantly different between the groups (NV p = 0.003; F = 10.366). The distomesial lengths of the scratches are also significantly different (p = 0.025; F = 5.456), and the correlation between density and length of scratches is, again, only significant for the early modern humans (p < 0.001; r = −0.754) and not for the Neandertals (p = 0.282; F = −0.277).

The indices show a tendency for the early modern humans to have higher NV/NT values (early modern mean = 0.376; Neandertal mean = 0.278) and lower NH/NT values (early modern mean = 0.176; Neandertal mean = 0.221) (Fig. 10.5); these differences would place the inferred early modern human diets closer to cold-adapted inhabitants (Inuits, Fuegians, etc.) and the Neandertals closer to mixed-diet modern hunter-gatherers (Khoisan, Andamanese, etc.), following Lalueza et al. (1996). The same pattern is observed when considering only MIS 3 individuals (Table 10.5).

To classify the Palomas specimens relative to these two samples, a linear discriminant analysis of the two samples (Neandertals and Upper Paleolithic modern humans) within MIS 3 was performed (Fig. 10.6). Only one function derives from the analysis. F1 is significantly different between both populations (p < 0.001; r = 0.840), and it is mainly correlated with the vertical scratch density (NV r = 0.412; Fig. 10.5) and the total mean density (NT r = 0.263; Fig. 10.4). Post hoc correct classification of the individuals is 92.9%, although it decreases to 73.8% after cross-validation. All four specimens from Palomas are classified as Neandertals (Fig. 10.6), with probabilities higher than 70% (Palomas 1: 99.4%; Palomas 29: >99.9%; Palomas 59: 99.8%; and Palomas 75: 72.8%).

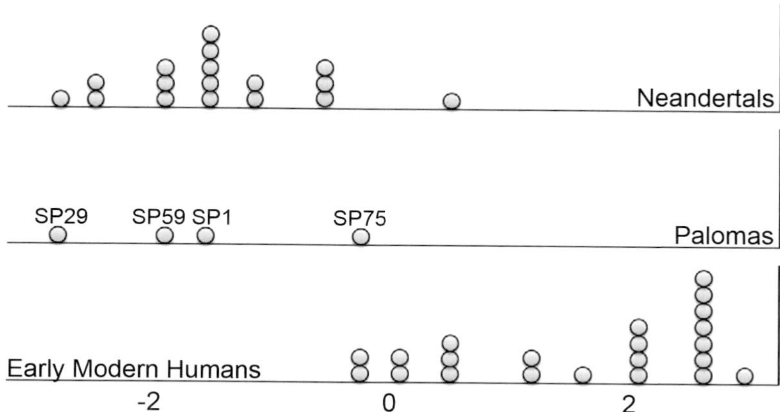

FIG. 10.6. Graphic representation of the linear discriminant analysis (LDA) function derived (p < 0.001; r = 0.840). The Neandertal values are provided above, and the early modern humans are below, with the four Palomas individuals in between. Microwear variables mostly related to F1 are NV (r = 0.412) and NT (r = 0.263).

Dentin Exposure Results

Palomas Postcanine Dentin Exposure

Twenty-two postcanine teeth deriving from five individuals provide analyzable dentin exposure (DE). However, all but three of the measureable teeth derive from Palomas 1 and 59, and of the remaining 19 measureable teeth, 16 are from Palomas 1. An additional 13 teeth can be scored, but they have insufficiently worn crowns to provide dentin exposure measures. Three out of the four P_3s provide DE values, although in these three cases it is lower than 10%. According to these percentages, the youngest to oldest individuals are Palomas 76 (0% DE), Palomas 59 (2.4% DE), Palomas 75 (3.6%), and then Palomas 1 (6.7%). Only one of the P_4s provides a DE value (Palomas 1), and the other 4 P_4s have no dentine exposure. On the P³ and P⁴, only Palomas 1 (both left and right) presents DE (35.8 and 32.3%, respectively, on the P³ and 28.6% and 13.8%, respectively, on the P⁴). On the M_1, 3 out of 5 teeth furnish DE measures, but 2 of them belong to Palomas 1 (55.6% on the left and 69.9% on the right); Palomas 59 has a value (10.9%) indicating a younger age. The two M¹s from Palomas 1 present DE values of 48.2% on the left and 46.6% on the right. On the M_2, only Palomas 80 has no dentine exposure. Palomas 1 has higher levels of DE (left and right: 35.0% and 33.0%, respectively); Palomas 59 furnishes a DE of 15.4%; and Palomas 29 one of 5.4%. On the M² and M³, only Palomas 1 retains teeth with measureable DE (40.9% on the left M²,

38.6% on the right M², 69.5% on the left M³, and 11.6% on the right M³). No M_3s are sufficiently intact. Of the three dm_2s, two (Palomas 70 and 88) provide no DE, but Palomas 83 provides a DE value of 5.4%.

Two of the individuals, Palomas 1 and 59, retain more than one tooth, and therefore the analysis of their dental series can be performed (Fig. 10.7). In the case of Palomas 59, the series includes the left P_3 to M_2. In this mandible, there is a reasonable relationship between the amount of dentin exposure and the dental eruption schedule for the P_3 to M_1, but the M_2 has an excessive amount of DE compared to the M_1. This abnormality of the M_2 with respect to the M_1, and also the premolars, has been described (Walker et al. 2011c), and it was explained as a possible supraeruption of this tooth, causing "greater occlusal forces and increased wear" (see also chapter 12). However, this abnormality does not imply a dysfunctional occlusion, as the occlusal plane was maintained. Palomas 1 retains 16 teeth with DEs, permitting a more thorough analysis (Fig. 10.7). First, concerning differences between dental arcades, the mandibular premolars have lower DE values than the maxillary ones. However, on the molars, the reverse pattern is evident on the M1s, but there is little difference across all four M2s. Regarding laterality, when only considering the maxillary molars, in all cases the left teeth possess higher DE than right maxillary ones. Yet the differences are small for the P³, M¹, and M². The M1s have higher DE than the rest of the teeth, except for the

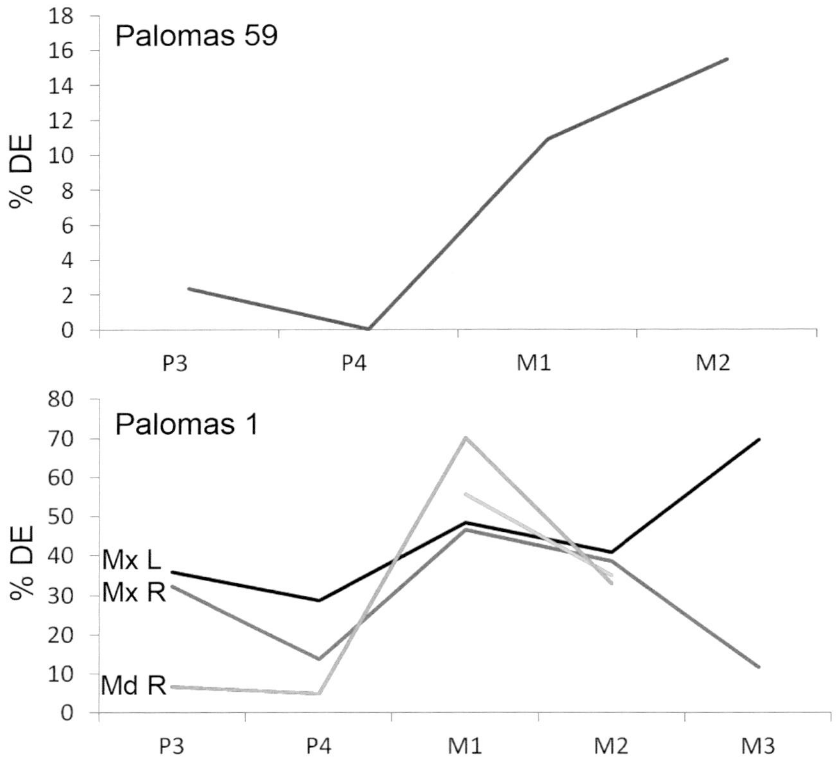

FIG. 10.7. Dentin exposure percentages (%DE) for the Palomas 59 left mandibular P_3 to M_2 dentition and the Palomas 1 available P3 to M3 maxillary (Mx R and Mx L) and mandibular (Md R) dentition. The Palomas 1 left M_1 and M_2 %DE are also indicated.

anomalous left M³; the P_3 to M2 pattern is therefore as would be expected following the eruption schedule. We lack DE measures for the mandibular M_3s, but the percentage of DE on the right M³ is considerably lower.

Comparative Pleistocene Dentin Exposure

In the Late Pleistocene dental arcades, a number of the individuals that include more than one analyzable tooth show great variability, due to either asymmetry or maxilla-mandible differences (Table 10.6). However, an analysis of this pattern shows that the individuals showing more than 10% difference are constrained to Švédův stúl (Ochoz) 1, Tabun 1, and La Quina 5, all of them Neandertals. In the case of La Quina 5, the only tooth distorting the degree of wear is the right M¹. However, the emergence schedule in La

Quina 5 is followed; the higher DE should therefore be considered normal within the individual, and in any case it is affecting both laterality differences or contrasts between the dental arcades. One can query whether it is related to the marked humeral diaphyseal asymmetry of that individual (Trinkaus et al. 1994). Tabun 1 and Švédův stúl 1, however, show marked differences (more than 10%) in three out of four teeth. In the Švédův stúl 1, P_3–M_1 show differences between 16–46% of DE, with the greatest difference in the P_3. Those differences are due to the consistently higher DE on the left side, and therefore clear laterality is found on this individual. It is probably related to the curious "step" in the occlusal plane across the right canine (Vlček 1969); absence of the maxillary dentition prevents further assessment of its wear pattern.

In the case of Tabun 1, differences greater than 10% occur on both premolars and the second molar. In the P_3s, the differences occur between dental arcades, and in the P_4s and M2s, where three P_4s and three M2s are measureable, marked differences occur for both laterality and between the dental arcades (but not obliquely), being higher than 20% in all the cases. It is difficult to explain why the M1s show differences lower than 10%, and the rest of the teeth are greater, especially since the M1 position will not allow contrasting pressures relative to the P_4s and/or the M2s. A similar result is obtained (although the differences are lower) on the Mauer 1 mandible. This Middle Pleistocene specimen shows differences in laterality lower than 1.5% in the M_2, whereas the M_1 and the M_3 surpass 7%.

The available Upper Paleolithic modern humans show relatively low values for these differences, as with the few available Middle Pleistocene specimens (Table 10.6). It is unclear why these few Neandertals exhibit such variable and marked differences across similar tooth positions in the degrees of dentin exposure from occlusal wear. There is something unusual in the Švédův stůl 1 dental wear, but the other specimens do not exhibit similarly unusual patterns. Indeed, despite her mandibular injury and posttraumatic facial remodeling, Dolní Věstonice 3 exhibits surprisingly symmetrical wear (Hillson 2006; Table 10.6).

Palomas 1 Dentin Exposure Differences

In the Palomas sample, this analysis can only be performed for Palomas 1 (Table 10.7). The results show considerable variability, with 8 of the 19 possible pairs surpassing the 10% level. Of note are those for the maxilla versus mandible for the right P3s and M1s and the asymmetry in the P_4s, the M_1s, and especially the M³s. The M³ asymmetry is difficult to explain, due to the fact that the M3 is the most distal tooth, and the laterality differences observed in the Palomas 1 M2s are lower than 3%. The differences are also elevated in the oblique left-right comparisons of the P3s and the P4s. As with specimens such as La Quina 5 and Tabun 1, it is unclear why the degrees of dentin exposure of Palomas 1 are so variable.

Dentin Exposure versus Buccal Microwear

Part of the variation in dentin exposure might be related to variation in the abrasiveness of the diet. A comparison of occlusal wear (DE) and microwear values shows that, in some teeth, there is a significant correlation between the variables (Table 10.8). Yet there is no clear pattern, and the individuals presenting a higher DE do not always present a higher microwear pattern. This is especially clear in the cases of the individuals with marked laterality or mandible-maxilla differences. For example, La Quina 5 presents a clear difference in occlusal wear, due principally to the right M¹, yet there is no indication that this tooth especially differs in its microwear signature from the rest of the teeth, nor is it the tooth with the higher density of scratches.

Discussion and Summary

The teeth from the Sima de las Palomas are poorly preserved. Even though the site has provided one of the larger fossil human postcanine dental samples, relatively few of those teeth can be analyzed for microwear, and a number are too damaged or encrusted for occlusal wear quantification. The enamel preservation of fossil teeth seems to be unrelated to chronology or geography (Pinilla 2012), but in this case, the brecciated sediments and fossil cleaning might have affected enamel preservation highly, producing erosion on the enamel surface of most of the teeth. In addition, the large number of immature dental remains (chapters 6 and 15) means that few teeth (other than those of Palomas 1) exhibit dentin exposure.

The variation observed within the site is quite large. However, the microwear variation is similar to that observed in other sites (for example, El Sidrón and Vindija) (Pinilla et al. 2011). Even though some authors have argued that the Neandertals had a stable diet through time and geography (for example, Bocherens et al. 2005), it is increasingly apparent that there was considerable dietary flexibility, evident in patterns of dental attrition (Pérez-Pérez et al. 2003; El-Zaatari et al. 2011; Fiorenza et al. 2011), dental calculus (Henry 2011; Henry et al. 2011, 2014; Salazar-García et al. 2013), inferences from the archeological record (Bar-Yosef 2004; Hardy 2010; Hardy and Moncel 2011), and buccal microwear.

It is possible that some of the variation is due to the stratigraphic separation of Palomas 1 from the other dental remains (chapter 2). However, the microwear results from Palomas might be merely normal internal variability. The limited sample size at the site (n = 4), together with the lack of a tendency in the comparative Neandertal sample, prevents inferring that a dietary shift occurred among the Neandertals during MIS 3. Nonetheless, when the Neandertal buccal microwear is compared to that of the early modern humans, the Neanderthals are to some extent distinct, with a correct classification rate after cross-validation of 73.8%. This difference mostly reflects more scratches together with more vertical scratches among the early modern humans relative to the Neandertals. In this context, the four Palomas individuals are classified as Neandertal with post hoc probabilities of more than 70% in all of the cases and more than 99% in three out of four cases. Even though more analyses are needed to understand what changed in the diet between those Neandertals and the Upper Paleolithic modern humans, these buccal microwear signatures indicate that, despite the substantial overlap in the individual values, significant differences between the groups existed.

It should be mentioned that the position of Palomas 59 in the buccal microwear variation may have been affected by its carious lesions. Although rare, carious lesions are present among Neandertals and early modern humans (chapter 12), but their effects on tooth wear are likely to be idiosyncratic, depending on lesion location and severity. Even though the presence of a carious lesion on Banyoles 1 is subject to debate (Lalueza et al. 1993b), and no laterality exists between left and right mandibular teeth (Pinilla 2012), the buccal microwear indices place Banyoles 1 among the most vegetarian individuals of the sample, supporting an association of the pathology with the ingestion of carbohydrates.

The buccal microwear variation is paralleled at an individual level by the differences across individual tooth positions in the percentages of dentin exposure of Palomas 1. These occur in terms of asymmetry within arcades and across the arcades. Yet similar levels of differential occlusal wear are present in other Neandertals, and it is unclear to what masticatory behaviors they can be attributed. None of them have evidence of dental agenesis, antemortem tooth loss, periapical lesions, facial trauma, or other factors that could have produced uneven occlusal wear. These unusual degrees of wear are joined by the greater M_2 than M_1 and premolar wear of Palomas 59 (Walker et al. 2011c; chapter 12).

TABLE 10.1. Sima de las Palomas postcanine teeth included in the microwear and dentin exposure analyses

SP NO.	TOOTH	SIDE	MICROWEAR PRESERVATION	% EXPOSED DENTIN
1a	P^3	Right	Erosion	32.34
1a	P^4	Right	Erosion	13.79
1a	M^1	Right	Preserved	46.59
1a	M^2	Right	Erosion	38.63
1a	M^3	Right	Erosion	11.62
1b	P^3	Left	Erosion	35.84
1b	P^4	Left	Erosion	28.64
1b	M^1	Left	Erosion	48.21
1b	M^2	Left	Erosion	40.89
1b	M^3	Left	Erosion	69.47
1c	P_3	Right	Erosion	6.73
1c	P_4	Right	Erosion	4.85
1c	M_1	Right	Erosion	69.94
1c	M_2	Right	Preserved	32.99
1c	M_3	Right	Erosion	damaged
1d	M_1	Left	Preserved	55.63
1d	M_2	Left	Preserved	35.01
1d	M_3	Left	Erosion	broken
29	M_2	Right	Preserved	5.38
36	dm^2	Left	Unerupted/Unworn	0.00
40	dm_1	Left	Erosion/Broken	—
45	P_3	Right	Erosion/Unworn	—
53	P^3	Left	Erosion/Broken	—
57	P_4	Right	Erosion	0.00
59	P_3	Left	Preserved	2.36
59	P_4	Left	Preserved	0.00
59	M_1	Left	Preserved	10.90
59	M_2	Left	Erosion	15.44
60	P^3	Right	Broken/Unworn	0.00
70	dm_2	Right	Erosion	0.00
72	M^1	Left	Broken/Erosion	—
75	P_3	Left	Preserved	3.56
76	P_3	Right	Erosion	0.00
78	P_4	Left	Unworn	0.00
80	M_2	Left	Erosion	0.00
81	M?	Left	Erosion	0.00
83	dm_2	Right	Erosion	24.55
84	M_1	Left	Erosion	0.00
87	P_4	Left	Erosion	0.00
88	dm_2	Left	Erosion/Matrix	0.00
94	P^4	Left	Erosion/Unworn	0.00

Note: The fourth column indicates whether the tooth was adequately preserved for buccal microwear and therefore analyzable, or its cause of exclusion (see chapter 6 for additional preservation information). The last column provides the surface area of exposed dentin as a percentage of total crown area (Fig. 10.1).

TABLE 10.2. Assignments of the Neandertal comparative specimens by climate, latitude, and MIS, bearing in mind imprecisions in the dating and the stratigraphic associations of some of the specimens

SITE	COLD VS. MILD	NORTH VS. SOUTH	MIS
Banyoles	Cold	South	4
Breuil	Mild	South	3
Cova Foradá	—	South	—
Figueira Brava	Mild	South	3
Guattari	Cold	South	3, 4
Kůlna	Mild	North	3
Malarnaud	Mild	North	4
Montgaudier	Cold	North	3
Le Moustier	Cold	North	3
La Quina	Cold	North	4
Saint Césaire	Mild	North	3
El Sidrón	Mild	South	3
Subalyuk	Cold	North	4
Švédův stůl (Ochoz)	Cold	North	4
Tabun	Mild	South	5
Vindija	Mild	South	3
Zafarraya	Mild	South	3

TABLE 10.3. Buccal microwear values for the Palomas individuals with adequate preservation of buccal surfaces for at least one tooth

MEASUREMENT	SP1	SP29	SP59	SP75
Number of horizontal scratches (NH)	24.5	11	31	42
Length of horizontal scratches (XH)	131.8	90.8	209.3	183.1
SD of length, horizontal scratches (SH)	74.8	73.8	164.2	159.0
Number of vertical scratches (NV)	46	17	18	22
Length of vertical scratches (XV)	134.5	96.5	166.8	130.9
SD of length, vertical scratches (SV)	85.5	38.0	85.2	113.1
Number of mesiodistal scratches (NMD)	34.5	25	25	83
Length of mesiodistal scratches (XMD)	116.0	120.8	166.0	152.4
SD of length, mesiodistal scratches (SMD)	70.5	53.8	80.4	129.5
Number of distomesial scratches (NDM)	30	19	45	50
Length of distomesial scratches (XDM)	107.7	106.0	158.0	120.5
SD of length, distomesial scratches (SDM)	53.1	37.1	94.2	98.1
Sum of the scratch densities (NT)	146.5	72	127	197
Mean of the scratch lengths (XT)	123.7	106.5	176.1	148.4
Mean of the standard deviations (ST)	73.1	50.5	129.1	128.8
Proportion of horizontal scratches (NH/NT)	0.167	0.153	0.244	0.213
Proportion of vertical scratches (NV/NT)	0.314	0.236	0.142	0.112

Note: See Table 10.1 for the included teeth, especially for Palomas 1 and 59.

TABLE 10.4. Means ± standard deviations of the buccal microwear values of the Neandertal comparative sample (n = 23) divided into north and south subsamples and into cold and mild subsamples

MEASUREMENT	NORTH	SOUTH	COLD[a]	MILD
	N = 11	N = 12	N = 7	N = 15
Number of horizontal scratches (NH)	44.9 ± 17.9	65.1 ± 36.8	50.2 ± 26.1	58.6 ± 33.8
Length of horizontal scratches (XH)	141.2 ± 24.2	129.5 ± 33.6	136.1 ± 21.4	132.3 ± 32.7
SD of length, horizontal scratches (SH)	97.4 ± 19.4	92.4 ± 27.8	94.8 ± 21.2	94.0 ± 26.1
Number of vertical scratches (NV)	76.2 ± 42.1	63.6 ± 36.3	52.7 ± 30.4	78.4 ± 41.6
Length of vertical scratches (XV)	138.4 ± 27.0	144.3 ± 47.7	129.5 ± 41.7	144.6 ± 37.4
SD of length, vertical scratches (SV)	100.8 ± 31.4	108.9 ± 53.4	85.1 ± 33.3	112.9 ± 46.9
Number of mesiodistal scratches (NMD)	66.2 ± 37.9	77.3 ± 28.8	57.7 ± 25.4	80.9 ± 34.4
Length of mesiodistal scratches (XMD)	114.2 ± 19.0	114.8 ± 22.0	113.5 ± 25.2	114.1 ± 18.8
SD of length, mesiodistal scratches (SMD)	80.3 ± 19.9	82.4 ± 15.6	77.5 ± 21.3	82.8 ± 16.3
Number of distomesial scratches (NDM)	39.2 ± 17.3	60.8 ± 19.5	43.6 ± 28.9	53.2 ± 17.6
Length of distomesial scratches (XDM)	116.8 ± 24.5	122.6 ± 24.8	116.1 ± 18.3	118.6 ± 25.1
SD of length, distomesial scratches (SDM)	77.2 ± 27.7	84.9 ± 25.9	86.1 ± 16.1	76.9 ± 29.8
Sum of the scratch densities (NT)	229.9 ± 95.5	269.4 ± 57.0	217.2 ± 90.8	269.6 ± 70.9
Mean of the scratch lengths (XT)	130.6 ± 21.9	129.0 ± 29.5	126.7 ± 22.5	128.7 ± 26.7
Mean of the standard deviations (ST)	97.1 ± 20.6	97.8 ± 24.9	92.9 ± 11.1	98.7 ± 26.6
Proportion of horizontal scratches (NH/NT)	0.195 ± 0.187	0.242 ± 0.646	0.231 ± 0.587	0.217 ± 0.476
Proportion of vertical scratches (NV/NT)	0.331 ± 0.441	0.236 ± 0.637	0.243 ± 0.334	0.291 ± 0.587

[a]Cova Foradá is not included.

TABLE 10.5. Means ± standard deviations of the buccal microwear values for the Neandertal and early modern human samples

MEASUREMENT	NEANDERTALS (N = 23)	EARLY MODERN HUMANS (N = 32)	MIS 4 NEANDERTALS	MIS 3 NEANDERTALS	E/MUP (MIS 3)	LUP (MIS 2)
Number of horizontal scratches (NH)	55.4 ± 30.5	48.9 ± 17.6	50.0 ± 21.8	57.4 ± 33.4	49.3 ± 18.3	48.0 ± 16.8
Length of horizontal scratches (XH)	135.1 ± 29.5	133.8 ± 30.4	129.7 ± 20.3	137.0 ± 32.4	126.8 ± 30.8	148.6 ± 25.0
SD of length, horizontal scratches (SH)	94.8 ± 23.7	96.5 ± 31.1	84.6 ± 21.2	98.4 ± 24.1	91.8 ± 30.4	106.3 ± 31.7
Number of vertical scratches (NV)	69.6 ± 38.8	104.8 ± 43.4	71.8 ± 35.1	68.9 ± 41.0	115.0 ± 46.2	83.2 ± 28.0
Length of vertical scratches (XV)	141.5 ± 38.4	142.9 ± 30.8	133.3 ± 33.2	144.4 ± 40.7	135.3 ± 29.8	159.0 ± 27.6
SD of length, vertical scratches (SV)	105.0 ± 43.5	112.0 ± 26.2	96.5 ± 24.3	108.0 ± 48.8	110.4 ± 27.5	115.3 ± 24.4
Number of mesiodistal scratches (NMD)	72.0 ± 33.2	62.7 ± 29.2	71.1 ± 36.2	72.3 ± 33.2	60.7 ± 29.1	67.0 ± 30.6
Length of mesiodistal scratches (XMD)	114.5 ± 20.2	117.4 ± 31.8	119.1 ± 22.6	112.9 ± 19.7	114.4 ± 36.9	123.8 ± 16.6
SD of length, mesiodistal scratches (SMD)	81.4 ± 17.4	87.0 ± 23.0	88.8 ± 12.6	78.8 ± 18.4	84.6 ± 25.0	92.0 ± 18.1
Number of distomesial scratches (NDM)	50.5 ± 21.2	59.1 ± 32.0	42.2 ± 19.5	53.4 ± 21.5	63.6 ± 27.8	49.5 ± 39.2
Length of distomesial scratches (XDM)	119.8 ± 24.3	115.0 ± 28.4	108.8 ± 17.5	123.7 ± 25.5	105.6 ± 22.2	134.6 ± 30.9
SD of length, distomesial scratches (SDM)	81.2 ± 26.4	83.7 ± 25.3	74.6 ± 16.7	83.5 ± 29.2	77.5 ± 20.6	96.7 ± 30.2
Sum of the scratch densities (NT)	250.5 ± 78.6	278.3 ± 67.6	249.2 ± 100.2	250.9 ± 73.2	290.4 ± 71.5	253.0 ± 53.1
Mean of the scratch lengths (XT)	129.8 ± 25.6	129.6 ± 25.0	124.0 ± 22.0	131.8 ± 27.1	122.1 ± 24.2	145.4 ± 19.3
Mean of the standard deviations (ST)	97.5 ± 22.4	103.9 ± 17.5	93.1 ± 10.8	99.0 ± 25.4	101.2 ± 17.5	109.5 ± 17.0
Proportion of horizontal scratches (NH/NT)	0.221 ± 0.388	0.176 ± 0.260	0.201 ± 0.218	0.229 ± 0.456	0.170 ± 0.257	0.190 ± 0.31
Proportion of vertical scratches (NV/NT)	0.278 ± 0.494	0.376 ± 0.642	0.288 ± 0.350	0.274 ± 0.561	0.396 ± 0.646	0.329 ± 0.52

Note: The samples are first pooled, then each is subgrouped by Marine Isotope Stages (MIS) 4 and 3 for the Neandertals and 3 and 2 for the early modern humans. The latter two stages, for the early modern humans, are equivalent to the Early/Mid Upper Paleolithic (E/MUP) and Late Upper Paleolithic (LUP) samples.

TABLE 10.6. Differences in dentin exposure values (DE, expressed as percent) for specimens providing antimeres and/or the same tooth from different arcades

	SPECIMEN		DIFFERENCE SIDE[a]	DIFFERENCE ARCADE[b]	DIFF. SIDE+ARCADE[c]
P3					
	Arago 2	MP	5.28	—	—
	La Quina 5	N	—	0.85	—
	Švédův stůl (Ochoz) 1	N	**46.90**	—	—
	Tabun 1	N	—	**34.14**	—
	Dolní Věstonice 3	UP	—	—	2.66
	Dolní Věstonice 14	UP	—	—	2.99
	Dolní Věstonice 15	UP	1.00	4.50	4.50
P4					
	Le Moustier 1	N	0	—	—
	La Quina 5	N	1.21	—	—
	Švédův stůl (Ochoz) 1	N	**16.80**	—	—
	Tabun 1	N	**28.24**	**23.50**	4.74
	Dolní Věstonice 13	UP	—	0	—
	Dolní Věstonice 15	UP	0	2.73	2.73
M1					
	Mauer 1	MP	7.85	—	—
	Montmaurin 1	MP	1.13	—	—
	Le Moustier 1	N	1.48	—	—
	La Quina 5[d]	N	2.49 (mand) / 26.87 (max)	5.97 (left) / 18.41 (right)	8.46 & **20.90**
	Švédův stůl (Ochoz) 1	N	**24.12**	—	—
	Tabun 1	N	6.42	—	—
	Dolní Věstonice 13	UP	—	1.39	—
	Dolní Věstonice 15	UP	1.51	—	—
	Mladeč 2	UP	0	—	—
	Pataud 1	UP	1.45	0.56	2.01
M2					
	Arago 2	MP	0	—	—
	Mauer 1	MP	1.21	—	—
	Montmaurin 1	MP	0	—	—
	La Quina 5	N	—	—	1.57
	Švédův stůl (Ochoz) 1	N	6.26	—	—
	Tabun 1	N	**35.61**	**40.61**	5.00
	Dolní Věstonice 13	UP	0	0	0
	Lachaud 3	UP	0	0	—
	Pataud 1	UP	0	—	—
	Předmostí A17088	UP	0	—	—
M3					
	Mauer 1	MP	9.45	—	—
	Montmaurin 1	MP	0	—	—
dm					
	Devil's Tower 1 dm1	N	—	—	3.53
	Devil's Tower 1 dm2		—	—	3.58

Note: Values equaling 0% difference are always related to the lack of DE for the individuals in at least one of the teeth analyzed (none of the individuals presented exactly the same wear). Differences of more than 10% are highlighted in boldface. The sample designations for the specimens are Middle Pleistocene (MP), Neandertal (N), and Upper Paleolithic (UP).

[a]Difference Side indicates that the dental arcade remains the same, and the differences are due to laterality (for example, left P^3 versus right P^3).

[b]Difference Arcade refers to individuals with the same side, but the teeth are from different dental arcades: maxillary versus mandibular (for example, left P^3 versus left P_3).

[c]Diff. Side + Arcade indicates that the individual provides teeth from different arcades and sides (for example, left P_3 and right P^3).

[d]The La Quina 5 M1s provide double values, as the four teeth are present and therefore all of the comparisons are made twice.

TABLE 10.7. Differences in dentin exposure values (DE) for Palomas 1

		DIFFERENCE SIDE[a]		DIFFERENCE ARCADE[b]		DIFF. SIDE + ARCADE[c]
P3	Maxilla	3.50	Right	**25.61**	Right-Left	—
	Mandible	—	Left	—	Left-Right	**29.11**
P4	Maxilla	**14.85**	Right	8.94	Right-left	—
	Mandible	—	Left	—	Left-Right	**23.79**
M1	Maxilla	1.62	Right	**23.35**	Right-left	9.04
	Mandible	**14.31**	Left	7.42	Left-Right	**21.73**
M2	Maxilla	2.26	Right	5.64	Right-left	3.62
	Mandible	2.02	Left	5.88	Left-Right	7.90
M3	Maxilla	**57.85**	Right	—	Right-left	—
	Mandible	—	Left	—	Left-Right	—

Note: See Table 10.1 for the raw values. Differences of more than 10% are highlighted in boldface.
[a]Difference Side indicates that the dental arcade remains the same, and the differences are due to laterality (for example, left P³ versus right P³).
[b]Difference Arcade refers to individuals with the same side, but the teeth are from different dental arcades: maxillary versus mandibular (for example, left P³ versus left P_3).
[c]Diff. Side + Arcade indicates that the individual provides teeth from different arcades and sides (for example, left P_3 and right P³). Right-Left indicates the maxillary right tooth DE minus that for the mandibular left tooth, and the reverse for Left-Right.

TABLE 10.8. Correlation coefficients and probability values for the teeth and buccal microwear variables for which the microwear and the dentin exposure percentage (DE) are correlated within each sample

	MEASUREMENT	CORRELATION COEFFICIENT (R)	PROBABILITY (P)
Early Modern Humans			
P4	ST	0.560	0.047
Neandertals			
P3	XV	0.690	0.040
	XDM	0.739	0.023
	XT	0.747	0.021
M1	XV	0.551	0.005
	SV	0.456	0.025
	XT	0.446	0.029
	ST	0.492	0.015
M3	NH	0.717	0.030
	NDM	0.751	0.020

Note: See Tables 10.3–10.5 for the full names of the measurements.

11

The Palomas Dental Remains

Nonmasticatory Wear

JOHN C. WILLMAN

THE USE OF TEETH AS TOOLS for non masticatory, manipulative behavior has long been a central tenet of Neandertal behavioral reconstructions and adaptive scenarios for the evolution of a suite of their craniofacial and dental traits (Brose and Wolpoff 1971; Smith 1983; Rak 1986; Trinkaus 1987; Smith and Paquette 1989; Antón 1990, 1994, 1996; Spencer and Demes 1993; Brace 1995; Dobson and Trinkaus 2002; O'Connor et al. 2005; Weaver 2009). In particular, the large anterior dentition, in terms of overall crown dimensions, mass-additive discrete morphology (features of the crown that increase its volume, such as labial convexity, shovel shape, and lingual tubercles), and root dimensions (Trinkaus 2004; Bailey 2006; Le Cabec et al. 2013), is regularly viewed as functionally adapted to high and/or repetitive dental loading and a means of maintaining functional proficiency throughout a lifetime of cumulative dental attrition. However, these traits are likely to be the ancestral condition for the Neandertals, since they are shared with Middle Pleistocene humans and overlap considerably with Middle and Upper Paleolithic early modern humans (Smith and Paquette 1989; Trinkaus 2004; Bailey 2006; Martinón-Torres et al. 2012; Le Cabec et al. 2013).

In addition to the morphological attributes of archaic dentitions, dental wear provides direct evidence for elevated rates of the use of the anterior dentition relative to the postcanine dentition through the analysis of overall occlusal wear (Trinkaus 1992; Bermúdez de Castro et al. 2003; Lozano et al. 2008; Clement et al. 2012) and of particular dental wear features (for example, enamel chipping, labial rounding, anterior nonocclusion, occlusal striations and gouging, labial cut marks, labial microwear and microwear texture, parafacets, subvertical grooves, and so on) (Ryan 1980; Brace et al. 1981; Lalueza-Fox and Frayer 1997; Henry et al. 2006; Lozano et al. 2008; Frayer et al. 2010; Hillson et al. 2010; Krueger and Ungar 2012; Volpato et al. 2012; Estalrrich et al. 2011; Estalrrich and Rosas 2013, 2015; Fiorenza and Kullmer 2013; Hlusko et al. 2013; Fiore et al. 2015).

Perhaps the most quintessential of Neandertal nonmasticatory behaviors is the so-called "stuff and cut" behavior (Brace 1975), typically attributed to the production of labial cut marks on the enamel of the anterior dentition (and occasionally on the premolars). This behavior is well known in the ethnographic literature, and a review of ethnographic sources shows variation in the type of tools used for cutting (for example, metal knives, stone tools, and bamboo) as well as cutting motions: oblique, horizontal, upward, and downward (Marshall and Gardner 1957; Weyer 1959; Woodburn and Hudson 1966; Semenov 1964; Merbs 1983; Uomini 2008, 2011). The orientations of instrumental striations may be indicative not only of the handedness of an individual but also of the behaviors practiced (Estalrrich and Rosas 2013; Fiore et al. 2015).

Since labial cut marks were first documented by Martin (1923) on the La Quina 5 dentition, a series of qualitative and quantitative descriptions of labial cut marks have been described for Early Pleistocene humans from Gran Dolina (Bruner and Lozano 2014); Middle Pleistocene humans from Boxgrove (Hillson et al. 2010), Broken Hill (Lalueza-Fox and Pérez-Pérez 1994), Mauer (Puech et al. 1987), Pontnewydd (Compton and Stringer 2012), and Atapuerca-SH (Bermúdez de Castro et al. 1988; Lozano-Ruiz et al. 2004; Lozano et al. 2008, 2009; Frayer et al. 2012); and Neandertal specimens from Angles sur l'Anglin (Patte 1960), Cova Negra (Bermúdez de Castro et al. 1988; Arsuaga et al. 2001b), Hortus (de Lumley 1973; Bermúdez de Castro et al. 1988; Estalrrich and Rosas 2015), Krapina (Lalueza-Fox and Frayer 1997; Fiore et al. 2015), Regourdou 1 (Volpato et al. 2012), Saint-Brais (Koby 1956), Shanidar (Trinkaus 1983), El Sidrón (Estalrrich and Rosas 2013, 2015), Spy (Estalrrich and Rosas 2015), Tabun (Lalueza-Fox and Pérez-Pérez 1994), and Vindija (Frayer et al. 2010). Some additional evidence comes from bioarchaeological studies of recent humans (Bax and Ungar 1999; Molnar 2008; Dinnis et al. 2014), but only labial cut mark presence has been documented for early modern human dentitions (Willman 2016).

Antemortem chipping has received comparatively less attention in the paleobiological literature, but some studies have addressed its occurrence in the Middle Pleistocene specimens from Atapuerca-SH (Lozano et al. 2008) and Neandertals (Lalueza-Fox and Frayer 1997; Estalrrich and Rosas 2015). Chipping provides data on unique events that record loading of the anterior dentition beyond the ability of the enamel to resist fracture. In addition, the analysis of Neandertal occlusal wear gradients has been underutilized with the exception of a few studies (Trinkaus 1992; Clement et al. 2012). Although anterior teeth are used for both ingestive and nonmasticatory behaviors, the relative wear on the anterior versus posterior dentition can reveal important variation in the relative use of the anterior dentition among diverse paleobiological groups.

Here, an analysis of relative anterior to posterior scaled occlusal wear gradients, labial cut marks, and enamel chipping is provided for the dental remains from Sima de las Palomas in comparison to available data for Middle Pleistocene archaic humans, Neandertals, Middle Paleolithic modern humans (MPMH), and Early/Mid Upper Paleolithic (E/MUP) humans. The publication of relevant raw data will allow the Palomas material to be more broadly compared in future studies in which researchers have an interest in the paleobiological implications of anterior tooth use for nonmasticatory behaviors among various groups of Pleistocene humans.

Materials and Methods

Macroscopic Observations and Scanning Electron Microscope Sample Preparation

All teeth included in this study were observed in person by the author. Initial assessments of crown and enamel preservation were made and scaled photographs using a Nikon D90 and macro lens were taken for reference. Occlusal wear was documented following Smith (1984) and checked against observations made previously (see chapter 6). The buccolingual crown dimensions in chapter 7 were used for the scaled macrowear gradient analysis.

All anterior teeth were cleaned with cotton-tipped applicators and 70% alcohol and then air-dried. Coltène President Plus Jet light-body polyvinylsiloxine (Coltène/Whaledent) was used for dental molding because of its accurate replication properties (Beynon 1987; Hillson 1992; Pérez-Pérez et al. 2003; Galbany et al. 2004b; Fiorenza et al. 2009). All molds were placed in dust-free plastic bags immediately after removal. Epo-tek 301 epoxy was used to make positive casts. Casts were mounted on aluminum pin stubs and sputter-coated with ≈20 nm of gold to ensure conductivity during scanning electron microscope (SEM) analysis.

A JEOL Neoscope JCM-5000 SEM with an accelerating voltage of 15 kV and working distance of 10–30 mm was used to microscopically examine taphonomic surface alterations and antemortem dental wear features on each dental cast. High-vacuum secondary electron emission and low-vacuum modes were used as needed. Both taphonomic and dental wear feature analyses began with a field of view that encompassed an entire occlusal or labial surface of each tooth before magnification was increased to identify features of variable size (Lozano et al. 2008).

Scaled Macrowear Gradients

Palomas 1 and 59 offer the most complete dentitions for assessing rates of anterior versus posterior occlusal macrowear scaled to buccolingual crown dimensions. This method uses the Smith (1984) wear score multiplied by the buccolingual (BL) crown dimension for each tooth. BL dimensions were made even if the root was functioning as the occlusal surface (Smith score 7–8) at the time of death. The sums of all anterior scaled tooth wear scores (although some analyses include only canines, see below) and sums of all postcanine scaled tooth wear scores are contrasted in bivariate plots. This method is the same as that of Trinkaus (1992), except that Smith (1984) wear scores are used rather than those of Molnar (1971), and crowns worn beyond maximum BL dimensions are included in the present analysis. Antimeres were averaged when present (wear and BL breadth) or the individual values were used if only one side was preserved. Maxillary and mandibular arcades are considered separately.

Individuals with all anterior teeth preserved *in situ* are not frequently found in Late Pleistocene samples, and the Sima de las Palomas sample is not an exception. Only anterior scaled wear scores for C^1 were available for Palomas 1, although a full set is available for the mandible. Palomas 59 preserves the only other relatively full set of mandibular teeth (C_1–M_2) for analysis. To increase sample sizes, analyses were first done using all postcanine teeth available, followed by a second analysis without M3s. The latter case allowed the inclusion in the comparative sample of individuals that lack fully erupted or occluding M3s as well as individuals with congenitally absent M3s.

The scores on each axis are a product of dental wear and crown (or root) size. The BL breadth of a tooth does not decrease dramatically with progressive wear. The exception is in older individuals that had functional dentitions despite being worn largely to root surfaces. However, changes in BL dimensions with age are also minimized by summing breadths across all available anterior or posterior teeth in an arcade. Since the width of a crown will not fluctuate too widely as the dimensions decrease under continual occlusal wear *in vivo*, a high value on either axis is going to be heavily influenced by occlusal wear. Larger values on one axis would indicate large BL dimensions and high wear and vice versa. Grade differences between individuals or groups are indicated by differences in relative slope.

Comparative data on Neandertals (Krapina 48, 50, 55, and 58; La Chaise BD 1; Ehringsdorf F 1010/69; Tabun 1; Saccopastore 2; Spy 1 and 2; Švédův stůl (Ochoz), Arcy-Hyène 8 and 9; Arcy-Bison P11.8; La Ferrassie 1; Genay 1; Monsempron 3; Le Moustier 1; Les Pradelles (Marillac) 4; Regoudou 1; La Quina 5; Saint-Césaire 1; Subalyuk 1; Banyoles 1; Valdegoba 1; Zafarraya 2; Shanidar 1, 2, and 6; Kebara 2; and Amud 1), Middle Paleolithic modern humans (MPMH: Skhul 4 and 5; Qafzeh 6, 7, 9, and 11), and Early/Mid Upper Paleolithic humans (E/MUP: Tianyuan 1; Brno 3; Dolní Věstonice 3, 13, 14, 15, and 16; Pavlov 1; Předmostí 1, 3, 4, 9, and 10; Wajak 2; Nahal 'En-Gev 1; Ohalo 2; Arene Candide IP; Barma Grande 2, 3, and 4; Ostuni 1; Tam Pa Ling 1; and Sunghir 2) were collected by the author from the original fossils and the literature.

Enamel Preservation and Postmortem Alteration

Assessments of taphonomic alterations of the enamel surfaces at Palomas are key to discriminating between antemortem and postmortem dental wear in the sample. Taphonomic alteration differentially affects deciduous teeth and unerupted and erupted permanent teeth

(Martínez and Pérez-Pérez 2004), yet postmortem abrasion, chemical erosion, and pristine enamel have distinct appearances that can be readily distinguished by microscopic observation (King et al. 1999; Pérez-Pérez et al. 2003; Martínez and Pérez-Pérez 2004). For these reasons all anterior teeth were examined microscopically to eliminate specimens that exhibit extensive postmortem enamel alterations that would bias results. Unerupted and erupting teeth that were not in functional occlusion, as well as the deciduous teeth of infants (Palomas 26, 39, 89, and 95), were examined prior to analyzing teeth worn *in vivo*. These teeth were largely subject to postmortem alterations, in contrast to teeth that were in functional occlusion during an individual's lifetime. As such, these teeth provide a baseline assessment of taphonomic surface alterations, since teeth in occlusion were subject to taphonomic alterations and *in vivo* dental wear through masticatory and nonmasticatory behaviors. These teeth are not described in detail here but provide context for taphonomic assessments of teeth that were in use at time of death.

Differential preservation of enamel is often found on a single crown. For some of these cases the presence or absence of chipping or cut marks on areas of well-preserved enamel can be documented, but taphonomic alterations on other areas of the tooth prevented full documentation of all possible antemortem wear features. Thus, all quantitative analyses may not be possible for a given tooth or feature. For example, postmortem enamel abrasion or fracture may destroy the start or end point of a labial cut mark, which eliminates the length measurement but may preserve a width and/or the dominant orientation.

Postmortem fracture is extensive among the Sima de las Palomas sample but can be easily distinguished from antemortem tooth chipping on the basis of the surface color of fractures and the presence or absence of rounded wear on the edges of fractures (Scott and Winn 2011). Given these potential preservation biases across the sample, in addition to differential enamel preservation biases for individual tooth surfaces, each type of data collected is listed in Table 11.1.

Labial Cut Marks and Enamel Chipping

Labial cut marks show a distinct morphological signature that assists in their identification and distinguishes them from dietary and taphonomically induced striations. Dietary microwear tends to be narrow (≤5 μm) in comparison to the much wider and longer striations of instrumental origin, which are often visible without magnification (Trinkaus 1983; Bermúdez de Castro et al. 1988; Lalueza-Fox and Pérez-Pérez 1994; Lalueza-Fox and Frayer 1997). Cutoffs for measuring the width of instrumental striations vary in the literature but are generally >10 μm. Cut marks frequently range from 20 μm to 40 μm, although some exceed 100 μm (Bermúdez de Castro et al. 1988; Lalueza-Fox and Pérez-Pérez 1994; Lalueza-Fox and Frayer 1997). In addition, experimental replication of labial cut marks shows a propensity for microstriations and hertzian cone fractures along the border of a cut mark, a V-shaped cross-section with microstriations in the groove, and patterned orientations reflecting handedness (Bromage and Boyde 1984; Bermúdez de Castro et al. 1988; Bromage et al. 1991; Lozano-Ruiz et al. 2004; Frayer et al. 2010; Estalrrich and Rosas 2013). *In vivo* wear creates a palimpsest of dietary and nondietary features and can round the edges of older cut marks (Frayer et al. 2010). However, the depth, width, and length of instrumental cut marks allow their identification even in the presence of some wear.

Length, width, and orientation are the most frequently analyzed variables in cut-mark assessment, although analyses focused on handedness may ignore length or width. Raw angle measurements are taken with the occlusal plane as a reference, but the cervix may be referenced to assist orienting canines with little occlusal wear. Raw measurements are subsequently condensed into four broad categories following Bermúdez de Castro and colleagues (1988): horizontal (0–22.5° or >157.5–180°), right oblique (>112.5–157.5°), vertical (>67.5–112.5°), and left oblique (>22.5–67.5°). Frayer and colleagues (2010, Fiore et al. 2015) have suggested the use of broader categories for determining orientation for right striations (95–175°) and left striations (5–85°), so these categories are also included to increase comparability between studies. In addition, the latter system is more sensitive for attributing handedness.

Length and width are more difficult to measure in the presence of extensive postmortem surface alterations, such as that seen in the Palomas sample. The end points of cut marks are frequently obscured by matrix, fracture, erosion, abrasion, or adhesive residue on enamel surfaces. As such, absolute length is not always measurable, even if a cut mark is clearly identifiable. In these cases no length measurement is taken, but orientation and width may still be measurable (Fig. 11.2A). While cut-mark widths are less affected by an attenuation of striation length due to postmortem damage, they are affected by other factors. Most significantly, cut marks are worn during the life of an individual through erosion by saliva (Frayer et al. 2010), the continued abrasion by particles in food, and the use of the teeth as tools. Postmortem alterations are also a concern. Widths are measured at the midpoint when full lengths are preserved or at a location that appears to reasonably record an average striation width when the true midpoint is obscured.

In sum, a conservative approach to the inclusion of teeth in analyses first takes postmortem taphonomic alterations into account, and it is followed by the exclusion of individual cut-mark measurements as dictated by local preservation on a tooth surface.

Antemortem Enamel Chipping

Antemortem chipping records individual events of high loads produced on hard objects that exceed the ability of enamel to withstand fracture (Constantino et al. 2010). The scoring system of Bonfiglioli et al. (2004:449) is used here to categorize chip size when present. A grade 1 chip indicates a slight crack or fracture ($\approx$0.5 mm), or a larger but superficial loss of enamel. Grade 2 chips are indicated by square, irregular lesions ($\approx$1 mm) with the enamel more deeply involved. A grade 3 chip is a fracture >1 mm that involves

enamel and dentine or a large, irregular fracture that could destroy the tooth.

Postmortem alterations—especially heavy pitting, abrasion, and enamel fracture—affect many occlusal surfaces in the Sima de las Palomas sample, and many individual teeth were eliminated from the chip frequency analysis for this reason.

Postmortem Taphonomic Alterations and Observable Dental Wear Features by Tooth

The enamel preservation status for each tooth is presented below. A summary of observable analytical data for each tooth is summarized in Table 11.1. Furthermore, Fig. 11.1 demonstrates several different forms of postmortem damage found in the sample, and examples of dental wear features can be seen in Fig. 11.2.

Palomas 1a (RC¹)

There is some enamel exfoliation and a large fracture on the mesiolabial crown near the cervix. Any striations of cultural origin on the remaining surface are obscured by glue. There is an antemortem occlusal chip on the distobuccal corner bordering the P³ contact facet.

Palomas 1b (LC¹)

Glue on the labial surface, similar to that of the Palomas 1a, obscures most of the surface and prevents an assessment of cut marks. The heavily abraded occlusal surface cannot be analyzed.

Palomas 1c (RI₁)

The labial surface is eroded with several areas of enamel exfoliation and a large postmortem fracture on the distal occlusal edge. Several striations are present, but postmortem alterations make it difficult to discern ante- or postmortem cause. Chipping cannot be assessed due to significant postmortem damage to the occlusal surface.

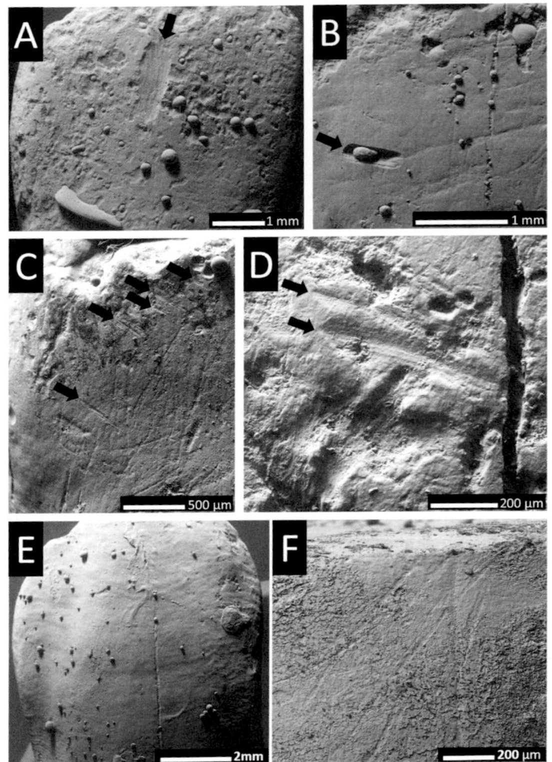

FIG. 11.1. Examples of taphonomic alterations to enamel at Sima de las Palomas. It should be noted that the spherical objects on some surfaces are preservational defects/inclusions (see especially, *A–C, E*). *A*: Palomas 89 is an unerupted tooth exhibiting extensive enamel exfoliation and a large furrow caused by postmortem abrasion (*arrow*). *B*: Palomas 79 shows enamel flaking in the upper left corner of the image and a large postmortem furrow with sharp edges on the lower left portion of the image (*arrow*). *C*: Arrows indicate the direction of a series of parallel striations on the labial surface of Palomas 97 (RI$_1$) caused by postmortem contact with abrasive material. This feature resembles damage by trampling. *D*: There are two wide furrows (*arrows*) on the labial surface of Palomas 19. They may be cut marks, but postmortem damage to the surrounding enamel is too extensive to be certain. *E*: extensive glue covering the surface of Palomas 26. Clumps and "ribbons" are clearly visible. *F*: A layer of glue covers the entire surface of Palomas 1 (RC$_1$). Cut marks can be seen underneath the glue, but length and width measurements are obscured.

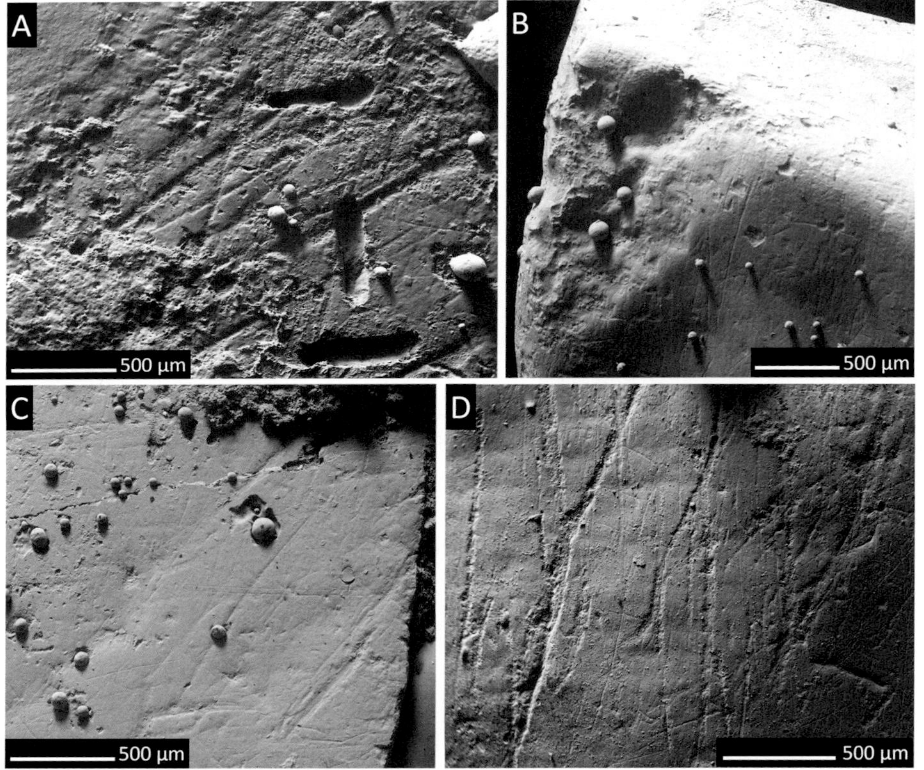

FIG. 11.2. Examples of dental wear features from Sima de las Palomas. *A*: Cut marks on the surface of Palomas 73 showing lengths that are clearly obstructed by taphonomic alterations but clear RO orientation and some measurable widths. *B*: Antemortem enamel chip on mesial-occlusal edge of Palomas 59 with clearly rounded edges (despite some postmortem exfoliation). *C*: Several broad to thin cut marks on the surface of Palomas 90, predominately RO with one H orientation. Fracture edge is on the right side of the image. *D*: Cut marks on the surface of Palomas 97 (LI$_1$) showing V to RO orientation of most striations.

Palomas 1c (RC₁)

Glue obscures much of the labial surface. Cut marks are visible under the layer of glue, with a particularly prominent concentration on the mesial edge near the occlusal surface (Fig. 11.1F). Unfortunately, length and width are not quantifiable because absolute dimensions are obscured by glue. However, many striations are large, and orientation is measurable. A very small chip (>0.5 mm) on the mesiobuccal edge of the interproximal facet is probably postmortem, and no other antemortem chips were observed.

Palomas 1d (LC₁)

The majority of the labial surface was fractured and lost postmortem. There is too little enamel for any analysis.

Palomas 18 (LC₁)

Extensive postmortem enamel fracture, postmortem microcracking, pitting, erosion, and glue blanket the labial surface, although some dietary microwear is present in isolated areas. Potential instrumental cut marks are visible, but postmortem damage has altered their morphology to such an extent that they do not meet conservative criteria for inclusion in this study (Fig. 11.1D). The occlusal surface is abraded, but there is no clear evidence of antemortem chipping.

Palomas 19 (LI₁)

Enamel fracture and abrasion are prominent on labial and occlusal surfaces. There is extensive postmortem enamel exfoliation in the superior third of the labial face, most significantly at midline but dissipating toward the mesial and distal edges. Inferior to that damage is an area of matrix (≈1 mm × 1 mm) and a large postmortem enamel fracture along the distal margin of the labial surface. The distal fracture is less problematic than the midline enamel loss, because cultural striations accumulate less frequently along the distal margins as one moves toward the cervix on a mandibular anterior tooth (personal observation). There is only modest adhesive residue, which is concentrated near the cervix. Importantly, several wide striations are partially overlaid by matrix, indicating an antemortem, instrumental origin. Data on cut-mark length and width were collected. No data on occlusal chipping were pursued due to the uneven occlusal enamel preservation.

Palomas 20 (LI₂)

There are isolated instances of enamel flaking elsewhere, but it is infrequent and clearly distinguished from antemortem wear. Dietary microwear is present across the labial surface, but there are no striations approaching 20 µm wide that would clearly indicate instrumental origin. Thus, there is no clear evidence for the presence of instrumental cut marks on the labial surface, but a layer of glue makes an assessment of complete absence untenable. Likewise, glue and postmortem abrasion prevent assessment of occlusal enamel chipping. No analyses were possible.

Palomas 21 (LI₁)

Both the labial and occlusal surfaces have large areas of postmortem enamel spalling, abrasion, and extensive glue covering portions of each surface. In isolated areas dietary microwear is present, but no striation measures approach 20 µm. Palomas 21 was excluded from dental wear feature analyses on the basis of poor preservation.

Palomas 24 (LI¹)

The labial enamel is extremely fragmentary, and in addition, matrix and glue obstruct most of the enamel surface that is free of postmortem fracture. The occlusal surface is heavily damaged as well. Palomas 24 was not included in dental feature analyses.

Palomas 31 (Ldc¹)

Most of the surface is obscured by postmortem alterations, including glue, matrix, erosion, and subtle enamel flaking. Dietary microwear is visible in some places; overall, the poor preservation rules out any dental wear feature analyses.

Palomas 34 (LI¹)

Several striations of possible cultural origin are present in addition to the presence of dietary microwear on isolated areas of the labial surface. However, the labial surface is badly eroded and exfoliated, and glue and matrix are widespread. Conservative criteria exclude the tooth from labial cut-mark analysis. The tooth has subtle occlusal wear with visible mammelons, but there is no sign of antemortem chipping.

Palomas 35 (LC¹)

There is heavy erosion in addition to glue on the labial surface. A large postmortem fracture in the center of the crown extends distally. The surface texture of the perikymata clearly demonstrates extensive chemical erosion of the labial surface. Cut-mark analysis is not possible. The occlusal enamel is generally well preserved. Several small, postmortem chips are missing, but the mild rounding of their edges is due to erosion rather than *in vivo* wear. A larger chip on the mesiolabial edge of the occlusal surface has a bright white color that stands out against the dull color of the rest of the enamel. This color indicates a postmortem fracture. No antemortem chipping was identified.

Palomas 44 (RC₁)

There is some postmortem erosion and diffuse enamel exfoliation, but dietary microwear indicates well-preserved enamel elsewhere. Two possible cut marks are present, but these features are eroded postmortem. Cut-mark analysis was not attempted due to the low number and questionable etiology of the cut marks. There are no antemortem chips.

Palomas 48 (RI²)

The enamel is heavily exfoliated and eroded, with glue and matrix obscuring most of the remaining enamel. Some cut marks may be present beneath the residue, but overall the surface is too degraded for analysis. The combination of abrasion and erosion has made it impossible to differentiate between antemortem and postmortem chipping, so this tooth was not included in analyses.

Palomas 54 (RC₁)

Only a small portion of mesial enamel, close to the occlusal surface and spanning a few millimeters vertically and laterally, is moderately well preserved. There is some evidence of dietary microwear in this area, but it is also somewhat eroded. No cut-mark analysis was possible. The occlusal surface is minimally worn, and there is no evidence of antemortem chipping.

Palomas 59 (LC₁)

Overall, the labial surface is well preserved, with minimal taphonomic alteration. The mesial edge, about halfway down the crown, exhibits some enamel exfoliation, but it interferes minimally with cut-mark analysis. Other alterations are few and easily differentiated from antemortem alterations. Dietary microwear is prevalent. The mesial corner of the occlusal surface exhibits a postmortem chip (Fig. 11.2B).

Palomas 73 (RI¹)

The labial surface is heavily abraded and eroded, with large portions of enamel missing postmortem. Isolated pieces of breccia still adhere to the enamel, and some glue is present. Despite extensive postmortem alteration, there are still a noteworthy number of antemortem cut marks on the labial surface that are clearly distinguished from postmortem wear (Fig. 11.2A). These instrumental striations are concentrated most heavily from midline to the mesial edge and continue within a few millimeters of the cervix. The distal side of the tooth is more heavily modified by postmortem alterations, so cut-mark analysis across the entire labial surface is not possible. Damage to the occlusal surface is too extensive for chipping to be assessed.

Palomas 74 (LC¹)

Chemical erosion of the labial surface is indicated by the heavily accentuated and exposed perikymata, especially on the distal portion of the crown. The distal portion is also most heavily affected by enamel exfoliation, although it is also found in a relatively even and diffuse distribution across the rest of the crown. These patches are easily differentiated from antemortem wear. The mesial half of the crown preserves some dietary microwear as well as instrumental striations, which are located where they would be expected on a canine (Bermúdez de Castro et al. 1988; Lalueza-Fox and Frayer 1997). The occlusal edge is somewhat exfoliated, but it is also relatively unworn. There is no evidence of antemortem chipping.

Palomas 79 (LI¹)

The distal portion of the labial surface is heavily fractured and missing postmortem. The fractures extend across the mesial face but to far less effect than on the distal side (Fig. 11.1B). Diffuse enamel exfoliation is seen elsewhere and easily distinguished from antemortem wear. Antemortem cut marks are numerous on the remaining enamel, although the absolute lengths of many individual cut marks are not measurable due to postmortem damage. Widths, some lengths, and orientations are measurable for remaining cut marks. The occlusal edge is heavily battered postmortem, so antemortem chipping could not be assessed.

Palomas 82 (LC₁)

The occlusal surface is heavily fractured postmortem, with extensions running down the midline of the labial surface. Chemical erosion, abrasion, and enamel exfoliation is also present. Cut marks are still visible, with the heaviest concentration on the mesial surface. Antemortem chipping could not be assessed.

Palomas 85 (Ldi²)

The labial surface is heavily eroded, and large portions of enamel are fractured and missing postmortem. No cut marks were visible on the labial surface, but the extent to which the enamel is altered postmortem removes it from consideration in dental wear feature analyses. Similarly, the occlusal rim of enamel is also heavily altered postmortem, preventing the analysis of enamel chipping.

Palomas 90 (LI¹)

The distal-occlusal quadrant of the labial surface is fractured and missing postmortem. Erosion and exfoliation are prevalent but diffusely spaced. Several large pieces of matrix still adhere to the labial surface. Extensive cut marks can still be recorded despite postmortem alterations (Fig. 11.2C). Too much occlusal surface is missing for an adequate assessment of antemortem enamel chipping.

Palomas 91 (LI₂)

The occlusal surface is heavily damaged due to postmortem processes and cannot be analyzed. The labial surface is also missing very large portions of enamel in several places. Postmortem chemical erosion accentuates the perikymata, especially toward the distal margins. Despite postmortem alterations, a considerable amount of dietary microwear is found on the labial surface. No dental feature analysis was possible.

Palomas 93 (Rdc₁)

Postmortem erosion and abrasion is extensive on this deciduous crown. No dental feature analysis was possible.

Palomas 97 (LI₁)

The labial face is extremely well preserved. Only a small amount of glue and matrix is present. Instrumental striations are present (Fig. 11.2D).

The labio-occlusal edge was observable, and there is no sign of antemortem chipping.

Palomas 97 (RI$_1$)

The labial surface has little postmortem enamel damage. Most damage is near the occlusal surface, near the mesial edge, which has modest enamel exfoliation. The distal edge has isolated abrasion near the occlusal surface that resembles trampling (Fig. 11.1C). A fine layer of residue adheres to the majority of the occlusal half of the labial crown and obscures some features. Cut marks are present across much of the surface. Definitive antemortem chipping is present on the labio-occlusal edge.

Palomas 97 (RI$_2$)

The labial face is roughly 70% intact. The entire occlusal surface and the majority of the distal edge are fractured and missing postmortem. Nevertheless, there are well-preserved instrumental striations on the remaining labial surface. Antemortem chipping could not be assessed.

Palomas 98 (RI$_2$)

The distal edge of the labial face is covered in matrix, there is a large postmortem fracture running the length of the crown near midline, and heavy enamel exfoliation affects the superior half of the crown. Some dietary microwear and a potential cut mark are visible, but there is too much postmortem damage to attempt quantification. Likewise, the occlusal surface is too damaged for the assessment of chipping.

Associations between Isolated Dental Elements

The possibility that multiple isolated teeth are associated with the same individual is an issue that can influence the group-level ratio of right- to left-handedness within the Palomas sample. Of the teeth with observable cut marks (Table 11.1), some associations between isolated elements can be made, while others can be ruled out.

The RC$_1$ of Palomas 1 is easily isolated from the rest of the sample, since there are no isolated RI$_2$s that had preserved cut marks, and the only isolated maxillary central incisors exhibit less wear than the upper canines associated with Palomas 1. Thus, the RC$_1$ of Palomas 1 can be considered a unique individual.

Given the high prevalence of linear enamel hypoplasia (LEH) formation in maxillary central incisors, maxillary canines, and mandibular canines (Goodman and Rose 1990), the location of LEH can assist in matching isolated dental elements. Thus, Palomas 73 and 79 can be paired with some certainty on the basis of similar occlusal wear, similarity in cut-mark orientation, proportions, and the width and location of LEH. Palomas 19 (LI$_1$) may also correspond to Palomas 73 and 79, as it exhibits a similarly broad LEH in the cervical third of the crown as well as matching dental wear (Smith score 3), although this association is less certain.

Similarly, two wide and very shallow hypoplasias are visible on Palomas 74 (LC^1) and 82 (LC$_1$), both of which are attributed to juveniles with minimal occlusal wear (chapter 6). Therefore, Palomas 74 and 82 represent another possible association.

Palomas 19 (LI$_1$) does not appear to correspond to Palomas 59 (LC$_1$) or 90 (LI^1), as the latter represent tooth types that are more susceptible to LEH but nonetheless lack the distinct LEH seen on Palomas 19. Palomas 59 and 90 do share similar occlusal wear, but little more can be said for associating them.

Possible, probable, and certain associations of isolated elements are clearly noted in Tables 11.2 and 11.4. A conservative estimate of the minimum number of individuals is used in the final interpretation of handedness among the Palomas Neandertals.

Scaled macrowear gradients

Figs. 11.3 to 11.7 show bivariate plots of scaled anterior versus posterior macrowear gradients for the Palomas 1 maxilla and mandible and the Palomas 59 mandible. The maxillary canine wear of Palomas 1 relative to the postcanine dentition follows the overall trend seen across the Neandertal, MPMH, and E/MUP human groups (Figs. 11.3 and 11.4). There is no major separation of any particular group in terms of slope (that is, trends in progressive wear). Palomas 1 is relatively unexceptional in this overall trend (Fig. 11.3), even when the M_3 is excluded to increase sample size (Fig. 11.4).

Mandibular anterior wear (I_1–C_1) relative to postcanine wear is somewhat low for Palomas 1 when compared to the overall Neandertal trend, and Palomas 1 groups more closely to the pattern seen among the MPMH and E/MUP sample (Figs. 11.5 and 11.6). This pattern occurs despite the clustering of Palomas 1 largely with the Neandertals (and Middle Pleistocene archaic humans) in comparisons of anterior versus posterior crown dimensions (Fig. 7.6). This position for Palomas 1 indicates modestly less anterior wear relative to its posterior wear. This trend continues when just the C_1 relative to P_3–M_2 is considered for Palomas 1 and the partial arcade of Palomas 59 (Fig. 11.7). Both Palomas arcades fall in line with the general trend seen in MPMH and E/MUP humans—less anterior wear relative to posterior wear—but do not stray far from the general Neandertal pattern.

Labial Cut Marks

A total of 11 teeth preserve enough evidence of labial cut marks to be included in analyses. All 11 provide data on striation orientation, 9 provide length data, and 10 provide data for striation width. Table 11.2 provides descriptive statistics for length and width measurements for each tooth and cumulative data on definitive, probable, and possible tooth associations. The number of each tooth type represented in the sample is small, which makes general length and width trends by tooth type or side difficult to assess in a meaningful way. However, comparisons of Palomas with published, group-level descriptive statistics from Krapina, Atapuerca-SH, and Native Australians (Table 11.3) show some interesting differences.

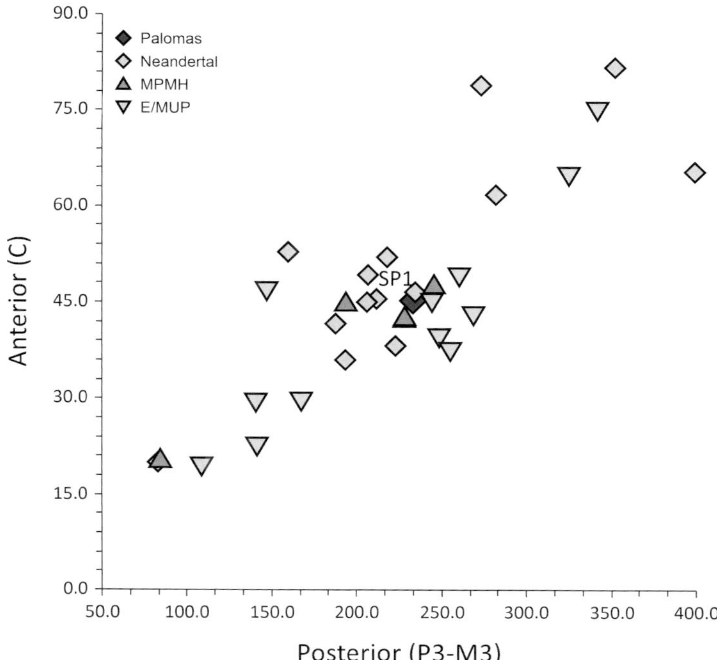

FIG. 11.3. Scaled anterior (C¹) versus posterior (P³–M³) wear gradients for the Palomas 1 maxilla, Neandertals, Middle Paleolithic modern humans (MPMH), and Early/Mid Upper Paleolithic humans (E/MUP).

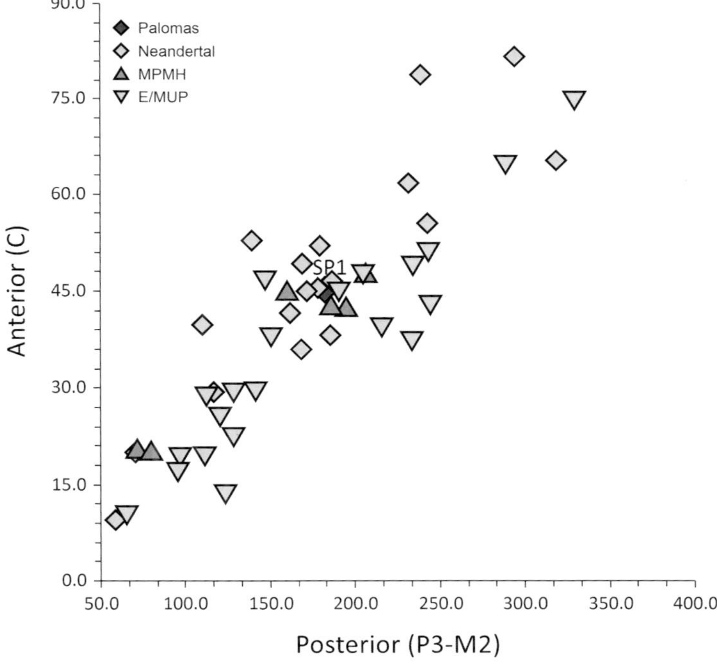

FIG. 11.4. Scaled anterior (C¹) versus posterior (P³–M²) wear gradients for the Palomas 1 maxilla, Neandertals, Middle Paleolithic modern humans (MPMH), and Early/Mid Upper Paleolithic humans (E/MUP).

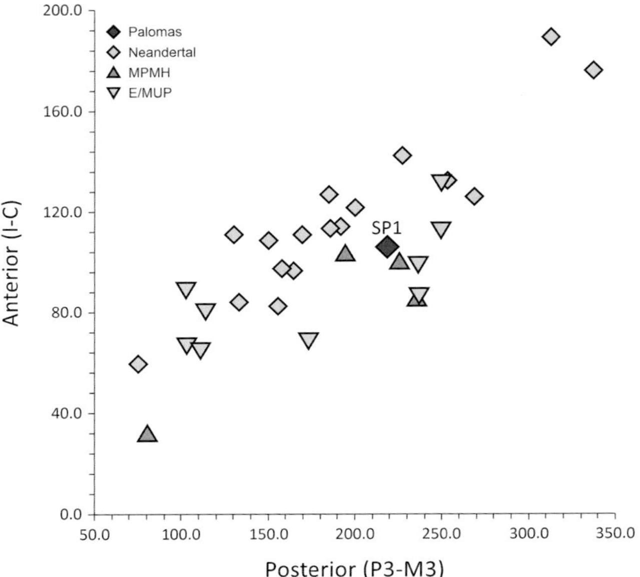

FIG. 11.5. Scaled anterior (I_1–C_1) versus posterior (P_3–M_3) wear gradients for the Palomas 1 mandible, Neandertals, Middle Paleolithic modern humans (MPMH), and Early/Mid Upper Paleolithic humans (E/MUP).

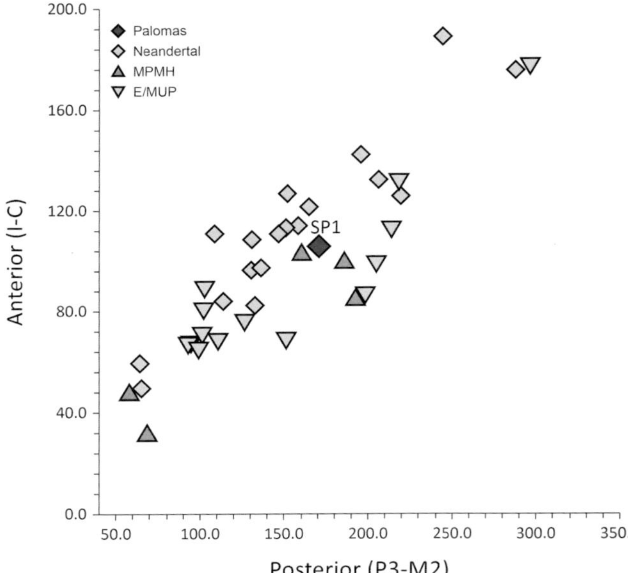

FIG. 11.6. Scaled anterior (I_1–C_1) versus posterior (P_3–M_2) wear gradients for the Palomas 1 mandible, Neandertals, Middle Paleolithic modern humans (MPMH), and Early/Mid Upper Paleolithic humans (E/MUP).

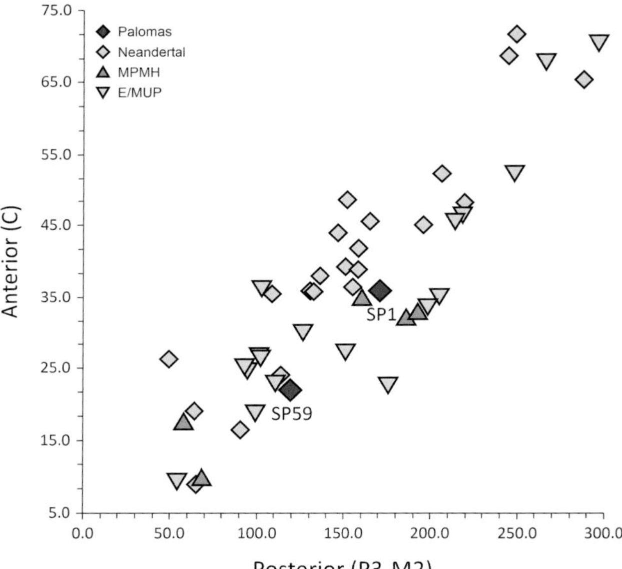

FIG. 11.7. Scaled anterior (C_1) versus posterior (P_3–M_2) wear gradients for the Palomas 1 and 59 mandibles, Neandertals, Middle Paleolithic modern humans (MPMH), and Early/Mid Upper Paleolithic humans (E/MUP).

Mean width stands out as being relatively low in the Palomas sample, but it is nonetheless within the range of variation seen in other groups. The low values may partially result from small sample size and/or erosion of cut-mark edges due to postmortem processes. The Palomas striations are also the shortest on average, but again, they exhibit substantial overlap with other groups.

Handedness is overwhelming right-dominated, whether the Bermúdez de Castro and colleagues (1988) or Frayer et al. (2010) classifications are used (Table 11.4). All maxillary teeth are either right-oblique (Bermúdez de Castro et al. 1988) or right-biased (Frayer et al. 2010), while a few mandibular teeth display tendencies toward vertical and horizontal orientation. This finding is not surprising in the context of other mandibular cut-mark analyses (for example, Volpato et al. 2012, Estalrrich and Rosas 2013; Fiore et al. 2015) and may relate to particular behaviors. All mandibular teeth are right-biased using the more sensitive analysis of handedness (Frayer et al. 2010).

Since it has been suggested that the frequencies

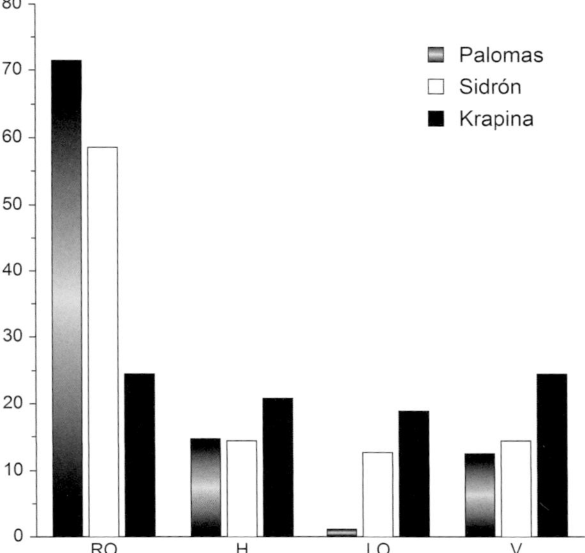

FIG. 11.8. Comparative frequencies of cutmark orientation (RO: right oblique; H: horizontal; LO: left oblique; V: vertical) for the maxillary teeth from Sima de las Palomas, El Sidrón, and Krapina. (Data from this study, Estalrrich and Rosas 2013 and Fiore et al. 2015)

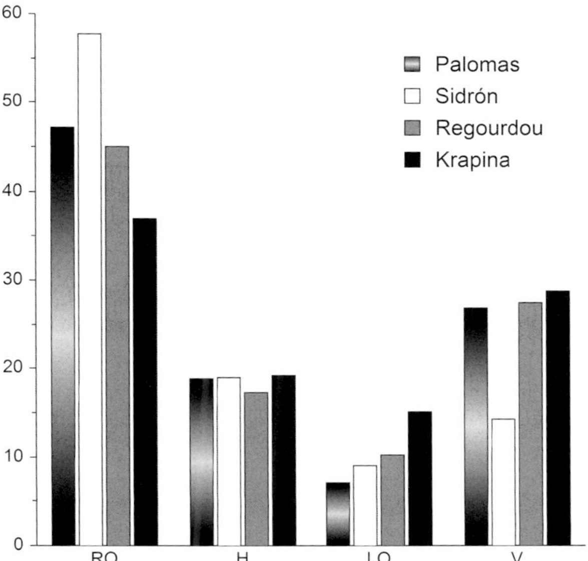

FIG. 11.9. Comparative frequencies of cutmark orientation (RO: right oblique; H: horizontal; LO: left oblique; V: vertical) for the mandibular teeth from Sima de las Palomas, El Sidrón, Regourdou 1, and Krapina. (Data from this study, Estalrrich and Rosas 2013, Volpato et al. 2012, and Fiore et al. 2015)

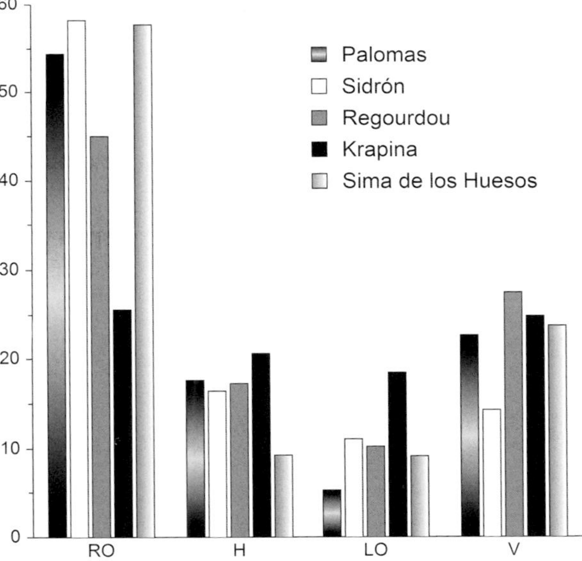

FIG. 11.10. Comparative frequencies of cutmark orientation (RO: right oblique; H: horizontal; LO: left oblique; V: vertical) for the combined maxillary and mandibular teeth from Sima de las Palomas, El Sidrón, Regourdou 1 (mandible only), Krapina, and Sima de los Huesos (Atapuerca). (Data from this study, Estalrrich and Rosas 2013, Volpato et al. 2012, Fiore et al. 2015, and Lozano et al. 2009)

of orientation categories could relate to specific behaviors (Estalrrich and Rosas 2013; Fiore et al. 2015), the frequencies of right-oblique, horizontal, left-oblique, and vertical orientations for Palomas compared to the published literature were plotted for sites with published data in Figs. 11.8 (maxillary teeth), 11.9 (mandibular teeth), and 11.10 (maxillary and mandibular teeth combined). The frequency data again show a proclivity toward right-oblique over left-oblique in maxillae, mandibles, and overall—this is expected. However, the horizontal and vertical categories are around 20% to 30% of total cut marks, respectively, whether examining maxillary, mandibular, or combined frequencies. Vertical cut marks are particularly prominent in mandibles, which may indicate different stuff-and-cut motions or behaviors. Maxillary and mandibular differences should be taken into consideration in future studies rather than pooling all cut marks from both jaws.

The data on handedness from Sima de las Palomas (Table 11.5) fit in well with the general trend toward right hand dominance in Middle and Late Pleistocene Archaic *Homo* (Lalueza-Fox and Frayer 1997; Lozano et al. 2009; Frayer et al. 2010, 2012; Hillson et al. 2010; Uomini 2011; Volpato et al. 2012; Estalrrich and Rosas 2013; Bruner and Lozano 2014; Fiore et al. 2015). Using the most conservative estimate of the number of individuals represented by associated and isolated anterior teeth at Palomas gives a ratio of 5:0 right-handed to left-handed individuals, making the ratio of right- and left-handedness among Middle and Late Pleistocene archaic humans 95:5 (54 right-handed and 3 left-handed individuals). If data from upper limb asymmetry in the absence of dental evidence are included (including the left-side dominant Palomas 96), the ratio is 94:6 (60 right-handed and 4 left-handed individuals). Note that, reassuringly, the three Neandertals providing both cut-mark and humeral asymmetry indications of handedness (La Quina 5, Regourdou 1, and Tabun 1) are all right-handed by both indicators. This overall handedness ratio is very close to the 9:1 ratio frequently invoked in the literature as evidence for archaic human language capacity (Lozano et al. 2009; Frayer et al. 2010, 2012; Fiore et al. 2015).

Antemortem Enamel Chipping

Antemortem chipping is relatively infrequent at Palomas. Poor occlusal preservation, in addition to the relatively low levels of occlusal wear, for many of the anterior teeth probably influences this trend. There are no clear relationships between severity of occlusal wear and chipping or maxillary versus mandibular chipping rates— again, it is likely an artifact of preservation and the young age at death for preserved teeth. Only 27% of teeth (3 of the 11 teeth) that could be analyzed exhibit chipping. One tooth exhibits multiple chips (Palomas 97 RI$_1$). Only chips of grades 1 and 2 are present.

The paucity of chipping is even more evident when compared to published data on El Sidrón, Spy, and Hortus (Estalrrich and Rosas 2015). Spy represents only two individuals and has rates of mandibular chipping (25%) that are close to those seen at Palomas (29%). However, when looking

at combined maxillary and mandibular chipping, there are clearly far fewer chips in the Palomas sample. Whether this is an artifact of preservation or an empirical reality of the sample is difficult to elucidate.

Discussion and Summary

The analysis of anterior dental wear among the Sima de las Palomas material provides new data on relative rates of anterior versus posterior wear gradients, labial cut marks, and enamel chipping. These results add new commentary on previously established trends while reinforcing patterns previously established.

The scaled occlusal macrowear analysis is not new (see Trinkaus 1992), but it has nonetheless not been used since its inception. The results from this study show that it is a particularly useful technique when dealing with incomplete dentitions (that is, the Palomas 1 maxilla and Palomas 59 mandible). Overall, there is little separation between groups in terms of the relative rates of anterior versus posterior dental wear. Neandertals do appear to have somewhat elevated rates of anterior dental wear, but there are considerable overlap and exceptions with the comparative MPMH and E/MUP human samples. The Palomas samples fit the Neandertal trend for the most part, but on the less worn end of the spectrum, they fall in line with trends seen among many of the MPMH and E/MUP humans. This is not particularly surprising in light of Clement and colleagues' work on occlusal wear gradients (Clement et al. 2012), which shows that some early modern and recent human samples have anterior wear relative to posterior wear that far exceeds that of their Neandertal sample. However, their analysis does not account for variation in tooth size. In that respect, scaled macrowear analyses are a useful complement to the established literature.

The labial cut-mark analyses reveal additional variation in terms of length and width of instrumental striations on the anterior teeth. The Palomas samples fall on the low end of the variation in both length and width of striations but are nonetheless encompassed within the published ranges of variation from various sites. Further studies may reveal temporal or ecogeographic patterning of such behaviors. It is particularly interesting to note that the absolute lengths of the oldest comparative samples, from Atapuerca-SH and Krapina, are longer on average than those of Palomas and suggest a possible temporal influence on variation. The lower incisors from Boxgrove have mean absolute lengths intermediate between those of Atapuerca-SH and Krapina, further reinforcing this possibility.

Handedness indicated by labial cut marks at Palomas was 100% right-handed for a conservatively estimated five individuals. This is not surprising, given the extensive analyses of handedness among archaic *Homo* and their focus on language capacities among Neandertals and their predecessors (Lozano et al. 2009; Frayer et al. 2010, 2012; Uomini 2011; Volpato et al. 2012; Estalrrich and Rosas 2013; Bruner and Lozano 2014; Fiore et al. 2015). In fact, laterality is known from the Early Pleistocene in the form of antebrachial asymmetry for KNM-WT 15000 (Weaver et al. 2001) and labial cut marks from one specimen from Gran Dolina (Bruner and Lozano 2014), which together suggest laterality is a hallmark of *Homo* since the Early Pleistocene.

A focus on determining handedness seems to overshadow an important aspect of variation in cut-mark orientation. For example, this study also found important differences in mandibular versus maxillary rates of horizontal and vertical cut marks that may indicate differences in the susceptibility of the teeth in upper or lower jaws to certain cut-mark patterns, perhaps even reflecting different tasks, such as scraping versus cutting. This is worth pursuing in light of research suggesting that the biomechanical properties of Neandertal

humeri may reflect habitual hide-scraping activities (Shaw et al. 2012). It is suggested here that the traditional system of Bermúdez de Castro and colleagues (1988) be used for its ability to detect subtle differences in orientation that may reflect task-related behaviors. In addition, the use of Frayer and colleagues' system (Frayer et al. 2010, Fiore et al. 2015) should also be used for its sensitivity to detecting handedness. However, more comparative data are needed before specific behaviors can be attributed to different patterns of cut-mark orientation.

The analysis of antemortem chipping was not particularly enlightening in the framework of Late Pleistocene nonmasticatory behavior. Taphonomic issues are likely to blame, but it may also be that the Palomas Neandertals were not loading the anterior dentition with frequent high loads. This could gain support from additional samples yielding similarly low levels of antemortem chipping. Thus, Palomas could be an anomaly in this regard or simply reflect taphonomic biases in the sample.

In summary, the analysis of nonmasticatory dental wear from Sima de las Palomas falls in line with what is already known about Neandertal nonmasticatory behavior, but more can still be gained from future studies. In particular, studies of anterior dental wear among Neandertals, and Pleistocene *Homo* more generally, have historically focused on within-sample variation or individual fossils with few, but notable, exceptions. This study attempted to expand upon these analyses but also provide raw data to facilitate comparative research on a broader scale. Broader comparisons are still sorely needed to understand Neandertal nonmasticatory behavior in light of temporal and ecogeographic variation, although much has been gained by recent renewed interest in this topic.

TABLE 11.1. Enamel chipping and labial cutmark analyses possible for each anterior tooth from Sima de las Palomas

SP NO.	TOOTH[a]	WEAR[b]	CHIPPING[c]	CUTMARKS[d]
1a	RC^1	4–5[e]	Present	n/a
1b	LC^1	4	n/a	n/a
1c	RI_1	5	n/a	n/a
1c	RC_1	4	Absent	O
1d	LC_1	4	n/a	n/a
18	LC_1	2	Absent	n/a
19	LI_1	3	n/a	L, W, O
20[f]	LI_2	5	n/a	n/a
21[f]	LI_1	5	n/a	n/a
24	LI^1	1–2	n/a	n/a
31	Ldc_1	5	n/a	n/a
34	LI^1	1	Absent	n/a
35	LC^1	4	Absent	n/a
44	RC_1	4	Absent	n/a
48	RI^2	3	n/a	n/a
54	RC_1	2–3	Absent	n/a
59	LC_1	3	Present	L, W, O
73	RI^1	4	n/a	W, O
74	LC^1	1	Absent	L, W, O
79	LI^1	3–4	n/a	L, W, O
82	LC_2	1	n/a	L, W, O
85	Ldi^2	5	n/a	n/a
90	LI^1	3	n/a	L, W, O
91	LI_2	3	n/a	n/a
93	Rdc_1	5	n/a	n/a
97	LI_1	1–2	Absent	L, W, O
97	RI_1	1–2	Present	L, W, O
97	RI_2	1–2	n/a	L, W, O
98	RI_2	4–5	n/a	n/a
Total teeth included in analyses:			11	11

[a]Only teeth that are clearly erupted and worn are considered.

[b]Scores based on Smith, 1984.

[c]Chipping analysis: Present, absent, or not applicable.

[d]Cutmark analysis: If cutmarks are present, the measurements taken are indicated as L (length), W (width), and O (orientation). Teeth too damaged to assess cutmarks are labeled n/a (not applicable).

[e]This tooth was scored as a 5 for scaled occlusal wear gradient analysis. Therefore, the average for the antimeres is 4.5.

[f]SP20 and 21 are included with the teeth of the Palomas 1 mandible for macrowear gradient analysis. As noted in chapter 6, they most likely derive from Palomas 1.

TABLE 11.2. Descriptive statistics for labial surface cutmark analysis by each tooth, individual, and associations between isolated teeth (all measurements in μm)

SP NO.	TOOTH	LENGTH			WIDTH		
		MEAN	STD DEV	N	MEAN	STD DEV	N
Maxilla							
73	RI^1	n/a	n/a	n/a	59.0	15.8	6
74	LC^1	549.9	389.5	13	28.9	12.1	13
79	LI^1	859.7	736.2	16	38.1	11.8	21
90	LI^1	506.7	274.2	32	25.4	11.7	48
73 and 79[a]		859.7	736.2	16	42.8	15.3	27
Mandible							
19	LI_1	1076.1	1142.1	16	34.8	18.8	18
59	LC_1	540.7	264.6	38	21.1	8.0	38
82	LC_1	602.6	315.8	14	33.9	19.2	16
97	LI_1	782.0	733.9	38	34.3	15.2	39
97	RI_1	634.7	463.3	37	36.1	17.6	38
97	RI_2	626.2	298.7	24	29.9	10.6	25
97 combined		689.2	555.5	99	33.9	15.2	102
Maxilla & Mandible							
19, 73, and 79[a]		967.9	951.6	32	39.6	17.0	45
74 and 82[a]		577.2	347.2	27	31.7	16.3	29
59 & 90[b]		525.2	267.6	70	23.5	10.4	86
Total, All Maxillary Teeth		608.5	477.3	61	31.3	15.0	88
Total, All Mandible Teeth		685.2	568.1	167	31.2	15.6	174

[a]Isolated teeth of probable association.
[b]Isolated teeth of possible association.

TABLE 11.3. Descriptive statistics for cutmark length and width at Sima de las Palomas compared to published data (all measurements in μm)

SAMPLE	LENGTH			WIDTH			SOURCE
	MEAN	STD DEV	N	MEAN	STD DEV	N	
Sima de las Palomas	664.9	559.1	228	31.2	15.4	262	This study
Krapina	739.1	563.7	1582	Range: 25.5–67.7			1, 2
Atapuerca-SH	1507.3	597.5	592	43.2	17.8	390	3
Native Australians	681.9	364.2	26	44.4	18.7	26	3

Sources: (1) Fiore et al. 2015; (2) Lalueza-Fox and Frayer 1997; (3) Lozano et al. 2008.

TABLE 11.4. Labial cutmark orientations for each tooth with measurable striations at Sima de las Palomas

| | | BERMÚDEZ DE CASTRO ET AL.[a] | | | | | | | | FRAYER ET AL.[b] | | | | | |
SP NO.	TOOTH	RO	H	LO	V	N	χ^2	P	DOMINANT ORIENTATION	RIGHT	LEFT	N	χ^2	P	DOMINANT ORIENTATION
Maxilla															
73	RI^1	5	1	0	0	6	11.33	0.010053	RO	6	0	6	6.00	0.014306	Right
74	LC^1	11	1	0	1	13	24.85	0.000017	RO	12	0	12	12.00	0.000532	Right
79	LI^1	17	2	0	2	21	35.57	<0.000001	RO	21	0	21	21.00	0.000005	Right
90	LI^1	30	9	1	8	48	39.17	<0.000001	RO	43	4	47	32.36	<0.000001	Right
73 and 79^c		22	3	0	2	27	46.63	<0.000001	RO	27	0	27	27.00	<0.000001	Right
Mandible															
1	RC$_1$	12	1	0	1	14	27.71	0.000004	RO	13	1	14	10.29	0.001341	Right
19	LI$_1$	8	2	0	8	18	11.33	0.010053	RO & V	14	1	15	11.27	0.000789	Right
59	LC$_1$	13	15	4	6	38	8.95	0.029999	H	21	7	28	7.00	0.008151	Right
82	LC$_1$	7	3	0	6	16	7.50	0.057558	RO	13	2	15	8.07	0.004509	Right
97	LI$_1$	11	3	0	25	39	38.44	<0.000001	V	25	2	27	19.59	0.000010	Right
97	RI$_1$	24	3	1	10	38	34.21	<0.000001	RO	36	1	37	33.11	<0.000001	Right
97	RI$_2$	25	13	10	1	49	24.06	0.000024	RO	23	2	25	17.64	0.000027	Right
97 combined		60	19	11	36	126	44.73	<0.000001	RO	84	5	89	70.12	<0.000001	Right
Maxilla and Mandible															
19, 73, and 79^c		30	5	0	10	45	46.11	<0.000001	RO	41	1	42	38.10	<0.000001	Right
74 and 82^c		18	4	0	7	29	24.66	0.000018	RO	25	2	27	19.59	0.000010	Right
59 and 90^d		43	24	5	14	86	37.07	<0.000001	RO	64	11	75	37.45	<0.000001	Right

[a]Bermúdez de Castro et al. (1998): RO = right oblique, H = horizontal, LO = left oblique, V = vertical, n = total observations. Chi-squared tests were are for striation counts with three degrees of freedom assuming equal proportions of each orientation category (that is, 25:25:25:25). P-values are reported as exact values to six decimal points. Dominant orientation is the result of Chi-squared test.

[b]Frayer et al. (2010): Right = >95°–175°, Left = >5°–85°. Total observations are fewer than when using Bermúdez de Castro et al. (1998) method in many cases because vertical (85°–95°) and horizontal (0°–5° or >175°–180°) are not considered. Chi-squared tests are for striation counts with one degree of freedom assuming equal proportions of each orientation category (50:50). P-values are reported as exact values to six decimal points. Dominant orientation is the result of Chi-squared test.

[c]Isolated teeth of probable association.

[d]Isolated teeth of possible association.

TABLE 11.5. The archaic human context for handedness at Sima de las Palomas[a]

FOSSIL INDIVIDUAL OR SITE	RIGHT	LEFT	SOURCE
Evidence from labial cutmarks:			
Sima de las Palomas			
Each tooth	11	0	This study
By individual (including probable associations)[b]	6	0	This study
By individual (conservative)[c]	5	0	This study
Atapuerca-SH	12	0	13
Boxgrove 2 & 3[d]	1	0	1
Broken Hill (Kabwe) 1	1	0	2
Cova Negra	1	0	3, 4
El Sidrón	11	0	5
Hortus	4	1	3, 6
Krapina	9	2	7
Mauer 1	1	0	8
La Quina 5[e]	1	0	3, 4
Regourdou 1[f]	1	0	9
Saint-Brais	1	0	3, 10, 11
Shanidar 2	1	0	3, 12
Tabun 1[e]	1	0	2
Vindija	4	0	14
Conservative total count based on cutmarks:	54	3	
Evidence from Neandertal upper limb asymmetry in the absence of labial cutmark data:			
Amud	1	0	15
La Chapelle-aux-Saints 1	1	0	16
La Ferrassie 1	1	0	16
Neandertal 1[g]	1	0	16
Palomas 96	0	1	17
Spy 2	1	0	16
Shanidar 4	1	0	15
Total count based on upper limb asymmetry:	6	1	

Observed Ratios	Right:Left
Cutmarks	95:5
Upper limb	86:14
Cutmarks and upper limbs	94:6

Sources: (1) Hillson et al. 2010; (2) Lalueza-Fox and Pérez-Pérez 1994; (3) Bermúdez de Castro et al. 1988; (4) Arsuaga et al. 2001; (5) Estalrrich and Rosas 2013; (6) de Lumley 1973; (7) Fiore et al. 2015; (8) Puech et al. 1987; (9) Volpato et al. 2012; (10) Koby 1956; (11) Willman pers. obs. of original fossil; (12) Trinkaus 1983; (13) Frayer et al. 2012; (14) Frayer et al. 2010; (15) Trinkaus and Churchill 1999; (16) Trinkaus et al. 1994; (17) chapter 13, this volume.

[a]The assessments of handedness for Sima de las Palomas are based on the more side-sensitive method of Frayer et al. (2010); see also Table 11.4.

[b]Isolated teeth treated as separate individuals (Palomas 59 and 90) in addition to individuals isolated with certainty (Palomas 1 and 97) and individuals based on the probable associations between isolated teeth (Palomas 19, 73, and 79 and Palomas 74 and 82).

[c]This is the most conservative estimate of the total number of individuals. It includes individuals isolated with certainty (Palomas 1 and 97), individuals based on the probable associations between isolated teeth (Palomas 19, 73, and 79 and Palomas 74 and 82), and an individual based on the possible associations between an isolated tooth and an arcade (Palomas 59 and 90).

[d]The Boxgrove 2 and 3 lower incisors are likely from the same individual and exhibit cut marks on both the crown and root, with statistically significant differences in orientation between each tissue, but the predominate pattern is one of right-handedness (Hillson et al. 2010).

[e]Additional support from humeral diaphyseal asymmetry (Trinkaus et al. 1994)

[f]Additional support from humeral diaphyseal asymmetry (Vandermeersch and Trinkaus 1995; Volpato et al. 2012).

[g]Posttraumatic or pathological influence on asymmetry. Also, due to an amputation, Shanidar 1 (Trinkaus 1983) was not considered.

TABLE 11.6. Assessment of antemortem enamel chipping for each tooth with well-preserved occlusal surfaces at Sima de las Palomas

SP NO.	TOOTH	WEAR[a]	CHIPPING SCORE[b]
1	RC¹	5	1
1	RC₁	4	0
18	LC₁	2	0
34	LI¹	1	0
35	LC¹	4	0
44	RC₁	4	0
54	RC₁	2	0
59	LC₁	3	2
74	LC¹	2	0
97	LI₁	1–2	0
97	RI₁	1–2	1 and 2

[a]Smith 1984.
[b]Bonfiglioli et al. 2004: 0 = observable but absent; 1, 2, or 3 = present.

TABLE 11.7. Antemortem enamel chipping at Sima de las Palomas compared to Neandertal data from Estalrrich and Rosas (2015)

GROUP	MAXILLARY		MANDIBULAR		TOTAL	
	PERCENT	COUNT	PERCENT	COUNT	PERCENT	COUNT
Sima de las Palomas	25%	1/4	29%	2/7	27%	3/11
El Sidrón	76%	28/37	86%	30/35	81%	58/72
Spy	100%	6/6	25%	1/4	70%	7/10
Hortus	79%	11/14	100%	2/2	81%	13/16

The Palomas Oral Paleopathology and Stress Indicators

12

SARAH A. LACY, JOSEFINA ZAPATA, A. VINCENT LOMBARDI,
JOHN C. WILLMAN, AND ERIK TRINKAUS

NEANDERTALS SUFFERED from the same oral health problems as those experienced by living people. Periodontal disease was widespread, antemortem tooth loss was present if not especially common, and caries was not as rare as initially thought (Heim 1976; Trinkaus 1983, 1985; Lalueza et al. 1993; Lebel and Trinkaus 2002a, 2002b; Walker et al. 2011c; Lacy et al. 2012; Lacy 2014b; Trinkaus et al. 2014a). These conditions were also present among other late archaic humans and early modern humans, varying more in frequency than in kind (for example, Sognnaes 1956; Tillier et al. 1995; Trinkaus and Pinilla 2009; Humphrey et al. 2014). Various Neandertals have been identified with infected alveolar cysts (El Sidrón 2; Mezzena 1), persistent deciduous teeth (El Sidrón 2 and 3; Le Moustier 1), and dental agenesis (Malarnaud 1) (Heim and Granat 1995; Ponce de León and Zollikofer 1999; Condemi et al. 2012; Dean et al. 2013), and these conditions continued into the Upper Paleolithic (Trinkaus et al. 2006, 2014a; Lacy 2014b). These largely recent oral paleopathological analyses of Late Pleistocene humans have helped to shift the understanding of their oral health from generalizations of being "ubiquitously good" to a more nuanced analysis of the temporal, regional, and demographic patterns. Few of the dentitions in the Late Pleistocene were a dentist's dream (Lacy 2014b).

There have also been a variety of analyses of Neandertal dental enamel hypoplasias (for example, Molnar and Molnar 1985; Ogilvie et al. 1989; Brennan 1991; Guatelli-Steinberg et al. 2004). These reflections of developmental disruption during crown calcification from nonspecific stress (Goodman and Rose 1990; Hillson and Bond 1997) are common among the Neandertals, with inferred frequencies varying among the analyses.

The Palomas dentoalveolar remains follow the same general pattern. Their preservation is biased towards individual isolated teeth (chapter 6), and the age-at-death profile of the sample is relatively young (chapter 15). These aspects of the sample are reflected in the diagnoses being heavily towards dental pathology; only in a few cases can periodontal (alveolar) disease and antemortem tooth loss be assessed, and they appear to have been absent. Individual remains from Sima de las Palomas nonetheless present dental caries, dental enamel hypoplasias, hypercementosis, and two possible alveolar lesions (see also Walker et al. 2011c). This chapter describes the few instances of oral pathology.

Materials and Methods

The teeth and mandibles were measured with digital calipers, photographed, and visually examined.

Radiographs were taken with a Nomad eXaminer portable x-ray generator and Digirex digital sensor with processing software (Dentamerica); additional observations were made on micro-CT images of the specimens, courtesy of P. Bayle and K. A. Robson Brown. Carious lesions were identified either visually or through radiographs, and they are scored following Hillson (2001). Periodontal disease was assessed using buccal and lingual cementoenamel junction to alveolar crest (CEJ-AC) distance measurements per tooth, when present (Lavigne and Molto 1995; Armitage 2004), as well as an ordinal scoring method for assessing interdental septa condition (Lacy 2014b, adapted from Costa 1982). Tooth loss was assessed visually by looking for evidence in an open alveolus of osteoblastic activity or resorption or for a fully healed alveolus with a missing tooth and some gap in the dental arcade (to differentiate it from dental agenesis or noneruption). Alveolar lesions were diagnosed from either an open sinus or from radiolucency in a radiograph (Dias and Tayles 1997). Dental enamel hypoplasias were diagnosed visually and confirmed with hand magnification (10x). They are positioned by the perpendicular distance at buccal midline from the middle of the disruption to the cementoenamel junction (Rose et al. 1985).

In this assemblage, postmortem erosion, chipping, and pitting of dental enamel are common, and therefore suspected carious lesions were differentiated from postmortem damage by the antemortem acidic erosion of caries. Suspected carious lesions were radiographed and/or micro-CT scanned to confirm their diagnosis.

The general assessment of the Palomas human oral pathology sample includes its placement in the larger context of oral pathologies in Late Pleistocene Neandertals and early modern humans from western Eurasia (Lacy 2014b). Those observations derive from personal observations of original remains, principally by Sarah A. Lacy.

Results

Table 12.1 lists each Sima de las Palomas num-

bered dentoalveolar specimen, indicating whether pathology is present and, if so, what kind. Below are descriptions for those remains that presented with pathology, or for which the preservation is problematic to the point that specific kinds of data could not be collected, with an explanation.

Palomas 1 Maxillae and Mandible Right and Left

Considering the nature of the reconstruction of the teeth in the alveolar bone and the associated distortions (Figs. 4.1, 5.2, and 5.3), it is not possible to get accurate CEJ-AC distances. No other pathology was observed.

Palomas 6 Mandible with C_1 to M_2 Roots Left

The roots of the Palomas 6 teeth are broken below the crowns, and the alveolar crests of the alveoli are damaged (Plate 5.5); therefore, its periodontal status cannot be reliably assessed. Based on radiographic imaging, there is a radiolucency around the root of the left mandibular canine (Plate 12.1). In the micro-CT slice, the nature of this radiolucency is difficult to detect; the trabecular bone is sparse, but no distinct cavity is observed. It is possible that an acute abscess was present, because osteoclastic activity is minimal, as opposed to the extensive bony resorption associated with a chronic granuloma or cyst. However, the lamina dura appears to be intact (the first aspect that is usually destroyed in any lesion), and no sinus was identified, as would be necessary for the diagnosis of a chronic abscess (Dias and Tayles 1997). As an acute abscess can be very difficult to diagnose osteologically, the radiolucency should remain undiagnosed.

Palomas 7 Mandible with C_1 and P_3 Crowns Left

The alveolar margin of the immature Palomas 7 mandibular fragment is not well preserved (Plate 5.9), and no erupted teeth are preserved; therefore, the periodontal health of the individual cannot be assessed.

Palomas 18 Mandibular C₁ Left

The canine crown (Plate 6.16) has a single furrow-type linear hypoplasia ≈5.6 mm from the cervix.

Palomas 19 Mandibular I₁ Left

The labial surface exhibits a broad and shallow linear enamel hypoplasia (Plate 6.17) located 2.7 mm from the cervix.

Palomas 23 Mandible with Partial I₁ to M₃ Right

The Palomas 23 alveolar crests are mostly damaged, and dental crowns are present only on the two molars (most of the M_2 and a mesial portion of the M_3) (Plate 5.7). The CEJ-AC distances could therefore not be reliably assessed for periodontal disease. What remains of the buccal alveolar margin, from the mid-M_1 to the distal M_3, appears to have been unaltered antemortem.

Palomas 25 Mandibular dm₁ Right

This lower right deciduous first molar has an occlusal carious lesion just mesial to the hypoconid (distobuccal cusp) in the crown of the tooth (Plate 12.2; see also Walker et al. 2011c). It is about 1.2 mm in diameter and reveals dentin in the floor of the lesion and on the mesial side, providing it with a score 5 occlusal carious lesion following Hillson's (2001) diagnostic criteria for archaeological dental remains. It is roughly circular and may have initially formed in a groove or a developmental pit. There is some persistent hard sedimentation in the lesion, but acidic etching of the enamel and the dentin surface within the lesion is still visible.

The tooth is too heavily mineralized to view the lesion clearly in a radiograph (Plate 12.3); the lesion can be localized, but not the demineralization of the surrounding enamel and dentin that would have been present in life, producing radiolucency. Even though there is some antemortem occlusal attrition revealing dentin, and postmortem damage in the form of a crack through the

crown that passes through the defect, the rounded contours, the position, and surface etching within the lesion suggest it is a true carious lesion despite the lack of radiographic confirmation.

Palomas 34 Maxillary I¹ Left

The Palomas 34 unerupted I¹ root (Plate 6.29) presents several strongly marked furrow-type linear cementum hypoplasia, and it exhibits a deformation on the distolabial surface at the middle of the root. The crown lacks clear hypoplastic defects.

Palomas 35 Maxillary C¹ Left

This maxillary canine crown (Plate 6.30) shows a broad and shallow furrow-type linear hypoplasia ≈1.6 mm from the cervical line, evident primarily as a step in the labial enamel.

Palomas 37 Maxillary I² Right

The unerupted Palomas 37 crown (Plate 6.32) shows evidence of four small furrow-type linear hypoplasia, ≈0.2 mm, ≈1.3 mm, ≈2.9 mm, and ≈7.4 mm from the cervical line.

Palomas 45 Mandibular P₃ Right

The root of this mandibular premolar exhibits mild hypercementosis (Plates 6.37 and 12.4), partially removed postmortem from the root. It is evident distally and lingually and in a radiograph of the tooth.

Palomas 49 Mandible with Deciduous Alveoli

There are no erupted teeth present in this mandible (Plate 5.10), and the alveoli are only preserved for the dm₁s and the left dm₂. This individual was about two and one-half years of age (chapter 5), and considering its preservation, periodontal health cannot be reliably assessed.

Palomas 59 Mandible with C₁ to M₂ Left

The Palomas 59 left mandibular corpus with five

teeth (Plate 5.11) presents caries and an unusual pattern of supraeruption of the M_2 (see also Walker et al. 2011c and discussions of occlusal wear in chapters 6 and 10). There is an interproximal carious lesion (score 5; Hillson 2001) on the mesial side of the left M_2 superior to the cervix of the crown (Plate 12.5). It can be seen somewhat visually and is also evident radiographically. There is a corresponding lesion on the distal interproximal side of the left M_1. Radiographically (Plate 12.5), there also appears to be a mesial interproximal lesion on the left M_1, but this could not be verified. The largest radiolucency seen on the radiograph at the cervix of the M_2 is an artifact of a gap in the buccal breccia and is therefore not a sign of antemortem pathology (Plate 12.5). Therefore, the only two lesions that appear likely to be carious are the more superior of the two mesial interproximal radiolucencies on the M_2 and the corresponding distal radiolucency on the interproximal M_1. Both would be considered score 5 contact area caries, as the dentin is involved, given the relative thinness of dental enamel on interproximal crowns (Hillson 2001).

Palomas 59 does not exhibit periodontal disease; all of the CEJ-AC distance averages per tooth are below 2 mm, excluding those that have postmortem damaged alveolar crests (Table 12.2). There appears to be some alveolar resorption along the molars, but it is minor (Plate 12.6). The second molar also shows greater wear than the first, whether scored ordinally (chapter 6) or especially by percentage of exposed dentin (Table 10.1). This is an uncommon and perhaps pathological pattern, given the occlusal eruption of the M_1 normally at least half a decade before the M_2. The most likely explanation of this differential occlusal wear derives from pathological changes to the apex of the M_2, causing some supraeruption of the tooth (Plate 12.6). There does not appear to be marked supraeruption of the M_2 relative to the other teeth, but this appears to be the only available explanation, as the opposing maxillary molars are not preserved. This supraeruption may have been caused by the radiopaque mass located between the apices of the molar's roots (Plate 12.6) that could be a cementoma (fibro-osseus

lesion from a periodontal ligament) (Underhill et al. 1992) or a calcified periapical cyst (Dias and Tayles 1997; Dias et al. 2007). (The term *calcified abscess* has been used [Walker et al. 2011c], but *calcified periapical cyst* is more appropriate given the size, smooth contours, and lack of a sinus.) Because there is no obvious source for pulpal inflammation or infection, a cementoma seems more probable than a periodontal cyst (Underhill et al. 1992). However, no differential diagnosis can be definitive based solely on the radiographic images.

Palomas 68 Maxillary P³ and P⁴ Right

The Palomas 68 P⁴ crown (Plate 6.52) has a linear hypoplasia around the upper third.

Palomas 70 Mandibular dm₂ Right

There is a large pit (1.4 mm wide) in the middle of the buccal shoulder of the crown (evident in the occlusal view of Plate 6.54) that is probably a hypoplasia or some other developmental defect.

Palomas 72 Maxillary M

The molar crown has a linear hypoplasia around the midcrown, ≈2.5 mm from the cervix.

Palomas 73 Maxillary I¹ Right

The labial portion of an I¹ crown shows evidence of two relatively broad furrow-type linear hypoplasia, ≈3.4 and ≈4.1 mm from the cervical line (Plate 6.57).

Palomas 79 Maxillary I¹ Left

The labial surface of the I¹ crown exhibits two shallow, narrow, and adjacent linear hypoplasia furrows, ≈2.4 and ≈2.8 mm from the cervical line.

Palomas 80 Mandible with M₂ and M₃

The alveolar margin is damaged at the left P_4 and left M_1, the left M_2 is in mideruption, and the left

M$_3$ is unerupted (Plate 5.12); therefore, periodontal disease cannot be assessed.

Palomas 87 Mandibular P$_4$ Left

There are five tiny hypoplastic pits on the buccal surface of the main cusp and cusplets near the occlusal surface of the crown of P$_4$ (Plate 6.67). They are distinct from the areas of postmortem loss of surface enamel and represent minor developmental defects.

Discussion

Considering that this assemblage is heavily dominated by immature individuals (80.7% of the dental remains; Table 15.2), the relatively low prevalence of oral pathology is not surprising. Most oral pathologies, with the exception of caries, increase with age in the Late Pleistocene sample (Lacy 2014b) as well as in modern human samples without access to oral hygiene (Ånerud et al. 1979; Loë et al. 1986; Neely et al. 2001, 2005; Ronderos et al. 2001; Oztunc et al. 2006; Eke et al. 2012). Caries is the only pathology specific to the young in Late Pleistocene humans (Lacy 2014b), and this pattern likely reflects increasing levels of dental attrition with age, which remove evidence of minor caries. This finding may also have been confounded by other behaviors, such as a differing diet in the young.

Caries is a relatively rare pathology in the Pleistocene, though it is not unknown among Middle Paleolithic humans (for example, Sognnaes 1956; Lalueza et al. 1993; Tillier et al. 1995; Trinkaus et al. 2000a; Lebel and Trinkaus 2002a; Trinkaus and Pinilla 2009; Lacy et al. 2012; Humphrey et al. 2014; Lacy 2014b) and among earlier humans (for example, Carter 1928; Brodrick 1948; Robinson 1952; Clement 1956; Grine et al. 1990; Lacy 2014a). For the Neandertals overall, the rate of carious teeth is 1.3% (9 out of 668) (Lacy 2014b). Even though this percentage is higher than that of previous estimate (0.5% of Walker et al. 2011c; 0.5% of Lanfranco and Eggers, 2012), it is still low in the context of recent humans practicing agriculture but not of high-latitude foragers. The caries per-tooth percentage for the Point Hope arctic sample is a comparable 1.0% (Lacy 2014b).

The Palomas caries likely reflect a regional pattern in caries prevalence, with the prevalence of caries clustering along the Mediterranean in the Late Pleistocene until the Upper Paleolithic, when caries appear farther north, although still sporadically (Lacy 2014b). The other cases of Neandertal caries—Amud 1 (Lacy 2014b); Aubesier 5 (Trinkaus et al. 2000a); Aubesier 12 (Lebel and Trinkaus 2002a); Banyoles 1 (Lalueza et al. 1993); Kebara 1 and 27 (Tillier et al. 1995); and Tabun 2 (Lacy 2014b), as well as those of the southwest Asian Middle Paleolithic modern humans (Sognnaes 1956; Lacy 2014b) and the east Asian Zhiren 2 (Lacy et al. 2012)—are all relatively southern and contrast with the absence of caries among contemporaneous fossils farther north. The Palomas 25 and 59 cases support this regional pattern and may reflect climatic and the resultant dietary differences, such as environmental prevalence of plant sugars (Kirschbaum 2004; Zheng et al. 2009). The regional differences may also relate to the mineral composition of water, such as decreases in groundwater fluoride concentration in closer proximity to the sea (Dunning 1953; Bang 1964), and therefore reflect proximity to the Mediterranean more than latitude *per se*. Regional clines in caries prevalence have been found for Mesolithic Europe (Meiklejohn et al. 1988; Frayer 1989), prehistoric India (Lukacs and Pal 1993), and prehistoric Japan (Fujita 2012). There is as well an inflection in the predominance of carbohydrates in hunter-gatherer diets in which they decrease markedly north of 40° latitude (Ströhle and Hahn 2011). The Sima de las Palomas, at 37°47'59" N, is well below 40° N. Periodontal disease and antemortem tooth loss did not present an interregional pattern (Lacy 2014b).

Walker and colleagues (2011c) questioned whether caries-causing bacteria such as *Streptococcus mutans* would have even been present in Neandertals. The oral microbiome for Neandertals has not been fully explored, but *S. mutans* was not present in dental calculus samples from Mesolithic (that is, preagricultural) Poland and only predominated after the industrial revolution (Adler et al. 2013; Warinner et al. 2014). *Porphyromonas gingivalis*, associated with periodontal disease, was present in this earlier sample, though (Adler et al. 2013). This pattern does not rule out that *S. mutans* or closely related species were present in at least some Neandertal populations (especially Mediterranean ones, see above), but its absence may explain the relatively low prevalence of the condition in the Middle Paleolithic of Europe.

The prevalence of dental enamel hypoplasias at Palomas is relatively low. Hypoplasia is a relatively common developmental defect in Neandertals, with estimates of its prevalence ranging from a third (Guatelli-Steinberg et al. 2004) to ≈75% of individuals (Molnar and Molnar 1985; Ogilvie et al. 1989), depending on which teeth (all or anterior only) were in the fossil sample analyzed and the identification methods used. Only 11 individuals out of 72 (15.3%, assuming that the isolated teeth represent separate individuals, but see considerations of Palomas 19, 73, and 79 in chapters 6 and 11) present the defect here (not including Palomas 74 and 82, whose defects are very minor). Alternatively, it is appropriate to consider only the crowns of the sufficiently well preserved anterior permanent teeth (incisors and canines), those teeth that are most commonly scored for hypoplasias. The five Palomas teeth with at least one defect (Palomas 18, 19, 37, 73, and 79) represent 26.3% of that sample (n = 19).

Possible alveolar abnormalities were identified by radiographs (Palomas 6 and 59) and therefore have no outward indicators, such as sinuses, with the exception of the possible supraeruption of the M_2 in Palomas 59. Palomas 59 did not have sufficiently pronounced dental attrition to expose the pulp of the tooth to pulpal infection and a resultant necrosis, so the source of an infection is difficult to identify, if indeed there was one. Excessive occlusal forces can produce reactive lesions (Underhill et al. 1992), but the opposing occlusal surfaces for both Palomas individuals are missing.

The only other unusual feature of the Palomas masticatory anatomy is the anterolateral projections of the Palomas 6 and 23 mandibles (chapter 5). They do not appear to be pathological. They do not conform to a normal anterior marginal tubercle, although the more modest one on the immature Palomas 7 mandible might represent such a feature.

Summary

Most forms of oral pathology (for example, tooth loss, oral infection or periapical lesions, caries) are relatively rare in Neandertals. All of these pathologies plus periodontal disease, except for caries, increase through the aging process, especially in the Late Pleistocene sample (Lacy 2014b). Because the Palomas sample is skewed towards younger individuals, and many individuals are represented by a single isolated tooth with little or no alveolar bone, it is not surprising that periodontal disease and tooth loss were not found in this relatively large human sample from a Late Pleistocene site. Caries was the only oral pathology in the Late Pleistocene found previously to be most common in the younger cohorts (Lacy 2014b), and this is further supported by the Palomas material (Palomas 25 being a child and Palomas 59 a young adult). Two possible periapical pathologies were identified (Palomas 6 and 59), as well as a number of instances of dental enamel hypoplasia (Table 12.1). Considering the prevalence of dental enamel hypoplasia in other surveys of Neandertals (Molnar and Molnar 1985; Ogilvie et al. 1989; Guatelli-Steinberg et al. 2004), the Palomas sample has a low prevalence of the developmental defect. The Palomas sample continues to support the assertion that oral pathology was relatively rare in younger Late Pleistocene humans and in all age categories when compared to recent human samples (Lacy 2014b).

TABLE 12.1. The Palomas dental and alveolar remains with presence/absence of pathological lesions

SP NO.	IDENTIFICATION	PATHOLOGY PRESENCE
1a	Maxilla right with C^1 to M^3	None
1b	Maxilla left with C^1 to M^3	None
1c	Mandible right with I_1, C_1 to M_3	None
1d	Mandible left with C_1, M_1 to M_3	None
6	Mandible corpus left with C_1 to M_2 roots	Periapical lesion?
7	Mandible corpus left with C_1 and P_3 germs	None
18	Canine mandibular left	Hypoplasia
19	Incisor 1 mandibular left	Hypoplasia
20	Incisor 2 mandibular left (probably SP1)	None
21	Incisor 1 mandibular left (probably SP1)	None
22	Premolar 3 mandibular left (probably SP1)	None
23	Mandible with left I_1 to right P_4 roots plus right M_1 to M_3	None
24	Incisor 1 maxillary left	None
25	Deciduous molar 1 mandibular right	Caries
26	Canine mandibular right	None
27	Deciduous canine	None
29	Molar 2 mandibular right	None
31	Deciduous canine mandibular left	None
33	Root fragment (mandibular incisor?)	None
34	Incisor 1 maxillary left	Hypoplasia
35	Canine maxillary left	Hypoplasia
36	Molar 1 maxillary left	None
37	Incisor 2 maxillary right	Hypoplasia
38	Molar 2 or 3 with mandibular fragment	None
39	Deciduous incisor 1 maxillary left	None
40	Deciduous molar 1 mandibular left	None
41	Incisor 1 mandibular right	None
42	Molar crown and root fragment	None
43	Incisor 2 maxillary left	None
44	Canine mandibular right	None
45	Premolar 3 mandibular right	Hypercementosis
46	Root fragment	None
47	Incisor 1 mandibular right	None
48	Incisor 2 maxillary right	None
50	Molar mandibular right	None
51	Molar 3 maxillary right with alveolus	None
53	Premolar 4 maxillary left	None
54	Canine mandibular right	None
55	Root, anterior maxillary tooth	None
57	Premolar 4 mandibular right	None
58	Molar 3 mandibular right	None
59	Mandibular body left from I_2 to M_3, with C_1 to M_2	Caries, pathological radicular apex
60	Premolar 3 maxillary right	None
61	Deciduous molar 1 right; mandibular fragment	None

SP NO.	IDENTIFICATION	PATHOLOGY PRESENCE
68	Premolars 3 and 4 right in maxillary fragment	Hypoplasia
69	Deciduous molar mandibular probably right fragment	None
70	Deciduous molar 2 mandibular right	Hypoplasia
71	Deciduous molar 1 maxillary left	None
72	Molar maxillary left fragment	Hypoplasia
73	Incisor 1 maxillary right labial crown	Hypoplasia
74	Canine maxillary left	None
75	Premolar 3 mandibular left	None
76	Premolar 3 mandibular right	None
78	Premolar 4 mandibular left	None
79	Incisor 1 maxillary left	Hypoplasia
80	Mandible left with molar 2 and molar 3 germ	None
81	Molar 1/2? mandibular left	None
82	Canine mandibular left	None
83	Deciduous molar 2 mandibular right	None
84	Molar 1 mandibular left	None
85	Deciduous incisor 2 maxillary left	None
87	Premolar 4 mandibular left	Hypoplasia
88	Deciduous molar 2 mandibular left with alveolar fragment	None
89	Incisor 2 mandibular right	None
90	Incisor 1 maxillary left	None
91	Incisor 2 mandibular left	None
93	Deciduous canine mandibular right	None
94	Premolar 4 maxillary left	None
95	Deciduous canine mandibular left	None
98	Canine mandibular right	None
99	Incisor 2? mandibular left	None
100	Molar mandibular left	None

TABLE 12.2. Dentoalveolar aspects of the Palomas 59 left mandibular dentition

	M_2	M_1	P_4	P_3	C_1
CEJ–AC Average[a]	1.6	1.3	0.0 (unbroken lingual side)	1.9	0.4 (unbroken lingual side)
Dental Wear Score[b]	4c	4b	3	2	3

[a]Cemento-enamel junction to alveolar crest height, in mm.
[b]Dental wear scores following Smith (1984).

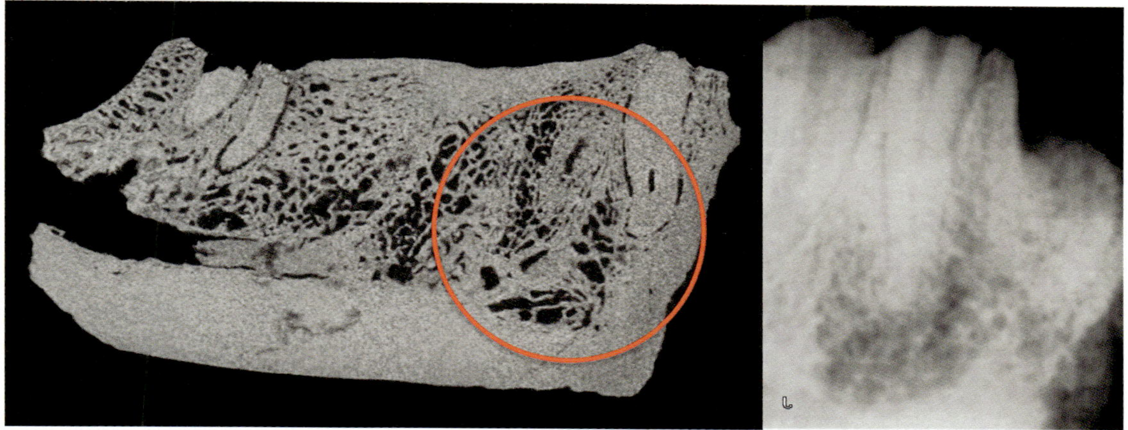

PLATE 12.1. Abnormalities of the Palomas 6 mandibular canine root. Micro-CT slice through the mandible with the region circled (*left*) and radiograph of the region showing the radiolucency around the root apex (*right*). (Micro-CT scan courtesy of P. Bayle)

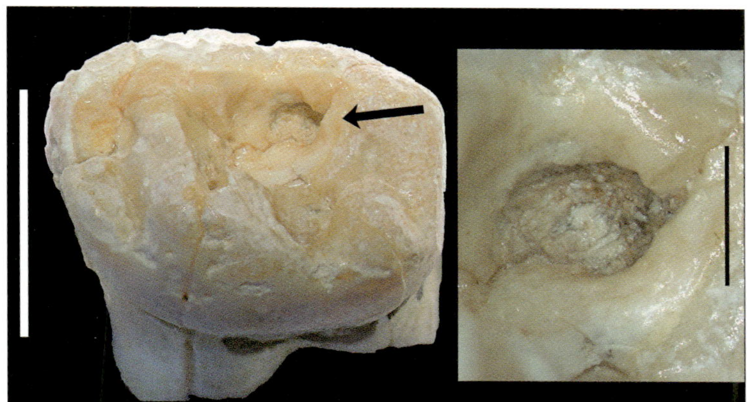

PLATE 12.2. Bucco-occlusal view of the Palomas 25 right dm_1 crown (*left*) with the carious lesion indicated (scale: 5 mm) and detail of the lesion (*right*). Scale: 1 mm

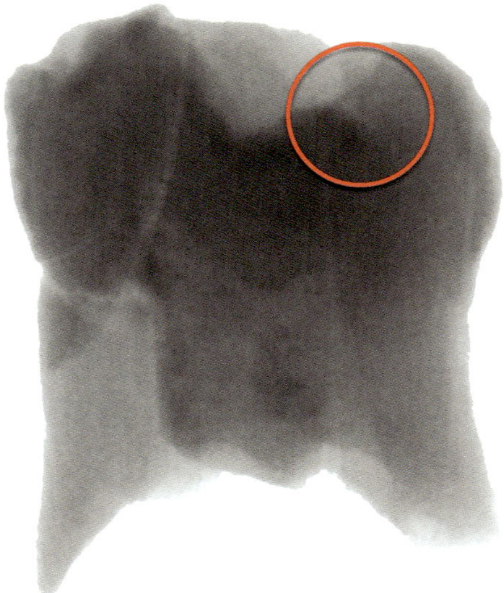

PLATE 12.3. Radiograph of the Palomas 25 right dm_1, with the area of the carious lesion indicated.

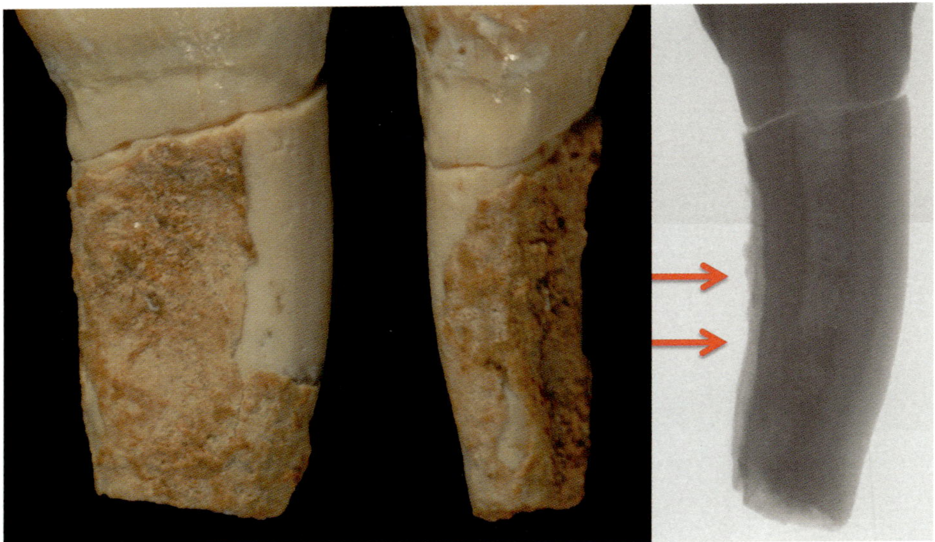

PLATE 12.4. Distal (*left*) and lingual (*middle*) views of the Palomas 45 mandibular P$_3$ root with the partially flaked off hypercementosis plus a buccolingual radiograph of the root showing the apposition of cementum (*right*). (Radiograph courtesy of P. Bayle)

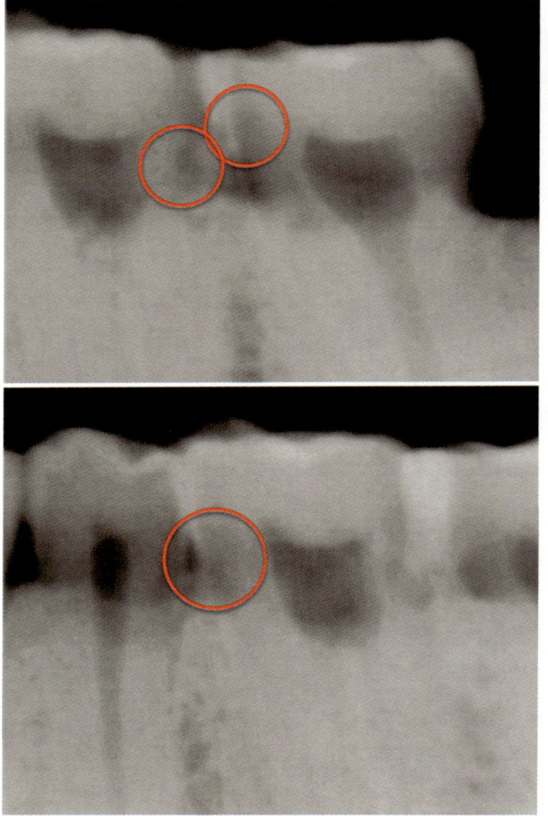

PLATE 12.5. Buccolingual radiograph of the Palomas 59 M$_2$ and M$_3$ with the areas of probable carious lesions circled (*above*), and buccolingual radiograph of the P$_4$ and M$_1$ with the probable mesial M1 carious lesion indicated (*below*).

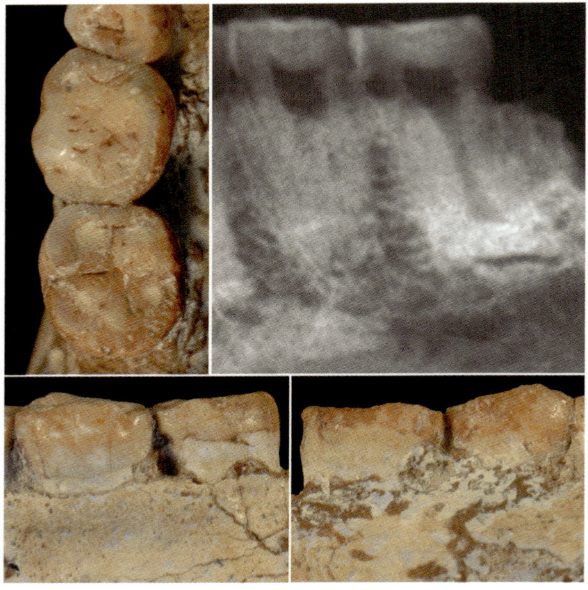

PLATE 12.6. The Palomas 59 left M$_1$ and M$_2$ in occlusal (*upper left*), buccal (*lower left*), and lingual (*lower right*) views plus a buccolingual radiograph of the M$_1$ and M2 showing the changes at the apex of the M$_2$ taurodont root. Not to the same scales.

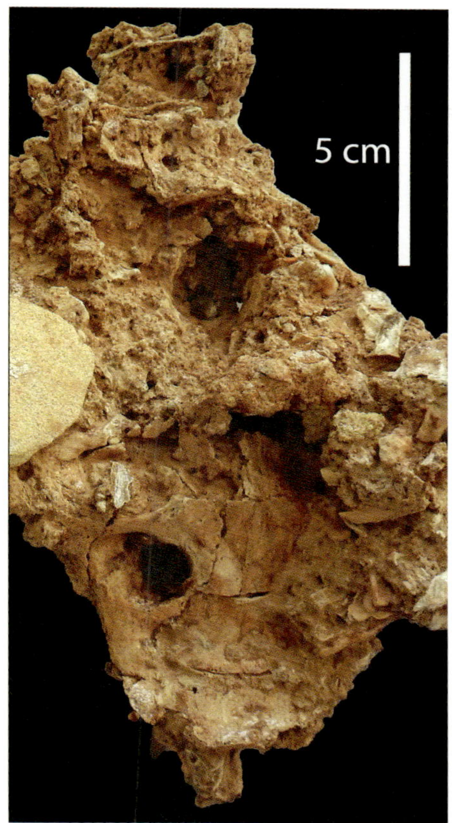

PLATE 13.1. The Palomas 92 caudal vertebral column, including (*from above*) the L2 (92kkk), L3 (92jjj), L4 (92iii), and L5 (92hhh) partial bodies and the ventral surface of the sacrum (92ss). The more complete vertebral bodies with intervertebral disk surfaces evident are the L2 and the L3. The more intact ventral sacral vertebrae are the S2 to S4, and there is a left ventral section of the S1 evident. Portions of the posterior ilia are crushed against the sacral alae.

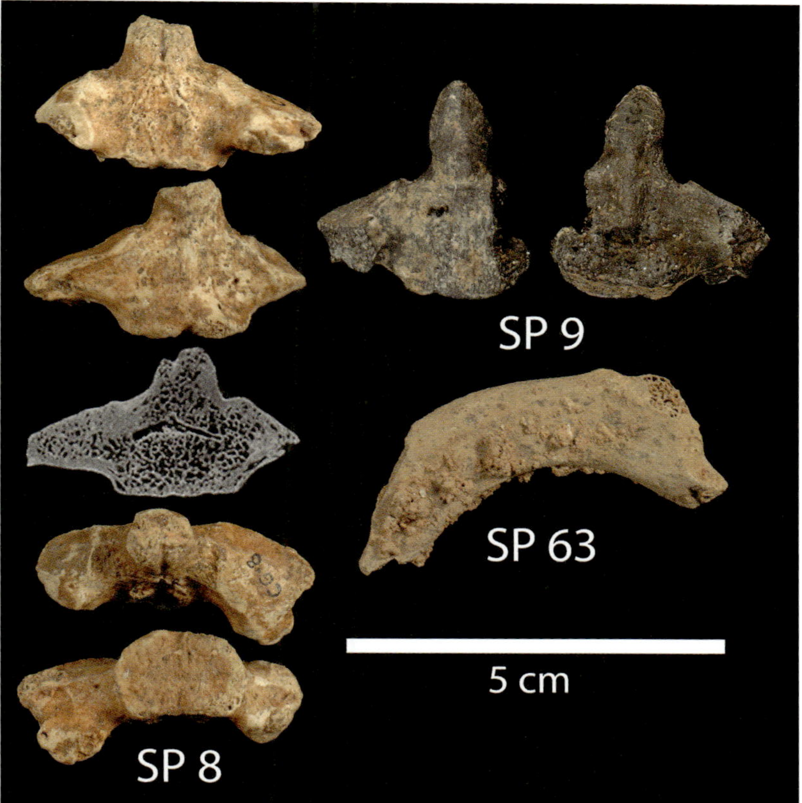

PLATE 13.2. Isolated axial remains from Palomas. Palomas 8: immature axis (C2) in (from above) dorsal, ventral, cranial, and caudal views, with a transverse micro-CT slice showing the partial fusion of the odontoid process onto the C2 body. Palomas 9: mature axis (C2) in dorsal (*left*) and ventral (*right*) views. Palomas 63: caudal view of left first rib.

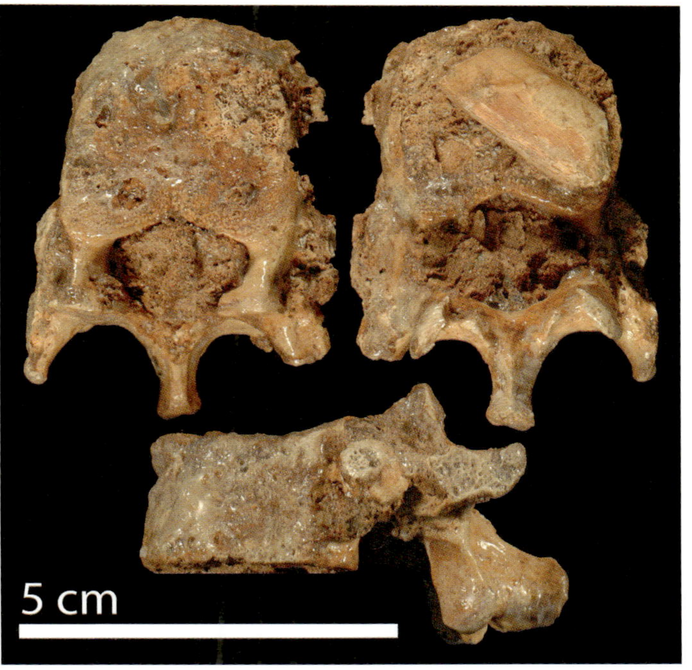

PLATE 13.3. The Palomas 92 (92ll) last thoracic vertebra (T12) in cranial (*left*), caudal (*right*), and left lateral (below) views.

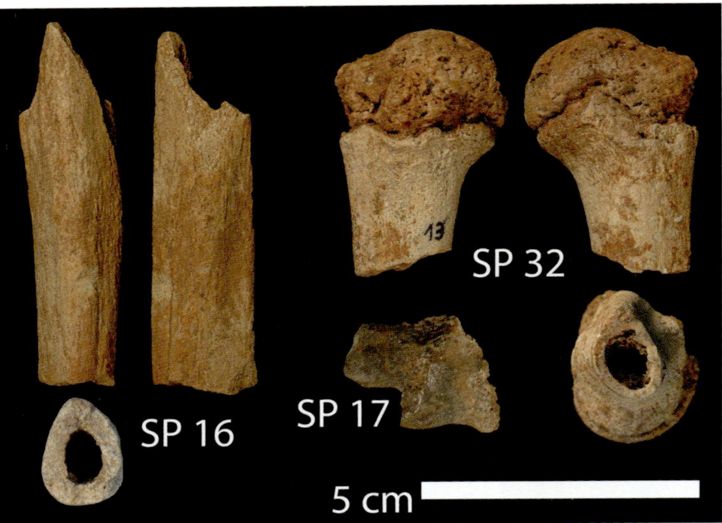

PLATE 13.4. Isolated partial humeri. Palomas 16: proximal left humerus in anterior (*left*), lateral (*right*), and distal (*below*) views. Palomas 17: distal left trochlea and medial epicondyle in anterior view. Palomas 32: proximal epiphysis and metaphyseal region of an immature right humerus in anterior (*left*), posterior (*right*), and distal (*below right*) views.

PLATE 13.5. Posteroanterior, mediolateral, and distal (≈75%) transverse micro-CT slices of the Palomas 32 proximal immature right humerus.

PLATE 13.6. Posterior view of the Palomas 92 right humerus (92ff) with the four sections aligned anatomically (*right*), plus the anterior view of the distal humerus with the attached proximal ulna and radius (*left*). There is a piece of faunal bone cemented to the mid-proximal posterolateral diaphysis, and the proximal ulna (92gg) is cemented around the trochlea. The length as assembled is approximate, given missing bone and matrix across the diaphyseal joins.

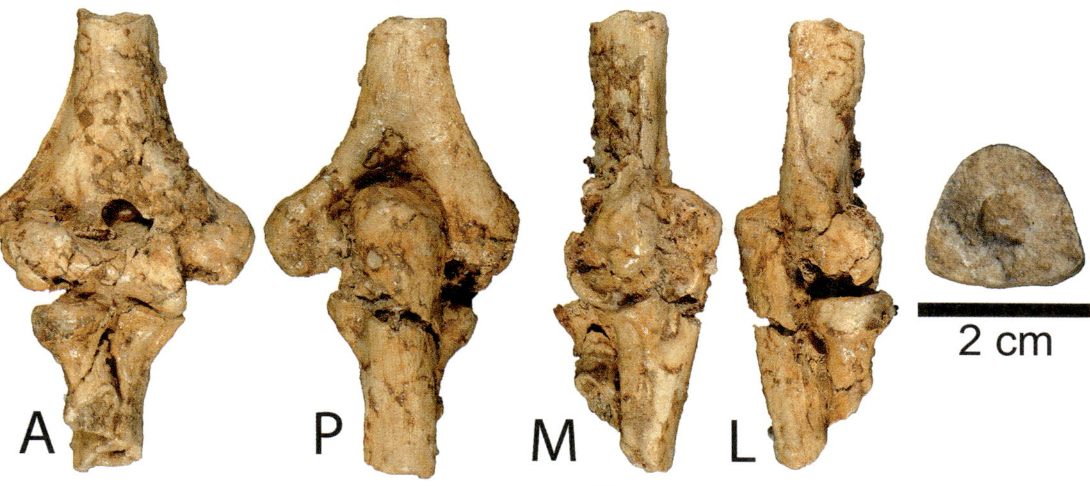

PLATE 13.7. The cemented-together right elbow in anterior (A), posterior (P), medial (M), and lateral (L) views. The proximal view of the middistal humeral diaphyseal cross-section, with anterior above, is provided enlarged relative to the other views.

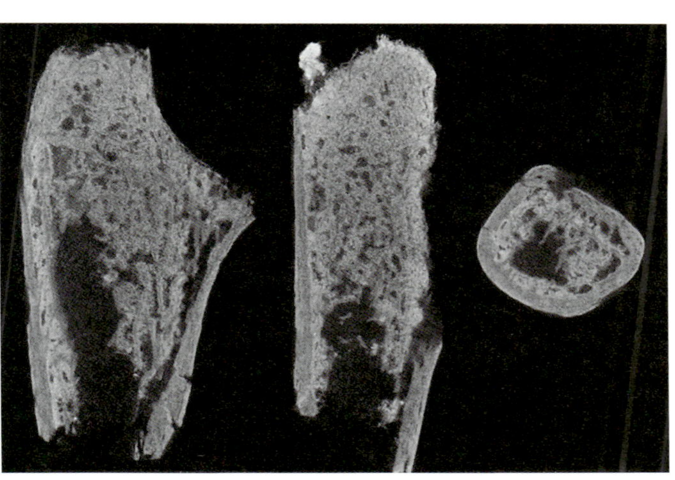

PLATE 13.8. Views of the Palomas 14 immature left proximal ulna and the Palomas 64 mature left proximal radial diaphysis. The arrow locates the proximal end of the interosseus crest. The bony growth into the Palomas 64 medullary cavity may be abnormal.

PLATE 13.9. Mediolateral, anteroposterior, and ≈75% diaphyseal cross-sections of the Palomas 14 immature left proximal ulna, from micro-CT slices.

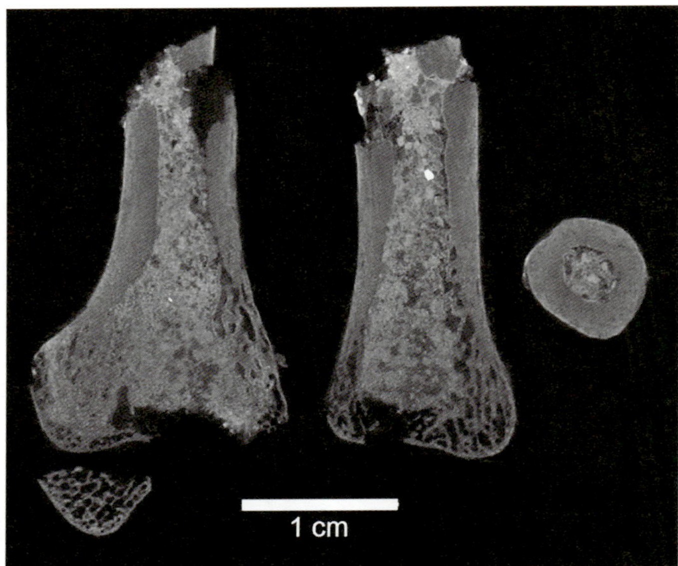

PLATE 13.11. Mediolateral, anteroposterior, and ≈25% diaphyseal cross-sections of the Palomas 92hh right distal ulna, from micro-CT slices.

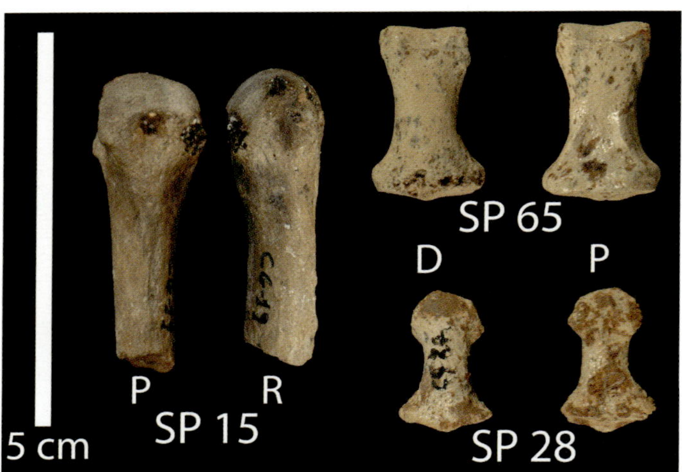

PLATE 13.12. Views of the Palomas isolated hand bones: the Palomas 15 right metacarpal 2 in palmar (P) and radial (R) views, and the Palomas 65 and 28 manual middle and distal phalanges 2–4 in dorsal (D) and palmar (P) views.

PLATE 13.10. The Palomas 92 right forearm and elbow in approximately medial view. The partially cemented assembly includes the distal humerus (92ff), three separated pieces of the ulna (92gg), and largely connected sections of the radius (92ii).

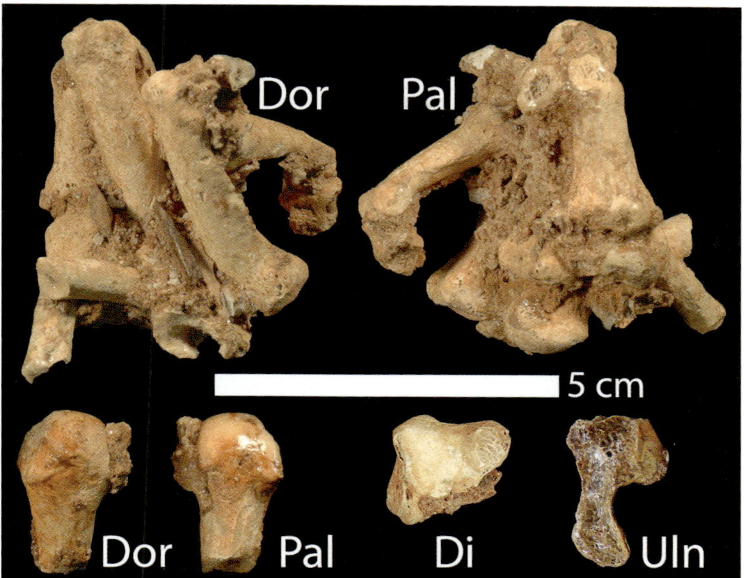

PLATE 13.13. Views of the Palomas 92 hand bones. *Above:* approximately dorsal and palmar views of the cemented left metacarpals and phalanges. *Below, left to right:* dorsal and palmar views of the right metacarpal 2 head, distal view of the right trapezium, and ulnar view of the left hamate bone.

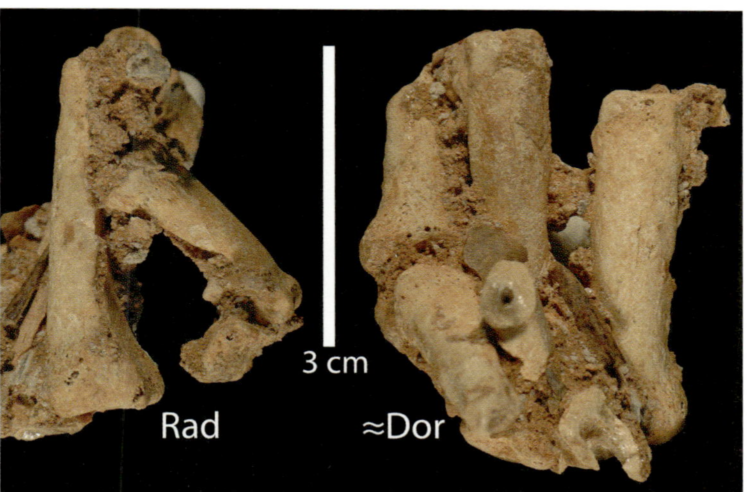

PLATE 13.14. Radial view of the Palomas 92 third ray manual phalanges (*left*) and approximately dorsal views of the proximal manual phalanges 3 to 5 (*right*).

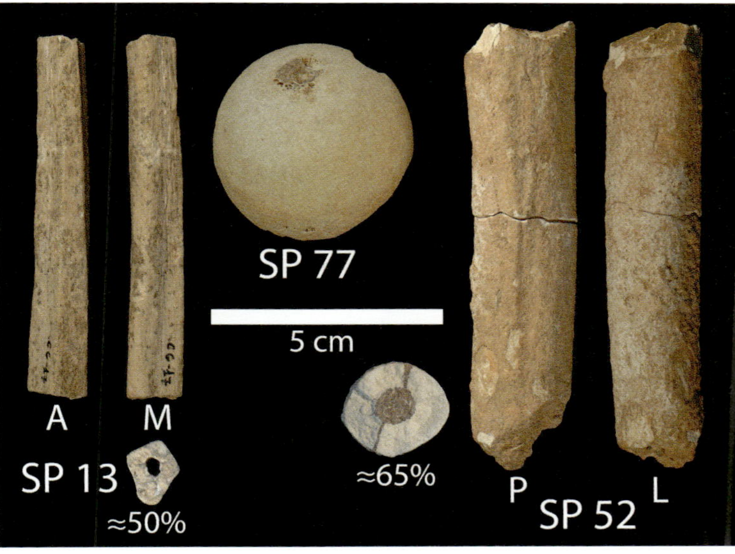

PLATE 13.15. Isolated lower limb remains from Palomas. *Left:* Palomas 13 left proximal fibular diaphyseal section in anterior (A) and medial (M) views, plus the approximately midshaft cross-section. *Center:* articular view of the Palomas 77 isolated femoral head. *Right:* posterior (P) and lateral (L) views of the Palomas 52 proximal femoral diaphysis, plus its ≈65% (midproximal) cross-section.

PLATE 13.16. The Palomas 92 femoral remains. *Left:* posterior view of the left femoral diaphysis. *Right:* lateral view of the right femur from the distal trochanters to the lateral condyle, with the proximal and distal sections placed adjacent to each other at their midshaft contact. *Below left:* the ≈65% (midproximal) diaphyseal cross-section of the right femur.

PLATE 13.17. Dorsal views of the Palomas 66 and 86 immature lateral proximal pedal phalanges and the Palomas 67 adult middle pedal phalanx.

Plate 13.18. Dorsal view of the Palomas 92 left pedal anterior tarsals, metatarsals, and phalanges.

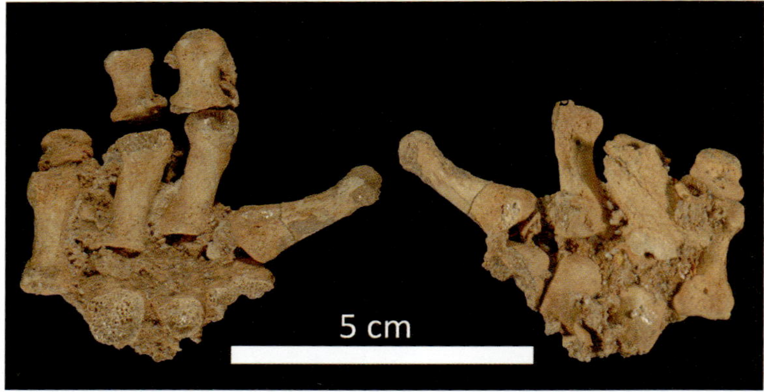

PLATE 13.19. Dorsal (*left*) and plantar (*right*) views of the Palomas 92 pedal phalanges. Given fossilization cementation, the proximal phalanx 2 is in approximately lateral and medial views.

The Palomas Postcrania 13

ERIK TRINKAUS, KATHARINE A. ROBSON BROWN, JON ORTEGA, AND KLÁRA KARAKOVÁ

ALONG WITH THE ABUNDANCE of dental remains, the Palomas Neandertal sample contains a number of postcranial elements. These remains include the seventeen isolated elements (most of them incomplete) found on the hillside, among the shaft debris, and within the Upper Cutting infilling (Table 13.1). They also include the associated partial skeleton of Palomas 92, which was variably crushed and cemented within the brecciated Conglomerate A (Tables 13.2, 13.4, 13.13, 13.26, and 13.30). Despite the conditions of these postcrania and the lack of stratigraphic provenience for ten of the isolated bones, these elements provide substantial morphological and paleobiological data for these southwestern European Neandertals.

There are also variably complete data available for the appendicular remains of Palomas 96. The description of Palomas 96 is not part of this monograph (chapter 3), given its current condition partially encased in breccia. There are nonetheless a number of measurements available from its long bones and hand remains, only a few of which have been previously published (Walker et al. 2011b; Trinkaus and Ruff 2012; Trinkaus et al. 2014b). These quantitative data are included in the tables, and they are presented in the graphical comparisons as appropriate in the context of the assessments of the other Palomas postcrania.

The Ages at Death of the Palomas Postcrania

Five of the isolated postcranial elements, Palomas 8, 14, 32, 66, and 86, derive from immature individuals. They range from relatively young bones (the Palomas 14 ulna) to ones at least later in the first decade in age (the Palomas 8 vertebra and the Palomas 32 humerus). More detailed assessments of their ages at death are provided with their separate descriptions.

The remainder of the isolated specimens appear to be fully mature, as indicated by either complete closure of the preserved epiphyses or synchondroses (Palomas 9, 15, 17, 28, 65, 67, and 77) or general size and morphology (Palomas 13 and 64). They could be late second decade in age or fully mature.

At the same time, three of the isolated Palomas postcranial specimens (Palomas 16, 52, and 63) have had their complete skeletal maturity queried (Walker et al. 2008). This uncertainty derives from the absence of epiphyseal regions for these diaphyseal sections and, especially, the perception of small dimensions for the specimens. It was therefore uncertain whether they could represent adolescent rather than fully adult individuals; certainly the abundance of immature individuals represented by the dental remains (chapters 6 and 15) and the presence of immature elements in the postcranial sample would make this interpretation feasible. However, the two fully adult associated partial skeletons from Palomas, Palomas 92 and Palomas 96, plus several of the clearly mature isolated Palomas elements, are also very small—at the bottom of the known Neandertal body size range (chapter 14). The Palomas 16, 52, and 63 elements are no smaller than those of these other adult Palomas remains (see comparisons below for Palomas 16 and 52). Moreover, aspects of their morphology conform to what is known for adult specimens (see below). They are therefore considered to be fully mature or nearly so; in either case, the morphometric comparisons to other mature remains are appropriate.

It is possible, and appropriate, given its preservation of multiple skeletal elements, to assess the age at death of Palomas 92 in detail here. All of its epiphyses—of the hand and foot bones, the femora, the right elbow, and the distal radius and ulna—are fully fused, and there is no evidence of the fusion lines for those epiphyses. Palomas 92 was therefore fully mature. In this context, the degree of ventral fusion of the sacral bodies permits a more precise age determination.

There is no indication of fusion between the S1 and S2, and that appears to have been the case at the S2/S3 juncture (Plate 13.1). They therefore conform to the stage 0 of Belcastro et al. (2008). The S3/S4 juncture conforms to their stage 1 (<50% fused), in that there is some indication of lateral fusion. The fusion line between S4 and the missing S5 appears to have been partially fused laterally, but crushing makes that determination tenuous. Based on the recent European samples of Belcastro et al. (2008), the lack of fusion at S1/S2 implies an age at death in the third decade, but the lack of fusion at S2/S3 indicates a probable age at death in the first half of the third decade, especially for females. The partial fusion of S3/S4 (and probably at S4/S5) also implies a very young adult age, again especially for females. A mixed Euro-American and Afro-American male sample (McKern and Stewart 1957) provides a similar age range. The Palomas 92 sacrum therefore indicates a very young adult age at death, most likely less than 25 years

of age. If the annular rings of the T12 (see below) were only partially fused, that would indicate an age at death closer to 20 years postnatal (McKern and Stewart 1957).

In this age-at-death context, it is possible to assess the suggestion (Walker et al. 2012; chapter 2) that Palomas 1 and 92 might represent the same individual, based on their similar stratigraphic positions and nonduplication of skeletal elements. Although its age assessment based on dental wear is less precise (chapters 6 and 15), the Palomas 1 facial remains are unlikely to come from an individual less than 30 years of age; therefore, Palomas 1 and 92 should be considered to derive from different individuals.

Because data for Palomas 96 are also included in the postcranial assessments, its age at death is also relevant. All of the sufficiently preserved appendicular epiphyses are fully fused, yet the proximal clavicle was completely unfused or had only minimal fusion (which condition is uncertain, given damage to the margins). A similar condition of the proximal clavicle indicates a probable age of less than 25 years if partially fused, less than or about 22 years if unfused (Webb and Suchey 1985). The ventral sacral bodies exhibit a lack of fusion (stage 0) of S1/S2, S2/S3, and S3/S4 and either no fusion or partial fusion (stage 0 or 1) for S4/S5. This modestly lesser stage of fusion than that of the Palomas 92 sacrum provides a maximum age at death of about 20 years and a probable one of less than 20 years (following Belcastro et al. 2008). The third molars were in full occlusal eruption, but root closure is not known. Together, these developmental indicators provide an age at death of about 20 years and probably slightly younger (see also Walker et al. 2011b).

Sex Assessment for the Palomas Postcrania

None of the isolated postcranial elements from Palomas or Palomas 92 provide sexually diagnostic pelvic remains; the only pieces of the pelvis known are the crushed pieces of ilium of Palomas 92 in matrix. Palomas 96, on the other hand, retains most of the pelvis, and it clearly indicates that Palomas 96 was female (Walker et al. 2011b). Given the general similarity in overall size of Palomas 92 and some of the other small postcrania (especially Palomas 16, 52, and 77) to Palomas 96, as well as their body sizes at the bottom of the Neandertal size range (chapter 14), it is likely that these elements also represent females. There are no substantially larger individuals represented by the Palomas postcrania, ones that could represent males, even though much larger individuals are represented by Neandertal postcranial elements from the Iberian Middle Paleolithic sites of Oliveira, El Sidrón, and Zafarraya. The Palomas mandibles are also among the smaller of the Neandertal ones in corpus dimensions, especially Palomas 6, 23, and 59 (Fig. 5.16). Yet it is tenuous to assign sex to Neandertals based solely on size (Trinkaus 1980, 2016c), and these Palomas specimens (other than Palomas 96) therefore are likely to be female but should remain indeterminate as to sex.

Postcranial Methods

All of the measurements are in millimeters unless otherwise indicated. Measurements with minor estimation are in parentheses. M numbers in the tables refer to the numbered measurement in the Martin system (Bräuer 1988). Measurements, even if estimated, are provided to the nearest 0.1 mm.

The cross-sectional parameters of the diaphyses provided (total and cortical areas and especially second moments of area) were primarily determined from natural (in fossilization) breaks of the diaphyses, the ones sufficiently perpendicular to the diaphyseal axis and with minimal damage so as to permit accurate representations of the cross-sections. Scaled photographs of them were projected enlarged onto a Summagraphics 1812 tablet and digitized, and the cross-sectional parameters were calculated using SLICE/SLCOMM (Nagurka and Hayes 1980; Eschman 1992). The values are the averages of three digitizations. In cases in which the axial orientation of the bone is approximate, the anatomically oriented second moments of area (I_x and I_y) are in parentheses or not included. Areas are provided to the nearest 0.1 mm^2; second moments of area are to the nearest mm^4.

A number of the postcranial bones were micro-CT scanned to provide internal morphological data for several of the isolated elements and to permit the measurement of the sets of Palomas 92 (and 96) manual and pedal phalanges that are cemented together. The scanning system is described in chapter 3. For the humeri, ulnae, and phalanges, the scanner was set at 100 kV and 100 μA with an Al/Cu filter to provide images with an isometric voxel resolution of 35 μm. For vertebrae the system was set to 92 kV 108μA for the same resolution. Analysis of the resulting reconstructed image files was achieved using CTAn (Version 1.9.2.5, Skyscan, Kontich, Belgium), Avizo v.6 software, and BoneJ (Doube et al. 2010), the last a plugin for ImageJ (National Institutes of Health).

For the upper limb diaphyseal sections derived from micro-CT scans, cross-sectional values were computed at 25%, 50%, and 75% of length for those specimens preserving full length. The positions were approximated for less complete specimens. For metaphyseal regions, a series of measures of the trabecular bone dimensions, densities, and orientations was determined. For some of the manual phalanges, whose external surfaces were obscured by breccia and the cementing of the bones to each other (Plate 13.13), measurements were derived from the micro-CT scans.

For trabecular bone (epiphyseal regions primarily in the case of the Palomas remains), the regions of interest for analysis were extracted from individual specimens by aligning the preserved regions into a plane defined by three type II anatomical landmarks (*sensu* Bookstein 1991). Alignment was followed by extraction of the largest possible spherical volumes of interest (VOI) about the bone's centroid. Trabecular tissue was initially segmented using a binary global threshold based on the frequency distribution of the gray values (CT numbers) in the scan. The threshold value was placed at the trough between the characteristic peaks for air and bone. It was found that there was overlap between the grayscale distribution of the trabeculae and that of the matrix within the trabecular space. In order to derive an optimum threshold, a curve representing the sum of two normal distributions (with means μ_1 and μ_2 and variances σ^2_1 and σ^2_2) was fit to the average grayscale distribution by altering the values of the normal parameters (μ_1, μ_2, σ^2_1, and σ^2_2) using a quasi-Newton iterative method (Solver, Microsoft Office Excel 2007). The two normal distributions were assumed to represent the trabeculae and the matrix in the trabecular space. There was a clear minimum between the two normal distributions, and this value was selected as the threshold (Robson Brown et al. 2014).

The following trabecular architectural measures were collected: bone volume fraction (BV/TV), trabecular thickness (TbTh), separation (TbSp), structure model index (SMI), and degree of anisotropy (DA). Bone volume fraction is the amount of bone per unit volume and was calculated by dividing the number of threshold white voxels (binarized solid) by the total number of voxels in the VOI. Trabecular thickness and separation were calculated by fitting the largest sphere possible within a trabecule or space between two objects (Hildebrand and Rüegsegger 1997). The diameter of this sphere represents the thickness of the trabecular element or space. SMI measures the proportion of rod- and plate-shaped trabeculae within 3D space. SMI values range from 0 for a structure made of plate-shaped trabeculae to 3 representing a primarily rod-shaped one. Structures with a mixture of plate and rod shapes have values between 0 and 3. Anisotropy measures the direction and size of the preferred orientation of trabeculae, calculated as a ratio between the maximum and minimum radii of the mean intercept length ellipsoid (Chen et al. 2013; Odgaard et al. 1997). A value of 1 reflects a completely anisotropic structure, while 0 reflects an isotropic one.

For the micro-CT-derived cross-sections, the following values were measured: anteroposterior diameter (AP), mediolateral diameter (ML), periosteal perimeter (PP); total bone area (BA), cortical bone area (CA), percent cortical area (% CA), and mean cortical thickness (CTh). BA was measured by selecting the entire slice as the region of interest. CA measurements were repeated three times, and the mean was calculated to account for any minor discrepancies when selecting the region of interest. CTh was calculated as the mean of cortical thickness taken in the medial, lateral, anterior, and posterior sections of the bone.

In the assessments of the humeri, given often marked bilateral diaphyseal (but not epiphyseal) asymmetry related to handedness among Pleistocene humans (Trinkaus et al. 1994, 2014a; Churchill and Formicola 1997; Cowgill et al. 2015; Trinkaus 2015), it is ideal to make comparisons only among dominant or nondominant sides for diaphyseal properties. However, none of the isolated Palomas arm bones or the Palomas 92 ones provide bilateral data on diaphyseal hypertrophy. Moreover, among all of the Neandertals (except Palomas 96) the right humeral diaphysis is stronger or the level of asymmetry is minimal. Therefore, in the humeral diaphyseal comparisons here, the Palomas 16 left humeral midshaft is compared to left humeri and the Palomas 92 right humeral midshaft and middistal shaft are compared to right humeri. The right-left values for Palomas 96 are reversed so that the comparisons will best approximate dominant versus nondominant ones, but they still permit the inclusion of unilaterally preserved ones.

Some degree of asymmetry in manual diaphyses has been documented among recent (Roy et al. 1994; Mays 2002) and Pleistocene (Trinkaus et al. 2014a) humans. However, the levels of manual diaphyseal asymmetry tend to be modest. When lower-limb diaphyseal asymmetry is assessed (it is usually assumed to be minimal), femoral diaphyseal asymmetry is also usually small. For these reasons the Palomas values are compared to both right and left values for these bones, averaged by individual if both sides are preserved, and have provided data.

The Palomas Isolated Vertebrae

Palomas 8: Axis (C2)—Immature

The bone retains the centrum and the articular facets without the tip of the odontoid process (Plate 13.2) (maximum breadth: 38.7 mm; maximum height: 21.4 mm). The immature status is based on the absence of the caudal annular ring and of the tip of the odontoid process. In addition, there is only partial (lateral) fusion of the odontoid process (from C1) onto the body of the axis (from C2) (Fig. 13.2). Given that (1) the facets/pedicles fuse to the sides of the odontoid process by 3–4 years postnatal, (2) those portions fuse to the centrum by 4–6 years postnatal, and (3) the tip of the odontoid process fuses to its base by about 12 years postnatal (Scheuer et al. 2000), the age at death of the specimen is best seen as between 6 and 12 years postnatal and probably toward the younger end of that range.

There is little of note on the specimen. Among Neandertal axes relative to those of recent humans (Heim 1976; Trinkaus 1983; Arensburg 1991; Gómez-Olivencia et al. 2013a, 2013b; Trinkaus 2016b), the primary differences relate to cervical spinous process hypertrophy, and these cannot be appropriately assessed on this immature specimen relative to those of other Neandertals.

Palomas 9: Axis (C2)—Mature

The bone retains the odontoid process, the cranial left articular facet, the left body, and the abraded caudal body (Plate 13.2). The maximum preserved height, from the dorsal C2/C3 intervertebral disk surface margin to the tips of the dens, is 30.3 mm.

The main features of note for Palomas 9 are related to the dens. It deviates to the left, evident in the dorsal view but especially in the ventral view (Plate 13.2). It is not clear how this might have related to its articulation with the anterior arch of the atlas or the positions of the apical ligament of the dens and the alar ligaments. Given their role in axial rotation of the head (Dalton 2011), it is likely that there was some minor compensation in the atlas and the occipital bone. None of the other Neandertal odontoid processes (the mature ones of La Ferrassie 1, Krapina 103 to 105, Regourdou 1, and Shanidar 2 or the immature ones of Palomas 8 and Krapina 102) exhibit a similar deviation. On Palomas 9, neither the articular facet for the anterior arch of the atlas nor the cranial left articular facet exhibit any subchondral or marginal degeneration.

The height of the dens, from the transverse tangent across the medial cranial facet margins, is 14.8 mm (Table 13.3); this value is among those of other Neandertal axes (Kebara 2: ≈15.0 mm; Krapina 103: 13.7 mm; Krapina 104: 15.0 mm; Regourdou 1: 15.6 mm; Shanidar 2: ≈18.0 mm) and well within a range of 11–18 mm provided by Stewart (1962) for recent human males. Two Middle Paleolithic modern humans, Qafzeh 9 and Skhul 5, provide modest heights of ≈13 and 12.5 mm (McCown and Keith 1939; Vandermeersch 1981).

The Palomas 92 Vertebrae

The Palomas 92 vertebral column is represented by a surprisingly intact last thoracic vertebra (T12), fragments of four lumbar vertebrae (L2 to L5) in matrix adherent to the sacrum, and the ventral surfaces of most of the S2 to S4 with pieces of the S1 and S5 (Table 13.2; Plates 13.1 and 13.3). The L2 and L3 provide approximate measurements of their body heights (≈23.0 and ≈24.0 mm, respectively), and the ventral sacrum provides ventral body heights of the S2 to S4 (22.0, 17.5, and 15.0 mm, respectively). The L2 also permits measurement of the external breadth across its caudal articular facets (29.3 mm). The T12 permits a larger series of measurements and observations (Table 13.3).

Palomas 92ll: Thoracic Vertebra (T12)

A largely complete thoracic vertebra is preserved for Palomas 92, with damage to the transverse process areas and variable marginal erosion and adherent sediment (Plate 13.3). The cranial articular facets are largely coronally oriented, and the caudal ones are more parasagittally oriented (Table 13.3); it is therefore identified as a T12.

The body is largely complete. The annular rings are present and fused onto the centrum for 12 mm around the dorsal half of the caudal body, especially on the right side. There are also traces of the annular rings on the cranial body adjacent to the spinal canal. Where present, the margins between the annular rings and the centrum are readily apparent, suggesting fusion shortly before death. The remaining body edges are eroded, and it is therefore unclear to what extent the annular rings were fully fused to the centrum. T12 annular rings are usually fully fused to the centrum by 20 years or shortly thereafter (McKern and Stewart 1957), so partial fusion would not be unusual, especially if Palomas 92 was closer to 20 years at death.

The neural arches are mostly present, as are the articular facets and the spinous process. The lateral transverse processes are damaged. The spinal canal retains matrix, and a fragment of bone is cemented to the caudal body. The dorsal end of the spinous process is eroded. All of the primary portions of the vertebra are fused. The maximum dorsoventral diameter is 58.9 mm, and the maximum transverse diameter across the broken transverse processes is 41.6 mm.

Palomas 92ss: Sacrum

The extremely fragile partial sacrum is preserved in a block of breccia, with the left side of the S1 ventral body and the full ventral bodies of the S2 to the S4 exposed, especially on the right side (Plate 13.1). The S5 body is absent, but more dorsal fragments are preserved. The right pelvic sacral foramen is fully evident, but the other foramina are absent or buried in breccia. The S1/S2, S2/S3, and S3/S4 ventral body fusion lines are readily apparent, as is the one between the S4 and the missing S5 body (see age at death assessment above). The promontory has been crushed and eroded, as are the lateral margins. Aside from providing S2 to S4 ventral body heights and age-related degrees of body fusion, the sacrum furnishes little data.

Vertebral Column Heights

As a measure of vertebral column length, given consistency in vertebral proportions along the column (Holliday 1995), the middle body heights of the Palomas 92 T12, L2, and L3 can be compared to the few comparative measurements available, all for male Neandertals and all presenting developmental or degenerative abnormalities (Trinkaus 1983, 1985, pers. obs.; Arensburg 1991). Note that middle heights, not including the annular rings, are usually less than the values for ventral or dorsal height if the cranial and caudal intervertebral disk surfaces are largely parallel. The Palomas 92 height for T12 (≈20.5 mm) is similar to those of Kebara 2 (22 mm) and Shanidar 3 (20 mm). The height for the L2 (≈23 mm) is close to the same dimension for Kebara 2 and Shanidar 3. The value of ≈24 mm for the Palomas 92 L3 is slightly higher than the measure of 20 mm for Shanidar 3 and 21 mm for Kebara 2 and La Chapelle-aux-Saints 1. These figures therefore suggest, but are insufficient to confirm, that the trunk length of Palomas 92 did not differ markedly from that of these other Neandertals.

It is also possible to estimate the ventral sacral height of Palomas 92 from the ventral body heights of the S2 to S4. A multiple regression based on recent human sacra provides a least squares equation of VenHt = 46.98 + (1.41 × S2Ht) + (1.58 × S3Ht) − (0.72 × S4Ht) (r^2 = 0.652, n = 25). The resultant ventral height for Palomas 92 is 94.9 ± 1.6 mm (95% CI: 91.5–98.2 mm). The modest level of correlation is largely due to the considerable variation in sacral curvature among recent humans. Sacral curvature cannot be assessed for Palomas 92, given postmortem crushing.

This estimate range for Palomas 92 is below the high value of ≈115 mm for Kebara 2 (Rak 1991) and the estimates of ≈107.5 mm and ≈102 mm for Shanidar 1 and 3 (Trinkaus 1983). The range encompasses the sacral height of Palomas 96 (82.4 mm), as is expected, since the S2 to S4 ventral body heights of Palomas 96 (22.9 mm, 17.1 mm, and 14.8 mm, respectively) are close to those of Palomas 92. None of the other Neandertal sacra (La Chapelle-aux-Saints 1, La Ferrassie 1, Regourdou 1, Subalyuk 1) provide a height measurement, but the presumably Middle Pleistocene Broken Hill E688 has a sacral height of 103 mm. The distorted Qafzeh 9 sacrum provides an estimated ≈98 mm (Vandermeersch 1981). The sacral craniocaudal dimension of the Palomas 92 is therefore below those of the three male Neandertals with at least estimates of sacral height (all are distorted and/or incomplete), and it is similar to that of the female Palomas 96.

The Palomas Costal Remains

Palomas 63: Rib 1 Left—Mature(?)

The bone preserves the neck and tubercle plus the shaft ventrally close to the costal cartilage surface (Plate 13.2). It has scattered hard matrix encrustations on its caudal surface and along the internal margin, and most of the cranial surface is obscured by matrix. The maximum preserved length is 51.8 mm.

There is no evidence of the scalene tubercle, and the cranial muscle insertions are covered by matrix. The two available measurements are its internal-external diameter near the middle of 12.6 mm and the craniocaudal thickness along the external margin of 5.7 mm. The thickness measurement is moderately high compared to measurements available for other Neandertals (Kebara 2: 3.8 mm; Shanidar 4: 4.1 mm; Krapina: 4.1 ± 0.4 mm, n = 6), but the breadth measurement is comparatively small (Kebara 2: 23.0 mm; Shanidar 4: 19.0 mm; Krapina: 13.9 ± 0.3 mm, n = 6). Together, these Neandertal first ribs provide ranges of 13.5 to 23 mm for the breadth and 3.5 to 4.7 mm for the thickness. The Palomas 63 first rib therefore appears to be relatively narrow but thick for a Neandertal.

Palomas 92: Rib Fragments

There are two small pieces of rib remaining for Palomas 92—92fff and 92ggg—but neither one provides morphological data.

The Palomas Humeri

The Palomas sample retains three isolated pieces of humeri—Palomas 16, 17, and 32—plus a largely complete right humerus of Palomas 92 (Tables 13.1 and 13.4). Together, they provide data principally on their diaphyses (Palomas 16 and 92) plus limited data on their distal epiphyseal regions. There are also data from the Palomas 96 humeral diaphyses and a distal epiphysis.

Palomas 16: Humerus Left—Mature(?)

Palomas 16 is a humeral diaphyseal section from midshaft (very close to the distal end of the deltoid tuberosity) to the middle of the pectoralis major tuberosity (Plate 13.4). The maximum preserved length is 68.2 mm. The possible immature status of the bone was based on its small size and thin cortical bone at the proximal end, but the midshaft relative cortical bone (percent cortical area = 76.7%) is normal for an adult (Late Pleistocene left midshaft %CA: 71.3 ± 10.5%, n = 26). It is considered in the comparisons as an adult, bearing in mind that it might represent a slightly younger individual.

Palomas 17: Humerus Left—Mature

The bone retains only the medial trochlea and the anterior medial epicondyle, which had been burned (Plate 13.4). The maximum mediolateral dimension is 32.5 mm.

Palomas 32: Humerus Right—Immature

This humerus consists of a complete proximal metaphysis with the surgical neck and the complete head and tubercle epiphysis cemented onto the metaphysis (Plates 13.4 and 13.5). The maximum preserved length is 46.0 mm, and its maximum breadth is 32.2 mm.

Given that the epiphysis is completely unfused and that it normally fuses during the middle to late adolescent years (Scheuer et al. 2000), this suggests a juvenile age (7–12 years) for this

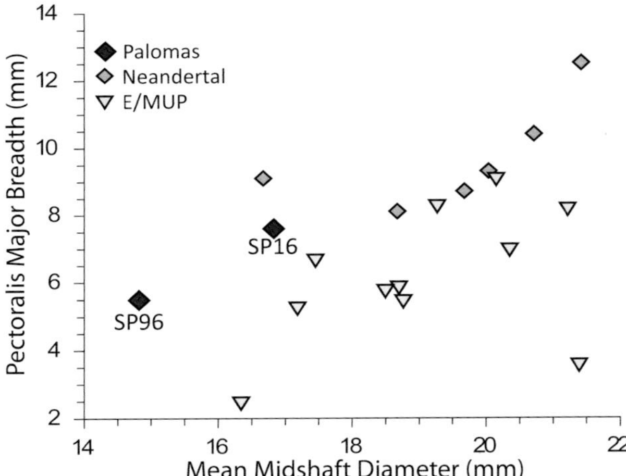

FIG. 13.1. Bivariate plot of pectoralis major breadth versus the geometric mean of the midshaft maximum and minimum diameters for Late Pleistocene left humeri. The value for Palomas 96 is for her nondominant right humerus; the other humeri are either left and nondominant or left and assumed to be nondominant. E/MUP: Early/Mid Upper Paleolithic modern humans.

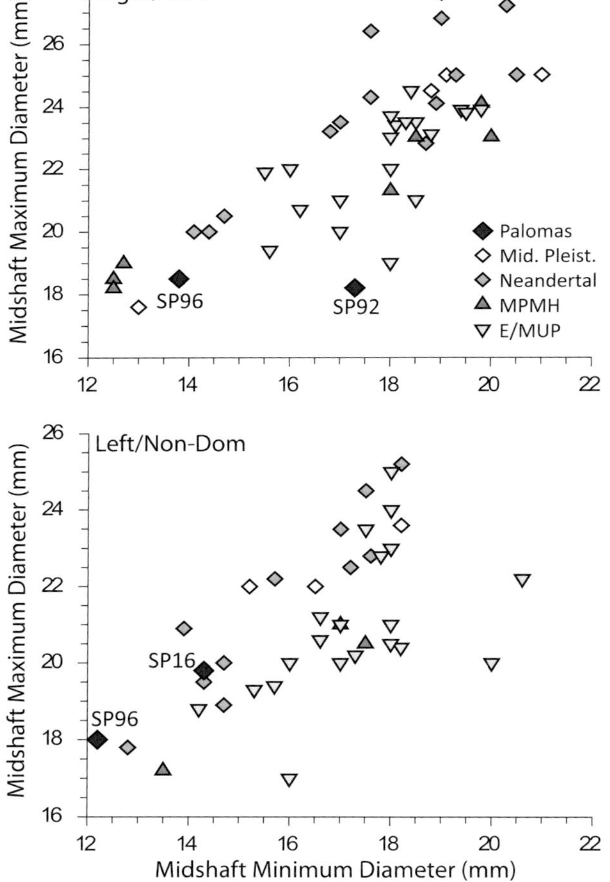

specimen. The degree of formation of the proximal epiphysis closely resembles that of an 8-year-old recent human (Scheuer et al. 2000: 282), supporting this age assessment.

Palomas 92ff: Humerus Right

The bone consists of four sections that together make up a largely complete right humerus (Plate 13.6); note that previously only the distal portion had been extracted from the breccia (Walker et al. 2011a). The proximal section includes most of the head, a lateral portion of the surgical neck, and pieces of diaphysis down to close to midshaft. The whole section is largely coated in thin matrix that, along with the breakage, obscures surface details. The section, however, appears to be close to its original length. The next two sections are cylindrical sections of diaphysis, one through midshaft and one from below midshaft to the middistal diaphysis. The distal portion retains the diaphysis from proximal of the supracondylar crests, through the metaphyseal region, to the distal articulations. The epicondyles are complete, the anterior capitulum has been pushed proximally, and the lateral trochlea is in small pieces. However, the breadths of the distal end (olecranon fossa, distal articular, and epicondylar) should be minimally altered.

To estimate the length of the Palomas 92 humerus, the lengths of the pieces along their posterior margins, where the matching broken edges are evident, were measured and summed. The pieces, from distal to proximal, measure 64, 46, 35, and 117 mm, respectively. The distal piece is measured from the distal trochlea to the proximal (middistal shaft) break, such that the sum of these lengths of 262 mm represents the

FIG. 13.2. Bivariate plots for the right (*above*) and left (*below*) humeral midshaft maximum versus minimum subperiosteal diameters. The right and left humeri of Palomas 96 are reversed, given her left-side dominance. The other specimens are either right-side dominant or assumed to be. Mid. Pleist.: Middle Pleistocene archaic humans; MPMH: Middle Paleolithic modern humans; E/MUP: Early/Mid Upper Paleolithic modern humans.

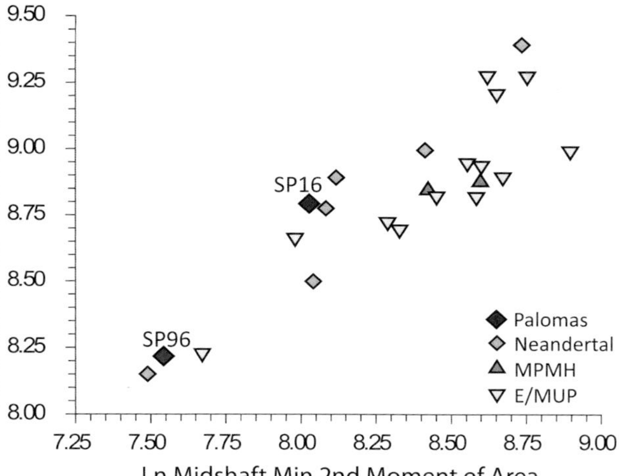

FIG. 13.3. Bivariate plot of the Palomas 16 and other Late Pleistocene left humeral midshaft maximum versus minimum second moments of area. The value for Palomas 96 is for her nondominant right humerus; the other humeri are either left and nondominant or left and assumed to be nondominant. MPMH: Middle Paleolithic modern humans; E/MUP: Early/Mid Upper Paleolithic modern humans.

maximum humeral length. Given the slightly eroded and modestly encrusted diaphyseal breaks, plus the crushing of the proximal section, the probable range of the original maximum length is 260–265 mm. However, given that the length estimate is employed for humeral diaphyseal scaling, as well as for assessments of body size (chapter 14), a more conservative range for maximum length would be 260–270 mm. This range for the Palomas 92 humeral maximum length is close to the value for Palomas 96 of 272 mm. Using a least squares regression based on recent human humeri, these length ranges provide humeral articular length estimates of 257–262 mm or, conservatively, 257–266 mm (HArtL = 0.976 × HMxL + 2.74, r^2 = 0.993, n = 86; SE_{est} = 1.8 mm / 0.7%).

Palomas Humeral Proximal Morphology

Although the head and surgical neck are substantially preserved for Palomas 92, damage to them

and breccia encrustation limit observations on them and the adjacent proximal diaphysis. The Palomas 32 immature proximal humerus (Plate 13.4) presents a rounded-head epiphysis with a prominent greater tuberosity. A broad and wide bicipital sulcus is evident on the anterior surface of the surgical neck. The bone is insufficiently preserved distally to assess humeral torsion. It nonetheless provides data for its trabecular dimensions and orientations. The cancellous architecture of the epiphyses appears relatively dense and isotropic (Plate 13.5; Table 13.8), whereas in the proximal metaphysis an anisotropic structure is formed of trabecular arcades. To quantify the cancellous bone architecture, volumes of interest representing maximum spheres were selected for the proximal metaphysis. Cross-sectional geometry of the specimen was analyzed at approximately the 75% level in a slice at a right angle to the bone's long axis (Plate 13.5).

Palomas Humeral Diaphyseal Morphology

The Palomas 16 partial proximal and midshaft diaphysis presents a distinct deltoid tuberosity, the proximal and distal ends of which extend across the postmortem breaks (Plate 13.4). There is an angled sulcus along the anterior margin of the tuberosity, evident in anterior view, and a broader sulcus along its posterior margin, evident in lateral view. The distal portion of a moderately rugose and distinctly bordered pectoralis major tuberosity is present. Its maximum preserved breadth of 7.6 mm may represent the original breadth or be slightly below it, depending on whether it broadened proximally of the postmortem break.

A pectoralis major breadth of 7.6 mm is the smallest one available for Neandertal left humeri (9.8 ± 1.5 mm, n = 7), but a plot of it relative to average diaphyseal diameters places it with the other Neandertal left humeri, along or above the upper margins of an earlier Upper Paleolithic sample (Fig. 13.1). Bearing in mind that the Palomas 16 tuberosity may have been slightly larger, and that scaling the tuberosities to diaphyseal dimensions minimizes possible contrasts (both should reflect shoulder region hypertrophy), the Palomas 16 individual was relatively robust, if modest in size. For comparison, the Palomas 96 right pectoralis major breadth (5.5 mm), from the nondominant arm, is well below other Neandertal left tuberosity breadths as well as the right ones (10.1 ± 1.8 mm, n = 13). Yet when scaled against its diaphyseal diameters (Fig. 13.1), it falls along the line of the Neandertal left humeri and above most of the Upper Paleolithic ones.

The midshafts of the Palomas humeri are variably ovoid (versus rounded), as is reflected in their subperiosteal diameters (Table 13.5; Fig. 13.2) and left second moments of area (Table 13.6; Fig. 13.3). For these comparisons, given left-side dominance, the Palomas 96 values are side-reversed. In general, Neandertal (and Atapuerca-SH Middle Pleistocene) humeri are more ovoid in their midshafts than are those of early modern humans, despite some overlap in the right humeral distributions and the rather ovoid proportions of several small Skhul right humeri. The Palomas 16 humerus and the two Palomas 96 humeri follow the archaic human pattern of having more ovoid midshaft contours, or greater maximum (anterolateral to posteromedial) diameters. However, the Palomas 92 right humerus is with the least ovoid of the early modern human right midshafts. Comparative data are more limited for midshaft second moments of area, and they are available only for Palomas 16 and 96 (Fig. 13.3). There is less scatter in the distributions, reflecting better quantification of the biomechanically relevant aspects of the humeral midshafts. The Neandertals tend towards having greater relative maximum second moments of area, and the two Palomas specimens are aligned with them, especially Palomas 16.

It is also possible to assess middistal diaphyseal hypertrophy for the Palomas 92 right humerus (Fig. 13.4). Although cross-sectional geometric parameters, in this case the polar moment of area as a general bending and torsional rigidity measure, should be scaled to body mass and bone length (Ruff 2000), they are compared here to both bone length alone (to maximize sample size) and to the product of the two. In both comparisons

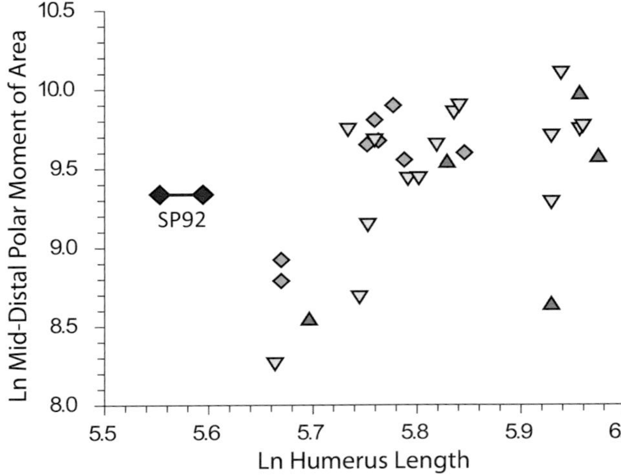

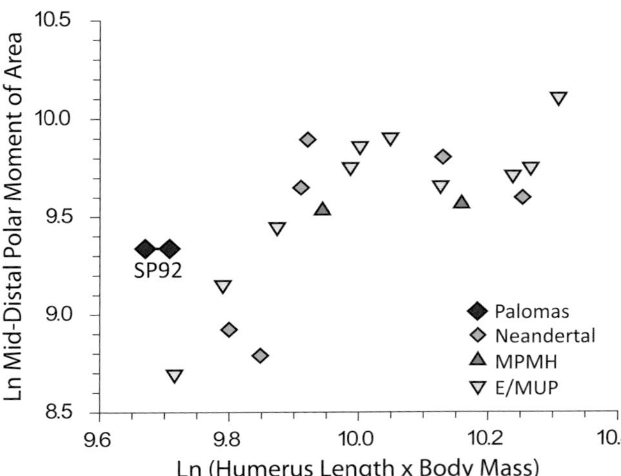

FIG. 13.4. Bivariate plots of humeral 35% polar moment of area versus humeral length (*above*) and humeral length times estimated body mass (*below*). The two values for Palomas 92 reflect the range of 260–270 mm for the humeral length estimate. MPMH: Middle Paleolithic modern humans; E/MUP: Early/Mid Upper Paleolithic modern humans.

there is little difference across the Late Pleistocene samples, despite some outliers in the comparison only to humeral length. Two values are provided for Palomas 92, one for each end of its humeral length estimate range (260 mm and 270 mm), and both of them provide substantially higher values for relative humeral rigidity than almost all of the other Late Pleistocene specimens. In order to bring the Palomas 92 values within the Neandertal range, its humeral length would have to be increased to ≈300 mm and/or its body mass estimate to ≈ 70 kg, both of which would be very unlikely. Palomas 92 therefore had a robust humeral diaphysis for its overall body size.

Palomas Humeral Distal Morphology

Palomas 17 and 92 provide information on their distal right humeri (Table 13.7), even though the former is very incomplete (Plate 13.4) and the latter is cemented to the proximal ulna and radius (Plate 13.7). Some additional data are available for the Palomas 96 right humerus, even though the articulations are crushed and embedded in matrix (Walker et al., 2011b).

None of the three Palomas medial epicondyles are deflected dorsally; a distinct dorsal deflection of the medial epicondyle is present in 69.2% (n = 13) of the Neandertals and 77.8% (n = 9) of the Atapuerca-SH Middle Pleistocene humeri, but none of the three sufficiently intact Middle Paleolithic modern humans (MPMH). The Palomas 92 distal humerus has a large septal aperture, evident in anterior view (Plate 13.7). It is not possible to determine whether Palomas 96 has one, but septal apertures are relatively common among Middle Paleolithic humans (Neandertals: 50.0%, n = 28; MPMH: 40.0%, n = 5). They are absent from Middle Pleistocene humeri (n = 5) and rare among earlier Upper Paleolithic (E/MUP) modern humans (8.0%, n = 25). They are also unusual among recent humans (pooled global sample: 6.7%, n = 3,791).

Neandertals and archaic *Homo* generally have been noted to have relatively thin lateral and especially medial pillars to the distal humerus, framing the olecranon fossa (Carretero et al. 1997; Yokley and Churchill 2006; Trinkaus et al. 2007; Bermúdez de Castro et al. 2012; Trinkaus 2012), whereas those of recent humans tend to be thicker. It remains unclear whether the difference is driven by variation in the pillar thicknesses or in the breadth of the olecranon fossa (Trinkaus et al. 2007). However, a plot of summed medial and lateral pillar thicknesses versus olecranon breadth (Fig. 13.5) separates most of the Middle and Late Pleistocene archaic humans from the early modern humans, the one exception being Shanidar 4.

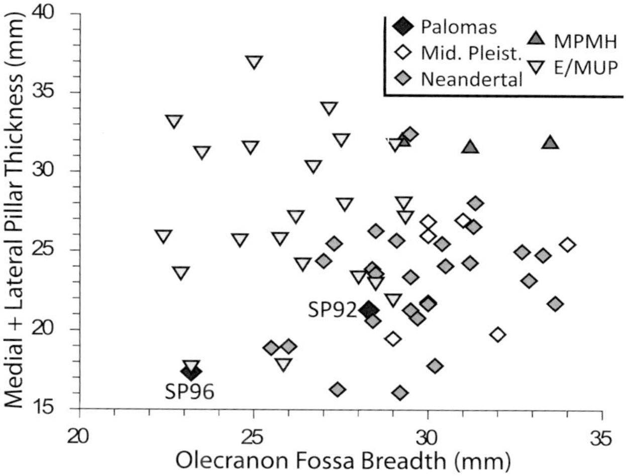

FIG. 13.5. Bivariate plot of the summed humeral medial and lateral pillar thicknesses versus olecranon fossa breadth for Palomas 92 and 96 and the comparative samples. Right and left values are included and averaged when both are available for a specimen. Mid. Pleist.: Middle Pleistocene archaic humans; MPMH: Middle Paleolithic modern humans; E/MUP: Early/ Mid Upper Paleolithic modern humans.

Palomas 92 falls in the middle of the Neandertal distribution, close to Shanidar 1 but nonetheless adjacent to a few E/MUP humeri. Palomas 96 is within the overlap zone of the Late Pleistocene samples, adjacent to Fanciulli (Grotte-des-Enfants) 6.

The Palomas Antebrachial Remains

Palomas 14: Ulna Left—Immature

Palomas 14 is the proximal end of a very immature left ulna, crushed in laterally but including an incomplete trochlear notch and truncated olecranon and partial proximal diaphysis. (Plates 13.8 and 13.9; Table 13.10). Maximum preserved length is 36.5 mm; maximum anteroposterior dimension is 19.8 mm.

Palomas 64: Radius Left—Mature

The bone preserves the proximal diaphysis and the tuberosity (Plate 13.8; Tables 13.11 and 13.12). The distal break is largely transverse at the proximal end of the interosseus crest. The proximal break is oblique anteromedioproximal to posterolaterodistal at the proximal end of the tuberosity. The tuberosity is complete. The surface of the bone is slightly powdery, and there is a hard but thin layer of mineral matrix on the external surface. The maximum preserved length is 67.3 mm.

The bone is externally free of abnormalities, but at the distal break there is an endosteal growth of bone along the anteromedial cortical bone, extending into the medullary cavity, with a trabecular bridging posteriorly (Plate 13.8). It is unclear whether this extra bone is a normal variant or a pathological growth.

Palomas 92gg and 92hh: Ulna Right

The principal portion of the Palomas 92 right ulna (Plate 13.10) consists of two proximal pieces, separated by 3–6 mm, cemented onto the distal humerus, plus four pieces of shaft. The most proximal piece consists of the olecranon with 26 mm of the dorsal surface; it is intact, but the articular surface is obscured. The next proximal section has the complete coronoid process and radial facet, plus ≈34 mm of the proximal shaft distal of the coronoid process. The proximal radius is cemented onto the radial facet and the proximolateral diaphysis. The larger diaphyseal section includes two pieces in matrix and still cemented to the radial diaphysis, the proximal end of which would have a reasonable join to the coronoid section were it separated from the radius. The last two pieces are small cross-sections of middistal shaft in matrix. It is principally the two proximal pieces, cemented to the humerus and radius, that provide data.

There is a separate complete head and part of the distal diaphysis (Palomas 92hh) (Plate 13.11). Side is based on the distal diaphyseal curvature and association with the right triquetral bone. The styloid process is absent, but it may be a piece of bone adherent to the dorsolateral head. There is no trace of the pronator quadratus tuberosity. The maximum preserved length is 30.7 mm. It has been possible to obtain cross-sectional parameters at ≈25% of bone length (Table 13.10).

Palomas 92ii: Radius Right

The bone (Plate 13.10) retains the head, the neck, and the dorsal corner of the tuberosity fused onto the humerus and ulna as though it were in full pronation, with the tuberosity pointing posterolaterally. There are then four pieces of diaphysis that make up a substantial portion of the bone, but they are variably in breccia, and the joins between them, although apparent, cannot be secured. The pieces can nonetheless be oriented anatomically and relative to each other.

Palomas 92bbb: Radius Left—Distal

The left radius is represented by the distal end of the diaphysis and the carpal articulation, partially crushed volarly and cemented to the proximal row of carpals.

Palomas Antebrachial Morphology

The fragmentary ulnae and radii from Palomas provide data for several discrete aspects of their morphology, but they do not permit assessment of overall lengths or shapes (Tables 13.9 to 13.12). These features involve ulnae trochlear notch orientation, radial tuberosity positioning, and radial interosseus crest projection. Additional head and neck metrics are available for the Palomas 92 right radius and for a few aspects of the Palomas 96 radii (Table 13.11).

Modern humans have been noted to have ulnar trochlear notches that are more proximally oriented than those of archaic humans, an orientation reflected in the greater volar projection of the coronoid process relative to the olecranon process (Solan and Day 1992; Churchill et al. 1996). The plot of coronoid versus olecranon height (Fig. 13.6) almost entirely distinguishes Neandertals from early modern humans, with one early modern specimen (Caviglione 1) approaching the Neandertal distribution. The small Middle Stone Age Klasies River Mouth ulna (Rightmire and Deacon 1991) appears intermediate, but it is close to the Neandertal line and below the early modern human one. Interestingly, the Palomas 92 right

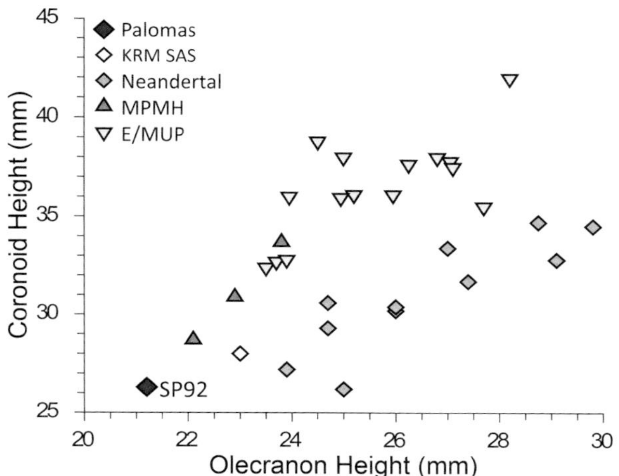

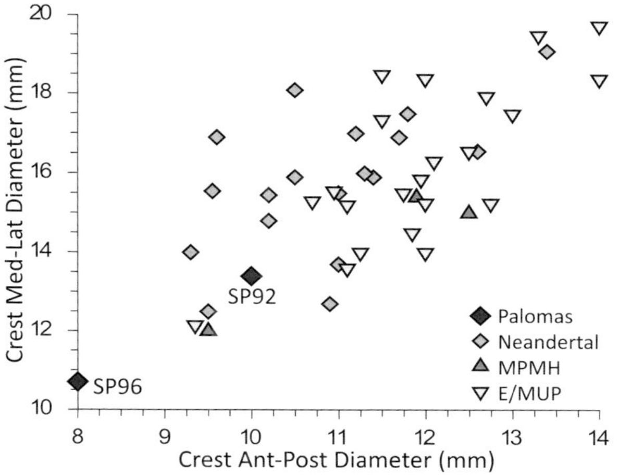

FIG. 13.6. Bivariate plot of ulnar coronoid versus olecranon anteroposterior heights, for Palomas 92 and Late Pleistocene comparative samples. The open symbol is for the Klasies River Mouth (KRM) SAS Middle Stone Age ulna. MPMH: Middle Paleolithic modern humans; E/MUP: Early/Mid Upper Paleolithic modern humans.

FIG. 13.7. Bivariate plot of the radial diaphyseal diameters at the interosseus crest maximum projection for Palomas 92 (and 96) and Late Pleistocene comparative samples. MPMH: Middle Paleolithic modern humans; E/MUP: Early/Mid Upper Paleolithic modern humans.

ulna has proportions that place it in line with the early modern humans. It is unclear whether this is a product of proportions that contrast with the Neandertals or a convergence of proximal ulnar proportions in the small body size range.

It has also been noted (Fischer 1906; Trinkaus and Churchill 1988) that archaic humans tend to have a directly medial orientation of the radial tuberosity when the arm is in full supination. The modern human pattern is an anteromedial orientation, such that the line of the interosseus crest passes in the posterior third of or behind the radial tuberosity. Palomas 92 and 96 exhibit the archaic pattern, in which the tuberosity is directly medial, as do 72.5% of the Neandertals (n = 20; note that Regourdou 1 is asymmetrical). This position is absent from Palomas 64, which exhibits a more anteromedial tuberosity position. Among early modern humans, only 8.3% of the MPMH (n = 6) and 6.3% of the E/MUP humans (n = 24) have the medial tuberosity orientation. One African later Middle Pleistocene specimen, Cave of Hearths 1, has the anteromedial position (Pearson and Grine 1997), but the more medial orientation

appears to be the ancestral human condition (Trinkaus and Churchill 1988).

The Palomas 92 right radius has a prominent interosseus crest with a teardrop-shaped cross-section, whereas the available left one of Palomas 96 is similarly teardrop-shaped in cross-section, but the crest is less prominent. Palomas 64 does not extend sufficiently distally for crest assessment, since only the proximal end of the interosseus membrane attachment is evident (Plate 13.8). A plot of mediolateral versus anteroposterior shaft diameters at the crest (Fig. 13.7) shows a tendency for the Neandertals to have relatively larger crests, but there is considerable overlap across the samples. Palomas 92 falls well within the overall Late Pleistocene distribution. The small size of Palomas 96 makes its proportional affinities difficult to assess, but it, too, appears to share its diaphyseal proportions with both the Neandertals and the early modern humans.

The Palomas Manual Remains

The Palomas remains include of three isolated

bones (a partial metacarpal 2 and two complete phalanges) (Table 13.1; Plate 13.12) plus the more complete series of hand bones for Palomas 92 (Table 13.13; Plates 13.13 and 13.14). There are also data from the hand skeletons of Palomas 96. However, most of the Palomas 92 left hand remains are cemented into two blocks of breccia, one of the radiocarpal region and the other of the ulnar metacarpals and phalanges. These bones are close to each other but not necessarily in their original anatomical connections. It has been possible to inventory them, based on morphology and preserved positions, to obtain some measurements on exposed portions and to extract additional data from micro-CT scans of the cemented blocks. The few right carpals and one metacarpal are separate.

The digits and sides assigned to the Palomas 15, 28, and 65 manual remains are based on their morphologies, and therefore those of the phalanges are approximate. The positions of the Palomas 92 metacarpals and phalanges, however, are based on their *in situ* positions in articulated and cemented partial hand skeletons. The same situation applies to the hand bones of Palomas 96 except for two partial distal phalanges. Therefore, the digit numbers given for the Palomas 92 (and 96) hand bones are mostly secure.

Palomas 15: Metacarpal 2 Right—Mature

The bone retains the complete head and distal half of the diaphysis of a second metacarpal (Plate 13.12). Digit number and side are determined by the oblique slope to the distal left head, hence the radial side of a metacarpal 2 head. The maximum preserved length is 39.0 mm.

Palomas 28: Manual Distal Phalanx 2–4—Mature

The bone is complete, with modest erosion to the apical tuft and thin matrix on portions of it (Plate 13.12). It is morphologically nonpollical, and the expanded apical tuft indicates that it is unlikely to be from the fifth digit. The maximum length is 19.5 mm.

Palomas 65: Manual Middle Phalanx 2–4—Mature

The phalanx is a complete bone with slight dorsal base abrasion (Plate 13.12); maximum length is 23.1 mm. Based on comparison of its articular length to those of Palomas 96 and other Neandertals, it likely derives from digit 2 (chapter 14).

Palomas 92vv: Scaphoid Left

Most of the scaphoid bone is present but cemented into a breccia block with the distal radius and adjacent carpal bones.

Palomas 92ww: Lunate Left

The partial bone is cemented into the breccia mass with the distal radius.

Palomas 92jj: Triquetral Right

Complete bone is present with abrasion to the ulnar side. Maximum length is 11.5 mm.

Palomas 92kk: Triquetral Left

A partial bone, crushed between the hamate, scaphoid, and distal radius in the left radiocarpal block.

Palomas 92lll: Pisiform Right

Partial bone with most of the facet side absent, and therefore principally the palmar side is preserved. It is insufficient to determine its projection (or dorsopalmar thickness). The maximum preserved (not original) dimensions are 11.1 mm long and 6.8 mm thick. The side is based on association with the right triquetral and ulnar head.

Palomas 92zz; Trapezium Right

The bone is an eroded right trapezium with the majority of the metacarpal 1 surface (Plate 13.13).

Palomas 92aa: Trapezium Left

A complete bone with minor edge abrasion to the radial metacarpal 1 facet and the edges of the scaphoid and trapezoid facets. There is breccia remaining within the palmar sulcus adjacent to the tubercle. There is an irregularity on the dorsal ulnar metacarpal 1 facet that is from fossilization erosion. Maximum radioulnar breadth is 20.1 mm.

Palomas 92bb: Trapezoid Left

A complete bone with minor edge abrasion and calcite adherent to the metacarpal 2 facet. Maximum dorsopalmar dimension is 9.6 mm.

Palomas 92aaa: Capitate Right

A heavily eroded core of the right capitate bone with little of the articular surfaces retained.

Palomas 92xx: Capitate Left

The largely complete bone is cemented between the scaphoid and hamate bones with exposure of the ulnar (hamate) facet and the ulnar side of the metacarpal 3 facet. There is at least proximal and dorsal bone loss and crushing.

Palomas 92yy: Hamate Left

The bone is cemented into the left radiocarpal breccia mass (Plate 13.13). It lacks most of its ulnar side, with exposed trabeculae, but it preserves what should be close to a midline profile through the body and to the ulnar side of the midline through the hamulus, with the dorsal body contours and the hamulus contours evident. The proximal angle is preserved but partly obscured, radially and ulnarly.

Palomas 92z: Metacarpal 2 Right

The head and the distal diaphysis with abrasion to the radial head (Plate 13.13). Maximum preserved length is 25.1 mm.

Palomas 92ccc: Metacarpal 2 Left

A fragment of the dorsoradial base, adherent to the distal trapezoid bone.

Palomas 92q: Metacarpal 3 Left

A section of the probably dorsal distal diaphysis and the beginning of the flare for the head, fused between the metacarpal 4 head and the proximal phalanx 3 base. Maximum preserved length is 18.2 mm; maximum preserved breadth is 9.9 mm.

Palomas 92r: Metacarpal 4 Left

A complete head and the distal two-thirds of the diaphysis, fused between the metacarpal 3, the proximal phalanx 3 and 4 bases, and the metacarpal 5 shaft. Maximum preserved length is 33.6 mm; maximum preserved breadth (head) is 11.5 mm.

Palomar 92s: Metacarpal 5 Left

The head missing the palmar surface and about two-thirds of the diaphysis. The bone is fused between matrix, the metacarpal 4, and the proximal phalanges 4 and 5. Maximum preserved length is 34.4 mm.

Palomas 92t: Manual Proximal Phalanx 3 Left

Complete bone, with the base fused to the proximal phalanx 4 and the metacarpals 3 and 4 and the distopalmar shaft fused to the middle phalanx 3 and matrix. Maximum preserved length is 41.0 mm.

Palomas 92u: Manual Proximal Phalanx 4 Left

Complete bone with the base partly buried in the adjacent metacarpal heads and proximal phalangeal bases. The palmar shaft is below matrix, the metacarpal 4 and the middle phalanx 4. Maximum preserved length is 39.0 mm.

Palomas 92v: Manual Proximal Phalanx 5 Left

The base and shaft fused to the metacarpal 5, the proximal phalanx 4, and the middle phalanx 4. Maximum preserved length is 27.4 mm.

Palomas 92w: Manual Middle Phalanx 3 Left

Complete bone, fused to the third proximal and distal phalanges. Maximum preserved length is 25.6 mm.

Palomas 92x: Manual Middle Phalanx 4 Left

The base fused to the proximal phalanges 4 and 5. Maximum preserved length is 8.4 mm.

Palomas 92y: Manual Distal Phalanx 3 Left

The base fused to the middle phalanx 3. Maximum preserved length is 9.0 mm.

Palomas 92uu: Manual Distal Phalanx 4 Left

The majority of the base of the phalanx, with loss of the dorsal surface and matrix on the palmar radial side, cemented adjacent to the head of the middle phalanx 4.

Morphology of the Palomas 92 Carpal Bones

There are portions of four right carpal bones and seven left ones preserved (and part of the left pisiform bone may be buried in the left radiocarpal block). Yet only two of them, the right trapezium and the left hamate bone, provide comparative information, even though the right trapezoid bone also provides metrics (Tables 13.14 and 13.15; Plate 13.13). The remainder of them are too damaged, too embedded in matrix, and/or appear to vary little from archaic and modern humans.

Neandertal trapezia have been noted to have metacarpal 1 facets that are sellar-shaped but dorsopalmarly relatively flat compared to those of recent humans, and this configuration is joined by the dorsopalmarly relatively straight, rather than concave, carpal facets of most of their first metacarpals (Trinkaus 1983, 2016a; Niewoehner 2001, 2006; Lorenzo et al. 2012). This relative flattening appears to be the ancestral configuration, given a similar morphology in the Early Pleistocene OH-7 (Trinkaus 1989; see also Marzke et al. 2010) and the Dinaledi archaic *Homo* hands (Kivell et al. 2015). However, it does not fully characterize the Middle Pleistocene Atapuerca-SH sample (Lorenzo 2007). These conclusions are evident in the dorsopalmar subtense-to-chord ratios of the metacarpal 1 facets of their trapezia (Fig. 13.8); note that data for a pooled recent human sample is included, given limited (n = 3) measurements for early modern humans. The Neandertals and OH-7 have low values, including one (Regourdou 1) with a subtense of zero, and the remainder fall well within the recent human variation.

In contrast to the remainder of the Neandertal sample, the Palomas 92 trapezium has a markedly convex metacarpal 1 facet. Its index of 23.1 is in the upper portion of the recent human

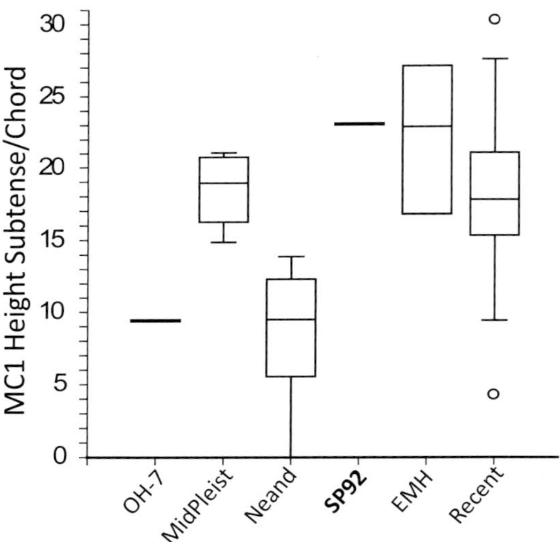

FIG. 13.8. Box plot of the dorsopalmar height subtense/chord index for the metacarpal 1 facet of the trapezium for Palomas 92 and the later Pleistocene comparative samples. The Early Pleistocene OH-7 immature specimen and a pooled recent human sample are included, the latter given the dearth of available early modern human data. Sample sizes: Middle Pleistocene (MidPleist), 10; Neandertals (Neand), 7; EMH (early modern humans), 3; recent humans, 92. (Atapuerca-SH data from C. Lorenzo pers. comm.)

distribution, in the middle of the early modern human values, above the range of the Middle Pleistocene sample, and far from the other Neandertal values. Ignoring the low value for Regourdou 1, the Palomas 92 index is 4.1 standard deviations from the Neandertal mean. The Palomas 92 trapezium therefore markedly extends the range of Neandertal variation and makes this feature in the Pleistocene to be one of distributional, rather

than absolute, differences. It should also be noted that the proximal left metacarpal 1 of Palomas 96 is distinctly dorsopalmarly concave, similar to those of recent humans, although damage to its palmar beak prevents accurate measurement of the dorsopalmar depth.

It is also possible to assess the relative dimensions of two of the carpal tuberosities that support the flexor retinaculum and associated tendons, the tubercle of the trapezium and the hamulus. Both are scaled against core dimensions of the bones, as available, bearing in mind that the metacarpal facet dimension of the trapezium may have been responding to habitual loads as well. The Palomas 92 and 96 hamuli, as with those of other Neandertals and Middle Pleistocene humans, are markedly projecting relative to those of early modern humans (Fig. 13.9). Interestingly, hamulus projection increases generally with hamate body size among early modern humans, whereas all of the Middle and Late Pleistocene archaic humans have projecting ones. Given the small body sizes of Palomas 92 and 96, their hamuli are relatively even more projecting than those of most other Neandertals, the exceptions being the La Ferrassie 2 and Tabun 1 females and the larger Shanidar 3 male.

The relative sizes of the trapezial tubercles of the Neandertals and their Middle Pleistocene predecessors are similarly large compared to an (albeit limited) early modern human sample (Fig. 13.9); the same pattern holds if recent human trapezia are included. In contrast to its hamulus, however, the trapezial tubercle of Palomas 92 is modest in size, similar to that of early modern humans. In this relative position it is joined by Shanidar 3; hence, both Shanidar 3 and Palomas 92 exhibit large hamuli but modest trapezial tubercles.

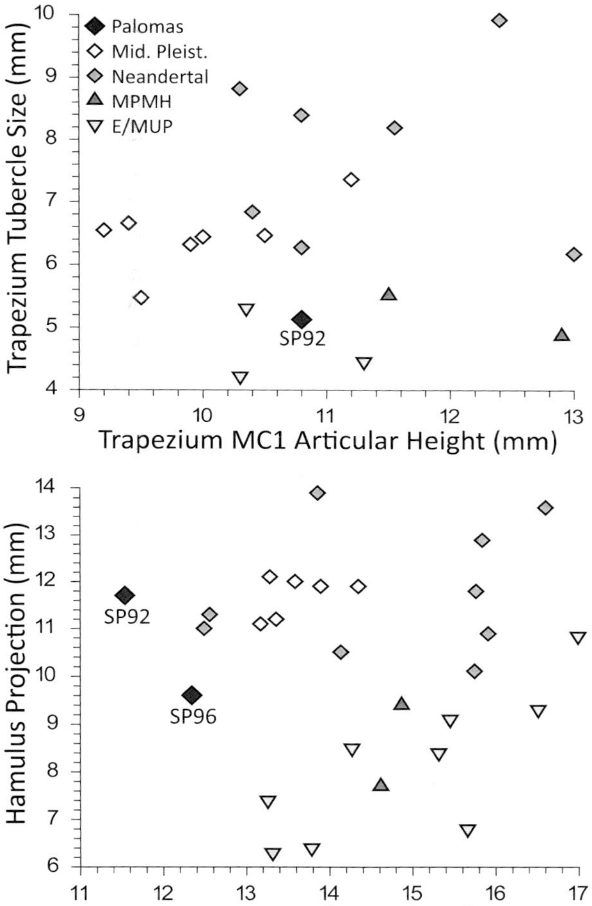

FIG. 13.9. Reflections of carpal palmar tuberosity hypertrophy for Palomas 92 and 96 and the comparative samples. *Above*: the geometric mean of the trapezial tuberosity dimensions versus metacarpal 1 articular height. *Below*: hamulus projection versus the geometric mean of the hamate body length and height. Mid. Pleist.: Middle Pleistocene archaic humans; MPMH: Middle Paleolithic modern humans; E/MUP: Early/Mid Upper Paleolithic modern humans. (Atapuerca-SH data from C. Lorenzo pers. comm.)

Palomas Metacarpal Morphology

The Palomas metacarpals consist of an isolated distal half of a second metacarpal (Palomas 15) and distal portions of the right second and left third to fifth metacarpals of Palomas 92 (Tables 13.12 and 13.16 to 13.18; Plates 13.13 and 13.14). There are also substantial portions of the right metacarpals of

Palomas 96 plus the left metacarpal 1 and 3, but the ulnar ones are mostly diaphyseal and distal as well. The distal articulations of the Palomas metacarpals conform entirely to the patterns of early and recent modern humans. The variations in ulnar digit carpometacarpal articulations within and between Late Pleistocene samples previously noted (Niewoehner et al. 1997; Trinkaus et al. 2014a; Ward et al. 2014) cannot be assessed on these metacarpals.

One feature of note is the percent cortical area (%CA) of the Palomas 15 and 92 metacarpals. The former, for a second metacarpal, is 80.3% (Table 13.17), and the latter, for a fourth metacarpal, is 94.7%. Comparative data for recent human second metacarpal %CAs (Pfeiffer and King 1983; Roy et al. 1994) place the Palomas 15 value in the middle of the recent human ranges of variation. The Palomas 92 metacarpal 4 value is, however, rather high, at least compared to metacarpal 2 data.

Palomas Manual Proximal and Middle Phalangeal Morphology

The Palomas sample preserves an isolated middle manual phalanx (Palomas 65), nine manual proximal phalanges from Palomas 92 and 96, and additional middle phalanges from the two associated skeletons (Tables 13.19 to 13.25; Plates 13.12 to 13.14). The digits for the Palomas 92 and 96 phalanges, as is rarely the case for Pleistocene human phalanges, are based on their *in situ* positions.

The Palomas 65 middle manual phalanx is a short and stout bone with a broadly flaring base (Plate 13.12). The insertions for the flexor digitorum superficialis tendons on the palmar midshaft are evident as modestly raised crests; they are smooth and lack the pronounced rugosities evident on some Neandertal ones, especially those from Krapina. The proximal phalanges from Palomas 92 (Plates 13.13 and 13.14) have prominent flexor sheath crests that project palmarly and are sufficiently radioulnarly expanded to place the minimum shaft breadths proximal on the diaphyses.

Neandertal phalanges have been noted to

have radioulnarly expanded epiphyses (Musgrave 1973; Semal et al. 2009). Given phalangeal preservation for Palomas 92 and 96 and the ability to assign proximal phalanges reliably to digit, the proximal maximum breadths of their third proximal phalanges are compared to their articular lengths (Fig. 13.10). There is only slight overlap between the Neandertal and early modern human phalanges; the one high outlier is Shanidar 6, with its markedly expanded base (Trinkaus

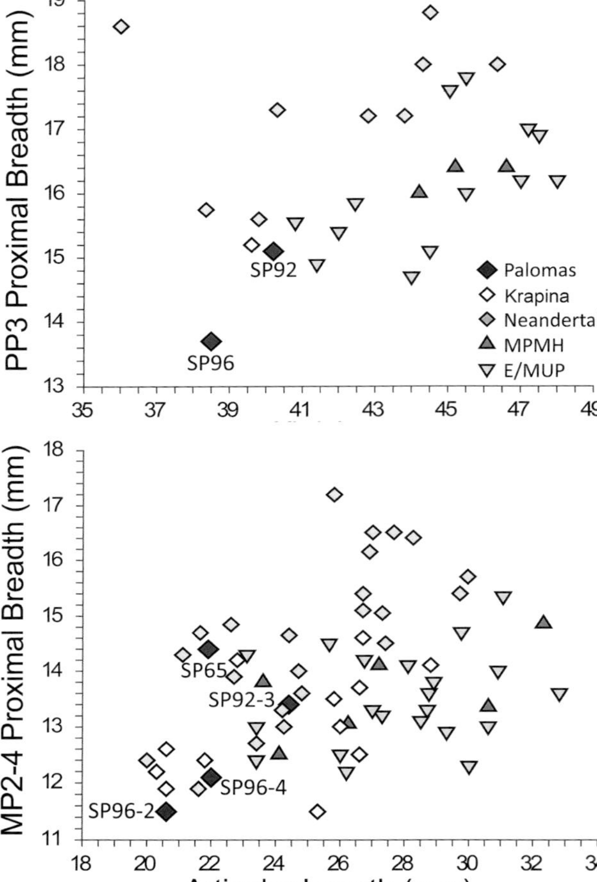

FIG. 13.10. Comparisons of proximal maximum breadth to articular length for proximal manual phalanx 3 (*above*) and pooled middle manual phalanges 2 to 4 (*below*) for Palomas 65, 92, and 96. Given the mixed nature of the Krapina phalanges, they are plotted separately from the other Neandertal ones. The additional numbers for the Palomas 92 and 96 middle phalanges indicate their digits. MPMH: Middle Paleolithic modern humans; E/MUP: Early/Mid Upper Paleolithic modern humans.

1983). Palomas 92 falls next to Krapina 204.5 and overlaps the E/MUP distribution. Palomas 96 is much smaller than the rest of the Late Pleistocene phalanges, and its proximal epiphyseal relative breadth is well within the early modern human distribution.

For the middle manual phalanges, given the uncertainties as to digit for the second to fourth rays for a number of Pleistocene specimens (especially the Krapina ones and Palomas 65, but not Palomas 92 or 96), the available data are pooled for digits 2 to 4, nonetheless averaging values from right and left pairs (Fig. 13.10). All of the Palomas middle phalanges are among the shorter of the Late Pleistocene ones. The one sufficiently complete phalanx from Palomas 92 and the two from Palomas 96 fall in the overlap of the late archaic and early modern human samples, but Palomas 65 is among the more robust (or radioulnarly expanded) of the Late Pleistocene middle manual phalanges.

Palomas Manual Distal Phalangeal Morphology

There is an essentially complete isolated distal phalanx from one of the middle three digits (Palomas 28) plus the bases of two distal phalanges of Palomas 92, both cemented to their middle phalanges (Table 13.25; Plates 13.12 and 13.14). Additional data are available for distal manual phalanges of Palomas 96, including the largely complete first, third, and fifth ones. Data are provided for all of the distal phalanges of Palomas 96 (Table 13.25), but the comparative considerations here involve the middle three digits, given the presence of Palomas 28.

Palomas 28 presents a relatively wide base with a prominent, proximally directed dorsal peak leading into a narrower shaft. Palmarly there is a concavity for the flexor digitorum profundus tendon, but it is not particularly rugose. Distally, the apical tuft is broad and rounded. Even though the distal and right margins of the tuft have been eroded, it is apparent that it was fully rounded and lacked the midline peak evident in most recent human distal hand phalanges. It also lacks any evidence of ungual spines; the presence of

ungual spines is variable within Pleistocene and recent humans and does not distinguish them from nonhuman primates (Susman 1998), but highly rounded tufts without even a radial or ulnar angulation are the more common configuration among archaic *Homo*. This pattern of a fully rounded apical tuft, at least for digits 2 to 4, is characteristic of Middle and Late Pleistocene archaic human distal phalanges 2 to 4 (none are known from the Early Pleistocene), being well documented for Neandertals and the Atapuerca-SH samples (Trinkaus 1983, 2016a; Vandermeersch 1991; Lorenzo et al. 2012).

It is difficult to provide an appropriate scaling for the breadth of the apical tuft, since early and recent modern humans tend to have relatively shorter distal phalanges than did the Neandertals (Villemeur 1994). The ancestral pattern of relative phalangeal lengths is unknown; the Dinaledi associated hand remains (Kivell et al. 2015) are intermediate between Neandertals and modern humans in relative ulnar phalangeal lengths. A comparison of apical tuft breadth to phalanx length (Fig. 13.11) highlights the absolutely wider tufts of the Atapuerca-SH and Neandertal distal phalanges, all of which are at or beyond the ranges of variation of the early modern human samples (the two <8.5 mm are Krapina 206.7 and Tabun 1). Yet, given the greater lengths of the archaic versus Upper Paleolithic phalanges, there does not appear to be a proportional difference between them; the same does not apply to the comparison of Neandertals to Middle Paleolithic modern humans, since the latter have relatively and absolutely narrower tufts than do most of the Neandertals.

Given these scaling considerations, the relative and absolute dimensions of the Palomas 28 apical tuft fall comfortably within the Neandertal variation. The Palomas 96 phalanx, when scaled against phalangeal length, is along the modest overlap zone of the Neandertals and early modern humans. However, if the Palomas 96 and other Late Pleistocene tuft breadths are scaled against humeral maximum lengths (Fig. 13.11), there is a clear separation of the Neandertals and early modern humans (although the east Asian early

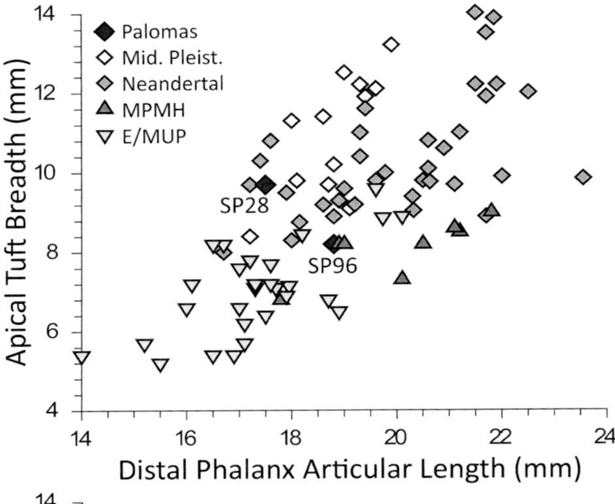

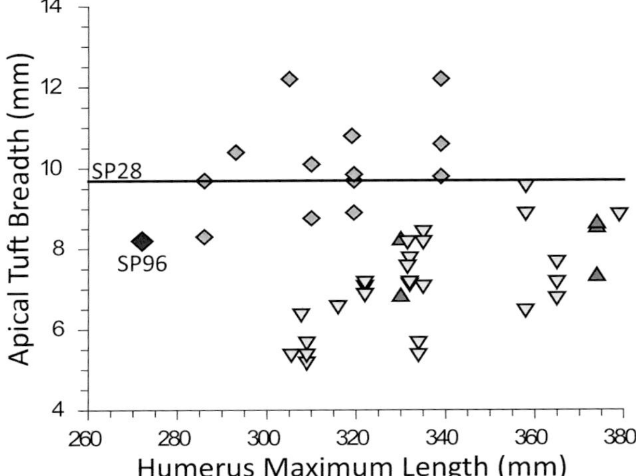

FIG. 13.11. Manual distal phalanx 2 to 4 apical tuft breadth versus phalanx articular length (*above*) and humeral maximum length (*below*). There is no associated humeral length for Palomas 28 (or the Atapuerca-SH phalanges), and therefore its apical tuft breadth is provided as a horizontal line, indicating its similarity to Neandertal dimensions and the very high humeral length that would be required for it to fall within the early modern human variation (see text). Note that multiple phalanges per individual are represented in the MDP 2–4 distribution, given difficulties in assigning digit for isolated phalanges or incomplete series of phalanges. Mid. Pleist.: Middle Pleistocene archaic humans; MPMH: Middle Paleolithic modern humans; E/MUP: Early/Mid Upper Paleolithic modern humans. (Atapuerca-SH data from C. Lorenzo pers. comm.)

modern Tianyuan 1 falls between the two distributions; Shang and Trinkaus 2010). Palomas 96 is among the Neandertals, albeit off of the small end of their variation. Given the generally small size of all of the Palomas postcrania (chapter 14) combined with the moderately large tuft breadth of Palomas 28, it would fall among the Neandertals unless its humeral length exceeded those of all known Neandertal humeri. Given that the size of the apical tuft reflects the overall dimensions of the distal finger and ungual pad (Mittra et al. 2007), these comparisons indicate relatively large terminal fingers for the Palomas individuals.

The Palomas Leg Remains

The isolated Palomas leg remains (Table 13.1) consist of a femoral head (Palomas 77), a midproximal diaphyseal section (Palomas 52), and a section of fibular diaphysis (Palomas 13). The first provides a body size indication, whereas the second in particular provides lower limb diaphyseal information.

Palomas 92 (Table 13.26) preserves pelvic remains, but they consist of fragments of the ilia preserved in breccia, either alone (92qq and 92rr) or in the same block as the ventral sacrum (92ss) (Plate 13.1). They provide little information. The right and left femora of Palomas 92 derive from several blocks of breccia. They are broken, cracked, and distorted. Yet three sections of the right femur fit together reasonably well with variably eroded breaks, permitting an assessment of overall length. The tibial fragments from Palomas 92 provide no morphological data.

In addition, Palomas 96 retains a largely complete pelvis, most of both of its femora, and most of the left tibia and fibula (Walker et al. 2011b). To the extent that they currently provide data, these remains are included in the tables and the morphometric comparisons.

Palomas 13: Fibula Left—Mature

The bone is a diaphyseal section from distal of the neck to approximately midshaft (Plate 13.15). The maximum preserved length is 80.7 mm.

Palomas 52: Femur Diaphysis Left

Palomas 52 retains two pieces of left midproximal to midshaft femoral diaphysis that fit cleanly together (Plate 13.15). The external surface is slightly eroded but intact, and the medullary cavity is filled with hard matrix. The proximal end has the proximal spreading of the muscle lines from the linea aspera, the beginning of the spiral line, and the distal portion of the gluteal buttress. It is distal of the lesser trochanter. The irregular distal break, based on linea aspera morphology and general contour, should be close to midshaft. The break in the middle of the piece should be close to the 65% cross-section based on the location of the proximal spread of the linea aspera. The clean break at ≈65% permits cross-sectional analysis of the diaphysis from the scaled photograph. The irregular distal break does not allow such transcription, but it permits approximation of the midshaft diameters (Table 13.27). Its maximum preserved length is 100.8 mm.

The bone could be adolescent or mature. Based on size, it is either a very small adult or an immature individual. Based on morphological estimates of the positions of the 65% and 80% cross-sections using cross-sectional morphology and especially posterior muscle markings, the distance between them is ≈50 mm. That estimate provides a biomechanical length (proximal neck to average of the distal condyles) of ≈333 mm, or basically 300–350 mm (the preserved landmarks are insufficient for a reliable length estimation for any scaling). This range is below any of the Neandertal adult femoral biomechanical lengths, the closest being Palomas 96 (360 mm), Shanidar 6 (≈366 mm), Palomas 92 (≈372 mm), La Ferrassie 2 (386 mm), and Tabun 1 (391 mm). Among early modern humans, it is closest to Nahal 'En-Gev 1 (365 mm). In addition, the well-marked muscle lines are similar to those of an adult, and the small

medullary cavity (or high relative cortical area: 84.7%; Table 13.28; Plate 13.15) suggests mature or late adolescent status (Ruff et al. 1994).

Palomas 77: Femur Head—Mature

Palomas 77 consists of a largely complete femoral head without any of its articular margins (Plate 13.15). There is a small area of surface bone on the exposed distal trabeculae (9.5 × 5.8 mm) that, on the basis of the fovea capitis orientation, should be inferior neck. Its side is indeterminate. The trabeculae show no signs of an epiphyseal surface or fusion line, so it is considered to be fully mature. The maximum diameter, which is approximately the anteroposterior diameter, is 45.6 mm. It does not derive from Palomas 92, since it comes from the Upper Cutting infilling, stratigraphically above the Conglomerate A.

Palomas 92mm: Femur Right

The right Palomas 92 femur consists of three pieces that fit together despite minor erosion along their transverse proximal diaphyseal join. They are described separately but are now one piece (Plate 13.16).

The more proximal piece (SP05H071B) is a badly crushed and twisted section of the trochanters and proximal diaphysis. The distal 18–19 mm of the diaphysis is intact with the full contour, and it fits well with the more distal SP05H002. The diaphysis is then compressed mediolaterally up to the lesser trochanter but with little length distortion. The spiral line and the gluteal tuberosity are evident for most of their lengths, but on twisted pieces. The greater trochanter is anteroposteriorly crushed and then twisted such that its mediolateral direction is now anteroposterior, with the medial side anterior with respect to the midproximal diaphysis. The anterior surface, digital fossa, and superior surface of the greater trochanter are largely present, but they are broken and distorted. Maximum preserved length is 159.5 mm.

The midproximal shaft section (SP05H002) consists of a short section of midproximal diaphysis, with the full cross-section exposed about the

60–65% position, then ≈28 mm of the full circumference distally of the transverse break. Further distally there is ≈30 mm of the anterolateral diaphysis continuing to an eroded distal end. The maximum preserved length is 62.1 mm.

These pieces then join to a distal femoral diaphysis and epiphysis piece from near midshaft to the condyles, with most of the lateral condyle preserved. The midshaft is compressed anteromedial to posterolateral. The popliteal area is intact posteriorly, cracked laterally, and compressed and covered in matrix anteriorly and medially. The medial condyle retains only a small section of its lateral (intercondylar) articular surface, 12 mm wide by 33.5 mm long. The lateral condyle is largely intact, with abrasion to the lateral edge of the articular condyle and to a strip along the rim of the condyle laterally. The lateral epicondyle is intact. The lateral patellar surface is present under matrix, and its preserved anterolateral edge is very close to the original margin. The sequence of pieces from the near midshaft break to the distal end fit tightly, with one crack having a 1 mm expansion. Maximum preserved length is 185 mm.

The resultant right femur of Palomas 92 is impressive mostly for its multitude of cracks and pieces of rock and sediment cemented tightly onto the subperiosteal bone (Plate 13.16). Despite these limitations it has been possible to sum the lengths of the individual sections, all of which would fit together cleanly if separated, to produce a reliable length from the lesser trochanter to the distal lateral condyle of 305 mm. From this measure the standard femoral lengths have been estimated using pooled recent human and Late Pleistocene femoral measurements (Table 13.27).

Palomas 9200: Femur Left

The left femur consists of two pieces, one proximal (SP05H107) and the other midshaft to distal diaphysis (SP05H071A) that join cleanly at ≈70% of biomechanical length (Plate 13.16). They are described separately, but their combined preserved length is 245.0 mm.

The more distal piece (SP05H071A) starts proximally with a clean break across the midproximal diaphysis (≈70%), just proximal of the juncture of the spiral line and the linea aspera. It then, going distally, has 23–24 mm of intact diaphyseal cross-section. Distally, it is anteroposteriorly compressed to the supracondylar region. There is little length distortion, except for what might be related to the twisting of the individual pieces. The most distal edge of the piece is posterolateral. Based on bone curvature and surface smoothness, plus comparisons to the more complete right femur, the distal extent should be close to the popliteal surface. The maximum preserved length is 185.0 mm.

The smaller proximal piece (SP05H107) originally consisted of fragments of the femoral head in a mass of faunal ribs and bone fragments, most of the medial and posterior femoral neck, which was pushed proximolaterally, and a portion of the proximal diaphysis distal to the juncture of the linea aspera and the spiral line. Most of the lateral surface (≈34 mm proximodistally) has been sheared off so that the only complete contour is at the distal lesser trochanter. A distal portion of the gluteal tuberosity, ≈27.5 mm long, is preserved in pieces of bone. Subsequent cleaning has removed the mass of broken femoral head and neck pieces and other bone to leave the diaphyseal section from the distal break (and join with SP05H071A) to the middles of the lesser trochanter and the gluteal tuberosity. The maximum preserved length is 64.0 mm.

Palomas Femoral Morphology

The comparative femoral morphology from the Sima de las Palomas entails the Palomas 92 gluteal tuberosity and the Palomas 52 and 92 diaphyseal characteristics, plus aspects of Palomas 96. The only articular elements preserved are the Palomas 77 head and the Palomas 92 lateral condyle, which furnish body size indications but little else of note. The measures of femoral diaphyseal hypertrophy are scaled against bone length times body mass, given the weight-bearing nature of the femora (Ruff 2000; Trinkaus and Ruff 2000).

The proximal diaphyses of both Palomas 92 femora provide portions of their gluteal tuberosities, providing part of the insertion for gluteus maximus. They are mildly rugose but have distinct margins. There is no sulcus in the right one and a shallow sulcus on the left side. They lack hypotrochanteric fossae. They are opposite prominent spiral lines, the left one of which is especially raised from the diaphyseal surface. The gluteal tuberosity breadths are 10.2 mm and ≈10.5 mm, values that are modest for a Neandertal (13.2 ± 2.1 mm, n = 16) but more elevated compared to those of early modern humans (MPMH: 9.0 ± 2.0 mm, n = 4; E/MUP: 9.9 ± 2.0 mm, n = 14). However, if their average value is scaled against body mass times bone length (Fig. 13.12), Palomas 92 is distinctly above the couple of early modern humans of similar body size and proportionately with those larger Neandertals that have relatively wide tuberosities.

In order to assess overall femoral diaphyseal hypertrophy, the scaled midshaft polar moments of area are compared across the Late Pleistocene samples (Fig. 13.13); note that no western Eurasian Middle Pleistocene femur currently provides cross-sectional parameters, a bone length, and a body mass estimate, although Middle Pleistocene data from elsewhere (Trinkaus and Ruff 2012) suggest that their levels of hypertrophy would be similar to those of the Neandertals. The femoral midshaft of Palomas 92 is too compressed to permit measurement of its midshaft cross-sectional parameters. However, it is possible to reliably estimate its midshaft polar moment of area (J) from its midproximal one (pooled Pleistocene *Homo* femora: ln50%J = 1.045 × ln65%J − 0.532, r² = 0.955, n = 42; Palomas 92: ln50%J = 10.663 ± 0.098 (42,745 ± 393 mm⁴), SE$_{est}$ = 0.92%). In the Late Pleistocene comparison, as documented elsewhere (Trinkaus and Ruff 1999, 2012), there is little difference across the Late Pleistocene samples in overall femoral (or tibial) diaphyseal hypertrophy. Palomas 92 and 96 are at the small end of the body size range, especially Palomas 96 (only Dolní Věstonice 3 is smaller), but they are well within the expected proportions of the other Late Pleistocene femora. Palomas 92, despite the estimations of its values, is close to the Tabun 1 female value. The Palomas femora therefore conform to the Neandertal (and Pleistocene *Homo*) level of femoral diaphyseal hypertrophy.

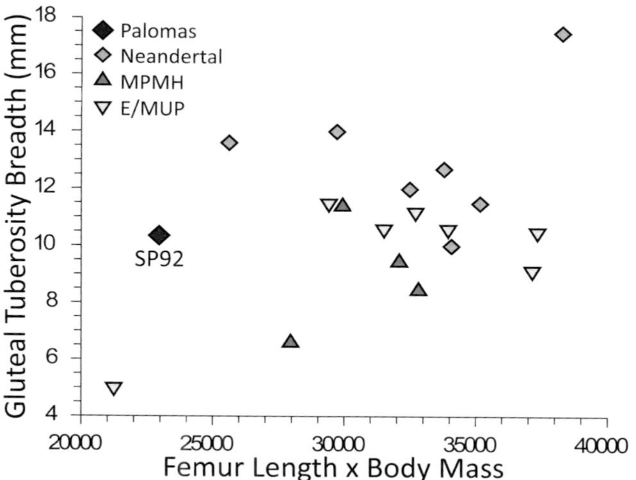

FIG. 13.12. Gluteal tuberosity breadth versus (biomechanical) femur length times body mass, for Palomas 92 and comparative Late Pleistocene samples. MPMH: Middle Paleolithic modern humans; E/MUP: Early/Mid Upper Paleolithic modern humans.

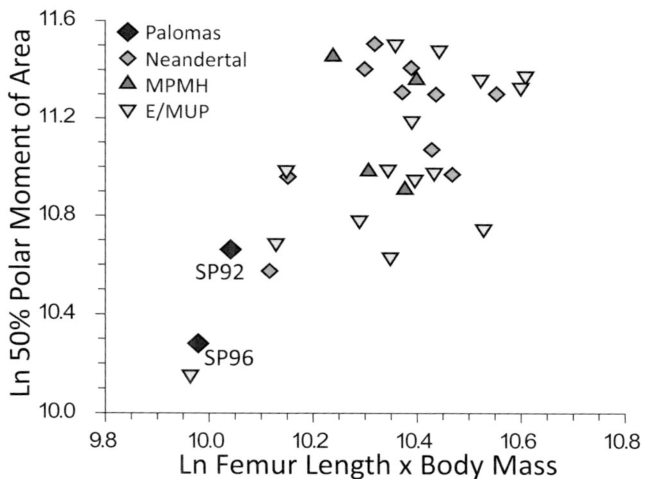

FIG. 13.13. The femoral midshaft (50%) polar moment of area versus (biomechanical) femur length times body mass for Palomas 92 and 96 and the Late Pleistocene comparative samples. Note that the Palomas 92 value is estimated from its 65% polar moment of area. MPMH: Middle Paleolithic modern humans; E/MUP: Early/Mid Upper Paleolithic modern humans.

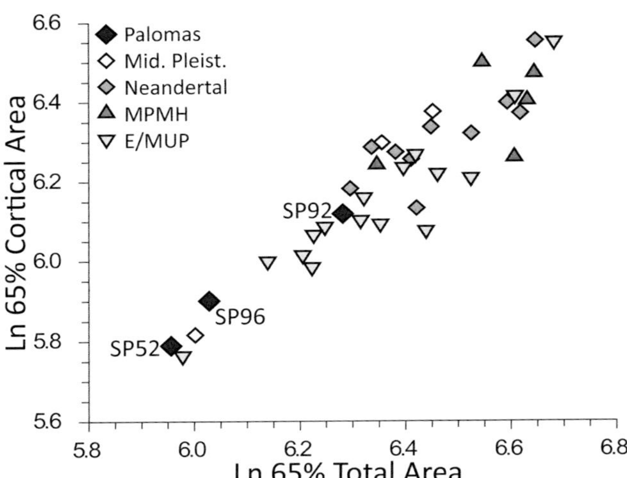

FIG. 13.14. Femoral mid-proximal (65%) cortical area versus total area for Palomas 52, 92, and 96 and the Middle and Late Pleistocene comparative samples. Mid. Pleist.: Middle Pleistocene archaic humans; MPMH: Middle Paleolithic modern humans; E/MUP: Early/Mid Upper Paleolithic modern humans.

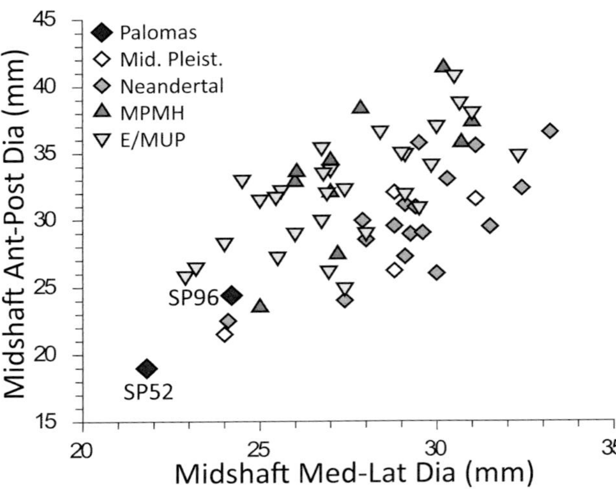

FIG. 13.15. Femoral midshaft anteroposterior versus mediolateral subperiosteal diameters for Palomas 52 and 96 versus the Middle and Late Pleistocene comparative samples. Note that the Palomas 52 values are estimated, given erosion to its femoral midshaft (Fig. 13.26). Mid. Pleist.: Middle Pleistocene archaic humans; MPMH: Middle Paleolithic modern humans; E/MUP: Early/Mid Upper Paleolithic modern humans.

A similar pattern is evident in relative cortical area in the midproximal diaphysis (Fig. 13.14). There is little difference across the Middle and Late Pleistocene samples. Palomas 92 is in the middle of the overall distribution. Palomas 52 and 96 are smaller, adjacent to the early Middle Pleistocene Gesher Benot Ya'acov 1 and the earlier Upper Paleolithic Dolní Věstonice 3.

The linea aspera of Palomas 52 converges to a single thin crest as it approaches midshaft, with a shallow and narrow sulcus along its medial side. On the Palomas 92 femora, it is a narrow crest ≈5 mm wide in the midproximal shaft, expanding to a flattened rugosity by midshaft, ≈10 mm wide. Neither one has any evidence of a pilaster. The Palomas 96 femora present a similar diaphyseal morphology.

The cross-sectional proportions can be assessed with both external diameters and, at the midproximal shaft, with second moments of area (Figs. 13.15 and 13.16). At midshaft, there is a well-documented distinction between archaic and early modern humans, more evident in second moments of area than in external diameters (Trinkaus and Ruff 2012), in which the archaic humans have relatively wider midshafts. This is evident in their midshaft

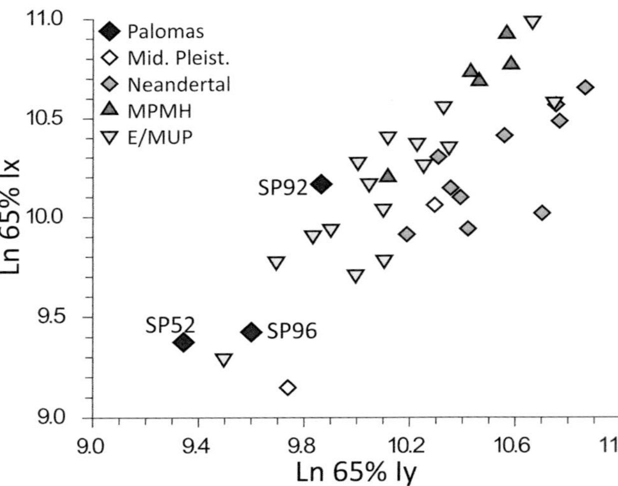

FIG. 13.16. The femoral mid-proximal (65%) anteroposterior second moment of area (I_x) versus the mediolateral one (I_y) for Palomas 52, 92, and 96 and the Middle and Late Pleistocene comparative samples. Mid. Pleist.: Middle Pleistocene archaic humans; MPMH: Middle Paleolithic modern humans; E/MUP: Early/Mid Upper Paleolithic modern humans.

external diameters (Fig. 13.15). Those of Palomas 96 and the estimates for Palomas 52 place them at the small end of the size range, especially for Palomas 52, in an overlap zone of archaic and early modern humans. There is a tendency for the anteroposterior to mediolateral proportion to increase with size (slope = 1.29), but not significantly so (95% CI: 0.85–1.72); related midshaft second moments of area provide similar results (slope = 1.32; 95% CI: 0.94–1.69). The relative positions of the Palomas proportions, as with the small ones for Gesher Benot Ya'acov 1 and Tabun 3, may reflect their small size as much as their affinities.

In the midproximal (65%) comparison of second moments of area (Fig. 13.16), the archaic human remains exhibit generally wider diaphyses than most of the early modern humans, as with the midshaft comparisons. All three of the Palomas femora, however, fall with the early modern humans in this feature, Palomas 52 and 92 more so than Palomas 96. They are close in size and shape to the midproximal femur of Dolní Věstonice 3 and separate from Gesher Benot Ya'acov 1; the small Tabun 1 and 3 femora do not provide midproximal proportions, and the other Neandertals are all larger at this position.

The Palomas 52 and 92 femora, joined by Palomas 96 for some of the comparisons, therefore largely conform to the archaic *Homo*/Neandertal pattern of femoral morphology. The main difference appears to be with respect to midproximal diaphyseal shape, in which they are relatively narrower or deeper than those of other later Pleistocene nonmodern humans.

Palomas Tibial Morphology

The fragments of tibia from Palomas 92 do not provide any morphological or morphometric data. However, there is an essentially complete left tibia from Palomas 96, albeit still partly in matrix (Walker et al. 2011b). Diaphyseal data are provided for it in Table 13.29 and are in Walker et al. (2011b) and Trinkaus and Ruff (2012). The other aspect of note is that the midshaft cross-section is amydaloid, lacking a distinct lateral longitudinal sulcus, and is therefore similar to that of most archaic human tibiae.

Palomas 13 Fibular Morphology

The Palomas 13 proximal fibular diaphysis (Plate 13.16) approaches the midshaft. Unlike most archaic human fibulae, it exhibits distinct longitudinal sulci on its sides, as opposed to mostly flat surfaces where the fibular sulci would be present in recent humans. The Palomas 96 left fibula has similar sulci along its more proximal portion, but near midshaft it has only one longitudinal sulcus, the other surfaces being convex in contour. Neither of these fibulae approaches the configuration seen in a number of early modern human fibulae, with deep longitudinal sulci (see Trinkaus 2006b).

In its cross-section near midshaft, Palomas 13 has a percentage cortical area of 91.3%. Comparative data are generally absent, but Palomas 96 provides one of 88.2% (Table 13.29).

The Palomas Pedal Remains

The Palomas sample contains three isolated pedal phalanges—two proximal ones of young individuals (Palomas 66 and 86) and a mature middle one (Palomas 67) (Tables 13.1, 13.34, and 13.36; Plate 13.17)—plus the articulated partial left foot of Palomas 92 (Tables 13.30 to 13.35; Plates 13.18 and 13.19). The Palomas 92 associated foot bones derive from several breccia blocks, and to varying degrees they remain cemented together. The two tarsals, the base of the metatarsal 1, the bases and diaphyses of metatarsals 2 to 4, all of metatarsal 5, and the medial hallucal sesamoid bone are in one unit (SP05H004; Palomas 92a to 92g, 92p). The remainder of the metatarsal 1 (SP05H008) connects to its base. The middle three metatarsal heads are joined in a block with the preserved proximal phalanges 2 to 5 and the fifth middle phalanx (SP05H005; Palomas 92c to 92f, 92h to 92k, 92n). There are two additional middle phalanges (SP05H007; Palomas 92l and 92m). The lateral hallucal sesamoid bone is fused to a fragment of the metatarsal 2 head (SP05H011; Palomas 92o). The Palomas numbers are by the anatomical bone, even if there is more than one piece for the bone and/or the pieces are joined to more than one cemented unit.

Palomas 66: Pedal Proximal Phalanx 2–5—Immature

The bone is complete but without any trace of the proximal epiphysis (Plate 13.17). The even and moderate waisting of the diaphysis relative to the articular/metaphyseal ends makes it unlikely to be manual (Pyle et al. 1971). The lack of evidence for proximal epiphyseal fusion makes it likely to be no older than mid-second-decade postnatal, and probably somewhat younger (Baker et al. 2005). Yet the overall length and the well-formed, if clearly subchondral, proximal and distal ends suggest a juvenile (7 to 12 years) or slightly older age at death. The maximum length is 15.0 mm.

Palomas 86: Pedal Proximal Phalanx 2–5—Immature

This heavily encrusted immature phalanx closely resembles Palomas 66 in shape, degree of maturity, and size (maximum length: 14.4 mm) (Plate 13.17). Despite the contrast in vertical position indicated by the difference in excavations spits (2g versus 2k; Table 13.1), they could derive from the same foot.

Palomas 67: Pedal Middle Phalanx 2–4 Left?—Mature

Palomas 67 is a complete bone with a thin layer of encrustation over the full surface (Plate 13.17). Given that the distal facet slopes to the left in dorsal view, assuming that the distal interphalangeal articulation would orient laterally, the bone becomes a left phalanx. The size and clear diaphysis separated from the epiphyses make it unlikely that it is a fifth digit phalanx. Maximum length is 15.5 mm.

Palomas 92a: Cuboid Left

The bone is complete with a vertical crack through the midcalcaneal facet. It is fused to the lateral cuneiform and metatarsals 4 and 5, obscuring the medial talar facets and the metatarsal facets. Maximum preserved (plantar) length is 30.3 mm.

Palomas 92b: Lateral Cuneiform Left

The bone is complete with abrasion to the dorsal navicular facet. It is fused to the cuboid bone and metatarsals 2 and 3. Maximum preserved length is 18.2 mm.

Palomas 92c: Metatarsal 1 Left

The bone is in two pieces. The lateral base is cemented to the plantar metatarsal 2, and the remainder of the bone is separate and has a hole in the lateral base. Maximum preserved length is 52.5 mm.

Palomas 92d: Metatarsal 2 Left

The bone consists of the base and the diaphysis with a hole in the medioplantar base. It is fused to metatarsals 1 and 3, the dorsal half of the head is fused to the metatarsal 3 head and the proximal phalanx 2, and there is a separate fragment of its head. Maximum preserved length is 63.3 mm.

Palomas 92e: Metatarsal 3 Left

The bone consists of the base and the shaft fused to the lateral cuneiform and the metatarsal 2 and 4 bases, plus the head fused to the metatarsal 2 and 4 heads and the proximal phalanx 3. Maximum preserved length is 53.9 mm.

Palomas 92f: Metatarsal 4 Left

The bone retains the base and the shaft with a small hole in the dorsal base. The base and diaphysis are cemented to the cuboid bone and to the metatarsals 3 and 5, and the head is fused to the metatarsal 3 head and the proximal phalanx 4. Maximum preserved length is 55.5 mm.

Palomas 92g: Metatarsal 5 Left

The bone is intact with distoplantar head abrasion. It is cemented to the cuboid bone and the metatarsal 4. Maximum preserved length is 71.3 mm.

Palomas 92h: Pedal Proximal Phalanx 2 Left

The complete bone is fused to the metatarsal 2 head. Maximum preserved length is 26.5 mm.

Palomas 92i: Pedal Proximal Phalanx 3 Left

The complete bone with dorsal head abrasion is attached to the metatarsal 3 head and the proximal phalanges 2 and 4. Maximum preserved length is 23.9 mm.

Palomas 92l: Pedal Proximal Phalanx 4 Left

The complete bone with dorsal head abrasion is buried between proximal phalanges 3 and 5 and a fragment of faunal bone. Maximum preserved length is 23.5 mm.

Palomas 92k: Pedal Proximal Phalanx 5 Left

The complete phalanx is fused to the metatarsal 5, the proximal phalanx 4, and the middle phalanx 5. Maximum preserved length is 21.1 mm.

Palomas 92l: Pedal Middle Phalanx 2–4 Left

The bone is complete with matrix on its base. Maximum preserved length is 13.1 mm.

Palomas 92m: Pedal Middle Phalanx 2–4 Left

The phalanx is intact. Maximum preserved length is 12.9 mm.

Palomas 92n: Pedal Middle Phalanx 5 Left

The phalanx is complete with lateral head abrasion, and it is fused to the proximal phalanx 5. Maximum preserved length is 8.7 mm.

Palomas 92o: Lateral Hallucal Sesamoid Bone Left

The complete sesamoid bone is attached to a metatarsal 2 head fragment. Maximum preserved length is 10.5 mm.

Palomas 92p: Medial Hallucal Sesamoid Bone Left

The bone was displaced and is cemented between the metatarsals 3 and 4. Maximum preserved length is 9.2 mm.

Palomas Pedal Morphology

There is little of note on the two Palomas 92 tarsals. Despite adhering matrix and the cementing of the lateral cuneiform bone against the cuboid bone, obscuring most of the medial surface of the cuboid bone, a navicular articular facet appears to be present on the more proximal portion of the dorsomedial side of the cuboid bone (Plate 13.18). Such facets (as opposed to a solely ligamentous connection) occur in 62.5% of Neandertals (n = 8), all three Middle Paleolithic modern humans, and 40.0% of earlier Upper Paleolithic modern humans. They occur in between ≈25% and ≈65% of individuals in recent human samples (Trinkaus 1975). The lateral cuneiform bone has a smooth and rounded tuberosity for the plantar ligaments.

The metatarsals are similarly smooth, with little definition of the dorsal lines for the interosseus muscles. It is not possible to assess whether a metatarsal 1/2 facet was present. The metatarsal 5 tuberosity is well developed. There is a modest degree of torsion in the fifth metatarsal (≈10°), suggesting a low but present transverse pedal arch (Morton 1922–24; Ward et al. 2011). The metatarsophalangeal articulations resemble those of recent humans, with distinctly proximodorsal orientations of the proximal articulations of the proximal phalanges (Plate 13.19).

The principal area of comparison concerns the proportions of the proximal pedal phalangeal diaphyses, given documented changes through Pleistocene humans (Trinkaus and Hilton 1996; Trinkaus 2005; Trinkaus and Shang 2008). Given differential loading on the middle three digits versus the fifth one (Stott et al. 1973) and difficulties in assigning digit number to the middle three phalanges unless they are in articulation or the series is complete, the comparisons pool the middle three digits and then consider the fifth ones separately. The Palomas 92 proximal pedal phalangeal shafts are compared to those from the Late Pleistocene samples, an Early Pleistocene *Homo* isolated phalanx from a middle digit (SKX 16699; Susman et al. 2001), and two recent human samples. The recent human samples are both reasonably robust, nonmechanized human samples, but they include one habitually unshod sample (Pecos Pueblo Amerindians, New Mexico) and one habitually shod sample (Point Hope Inuits, Alaska) (from Trinkaus 2005).

For the middle three phalanges, almost all of the pre–Upper Paleolithic phalanges have shaft breadths that exceed their shaft dorsoplantar heights (Fig. 13.17); the exceptions are two from Amud 15 and one each from Qafzeh 9 and Skhul 4. The two Palomas 92 phalangeal midshafts are distinctly wider than high, falling above the Neandertal mean and the MPMH range. They are, however, matched by some more recent modern humans. In the fifth toe (Fig. 13.17), most of the Pleistocene and recent human proximal phalanges are wider than high at midshaft, but this applies in particular to the pre–Upper Paleolithic ones. Palomas 92 is again among the ones with wider diaphyses.

It is also possible to scale their midshaft rigidities (modeled as solid beams and computed from the diaphyseal diameters using an ellipse formula; O'Neill and Ruff 2004) against phalangeal length (Fig. 13.18). For the middle three digits, the Middle Paleolithic (and one Early Pleistocene phalanx) have generally more robust phalanges than the Upper Paleolithic ones, despite some overlap in the distributions. The same difference is found in habitually unshod versus shod robust recent humans, and the Middle/Upper Paleolithic difference has

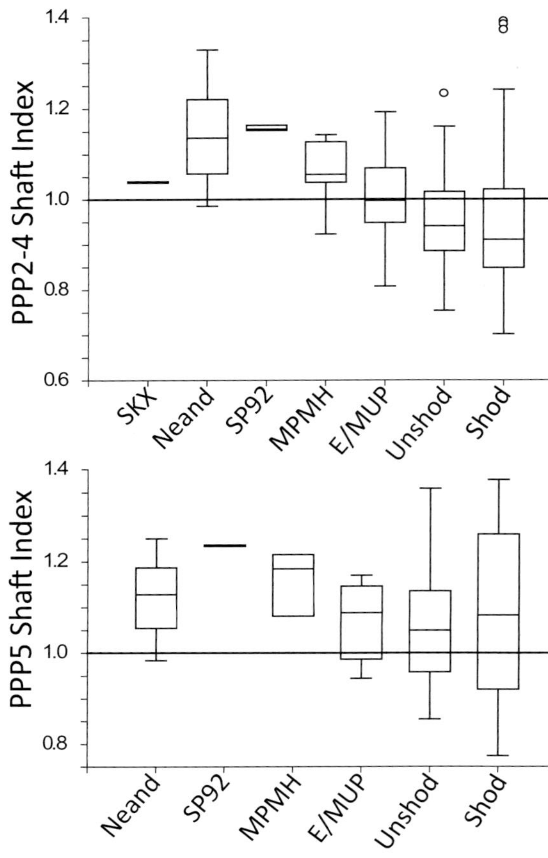

FIG. 13.17. Box plots of the index of pedal phalangeal midshaft breadth versus height for Palomas 92, Pleistocene *Homo* remains, and two recent human samples. SKX: SKX 16699 isolated phalanx; MPMH: Middle Paleolithic modern humans; E/MUP: Early/Mid Upper Paleolithic modern humans; Unshod and Shod: Pecos Pueblo AmerIndian and Point Hope Inuit recent human samples.

been interpreted in terms of a marked increase in the use of footwear with the emergence of the Upper Paleolithic (Trinkaus 2005). A similar contrast appears to have been present in eastern Eurasia, as indicated by the gracile Tianyuan 1 phalanx and the robust Denisova one (Trinkaus and Shang 2008; Mednikova 2011). The two Palomas 92 phalanges are largely with the Neandertals, especially the shorter one from the third toe. There are far fewer fifth proximal phalanges available, and one would expect little difference across the samples, given the reduced ground reaction forces on the little toe (Stott et al. 1973). Yet the same pattern, including the relative position of the Palomas 92 phalanx, occurs (Fig. 13.18).

It therefore appears that Palomas 92 conforms to the Middle Paleolithic (and apparently earlier Pleistocene) pattern of going habitually unshod. The relatively wider diaphyses of these pedal phalanges are likely related to differential mediolateral hypertrophy of the diaphyses, given the trussing nature of the extrinsic digital flexor tendons (Trinkaus and Hilton 1996).

The Palomas Postcranial Paleopathology

Palomas 92 and the isolated postcranial remains from the Sima de las Palomas are remarkably free of pathological lesions. This may be due in part

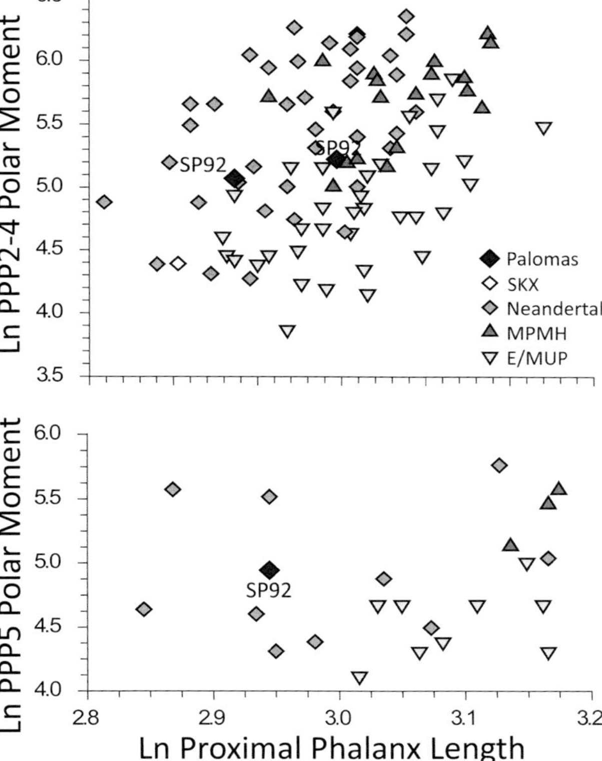

FIG. 13.18. Pedal proximal phalangeal midshaft rigidity: ln polar moments of area versus ln proximal phalangeal articular lengths for Palomas 92, Neandertals, Middle Paleolithic modern humans (MPMH), and Early/Mid Upper Paleolithic modern humans (E/MUP), plus the Swartkrans SKX 16699 Early Pleistocene phalanx. Note that multiple phalanges per individual are represented in the PPP 2 to 4 distribution, given difficulties in assigning digit to ray for isolated phalanges or incomplete series of phalanges.

to the dearth of preserved articulations that might reveal osteoarthritis, the young adult age at death of Palomas 92, and the presence of immature remains in the sample. Yet there is no evidence of either trauma or periosteal reactions. None of the muscular insertions reveal enthesopathies.

The only possibly pathological alteration is the bony growth into the midproximal medullary cavity of the Palomas 64 radius (Plate 13.8). Its etiology is unclear, and it does not appear to have affected the subperiosteal bone. It would probably not have been observed had the bone been intact.

Summary

The Palomas sample provides seventeen isolated postcranial elements, five of which are distinctly immature, plus an associated partial skeleton with upper limb, lower limb, and some axial remains. These remains are joined, in part, by the associated partial skeleton of Palomas 96. The overall pattern of these remains conforms generally to that of the Late Pleistocene Neandertals, to the extent that aspects of those Neandertals contrast with those of early and recent humans.

The Palomas postcrania are generally small (see chapter 14), which makes some of the implications of their postcranial proportions difficult to assess. Yet they are distinctly Neandertals (or more generally, archaic *Homo*) in their humeral diaphyseal and pectoralis major tuberosity proportions, the proportions of the humeral pillars of at least Palomas 92, the ulnar trochlear notch orientation of Palomas 92, the radial tuberosity orientations of two of the three proximal radii, their large hamuli, the epiphyseal breadths of the Palomas 65 middle manual phalanx but not those of Palomas 92 or 96, the apical tuft dimensions of their manual distal phalanges, and the absence of a pilaster on their femora. These features are joined by the scapular axillary border configuration, the pubic bone mediolateral attenuation, and tibial diaphyseal cross-sectional shape of Palomas 96 (Walker et al. 2011b).

At the same time, there are a few features in which they do not conform to the general Neandertal morphological pattern. The metacarpal articular facet of the Palomas 92 trapezium lacks the dorsopalmar flattening of other Neandertals. The Palomas 92 and 96 proximal hand phalanges lack proximal epiphyseal broadening. The Palomas 52, 92, and 96 midproximal femoral diaphyses are narrower and/or deeper than those of other archaic *Homo*, and the same appears to apply to their midshafts.

The isolated postcranial remains from the Sima de las Palomas, plus the Palomas 92 partial skeleton, despite incompleteness, damage, and cementation, provide considerable morphological information. They reinforce the affinities of the sample, evident in the mandibular and dental remains, to the Neandertals, yet they serve to extend the Neandertal range of morphological variation.

TABLE 13.1. Inventory of the isolated human postcranial remains from the Sima de las Palomas

SP NO.	OLD NO.	IDENTIFICATION	MATURITY	PROVENIENCE	DISCOVERY DATE
8	CG-8	Axis (cervical vertebra 2)	immature	Hillside rubble	1993
9	CG-9	Axis (cervical vertebra 2)	mature	Hillside rubble	1993
13	CG-17	Fibula diaphysis left	mature	Hillside rubble	1994
14	CG-18	Ulna proximal left	immature	Upper Cutting level 2b	9 July 1994
15	CG-19	Metacarpal 2 diaphysis and head right	mature	Main chamber rubble	1994
16	CG-20	Humerus diaphysis left	(mature)[a]	Hillside rubble (?)	1994
17	CG-21	Humerus trochlea and epicondyle left	mature	Hillside rubble	1994
28	CG-24	Manual distal phalanx 2–4	mature	Upper Cutting level 2g	6 July 1995
32	CG-55 #13	Humerus proximal diaphysis/ epiphysis right	immature	Upper Cutting level 2i	10 July 1995
52	—	Femur diaphysis left	(mature)[a]	Hillside rubble	2 August 1997
63	CG-84	Rib 1 left	(mature)[a]	Hillside rubble	23 July 1999
64	CG-82	Radius diaphysis left	mature	Hillside rubble	23 July 1999
65	CG-75 #9	Manual middle phalanx 2–4	mature	Shaft scaffolding tower	9 August 1999
66	CG-76 #10	Pedal proximal phalanx 2–5	immature	Upper Cutting level 2k	31 July 2000
67	CG-98	Pedal middle phalanx	mature	Upper Cutting level IA	29 July 2001
77	CG-93 (code22)	Femur head	mature	Upper Cutting level 2b	28 July 2003
86	—	Pedal proximal phalanx 2–5	immature	Upper Cutting level 2g	23 July 2004

[a]The Palomas 16, 52, and 63 elements are listed as (mature), indicating that their mature status is not secure. All appear morphologically mature, but they lack epiphyseal regions to confirm that diagnosis. All are small, but they are not small relative to the dimensions of Palomas 92 or to those of Palomas 96 (Walker et al. 2011b). They are analyzed as though they are mature.

TABLE 13.2. Inventory of the axial remains of the young adult Palomas 92 partial skeleton

SP NO.	FIELD NO.	IDENTIFICATION	EXCAVATION DATE	COMMENTS
92ll	SP05H036	Thoracic verterbra T12	1 August 2005	≈ Complete
92kkk	SP05H085	Lumbal vertebra L2	4 August 2005	Incomplete
92jjj	SP05H085	Lumbal vertebra L3	4 August 2005	Incomplete
92iii	SP05H085	Lumbal vertebra L4	4 August 2005	Incomplete
92hhh	SP05H085	Lumbal vertebra L5	4 August 2005	Incomplete
92ss	SP05H085	Sacrum + ilium fragments	4 August 2005	Incomplete
92fff	SP04H307	Rib	24 July 2004	Fragment
92ggg	SP04H307	Rib	24 July 2004	Fragment

Note: The elements are organized anatomically; hence, the SP numbers, field numbers, and dates of excavation are not in sequence. All are from Upper Cutting level 2h.

TABLE 13.3. Morphometrics of the Palomas vertebral remains, in millimeters and degrees

MEASUREMENT	PALOMAS 8	PALOMAS 9	PALOMAS 92
	C2 IMMATURE	C2	T12
Cranial facet angle[a]	—	—	right: 85° lateral left: 87° lateral
Caudal facet angle[a]	—	—	right: 34° left: 43°
Dorsoventral diameter	—	—	(59.0)
Cranial facets external breadth	—	—	31.7
Cranial facets internal breadth	—	—	12.4
Caudal facets external breadth	—	—	28.2
Caudal facets internal breadth	—	—	17.0
Canal dorsoventral diameter (M-10)	—	—	20.4
Canal transverse diameter (M-11)	—	—	19.0
Cranial body depth (M-4)	—	—	(24.0)
Caudal body depth (M-5)	12.0	—	(25.6)
Cranial body breadth (M-7)	—	—	32.0
Caudal body breadth (M-8)	17.8	—	(33.1)
Body middle height (M-3)	—	—	(20.5)
Spinous process length[b]	—	—	17.7
Spinous process angle[c]	—	—	127°
Lateral body height	15.0	15.5	
C2 maximum ventral height	—	30.3	
C2 mid body height	—	29.3	
C2 odontoid height (M-1a)	—	14.8	
C2 odontoid breadth	—	8.5	
C2 odontoid depth	—	9.3	

Note: Dimensions in parentheses are estimated.
[a]The facet angles are relative to the sagittal plane.
[b]The distance from the cranial ventral (interlaminal) margin to the process dorsal tip.
[c]The angle between the spinal process midline and the cranial body plane.

TABLE 13.4. Inventory of the shoulder and arm remains of the young adult Palomas 92 partial skeleton

SP NO.	FIELD NO.	IDENTIFICATION	PROVENIENCE	EXCAVATION DATE	COMMENTS
92eee	SP05H303+SP04H307	Scapula	Upper cutting level 2h	24 July 2004/4 August 2005	2 fragments
92ff	SP05H016+ SP04H301/302/306	Humerus right	Upper cutting level 2h/ SEL 2d	24 July 2004/ 1 August 2005/ 4 August 2005	5 sections
92gg[a]	SP05H016+SP04H304	Ulna right proximal and shaft	Upper cutting level 2h	9 August 2004/1 August 2005	3 sections
92hh	SP05H016	Ulna right distal	Upper cutting level 2h	1 August 2005	Incomplete
92ii[a]	SP05H016+SP04H305	Radius right	Upper cutting level 2h	9 August 2004/ 1 August 2005	4 sections
92bbb	SP05H017	Radius left	Upper cutting level 2h	1 August 2005	Incomplete

Note: The elements are organized anatomically, hence the SP numbers, field numbers, and dates of excavation are not in sequence.
[a]Portions of the right radial and ulnar diaphyses were formerly (Walker et al. 2011a) identified as fibular and cataloged as Palomas 92pp. Subsequent cleaning of matrix permitted their correct identification and articulation with the Palomas 92 right elbow region (Fig. 13.10).

TABLE 13.5. Lengths, proximal measurements, and diaphyseal dimensions of the Palomas humeri (including Palomas 96), in millimeters

MEASUREMENT	PALOMAS 16 LEFT	PALOMAS 32 RIGHT (IMMATURE)	PALOMAS 92 RIGHT	PALOMAS 96 RIGHT	PALOMAS 96 LEFT
Maximum length (M-1)[a]	—	—	(260–270)	272.0	—
Articular length (M-2)[a]	—	—	(257–266)	(270.0)	—
Head anteroposterior diameter (M-9)	—	25.5[b]	—	—	—
Proximal epiphysis mediolateral diameter	—	32.2[b]	—	—	—
Surgical neck maximum diameter	—	18.3	—	—	—
Surgical neck minimum diameter	—	12.2	—	—	—
Midshaft maximum diameter (M-5)	19.8	—	(18.2)	18.0	18.5
Midshaft minimum diameter (M-6)	14.3	—	(17.3)	12.2	13.8
Pectoralis major breadth	≥7.6	—	—	5.5	—
Deltoid tuberosity breadth	(8.5)	—	—	—	—
Mid-distal diaphyseal anteroposterior diameter	—	—	17.6	—	—
Mid-distal diaphyseal mediolateral diameter	—	—	18.7	—	—
Mid-distal diaphyseal circumference	—	—	61.0	—	—
Midshaft cortical thickness—anterior	5.5	—	—	4.6	5.8
Midshaft cortical thickness—posterior	3.8	—	—	5.4	6.4
Midshaft cortical thickness—medial	3.6	—	—	4.1	4.1
Midshaft cortical thickness—lateral	3.4	—	—	4.1	4.2
Mid-distal cortical thickness—posterior	—	—	5.2	—	—
Mid-distal cortical thickness—anterior	—	—	6.0	—	—
Mid-distal cortical thickness—anteromedial	—	—	4.9	—	—
Mid-distal cortical thickness—anterolateral	—	—	4.8	—	—

Note: Dimensions in parentheses are estimated.
[1] See chapter 13, p. 191, for the Palomas 92 humerus length estimates.
[2] Epiphyseal diameters.

TABLE 13.6. Cross-sectional parameters for the Palomas humeri at the specified locations

PARAMETER	PALOMAS 16 ≈50% LEFT	PALOMAS 92 ≈35% RIGHT	PALOMAS 96 ≈50% RIGHT	PALOMAS 96 ≈50% LEFT
Total area (mm²)	242.9	264.4	183.3	203.5
Cortical area (mm²)	186.4	233.0	160.5	171.3
Anteroposterior second moment of area (mm⁴) (I_x)	(6,591)	5,823	3,130	(4,626)
Mediolateral second moment of area (mm⁴) (I_y)	(3,068)	5,544	2,462	(2,253)
Maximum second moment of area (mm⁴) (I_{max})	6,593	5,834	3,705	4,629
Minimum second moment of area (mm⁴) (I_{min})	3,066	5,533	1,887	2,250
Polar moment of area (mm⁴) (J/I_p)	9,659	11,367	5,592	6,879

Note: The humeri were digitized from scaled photographs of the distally (Palomas 16 and 96 right) and proximally (Palomas 92ff and 96 left) facing diaphyseal breaks. The Palomas 16 and Palomas 96 left axial orientations are approximate, and hence Ix and Iy are in parentheses; the Palomas 92 and 96 right orientations are based on the distal articulation.

TABLE 13.7. Distal epiphyseal and metaphyseal measurements and discrete traits of the Palomas humeri (including Palomas 96), in millimeters

MEASUREMENT	PALOMAS 17	PALOMAS 92	PALOMAS 96
	LEFT	RIGHT	RIGHT
Epicondylar breadth (M-4)	—	58.5	(52.7)
Distal articular breadth (M-12a)	—	(38.8)	—
Medial trochlear ant-post diameter (M-13)	—	21.7	—
Mid minimum trochlear ant-post diameter (S-2)[a]	13.2	(11.6)	—
Olecranon fossa breadth (M-14)	—	28.5	23.2
Medial pillar thickness (S-12)	—	7.7	6.0
Lateral pillar thickness (S-13)	—	13.6	11.4
Septal aperture presence	—	present	—
Septal aperture mediolateral diameter	—	(8.6)	—

Note: Dimensions in parentheses are estimated.

[a]S-# refers to the equivalent measurement in Senut (1981).

TABLE 13.8. Microarchitecture and proximal diaphyseal properties of the Palomas 32 immature proximal humerus (Plate 13.5)

Trabecular thickness (TbTh) (mm)	0.318
Trabecular separation (TbSp) (mm)	0.356
Total orientation	27.71
% Trabecular orientation 0°–15°	30%
% Trabecular orientation 90°–105°	15%
Anteroposterior external diameter (mm)	18.7
Mediolateral external diameter (mm)	15.8
Subperiosteal perimeter (PP) (mm)	63.2
Total bone area (BA/TA) (mm²)	159.9
Cortical area (CA) (mm²)	67.8
Percent cortical area (%CA)	42.4%
Mean cortical thickness (CTh) (mm)	1.3

TABLE 13.9. Measurements of the Palomas 92 portions of the right ulna, in millimeters

Olecranon height (M-7)	21.2
Coronoid height (S-3)[a]	26.3
Midshaft maximum diameter	12.9
Midshaft minimum diameter	10.2
Distal minimum shaft anteroposterior diameter	8.5
Distal minimum shaft mediolateral diameter	8.6
Head breadth (MCH-3)[b]	16.5
Maximum head diameter	17.1
Distal maximum depth	15.7
Mean head cortical thickness	0.3

[a]S-# refers to the equivalent measurement in Senut (1981).

[b]MCH-# refers to the equivalent measurement in McHenry et al. (1976).

TABLE 13.10. Cross-sectional measurements of the Palomas 14 immature proximal ulna and the Palomas 92 mature distal right ulnar diaphysis

MEASUREMENT	PALOMAS 14	PALOMAS 92
	≈75%	≈25%
Anteroposterior external diameter (mm)	13.9	8.9
Mediolateral external diameter (mm)	12.7	8.2
Subperiosteal perimeter (PP) (mm)	189.9	33.6
Total bone area (BA/TA) (mm²)	105.4	56.8
Cortical area (CA) (mm²)	37.8	45.0
Percent cortical area (%CA)	35.8%	79.3%
Mean cortical thickness (CTh) (mm)[a]	1.7	2.5

Note: Measurements are at ≈75% and ≈25%, respectively, of diaphyseal length. Measures are derived from the micro-CT scans of the elements (Plates 13.9 and 13.11).

[a]The average of the cortical thicknesses taken at the anterior, posterior, medial, and lateral sections of the bone.

TABLE 13.11. Measurements and discrete observations of the Palomas 64, 92, and 96 radii, in millimeters

MEASUREMENT	PALOMAS 64		PALOMAS 92		PALOMAS 96
	LEFT	RIGHT	LEFT	RIGHT	LEFT
Head-neck length (M-1A)	—	31.6	—	—	—
Head diameter (≈anteroposterior) (M-5 (1), S-6)[a]	—	19.8	—	—	—
Neck anteroposterior diameter (M-5 (2), S-9)	—	10.0	—	—	—
Neck mediolateral diameter (M-4 (2))	—	10.3	—	—	—
Tuberosity length (S-1)	20.8	(20.0)	—	—	—
Tuberosity breadth (S-4)	13.3	12.0	—	—	—
Tuberosity projection (S-8)	16.1	—	—	—	—
Tuberosity position[b]	2	3	—	3	—
Proximal shaft anteroposterior diameter[c]	12.7	10.8	—	—	—
Proximal shaft mediolateral diameter	11.9	11.9	—	—	—
Proximal shaft circumference	39.5	—	—	—	—
Crest maximum diameter (≈mediolateral) (M-4)	—	(13.4)	—	—	10.7
Crest minimum diameter (≈anteroposterior) (M-5)	—	(10.0)	—	—	8.0
Distal maximum breadth (M-5 (6))	—	—	—	—	(26.5)
Distal ulnar proximodistal length	—	—	9.1	—	—

Note: Dimensions in parentheses are estimated.
[a]S-# indicates a measurements defined in Senut (1981).
[b]Anteromedial (2) and medial (3) relative to the plane of the interosseus crest, following Trinkaus and Churchill (1988).
[c]Shaft dimensions between the tuberosity and the interosseus crest (Trinkaus 1983).

TABLE 13.12. Cross-sectional parameters of the Palomas 64 mid-proximal left radial diaphysis and the Palomas 92 midshaft right ulnar diaphysis

PARAMETER	PALOMAS 64		PALOMAS 92
	RADIUS LEFT ESTIMATED ORIGINAL	RADIUS LEFT WITH MEDULLARY BONY GROWTH	ULNA RIGHT
Total area (mm²)	110.0	110.0	100.2
Cortical area (mm²)	83.3	96.7	91.2
Anteroposterior second moment of area (mm⁴) (I_x)	813	871	—
Mediolateral second moment of area (mm⁴) (I_y)	979	1,030	—
Maximum second moment of area (mm⁴) (I_{max})	989	1,031	933
Minimum second moment of area (mm⁴) (I_{min})	803	870	734
Polar moment of area (mm⁴) (J/I_p)	1,792	1,901	1,667

Note: The measurements are from scaled photographs of the diaphyseal breaks at approximately the midproximal and midshaft diaphyses. The values for Palomas 64 delete (original) and then include a bony growth into the medullary cavity from the anteromedial endosteal surface (Plate 13.8). The Palomas 64 axial orientation is based on the interosseus crest; the Palomas 92 piece is considered without parasagittal orientation.

TABLE 13.13. Inventory of the hand remains of the young adult Palomas 92 partial skeleton

SP NO.	FIELD NO.	IDENTIFICATION	EXCAVATION DATE	COMMENTS
92vv	SPo5H017	Scaphoid left	1 August 2005	≈ Complete
92ww	SPo5H016/017	Lunate left	1 August 2005	2 fragments
92jj	SPo5H016	Triquetral right	1 August 2005	≈ Complete
92kk	SPo5H017	Triquetral left	1 August 2005	Incomplete
92lll	SPo5H016/017	Pisiform right	1 August 2005	2 fragments
92zz	SPo5H017	Trapezium right	1 August 2005	≈ Complete
92aa	SPo5H100	Trapezium left	4 August 2005	≈ Complete
92bb	SPo5H100	Trapezoid left	4 August 2005	≈ Complete
92aaa	SPo5H017	Capitate right	1 August 2005	Incomplete
92xx	SPo5H017	Capitate left	1 August 2005	Incomplete
92yy	SPo5H017	Hamate left	1 August 2005	Incomplete
92z	SPo5H013	Metacarpal 2 right	1 August 2005	Incomplete
92ccc	SPo5H100	Metacarpal 2 left	4 August 2005	Incomplete
92q	SPo5H009/021/032	Metacarpal 3 left	31 July 2005	3 fragments
92r	SPo5H009/104	Metacarpal 4 left	31 July 2005	2 fragments
92s	SPo5H009	Metacarpal 5 left	31 July 2005	Incomplete
92ee	SPo5H104	Metacarpal 3, 4, or 5 right	4 August 2005	Missing
92cc	SPo5H020	Metacarpal proximal epiphysis	1 August 2005	Pulverized
92t	SPo5H009	Manual proximal phalanx 3 left	31 July 2005	Complete
92u	SPo5H009	Manual proximal phalanx 4 left	31 July 2005	Complete
92v	SPo5H009	Manual proximal phalanx 5 left	31 July 2005	Incomplete
92w	SPo5H009	Manual middle phalanx 3 left	31 July 2005	Complete
92x	SPo5H009	Manual middle phalanx 4 left	31 July 2005	Complete
92y	SPo5H009	Manual distal phalanx 3 left	31 July 2005	Incomplete
92uu	SPo5H009	Manual distal phalanx 4 left	31 July 2005	Incomplete

Note: The elements are organized anatomically, hence the SP numbers, field numbers, and dates of excavation are not in sequence. All are from the Upper Cutting level 2h.

TABLE 13.14. Measurements of the Palomas 92 left trapezium (92aa) and trapezoid (92bb) bones, in millimeters

MEASUREMENT	TRAPEZIUM	TRAPEZOID
Maximum length (M-1)		9.5
Maximum breadth (M-2)	20.1	10.4
Maximum height (M-3)	12.5	9.6
Maximum thickness	(11.9)	
Dorsal breadth	16.6	
Metacarpal articular height (—; M-11c)	11.0	14.5
Metacarpal articular breadth (M-4; M-10c)	(15.6)	10.5 (dorsal)
Scaphoid articular height (M-5a)	8.5	12.4
Scaphoid articular breadth (M-6; M-4c)	8.9	9.3
Trapezoid articular height (M-9)	7.5	
Trapeziod articular breadth (M-8)	9.5	
Metacarpal 1 articular height	10.8	
Metacarpal 1 articular subtense[a]	2.5	
Trapezium articular height (M-7c)		13.7
Trapezium articular breadth (M-6c)		8.4
Capitate articular height (M-9c)		13.4
Capitate articular breadth (M-8c)		5.3
Scaphoid-trapezoid articular angle	(129°)	
Trapezium tubercle length	(9.6)	
Trapezium tubercle thickness	3.3	
Trapezium tubercle projection	4.3	

Note: Martin numbers (M-#) are provided for the trapezium and then the trapezoid, as appropriate. Dimensions in parentheses are estimated.
[a]Maximum subtense from the articular height to the most distally projecting point on the dorsopalmar midline of the metacarpal 1 facet (Trinkaus 1989).

TABLE 13.15. Measurements of the Palomas 92 left hamate (92yy) and the Palomas 96 right and left hamate (96ak and 96ba) bones, in millimeters

MEASUREMENT	PALOMAS 92	PALOMAS 96	
	LEFT	RIGHT	LEFT
Maximum height (M-3)	22.1	20.6	(21.0)
Maximum length	14.5	13.8	13.7
Articular length (M-1)	(12.8)	13.6	—
Body height	10.4	11.2	11.2
Hamulus projection (M-5)	11.7	9.4	9.8
Hamulus length (M-14)	8.3	—	9.4
Hamulus thickness (M-15)	—	—	5.6

Note: The Palomas 92 hamate is partial and partially embedded in matrix, with the loss of most of the ulnar side of the bone. It is not apparent if the body surfaces extend to their full original extents; they are likely to be very close (within 1 mm). The hamulus margins are present. Dimensions in parentheses are estimated.

TABLE 13.16. Measurements of the Palomas 15 right metacarpal (MC) 2, the Palomas 92 left metacarpal bones, and the Palomas 96 right and left metacarpals, in millimeters

MEASUREMENT	PALOMAS 15 MC-2 R	PALOMAS 92CCC MC-2 L	PALOMAS 92R MC-4 L	PALOMAS 92S MC-5 L	PALOMAS 96T MC-1 R	PALOMAS 96U MC-2 R	PALOMAS 96V MC-3 R	PALOMAS 96RR MC-3 L	PALOMAS 96X MC-4 R
Maximum length	—	—	—	—	40.0	—	—	—	(48.0)
Articular length	—	—	—	—	38.0	(56.0)	—	—	(48.0)
Midshaft height	9.6	—	6.4	6.0	—	—	7.2	7.3	—
Midshaft breadth	7.2	—	6.3	6.3	9.5	6.2	6.5	6.6	—
Opponens breadth[a]					11.9				
Prox max height	—	—	—	—	—	—	—	—	10.2
Prox max breadth	—	—	—	—	—	—	—	—	10.7
Head height	13.0	12.5	12.3	—	—	—	—	12.0	—
Head dorsal breadth	13.5	(12.8)	—	—	—	12.7	13.0	13.0	—
Head palmar breadth	13.5	(13.5)	11.5	—	—	—	—	11.3	—

Note: Dimensions in parentheses are estimated.

[a]Radioulnar breadth across the shaft and the opponens pollicis crest at the maximum radial projection of the opponens crest.

TABLE 13.17. Midshaft cross-sectional parameters of the Palomas 15 and 92 metacarpals (MC), from scaled photographs of natural breaks

PARAMETER	PALOMAS 15 MC-2	PALOMAS 92R MC-4
Total area (mm²)	47.2	(32.2)
Cortical area (mm²)	37.9	(30.5)
Anteroposterior second moment of area (mm⁴) (I_x)	209	(90)
Mediolateral second moment of area (mm⁴) (I_y)	141	(77)
Maximum second moment of area (mm⁴) (I_{max})	213	(93)
Minimum second moment of area (mm⁴) (I_{min})	136	(74)
Polar moment of area (mm⁴) (J/I_p)	350	(167)

Note: Orientations are based on the metacarpal heads. Palomas 92r has required minor estimation of the damaged subperiosteal contour, hence the parentheses. Dimensions in parentheses are estimated.

TABLE 13.18. Distal trabecular parameters for the Palomas 92 left metacarpals (MC) 4 and 5

PARAMETER	PALOMAS 92R MC-4	PALOMAS 92S MC-5
Bone volume / tissue volume (Bv/Tv) (%)	33.25	31.80
Trabecular bone pattern factor (Tb.Pf) (mm⁻¹)	−7.26	−6.04
Trabecular thickness (Tb.Th) (mm)	0.372	0.365
Trabecular number (Tb.N) (mm⁻¹)	0.893	0.872
Trabecular separation (Tb.Sp) (mm)	1.062	0.851
Degree of anisotropy (DA)	1.155	1.187

TABLE 13.19. Measurements of the Palomas 92 left manual proximal phalanges (MPP), in millimeters

MEASUREMENT	PALOMAS 92T MPP-3	PALOMAS 92U MPP-4	PALOMAS 92V MPP-5
Maximum length (M-3)	41.0	39.0	—
Articular length	40.2	37.7	—
Shaft height	5.8	5.5	7.0
Shaft breadth	9.8	8.1	8.4
Proximal maximum height	11.7	10.8	10.0
Proximal maximum breadth	14.8	12.2	—
Proximal articular height	9.8	8.8	—
Proximal articular breadth	10.8	10.4	—
Distal height	6.8	—	—
Distal breadth	(11.5)	—	—

Note: Dimension in parentheses is estimated.

TABLE 13.20. Measurements of the Palomas 96 manual proximal phalanges (MPP), in millimeters

MEASUREMENT	SP96Y MPP-1 R	MPP-1 L	SP96W MPP-2 L	SP96AA MPP-3 R	SP96WW MPP-3 L	SP96YY MPP-5 L
Maximum length (M-3)		—	34.9	39.6	—	30.4
Articular length	(26.0)	—	33.3	38.5	—	(29.0)
Shaft height	5.7	—	4.6	5.6	5.7	4.1
Shaft breadth	7.9	—	7.9	8.5	8.3	7.2
Proximal max height	10.3	10.3	10.5	11.8	11.8	—
Proximal max breadth	12.7	—	(13.3)	—	13.7	12.2
Prox articular height	(9.2)	—	(9.3)	—	10.5	—
Prox articular breadth	(11.6)	11.9	(12.0)	—	12.2	9.9
Distal height	—	—	6.4	—	6.9	5.2
Distal breadth	—	—	(11.3)	—	11.4	9.3

Note: Dimensions in parentheses are estimated.

TABLE 13.21. Diaphyseal cortical thickness measurements for Palomas 65, 92, and 96 proximal (MPP) and middle (MMP) manual phalanges, in millimeters

MEASUREMENT	PALOMAS 65 MMP-2, 3, OR 4	PALOMAS 92T MPP-3	PALOMAS 92U MPP-4	PALOMAS 92W MMP-3	PALOMAS 96YY MPP-5
Proximal (75%)					
Dorsal	0.47	0.74	1.28	1.02	1.00
Palmar	0.29	0.99	1.28	0.82	1.15
Radial	0.33 rt	1.24	1.20	0.70	1.46
Ulnar	0.33 lt	0.87	1.32	0.59	1.41
Midshaft (50%)					
Dorsal	1.07	1.98	2.32	1.41	1.81
Palmar	0.74	1.65	1.65	1.56	1.51
Radial	0.70 rt	1.82	1.49	1.37	1.84
Ulnar	0.78 lt	1.86	1.65	1.41	2.21
Distal (25%)					
Dorsal	0.62	2.07	1.57	1.54	1.61
Palmar	0.33	2.19	1.20	2.11	1.66
Radial	0.87 rt	1.90	1.74	2.42	2.76
Ulnar	0.54 lt	2.03	20.3	1.80	2.51

Note: Measurements are taken at 75% (proximal), 50% (midshaft), and 25% (distal) of length. The Palomas 92 and 96 phalanges are all left based on *in situ* articulations, such that radial and ulnar sides can be determined. The Palomas 65 isolated phalanx cannot be sided; its radial and ulnar thickness are therefore labeled as right (rt) and left (lt), respectively.

TABLE 13.22. Proximal and distal trabecular parameters for the Palomas 92 and 96 left proximal manual phalanges (MPP)

PARAMETER	PALOMAS 92T MPP-3	PALOMAS 92U MPP-4	PALOMAS 92V MPP-5	PALOMAS 96YY MPP-5
Proximal				
Bone volume / tissue volume (Bv/Tv) (%)	29.10	29.40	—	31.64
Trabecular bone pattern factor (Tb.Pf) (mm^{-1})	−2.51	−3.99	—	−1.15
Trabecular thickness (Tb.Th) (mm)	0.426	0.492	—	0.794
Trabecular number (Tb.N) (mm^{-1})	0.842	0.597	—	0.399
Trabecular separation (Tb.Sp) (mm)	0.683	0.779	—	0.819
Degree of anisotropy (DA)	1.436	1.392	—	1.559
Distal				
Bone volume / tissue volume (Bv/Tv) (%)	27.50	32.00	36.30	27.52
Trabecular bone pattern factor (Tb.Pf) (mm^{-1})	−3.89	−5.40	−7.34	−2.80
Trabecular thickness (Tb.Th) (mm)	0.530	0.723	0.556	0.658
Trabecular number (Tb.N) (mm^{-1})	0.518	0.441	0.652	0.418
Trabecular separation (Tb.Sp) (mm)	1.402	1.229	1.121	1.314
Degree of anisotropy (DA)	1.458	1.267	1.278	1.206

TABLE 13.23. Measurements of the Palomas 65, 92, and 96 middle manual phalanges (MMP), in millimeters

MEASUREMENT	PALOMAS 65 MMP-2, 3, OR 4	PALOMAS 92W MMP-3 L	PALOMAS 92X MMP-4 L	PALOMAS 96CC MMP-2 R	PALOMAS 96DD MMP-3 R	PALOMAS 96ZZ MMP-3 L	PALOMAS 96EE MMP-4 R	PALOMAS 96 MMP-4L	PALOMAS 96AAA MMP-5 L
Maximum length	23.1	25.6	—	21.3	26.4	25.8	22.9	—	17.2
Articular length	21.9	(24.4)[a]	—	20.6	25.2	24.8	(22.0)[a]	—	(17.0)[a]
Midshaft height	5.8	5.8	—	4.3	—	4.5	—	4.5	—
Midshaft breadth	8.3	8.7	—	7.1	7.5	7.2	—	—	—
Proximal max ht	10.4	—	—	8.2	—	8.8	—	—	—
Proximal max br	14.4	13.4	(12.0)	11.5	—	—	—	12.1	—
Proximal artic ht	8.8	—	—	8.0	—	8.3	—	—	—
Proximal artic br	13.6	—	—	11.0	—	—	—	—	—
Distal height	6.1	5.5	—	4.3	—	4.9	—	—	—
Distal max breadth	12.1	10.4	—	9.4	—	10.1	—	—	—

Note: The Palomas 92 and 96 MMPs are identified as to digit and side based on their *in situ* articulated positions. Dimensions in parentheses are estimated.

[a]The articular lengths of the Palomas 92 manual middle phalanx 3 and the Palomas 96 middle phalanges 4 right and 5 left were estimated from their maximum lengths using a least squares regression based on Late Pleistocene middle phalanges 2 to 4 (N = 32) (ArtLen = 0.880 × MaxLen + 1.9; r^2 = 0.979; Palomas 92w MMP-3: 24.4 ± 0.4 mm; SE_{est} = 1.6%; Palomas 96ee MMP-4: 22.0 ± 0.5 mm; SE_{est} = 2.1%; Palomas 96aaa MMP-5: 17.0 ± 0.5 mm; SE_{est} = 3.0%).

TABLE 13.24. Proximal and distal trabecular parameters for the Palomas 65 and 92 middle manual phalanges (MMP)

PARAMETER	PALOMAS 65 MMP-2, 3, OR 4	PALOMAS 92W MMP-3	PALOMAS 92X MMP-4
Proximal			
Bone volume / tissue volume (Bv/Tv) (%)	20.52	22.80	—
Trabecular bone pattern factor (Tb.Pf) (mm⁻¹)	0.927	−1.155	—
Trabecular thickness (Tb.Th) (mm)	0.447	0.794	—
Trabecular number (Tb.N) (mm⁻¹)	0.793	0.399	—
Trabecular separation (Tb.Sp) (mm)	0.791	0.819	—
Degree of anisotropy (DA)	1.713	1.559	—
Distal			
Bone volume / tissue volume (Bv/Tv) (%)	18.75	32.45	34.60
Trabecular bone pattern factor (Tb.Pf) (mm⁻¹)	0.997	−5.806	−6.53
Trabecular thickness (Tb.Th) (mm)	0.550	0.365	0.445
Trabecular number (Tb.N) (mm⁻¹)	0.341	0.889	0.776
Trabecular separation (Tb.Sp) (mm)	1.079	0.860	0.881
Degree of anisotropy (DA)	1.450	1.297	1.323

TABLE 13.25. Measurements of the Palomas 28, 92, and 96 manual distal phalanges (MDP), in millimeters and degrees

MEASUREMENT	PALOMAS 28 MDP-2, 3, OR 4	PALOMAS 92Y MDP-3	PALOMAS 96DDD MDP-1	PALOMAS 96HH MDP-3	PALOMAS 96BBB MDP-2, 3, OR 4	PALOMAS 96 MDP-2, 3, OR 4CCC	PALOMAS 96FF MDP-5
Maximum length	19.5	—	24.2	19.6	—	18.5	17.3
Articular length	17.5	—	22.5	18.8	—	18.1	16.4
Midshaft height	4.1	—	(3.6)	4.0	—	—	3.6
Midshaft breadth	6.1	—	8.3	6.1	—	—	5.5
Proximal max height	7.5	—	6.3	6.2	—	—	5.4
Proximal max breadth	12.1	11.2	13.2	10.6	—	10.0	9.3
Proximal artic height	6.5	—	5.8	5.5	—	—	5.1
Proximal artic breadth	10.7	—	11.7	9.2	—	9.1	8.8
Distal max breadth	9.7	—	(9.5)	8.2	7.8	—	7.5
Ungual angle[a]			8°				

Note: Dimensions in parentheses are estimated.

[a]The angle between the longitudinal axis (midbase to midapical tuft) and the perpendicular to the proximal articular breadth tangent. A positive angle indicates an ulnar deviation of the distal bone, the normal condition and one used to side distal pollical phalanges. The Palomas 96 value is close to the mean of a Neandertal sample (7.2° ± 3.1°, n = 13) but above that of an early modern human one (2.7° ± 1.3°, n = 10). The two distributions bracket the central tendency of a pooled recent human sample (4.4° ± 1.6°, n = 70).

TABLE 13.26. Inventory of the lower limb remains of the young adult Palomas 92 partial skeleton

SP NO.	FIELD NO.	IDENTIFICATION	EXCAVATION DATE	COMMENTS
92qq	SP05H010	Pelvic fragment – ilium	31 July 2005	Incomplete
92rr	SP05H040	Pelvic fragment – ilium	1 August 2005	Incomplete
92ss	SP05H085	Ilium fragments + sacrum	4 August 2005	Incomplete
92mm	SP05H002 /071B/014	Femur right	29 July/1/4 August 2005	2 sections
9200	SP05H071A/107	Femur left	4/7 August 2005	3 sections
92ddd	SP05H001/077	Tibia (right or left?)	29 July/4 August 2005	4 fragments

Note: The elements are organized anatomically, hence the SP numbers, field numbers, and dates of excavation are not in sequence. All are from the Upper Cutting level 2h except the SP05H107 femur section, which is from level 2k.

TABLE 13.27. Osteometric measurements of the Palomas 52, 77, 92, and 96 femora, in millimeters

MEASUREMENT	PALOMAS 52	PALOMAS 77	PALOMAS 92		PALOMAS 96	
	LEFT	—	RIGHT	LEFT	RIGHT	LEFT
Maximum length (M-1)	—	—	(397.0)[a]	—	(395.0)[d]	—
Bicondylar length (M-2)	—	—	(394.0)[a]	—	(391.5)[d]	—
Biomechanical length[b]	—	—	(372.0)[a]	—	360	—
Femur head ant-post diameter (M-19)	—	45.6	(44.2)[c]	—	—	43.0
Femur head sup-inf diameter (M-18)	—	—	(44.3)[c]	—	—	
Gluteal tuberosity breadth	—	—	(10.5)	10.2	—	—
Subtrochanteric ant-post diameter (M-10)	19.2	—	—	—	—	—
Subtrochanteric med-lat diameter (M-9)	25.5	—	—	—	—	—
Mid-proximal anteroposterior diameter	21.1	—	28.2	—	—	—
Mid-proximal mediolateral diameter	23.1	—	25.2	—	—	—
≈Midshaft anteroposterior diameter (M-6)	(19.0)	—	—	—	24.4	—
≈Midshaft mediolateral diameter (M-7)	(21.8)	—	—	—	24.2	—
Lateral condylar depth (M-22)	—	—	60.0[c]	—	—	—

Note: Proximodistal locations of the diaphyseal diameters are approximate, given the preservation of the femora. Palomas 52 may be immature, given its small dimensions. Palomas 92 and 96 are mature despite their small dimensions. Dimensions in parentheses are estimated.

[a]Femoral lengths were estimated using the preserved portions of the right femur. From the distal lesser trochanter to the distal end of the proximal sections is 141 mm. By lining up the more proximal linea aspera with the more distal one, the best fit is on the anterolateral extension of proximal sections along an irregular proximal break in the distal piece. The proximal edge of the distal midshaft has thin matrix on it, but it provides a good contact for a minimal length estimate for the bone. From the contact point at midshaft to the distal lateral condyle is 165 mm. Minus 1 mm for the expansion crack gives 164 mm. Added to the proximal piece provides a length of 305 mm from the distal lesser trochanter to the distal lateral condyle (the "SP92Len").

A sample of recent human femora plus three Late Pleistocene femora (N = 40) provides least squares estimates of:

Biomechanical length = (1.12 × SP92Len) + 31.4; r^2 = 0.901; Palomas 92 Biomech Len: 372.3 ± 9.7 mm; SE_{est}: 2.6%.

Bicondylar length = (1.17 × SP92Len) + 37.3; r^2 = 0.900 ; Palomas 92 Bicond Len: 393.9 ± 10.2 mm; SE_{est}: 2.6%.

Maximum length = (1.17 × SP92Len) + 40.7; r^2 = 0.902 ; Palomas 92 Max Len: 397.0 ± 10.0 mm; SE_{est}: 2.5%.

[b]The distance from the proximal neck to the average of the distal condyles measured parallel to the diaphyseal axis (Ruff and Hayes 1983).

[c]The lateral condyle depth of Palomas 92 measures 59.5 mm, and it sustained only trivial marginal erosion. It is therefore rounded off to 60 mm. The anteroposterior and superoinferior diameters of the femoral head were estimated from the lateral condylar depth (posterior lateral condyle to anterior lateral patellar surface margin; 60 mm) using least squares regressions based on a pooled recent human (N = 41) and Late Pleistocene fossil human (N = 6) sample.

Anteroposterior diameter = 0.641 × LatCondDep + 5.8, r^2 = 0.881; estimated diameter: 44.2 ± 1.6 mm; SE_{est} = 3.6%.

Superoinferior diameter = 0.620 × LatCondDia + 7.1, r^2 = 0.849; estimated diameter: 44.3 ± 1.8 mm; SE_{est} = 4.1%.

[d]The bicondylar length was incorrectly provided as the maximum length in Walker et al. (2011b). Both are estimated from the maximum trochanteric length of 374 mm.

TABLE 13.28. Cross-sectional geometric parameters of the Palomas 52 left femur, the Palomas 92mm right femur, and the Palomas 96 right femur

PARAMETER	PALOMAS 52 ≈65%	PALOMAS 92 ≈65%	PALOMAS 96 ≈50%	PALOMAS 96 ≈65%
Total area (TA)	386.1	534.4	428.5	414.9
Cortical area (CA)	326.9	454.2	283.5	365.5
Ant-post second moment of area (I_x)	11,800	25,935	14,158	12,402
Med-lat second moment of area (I_y)	11,423	19,222	15,017	14,771
Maximum second moment of area (I_{max})	12,294	26,499	16,739	14,960
Minimum second moment of area (I_{min})	10,929	18,657	12,437	12,213
Polar moment of area (J/I_p)	23,223	45,157	29,175	27,173

Note: Each of the parameters were taken at approximately 65% (midproximal) of the biomechanical lengths of the bones plus ≈50% of length for Palomas 96. Orientations are based on placing the linea aspera midposterior. Areas are in mm², and second moments of area are in mm⁴.

TABLE 13.29. Osteometric and cross-sectional dimensions of the Palomas 13 and 96 left fibulae at approximately midshaft and the Palomas 96 left tibia at middistal (35%) and midshaft (50%)

DIMENSION	PALOMAS 13 FIBULA ≈50%	PALOMAS 96 TIBIA ≈35%	PALOMAS 96 TIBIA ≈50%	PALOMAS 96 FIBULA ≈50%
Maximum length (M-1a)	—	304.0		—
Lateral total length (M-1)	—	301.5		—
Lateral articular length (M-2)	—	285.0		—
Medial retroversion angle (M-12)	—	(22.0°)		
Maximum diameter (M-2; M-8)	14.6	—	26.6	—
Minimum diameter (M-3; M-9)	10.7	—	(20.2)	—
Total area (TA)	102.6	364.9	420.3	89.8
Cortical area (CA)	93.7	274.6	332.4	79.2
Maximum second moment of area (I_{max})	1,137	11,488	19,375	731
Minimum second moment of area (I_{min})	647	8,912	9,485	580
Polar moment of area (J/I_p)	1,784	20,400	28,860	1,311

Note: I_x and I_y are not included, as the orientations of the cross-sections are approximate. Measurements are in millimeters, areas are in mm², and second moments of area are in mm⁴. Dimensions in parentheses are estimated.

TABLE 13.30. Inventory of the left pedal remains of the young adult Palomas 92 partial skeleton

SP NO.	FIELD NO.	IDENTIFICATION	EXCAVATION DATE	COMMENTS
92a	SP05H004	Cuboid	31 July 2005	Complete
92b	SP05H004	Lateral cuneiform	31 July 2005	Complete
92c	SP05H004/008	Metatarsal 1	31 July 2005	3 fragments
92d	SP05H004/005/011	Metatarsal 2	31 July 2005	3 fragments
92e	SP05H004/005	Metatarsal 3	31 July 2005	2 fragments
92f	SP05H004/005	Metatarsal 4	31 July 2005	2 fragments
92g	SP05H004	Metatarsal 5	31 July 2005	≈ Complete
92h	SP05H005	Pedal proximal phalanx 2	31 July 2005	Complete
92i	SP05H005	Pedal proximal phalanx 3	31 July 2005	≈ Complete
92j	SP05H005	Pedal proximal phalanx 4	31 July 2005	≈ Complete
92k	SP05H005	Pedal proximal phalanx 5	31 July 2005	Complete
92l	SP05H007	Pedal middle phalanx 3	31 July 2005	Complete
92m	SP05H007	Pedal middle phalanx 4	31 July 2005	Complete
92n	SP05H005	Pedal middle phalanx 5	31 July 2005	≈ Complete
92o	SP05H011	Lateral hallucal sesamoid	31 July 2005	Complete
92p	SP05H004	Medial hallucal sesamoid	31 July 2005	Complete
92tt	SP05H007	Pedal distal phalanx 3	31 July 2005	Incomplete

Note: The elements are organized anatomically; hence, the SP numbers, field numbers, and dates of excavation are not in sequence. All are from the Upper Cutting level 2h.

TABLE 13.31. Measurements and a discrete observation of the Palomas 92 left cuboid (92a) and lateral cuneiform (92b) bones, in millimeters

CUBOID		LATERAL CUNEIFORM	
Plantar maximum length	30.3	Maximum height	24.9
Dorsal mid articular length	21.7	Dorsal length (M-1)	18.2
Dorsal lateral articular length	12.0	Maximum breadth	13.7
Dorsal medial articular length	20.6	Navicular articular height	16.2
Maximum height	25.8	Navicular articular breadth	11.7
Metatarsal 4/5 articular height	16.8		
Navicular facet	(present)		

TABLE 13.32. Measurements of the Palomas 92 (92c to 92g) left metatarsal bones (MT), in millimeters and degrees

MEASUREMENT	MT-1	MT-2	MT-3	MT-4	MT-5
Maximum length (M-1)	—	(72.0)	—	—	71.3
Mid articular length	53.0	68.5	—	—	65.2
Medial articular length					60.7
Lateral articular length					66.3
Shaft height (M-4)	—	8.9	—	—	10.0
Shaft breadth (M-3)	—	7.2	—	—	7.5
Proximal max height	—	20.3	—	—	—
Proximal max breadth	—	15.6	12.0	15.6	—
Tuberosity height					15.3
Torsion angle			—	—	(10°)

Note: Dimensions in parentheses are estimated.

TABLE 13.33. Measurements of the Palomas 92 (92h to 92k) left pedal proximal phalanges (PPP), in millimeters

MEASUREMENT	PPP-2	PPP-3	PPP-4	PPP-5
Maximum length	26.5	23.9	23.5	21.1
Articular length (M-1a)	24.4	21.6	(20.1)	19.0
Shaft height (M-3)	6.1	5.9	—	5.5
Shaft breadth (M-2)	7.1	6.8	(7.0)	(6.8)
Proximal max height (M-3a)	10.6	—	—	10.4
Proximal max breadth (M-2a)	11.6	—	—	—
Proximal artic height	—	—	—	9.3
Proximal artic breadth	—	—	—	8.8
Distal height	6.0	—	—	6.4
Distal breadth	8.9	9.0	9.2	9.7

Note: Digit and side are based on *in situ* articulation. Dimensions in parentheses are estimated.

TABLE 13.34. Measurements of the Palomas 67 pedal middle phalanx (PMP) and the Palomas 92 left middle and distal pedal phalanges (PMP and PDP), in millimeters

MEASUREMENT	PALOMAS 67 PMP-2–4	PALOMAS 92L PMP-3	PALOMAS 92M PMP-4	PALOMAS 92N PMP-5	PALOMAS 92TT PDP-3
Maximum length	15.5	12.9	13.1	8.7	—
Articular length (M-1a)	12.8	12.3	12.3	8.2	—
Shaft height (M-3)	6.2	(3.4)	4.0	3.7	—
Shaft breadth (M-2)	9.4	6.5	6.7	8.3	—
Proximal max height (M-3a)	10.4	7.3	7.3	6.7	—
Proximal max breadth (M-2a)	12.2	9.8	9.6	9.7	8.1
Proximal articular height	8.6	6.1	5.9	—	—
Proximal articular breadth	11.6	8.9	9.0	—	—
Distal height	5.8	3.6	3.8	4.0	—
Distal breadth	6.1	8.3	8.5	8.6	—

Note: Dimensions in parentheses are estimated.

TABLE 13.35. Measurements of the Palomas 92 hallucal sesamoid bones, in millimeters

MEASUREMENT	92P: MEDIAL	92O: LATERAL
Length	9.2	10.5
Breadth	6.5	7.0
Thickness	—	4.5

TABLE 13.36. Measurements of the Palomas 66 and 86 immature pedal proximal phalanges, in millimeters

MEASUREMENT	PALOMAS 66	PALOMAS 86
Maximum intermetaphyseal length	15.0	—
Midshaft height	3.8	3.6
Midshaft breadth	5.0	5.9
Proximal metaphyseal breadth	8.7	—
Distal height	4.1	—
Distal breadth	6.8	—

ERIK TRINKAUS

THE DISCUSSIONS of the Palomas human remains, and particularly those regarding the postcrania, have noted the diminutive dimensions of
these humans. These observations include the
supraorbital tori (chapter 4), the mandibles (chapter 5), some of the teeth (chapter 7), the appendicular articular and diaphyseal dimensions, and
especially the long-bone lengths of Palomas 92
and 96 (chapter 13). Even though many of these
assessments are evident in bivariate plots concerning skeletal proportions, and the dental ones
in terms of crown diameters, it is appropriate to
reevaluate some of these skeletal dimensions in
the context of Middle and Late Pleistocene western Eurasian human remains. It is also relevant
to assess, to the extent possible, the body proportions of Palomas 92, with comments on the more
complete data available for Palomas 96.

Body Length

The best indicators of body length (or stature) for
which the Palomas individuals provide data are
the lengths of the humeri and femora. Reliable
ones are available only for the two associated skeletons from Conglomerate A, Palomas 92 and 96
(Tables 13.5 and 13.27). Both of the measurements
are strongly correlated with their original statures (see references in Table 14.1). Comparisons

of these proximal limb segment lengths to later
Pleistocene samples (Fig. 14.1) place them at or
below the bottoms of those distributions. They
are well below all of the other humeral lengths,
and their femoral lengths are matched only by the
estimate for the early Middle Pleistocene Gesher
Benot Ya'acov 1 and the values for the earlier
Upper Paleolithic Nahal 'En-Gev 1 and Předmostí
1. With respect to the Neandertal sample, Palomas 92 and 96 are ≈2.9 and ≈2.5 standard deviations, respectively, below the Neandertal mean
for humeral length (309.3 ± 15.1 mm, n = 12). For
the femur they are, respectively, ≈2.1 and ≈2.2
standard deviations below the Neandertal mean
(440.5 ± 22.5 mm, n = 12).

The Palomas 92 partial skeleton also provides articular length measurements for three of
the metatarsals: the first (53.0 mm), the second
(68.5 mm), and the fifth (65.2 mm) (Table 13.32).
Given reasonably strong correlations of foot bone
lengths with stature (Byers et al. 1989; Pablos
et al. 2013), they should provide an indication of
relative body length, even though comparative
Late Pleistocene samples are small. With respect
to other Neandertals, the Palomas 92 metatarsal
1 length is between those of the Tabun 1 and La
Ferrassie 2 females and below those of four other
individuals (La Ferrassie 1, Kiik-Koba 1, Krapina
245, and Shanidar 1) (Fig. 14.2). The metatarsal 2 length exceeds only those of La Ferrassie

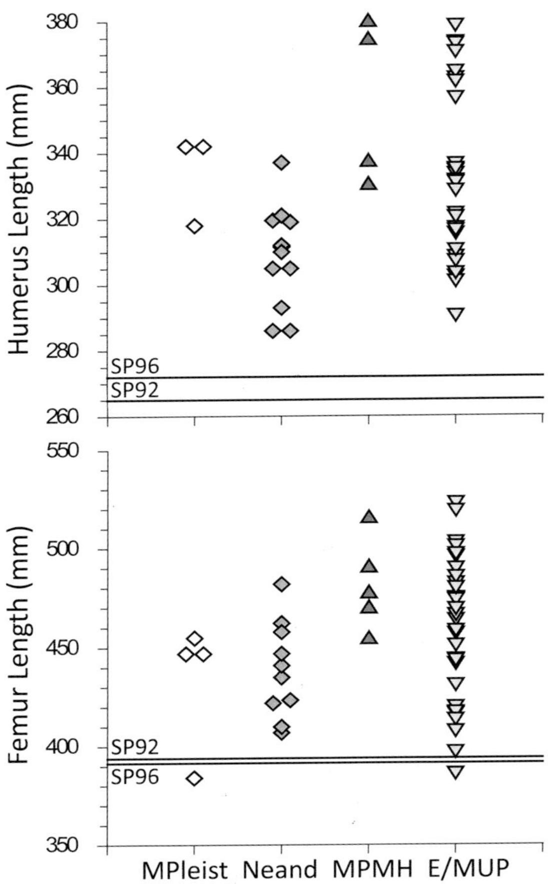

FIG. 14.1. Humerus maximum and femoral bicondylar lengths for Palomas 92 and 96 (using the modal estimates for their lengths) versus values for Middle Pleistocene (MPleist), Neandertal (Neand), Middle Paleolithic modern human (MPMH), and earlier Upper Paleolithic modern human (E/MUP).

2 and Shanidar 6, and the metatarsal 5 length is between the low values for La Ferrassie 2 and Shanidar 8 and the remainder of the Neandertal values. These limited comparisons confirm the humeral and femoral assessments placing the body length of Palomas 92 among the shortest of the Neandertals.

The only other Palomas specimens for which an appendicular indication of body length can be assessed are the manual middle and distal phalanges of Palomas 28 and 65. There are difficulties in assessing body length from manual phalanges; correlations between finger lengths and stature are similar to those using long bone

lengths (Musgrave and Harneja 1978; Meadows and Jantz 1992; Sen et al. 2014), but there have been changes in the relative lengths of the manual phalanges across Pleistocene humans (Villemeur 1994). The principal changes in phalangeal length proportions involve the proximal versus distal ones, in which the proximal ones are generally shorter and the distal ones longer among the Neandertals relative to modern humans; mean values for the isolated Atapuerca-SH phalanges provide proportions similar to those of the Neandertals (Trinkaus 2016a), and the undated archaic *Homo* Dinaledi hand (Kivell et al. 2015) furnishes ulnar digit proportions between those of the Neandertals and early modern humans. It is nonetheless possible to compare the lengths of the Palomas 65 middle phalanx (MMP) and the Palomas 28 distal phalanx (MDP), plus those of Palomas 92 and 96, to those of middle and distal phalanges from the middle three digits, thereby providing body length inferences for additional individuals at Palomas.

In manual middle phalangeal comparisons, however, digit assignment within rays 2 to 4 is only possible for associated series of middle phalanges (3 > 4 > 2) and not reliable for distal ones. The phalangeal lengths are therefore pooled, only averaging right and left values for those Late Pleistocene

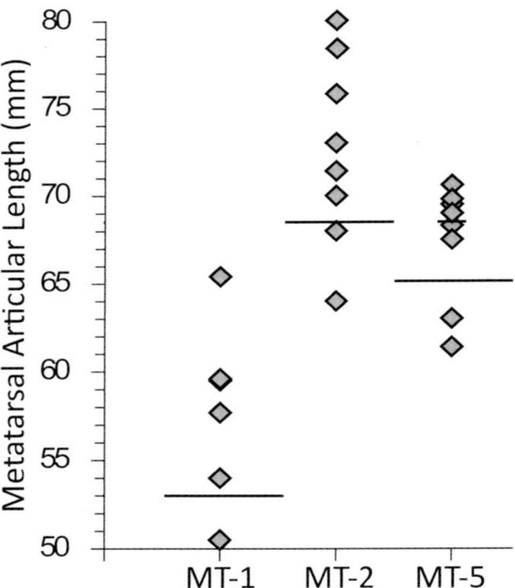

FIG. 14.2. Palomas 92 metatarsals 1, 2, and 5 articular lengths versus those available for the Neandertals.

individuals that preserve antimeres. The Krapina Neandertal phalanges are plotted separately, given their lack of association, as with the Atapuerca-SH sample. In the resultant comparison (Fig. 14.3), Palomas 65 and the Palomas 96 MMP-2 are at the bottoms of the variation, matched only by a couple of isolated Atapuerca-SH and Krapina phalanges and approached by phalanges of La Ferrassie 2 and Shanidar 3. This comparison strongly suggests that Palomas 65 derives from the second digit and had a body size similar to that of Palomas 96. The Palomas 92 MMP-3 is close to the average value for the same digit of Palomas 96. Among the Neandertal MMP-3s, it only exceeds the values of La Ferrassie 2 (24.3 mm) and Tabun 1 (24.2 mm) and is approached by that of Regourdou 1 (25.4 mm).

The sufficiently complete manual distal phalanges, Palomas 28 and two from Palomas 96, have lengths in the middles of the Middle Pleistocene and Upper Paleolithic samples but are among the smaller of the Middle Paleolithic ones (Fig. 14.3; see also Fig. 13.11). The Palomas 28 phalanx, with a length of 17.5 mm, is small for a Neandertal distal phalanx, exceeding only Krapina 206.7 and one each from La Ferrassie 2 and Shanidar 4. Its length is greater than most of the Upper Paleolithic ones (median: 17.4 mm), but that is a product more of relative phalangeal lengths than of stature.

Neither of the Palomas skeletons is sufficiently preserved to estimate trunk length (or skeletal trunk height). The more caudal vertebrae of Palomas 92 appear to have relatively high bodies, similar to those of a few male Neandertals. However, the sacral lengths of both Palomas partial skeletons are considerably smaller (chapter 13).

These various indicators of body length for the Palomas 92 and 96 partial skeletons and the two isolated manual phalanges indicate that they were among the shortest of the known Neandertals. They are most similar to the La Ferrassie 2 and Tabun 1 females and the probably female Shanidar 6 and 8 individuals. Most important, their modal humeral and femoral lengths are >2 standard deviations below the Neandertal means and outside of the known range of the other Neandertal values. Indeed, among Middle and Late

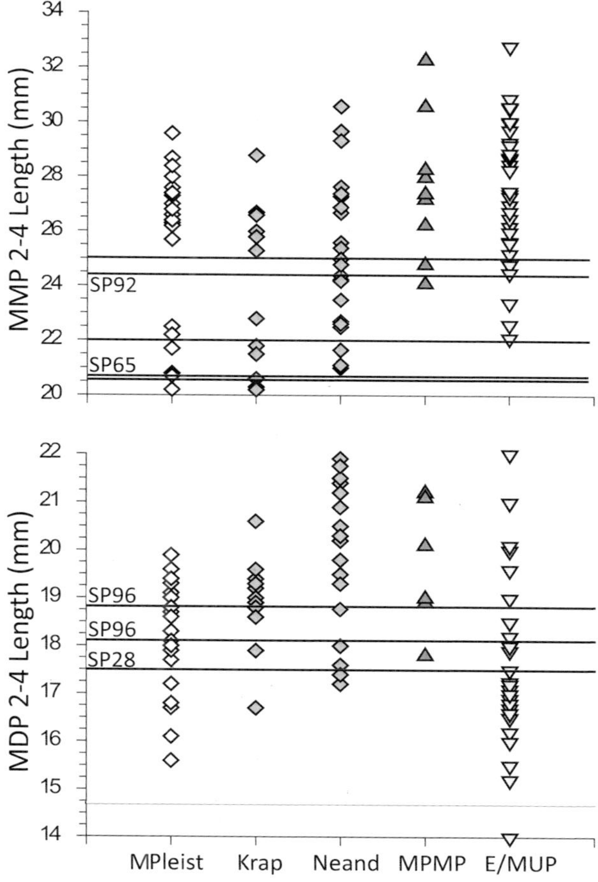

FIG. 14.3. Manual middle phalanx (MMP) 2–4 articular lengths for Palomas 65, 92, and 96 (*above*) and manual distal phalanx (MDP) 2–4 articular lengths for Palomas 28 and 96 (*below*) versus Middle Pleistocene (MPleist), Krapina (Krap), early last glacial Neandertal (Neand), Middle Paleolithic modern human (MPMH), and earlier Upper Paleolithic modern human (E/MUP) values. The unlabeled lines are for the Palomas 96 phalanges. For Palomas 96 and a number of the Late Pleistocene specimens, multiple values from the same individual are included; the numbers of individuals are 11 for the Neandertals, 3 for the MPMH, and 20 for the E/MUP. Right and left lengths, when available for a given digit of an individual, were averaged to a single value for that digit (including the MMP-3s of Palomas 96). The level of variation is exacerbated by including phalanges from digits 2 to 4, given the difficulties in assigning digit to Palomas 28 and 65, other isolated phalanges, and incomplete associated series.

Pleistocene western Eurasians, only two individuals have shorter long bones: the early Middle Pleistocene Gesher Benot Ya'acov 1 and the later Mid Upper Paleolithic Nahal 'En-Gev 1. The latter is mature (Arensburg 1977), but the former is only a diaphysis (Geraads and Tchernov 1983) and may be immature.

Stature Estimation

It is appropriate in this context to estimate the statures of Palomas 92 and 96 from their lower limb bone lengths, the femora of both individuals, and the tibia of the latter. There are diverse formulae available, based on cadavers or through the reconstruction of stature using the Fully technique (Table 14.1 and references therein). Given ecogeographic variation in recent and Late Pleistocene human body proportions, including both limb segment lengths to trunk length (Holliday 1995) and segment lengths within limbs (Trinkaus 1981), it should be best to match the Palomas body proportions to those of the recent human samples providing the regression formulae. There are also differences in proportions between males and females. For these reasons, stature was estimated for Palomas 92 and 96 using a range of reference samples and using female-based formulae (given the female sex of Palomas 96 and the likely female nature of Palomas 92, given its very small body size). The resultant mean estimates (Table 14.1) provide a range of ≈148 cm to ≈152 cm for both individuals from their femoral lengths. The tibia of Palomas 96, however, provides a lower range, mostly between ≈142 cm and ≈146 cm, with one higher value. The contrast between the femoral and tibial estimates for Palomas 96 is due to her low crural index of 77.7; it is unclear why the Native American arctic sample, with low crural indices, does not provide more similar femoral and tibial values for Palomas 96.

These estimates indicate that the original statures of these two Palomas individuals were in the vicinity of 150 cm. This value is below the female mean statures of 93.5% (n = 31) midlatitude and northern Eurasian (>30° N) modern human samples, the two shorter ones being of arctic Siberian natives (Gustafsson and Lindenfors 2009). Palomas 92 and 96 were therefore of short stature relative to both other Neandertals and higher-latitude modern humans.

Upper Limb Dimensions

There are limited dimensions of the Palomas upper limbs that are preserved, provide body size indications, can be compared across Middle and Late Pleistocene samples, and include specimens other than Palomas 92 and 96. These measures of size include left humeral midshaft dimensions, second metacarpal head diameters, and (for Palomas 92) humeral distal articular breadth. The more relevant ones are the articulation dimensions, because they are influenced by laterality less than diaphyseal measures (Trinkaus et al. 1994; Auerbach and Ruff 2006).

The midshaft cross-sectional parameters for Palomas 16 and 96 humeri place the former towards the middle of the Neandertal nondominant side humeri and the latter at its lower limit (Fig. 13.9). The articular diameters of the Palomas 15 and 92 second metacarpal heads (Table 13.16) are small relative to a modest Neandertal sample: 13.0 mm and 12.5 mm versus 15.8 ± 1.5 mm (n = 8) for height and 13.5 mm and ≈13.5 mm versus 16.2 ± 0.9 mm (n = 6) for breadth.

The humeral distal articular breadth of Palomas 92 places it at the bottom of the comparative samples' size ranges (Fig. 14.4). Two Neandertals provide smaller dimensions (Shanidar 6 and Tabun 1), and one (Krapina 166) is modestly larger. The Mid Upper Paleolithic values for Nahal 'En-Gev 1 and Předmostí 10 are slightly higher. To the extent that this cubital dimension reflects overall body size and arm loading regimes, it reinforces the small size of Palomas 92 for a later Pleistocene human.

Body Mass

The best skeletal reflections of body mass are the articulations of the lower limb as well as the cross-sectional areas of the femora and tibiae

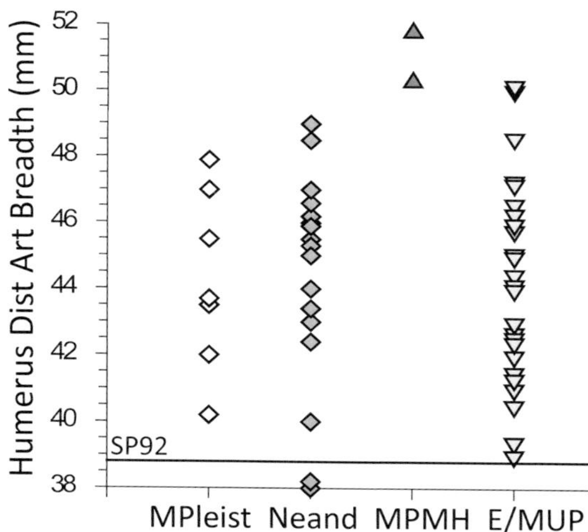

Fig. 14.4. Humerus distal articular breadth for Palomas 92 versus values for Middle Pleistocene (MPleist), Neandertal (Neand), Middle Paleolithic modern human (MPMH), and earlier Upper Paleolithic modern human (E/MUP).

can be estimated for many Pleistocene humans, including Palomas 92 and 96 (Table 14.1), despite ecogeographical body proportion variation (see below). However, the latter is poorly known for pre–Upper Paleolithic humans (Trinkaus 2011b), is unknowable for Palomas 92 and the isolated Palomas remains, but may eventually be known for Palomas 96 (Walker et al. 2011b).

For these reasons, body mass is estimated, following the protocol of Ruff (2010) and using the recent human formulae of Ruff et al. (1991), McHenry (1994), and Grine et al. (1995), from femoral head diameters for three Palomas individuals (Table 14.2). In addition, skeletal proxies for body mass are compared across the Middle and Late Pleistocene comparative samples (Figs. 14.5 and 14.6).

The mean estimates of body mass for Palomas 77, 92, and 96 from their directly measured or estimated femoral head diameters (Table 13.27) are provided in Table 14.2. Two of the regression formulae are based on pooled sex samples, but the analysis by Ruff et al. (1991) provides separate ones for females and males. Since Palomas 96 is female, it is the average using the female estimate (59.9 kg) that is most appropriate. Technically, the estimates averaging the female and male values should be used for Palomas 77 and 92 (64.9 and

(Ruff et al. 1993; Auerbach and Ruff 2004). These biomechanical bases for body mass assessment reflect the habitual loads on the weight-bearing lower limbs, since they provide resistance to those loads through the surface areas of the synovial joint hyaline cartilage and subchondral bone, the trabecular volumes of the epiphyseal regions, and the volume of diaphyseal bone resisting axial loads. The majority of those loads should be generated by body mass, directly from gravity and indirectly from body momentum. They are also augmented by muscular contractions. As a result, comparative assessments of Pleistocene estimated body masses, or the use of lower limb skeletal measures as a proxy for it, assume similar levels of habitual activity across the samples in question. Current analyses of Pleistocene human postcrania (for example, Holt 2003; Cowgill 2010, Trinkaus and Ruff 2012; Chirchir et al. 2015; Trinkaus 2015) strongly support such an assumption, even though the scaling of lower-limb hypertrophy is largely based on such body mass estimates (Trinkaus and Ruff 2000). It is also possible to estimate body mass using a geometric model of stature and body breadth (Auerbach and Ruff 2004; Ruff et al. 2012), but it requires reliable estimation of stature and a bi-iliac breadth. The former

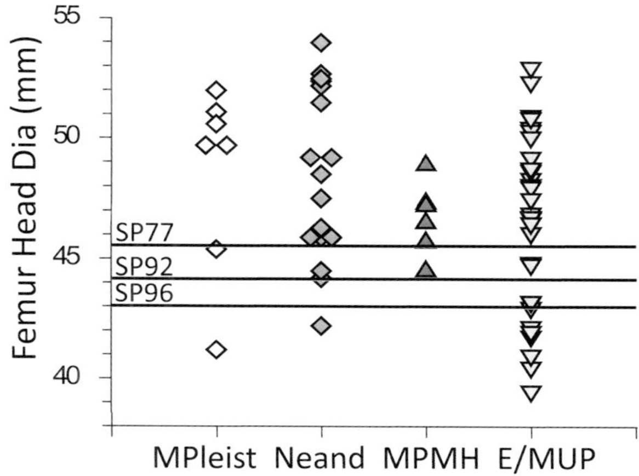

Fig. 14.5. Femur head diameters for Palomas 77, 92, and 96 versus values for Middle Pleistocene humans (MPleist), Neandertals (Neand), Middle Paleolithic modern humans (MPMH), and earlier Upper Paleolithic modern humans (E/MUP).

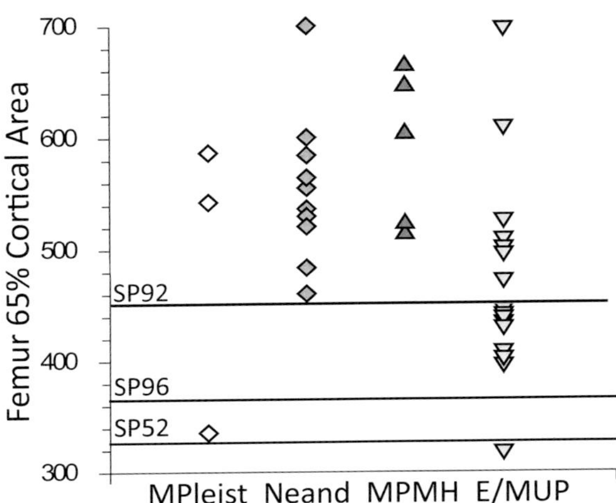

FIG. 14.6. Femur mid-proximal (65%) cortical areas for Palomas 52, 92, and 96 versus values for Middle Pleistocene humans (MPleist), Neandertals (Neand), Middle Paleolithic modern humans (MPMH), and earlier Upper Paleolithic modern humans (E/MUP).

61.7 kg, respectively), since they lack diagnostic pelvic remains. Yet their small body dimensions relative to other Neandertals would strongly argue for their being female, which would provide slightly higher mean estimates. In either case, the differences between the two sets of values are well within the estimation errors of the individual regression formulae, so it makes little difference.

These mean body mass estimates from the Palomas femoral heads are similar to those of three of the four pelvically sexed Neandertal females; the unsexed Krapina 208 hip bone and the Krapina 214 femur provide similar estimations. Yet these estimates are also close to the values of one definite male (Kebara 2) and one probable male (Regourdou 1) (Table 14.3).

If the femoral head diameters are compared across the Middle and Late Pleistocene samples (Fig. 14.5), the three Palomas specimens similarly fall among the smaller of the Neandertals as well as with the smaller of the Middle Pleistocene and the Middle Paleolithic modern human distributions. Their dimensions, however, are closer to the middle of the earlier Upper Paleolithic modern human sample (median: 47.5 mm;

46.7 ± 3.6 mm, n = 33). A more marked comparison is evident in the assessment of femoral cross-sectional cortical area at the midproximal (65%) level (the level for which three Palomas specimens provide data) (Fig. 14.6). Palomas 92 is at the bottom of the Middle Paleolithic range and well within the earlier Upper Paleolithic one, but Palomas 52 and 96 are much lower. Only two femora match the low values of these smaller Palomas individuals: the earlier Middle Pleistocene Gesher Benot Ya'acov 1 and the Upper Paleolithic Dolní Věstonice 3.

These different skeletal indicators, along with the body mass estimations, support what is evident in the assessments of body lengths. The Palomas individuals were small, at the low end of the Neandertal and general later Pleistocene ranges of variation. Moreover, although there are multiple indicators for Palomas 92 and 96, the isolated elements providing data relating to body mass follow the same pattern.

Facial Size

The craniomandibular remains from the Sima de las Palomas are all fragmentary, and even the largely complete one of Palomas 96 was extensively crushed and abraded. Yet four adult mandibular corpori provide a general indicator of overall size. The only one that approximates overall length, Palomas 1, lacks the condylar area and was extensively broken, with pieces rearranged *in situ*. Visual reconstruction of the mandible (Plate 5.4) provides a very crude superior length (condyles to infradentale) estimate in the vicinity of ≈110 mm. This value is in the middle of the Neandertal range of variation (100–122 mm, 109.5 ± 6.5 mm, n = 16), serving only to establish that the facial length of Palomas 1 was not exceptional for a Neandertal. The other measurement reflecting facial size is lateral corpus height at the mental foramen (Fig. 14.7). Even though the few Middle Paleolithic modern humans have high values, there is little difference across the other three samples. Palomas 1 has the highest of the Palomas corpori, its measure of ≈30.5 mm falling

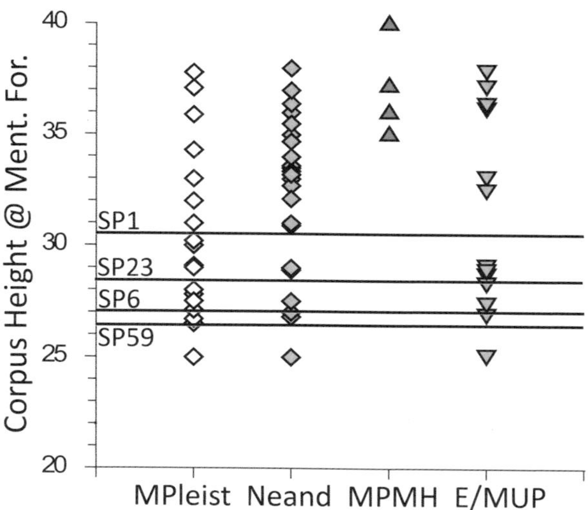

FIG. 14.7. Mandibular corpus heights at the mental foramen (in millimeters) for Palomas 1, 6, 23, and 59 versus values for Middle Pleistocene humans (MPleist), Neandertals (Neand), Middle Paleolithic modern humans (MPMH), and earlier Upper Paleolithic modern humans (E/MUP).

modestly below the Neandertal mean (32.5 ± 3.5 mm, n = 28). Palomas 6 and 59 (≈27.0 and 26.4 mm), however, are among the mandibles with the shortest corpori: Krapina 57, La Naulette 1, and Tabun 1. The Palomas 23 height (28.4 mm) is between these values. The sample of Palomas corpus heights is nonetheless significantly different from the Neandertal sample (t-test p = 0.005). The small body size of the southeastern Iberian Neandertals therefore appears to be reflected as well in their mandibular dimensions.

Body Proportions

An assessment of body proportions is only possible for Palomas 92, joined by the more complete appendicular remains of Palomas 96. The former provides estimates of humeral and femoral lengths and femoral head diameter. The latter furnishes lengths for all four limb segments and the clavicle plus femoral head diameter.

Given the preservation of Palomas 92, humeral and femoral lengths (as reflections of limb lengths) have been plotted against femoral head diameter

(reflecting body mass) (Fig. 14.8). In both of the distributions there is modest overlap of the Neandertals and the earlier Upper Paleolithic modern humans, with the former having on average relatively shorter proximal limb segments. The Middle Paleolithic modern humans have the relatively longest humeri and femora, distinct from those of the Neandertals. The proportions of humeral length to femoral head diameter of both Palomas

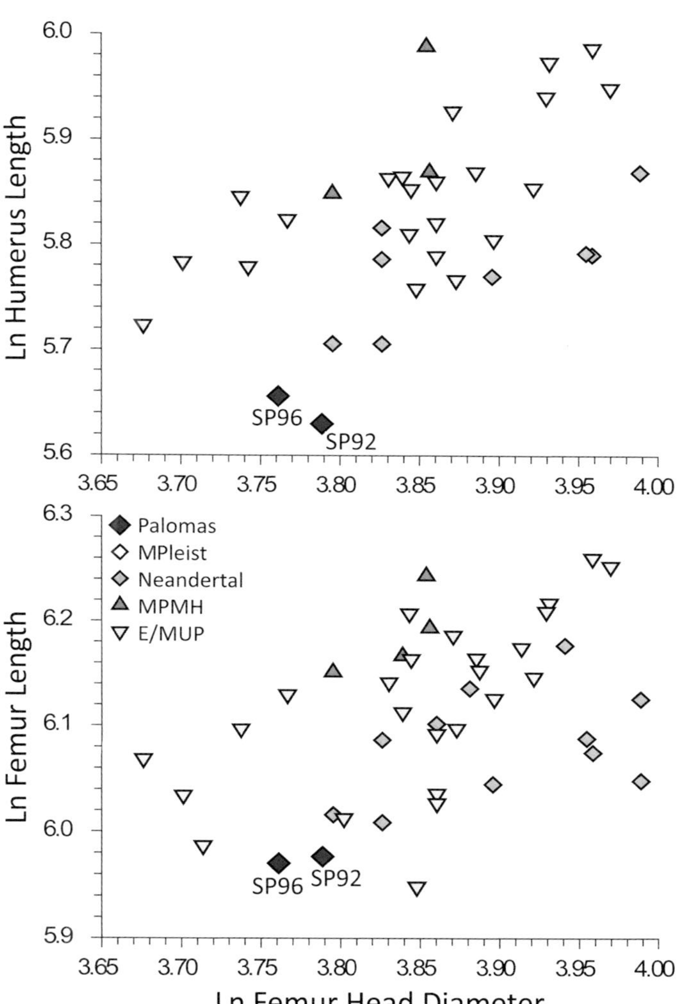

FIG. 14.8. Comparisons of humerus maximum and femoral bicondylar lengths (reflecting proximal limb segment lengths) to femoral head diameters (reflecting body core size) for Palomas 92 and 96 versus values for Middle Pleistocene humans (MPleist), Neandertals, Middle Paleolithic modern humans (MPMH), and earlier Upper Paleolithic modern humans (E/MUP).

partial skeletons are similar to those of the Neandertals, continuing their distribution to a smaller body size. The measurements for Palomas 96 are secure. Even if the estimates for Palomas 92 were shifted 10% for each dimension, it would remain closer to the other Neandertals and separate from the early modern humans. In the femoral comparison, both Palomas femoral lengths are relatively short but overlap both the Neandertal and the Upper Paleolithic modern human variation.

To these considerations can be added the limb proportional indices of Palomas 96 (residuals from bivariate distributions provide similar results [Walker et al. 2011b; Trinkaus et al. 2014b]). Its brachial and crural indices of 75.7 and 77.7 are among those of the Neandertals (brachial: 74.8 ± 3.0, n = 10; crural: 79.0 ± 2.2, n = 7). The Neandertal brachial indices overlap extensively with those of the Upper Paleolithic modern humans (77.8 ± 2.4, n = 20) but are still significantly lower (p = 0.016). The former's crural indices are substantially lower than those of the latter (84.8 ± 2.0, n = 20) (p = 0.0001). The Middle Paleolithic modern humans provide three divergent brachial indices (65.7, 76.4, and 81.3) and one very high crural index (89.6).

Palomas 96 also has a high claviculohumeral index (≈52.5), similar to those of the few other Neandertals providing it (La Ferrassie 1: 52.1; Kebara 2: 51.6, Regourdou 1: 51.9; Shanidar 3: ≈51.7). It is at the top of the earlier Upper Paleolithic modern human range (46.4 ± 3.6, n = 19), and the Middle Paleolithic modern human values (40.7, 45.1, and 49.7) are similar to the Upper Paleolithic ones. However, when clavicle length is scaled to body mass, the differences across these samples disappear, and Palomas 96 falls with the other Late Pleistocene (and recent) humans (Trinkaus et al. 2014b). It is the modest humeral lengths of Palomas 96 and other Neandertals (Fig. 14.8) that produce their high claviculohumeral indices. Of the three Upper Paleolithic ones with high indices, Mittlere Klause 1 has a relatively short humerus, whereas Dolní Věstonice 16 and especially Sunghir 1 have long clavicles.

Summary and Discussion

These considerations of the body size and proportions of the Palomas Neandertals indicate that the body proportions of the Palomas Neandertals closely match those seen in more northern European Neandertals and southwest Asian ones. These body proportions are extensively evident in the long bones of Palomas 96, and they are apparent in the less complete remains of Palomas 92. They are reflected in their humeral and femoral lengths relative to body mass (or femoral head diameter), and in the brachial, crural, and claviculohumeral indices of Palomas 96.

At the same time, most of the Palomas remains cluster among the smallest of the other Neandertals, in particular the pelvically sexed female La Ferrassie 2 and Tabun 1 partial skeletons. In addition to Palomas 92 and 96, , this small body size can be inferred from the isolated Palomas 15, 28, 52, 64, 65, and 77 skeletal elements. To the extent that mandibular corpus dimensions reflect overall body size, Palomas 6, 23, and 59 can be added to this group of small individuals. A couple of the elements are somewhat larger, including Palomas 1 and 16, but they nonetheless fall in the middles of the Neandertal ranges of variation. None of them approach the dimensions seen in the larger Neandertals, including the Iberian ones represented by Oliveira 7, El Sidrón 1609, and Zafarraya 1.

There are two interpretations of this generally small size of the Palomas remains. They could represent a physically small population (or local lineage, since they span considerable time; see chapter 2). Alternatively, the sample could have a disproportionate number of females. Palomas 96 is pelvically sexed as female, more reliably than any other archaic *Homo* specimen (Walker et al. 2011b). The others can only be sexed based on size. Despite the caveats in chapter 13 (see also Trinkaus 1980), it is tempting to assign probable sex to the various Palomas remains. If one were to do this, the small ones (Palomas 6, 15, 23, 28, 52, 59, 64, 65, 77, 92, and 96) would make up the "female" sample, and the other two (Palomas 1 and 16) would become "male" (even though they are not very large compared to pelvically sexed male Neandertals elsewhere). The resultant sex ratio is rather different from an expected 50:50 proportion (binomial p = 0.010). If the two supraorbital tori (Palomas 11 and 12) are added as "female" and "male," respectively, given their contrasting thicknesses, the difference decreases slightly (p = 0.014). It may well be that the generally small size of these Iberian Neandertals is a combination of an overrepresentation of females in the sample combined with generally small dimensions for these people.

TABLE 14.1. Stature estimates for Palomas 92 and 96

COMPARATIVE GROUP	PALOMAS 92 FEMUR	PALOMAS 96 FEMUR	PALOMAS 96 TIBIA	SOURCES
Europeans (North)	150.3	149.5	145.7	1
Europeans (South)	150.3	149.5	142.4	1
Euro-Americans	152.1	151.5	149.0	2
Native Americans (Arctic)	148.7	148.2	144.9	3
Native Americans (Temperate)	150.0	149.3	142.3	3
Native Americans (Great Plains)	151.2	151.4	143.8	3
Native Americans (Mesoamerica)	150.0	149.5	144.0	4
Egyptians	149.9	149.3	143.1	5
Afro-Americans	150.5	149.9	146.5	2

Sources: (1) Ruff et al. (2012); (2) Trotter and Gleser (1952); (3) Auerbach and Ruff (2010); (4) Genovés (1967); (5) Raxter et al. (2008).
Note: Stature estimates are based on the lengths of their femora and the tibia of Palomas 96 using stature formulae for recent human females, in centimeters.

TABLE 14.2. Mean body mass estimations from femoral head diameters for Palomas 77, 92, and 96

	PALOMAS 77 FEMALE?	PALOMAS 92 FEMALE?	PALOMAS 96 FEMALE
Femur Head Diameter (mm)	45.6	≈44.2	43.0
Body Mass Estimations (kg)			
McHenry (1994)	62.2	59.1	56.4
Ruff et al. (1991) Female	68.0	64.9	62.3
Ruff et al. (1991) Male	63.1	56.6	56.7
Ruff et al. (1991) Average	65.5	62.3	59.5
Grine et al. (1995)	67.0	63.8	61.1
Average Mass (male/female)[a]	64.9	61.7	59.0
Average Mass (female)[b]	65.7	64.7	59.9

Note: These mass estimates use the extant human regression formulae of Ruff et al. (1991), McHenry (1994), and Grine et al. (1995) (see also Auerbach and Ruff, 2004), following the protocol of Ruff (2010) to average the three results for femoral head diameters between 38 and 47 mm.
[a]Employing the average of the female and male results from Ruff et al. (1991).
[b]Employing the female results from Ruff et al. (1991).

TABLE 14.3. Comparative femoral head diameters and mean body mass estimates for Palomas 77, 92, and 96 and other Neandertals

SPECIMEN	FEMORAL HEAD DIAMETER (MM)[1]	BODY MASS ESTIMATE (KG)
Palomas 77	45.6	64.9
Palomas 92	44.2**	61.7
Palomas 96	43.0	59.9
Females		
La Ferrassie 2	45.9	66.4
Krapina 209	44.2*	62.6
Prince 1	49.2*	75.5
Tabun 1	44.5	63.3
Males		
Amud 1	51.5*	79.0
La Chapelle-aux-Saints 1	52.4	81.1
Feldhofer 1	52.2	80.6
La Ferrassie 1	54.0	84.9
Kebara 2	45.9*	64.8
Krapina 207	42.6*	57.1
Regourdou 1	45.9*	64.8
Shanidar 4	49.2	73.4
Uncertain Sex		
Fond-de-Forêt 1	48.5**	72.9
Krapina 208	46.3*	66.5
Krapina 213	52.7	80.9
Krapina 214	44.2	61.2
Shanidar 5	47.5	70.6
El Sidrón 1609	52.5	82.1
Spy 2	54.0	85.5

Note: Comparative Neandertal data from Trinkaus and Ruff (2012), updated as appropriate from Plavcan et al. (2015). The Palomas 77 and 92 values assume that their sexes are uncertain. The single asterisk * indicates that the diameter is estimated from its acetabular dimensions; the double asterisk ** indicates that it is estimated from its femoral condylar depth.

15

The Palomas Mortality Patterns

ERIK TRINKAUS

THERE HAVE BEEN SEVERAL attempts to assess the mortality profiles of Middle and Late Pleistocene humans, in part to evaluate paleodemographic implications but also to provide insight into the plethora of behavioral, mortuary, and taphonomic processes that go into the mortality age profiles evident in individual site assemblages and in pooled regional and temporal samples (for example, Wolpoff 1979; Trinkaus 1995, 2011a; Bocquet-Appel and Arsuaga 1999; Streeter et al. 2001; Zilhão and Trinkaus 2002b; Caspari and Lee 2004; Martinón-Torres et al. 2012; Trinkaus et al. 2014a). With these issues in mind, the age-at-death distribution of the Palomas sample has been assessed. It is based principally on the abundant dental remains, since they provide both the largest sample for the Palomas Neandertals and the best indicators of their ages at death. The three associated skeletons (Palomas 92, 96, and 97) are then added to the dental/mandibular sample.

The descriptions of the individual teeth or series in alveoli (chapter 6), as well as the mandibles (chapter 5), provide assessments of the ages at death per tooth or per dental arcade, based on dental calcification, overall occlusal wear, and dental eruption or occlusion. For the immature individuals, preference is given to dental (crown and root) calcification, but all are assessed with respect to their general degrees of occlusal wear, presence of interproximal wear facets, and (for deciduous teeth) possible root resorption. For the fully formed and erupted permanent teeth, plus the mature mandibles, the assessment is based largely on occlusal wear. Maturity status relies principally on full M3 eruption indicated by M3 position and/or the presence of a distal M2 interproximal facet.

It is also possible to assign approximate ages at death to some of the isolated postcranial remains (chapter 13). However, with the exception of the associated skeleton of Palomas 92 (and those of Palomas 96 and 97), it is unclear whether these postcranial elements (as well as the isolated cranial ones) derive from the same individuals as the isolated teeth. Considerations of the mortality pattern of the Palomas sample is therefore primarily concerned with the dental remains plus the three associated skeletons. For them, Palomas 92 represents a very young adult (chapter 13), Palomas 96 was similar in age at her death (chapter 13; Walker et al. 2011b), and Palomas 97 is from a juvenile (Walker et al. 2012).

It has been possible, based on developmental status, degree of occlusal wear, labial wear patterns, and enamel hypoplasias, to associate a few of the anterior teeth (chapter 11). The most likely association, that of Palomas 73 and 79 I's, does not affect the counts, since Palomas 73 is too incomplete to provide an independent age assessment. If the apparently slightly younger Palomas 19 I$_1$ belongs with them (categorized as juvenile versus adolescent), it would slightly modify the proportions of these immature age groups. The less probable association of Palomas 74 and 82 would reduce the number of juveniles by one but change the overall distribution little.

Age Intervals

There are a variety of ways in which researchers have divided the years of growth into those of infants, children, juveniles, adolescents, and then adults (see Scheuer et al. 2000). The stages used here (Table 15.1) generally follow those of Bogin (1997), whose stages are based on the growth and development of major biological systems and behavioral independence. Given the nature of the fossil remains, dental formation criteria are added to those of Bogin.

Infancy (birth to about 2 years) is the period of total dependence on adults, and it corresponds approximately to the formation of the deciduous dentition. Childhood (about 3 to 6 years) extends to the advent of possible survival without adult feeding and protection and to the beginning of the permanent dentition in occlusion. The juvenile period (about 7 to 12 years) is the period of some independence prior to puberty, and it ends with all but the M3s in occlusion. Adolescence (about 13 to 18 years) is then the period from the beginning of puberty to the attainment of (most) skeletal maturity and behavioral adulthood.

Adulthood is also separated into young versus old adults, with a dividing age at about 40 years. This age represents the approximate transition between prime-age (in physical and reproductive terms) adults and older individuals. The dividing age is approximate, given both individual variation in the aging process and increasing uncertainties with age-at-death assignments to older skeletal remains. With respect to the fossil remains, specimens aged near 40 years are placed in the older category (Table 15.3).

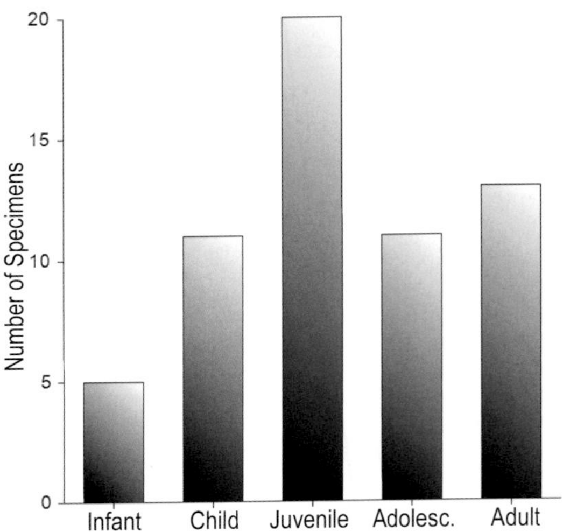

FIG. 15.1. Distribution of Palomas dental, mandibular, and associated skeleton remains by age category (see Table 15.1). Note that some duplication of individuals is undoubtedly present.

A Dominance of Immature Individuals

Given both the number of deciduous teeth and the number of incompletely formed dental roots, one has the impression of a sample dominated by immature individuals. Indeed, of the 60 isolated teeth and mandibles, plus the associated skeletons, 47 (78.3%) come from immature individuals (Table 15.1; Fig. 15.1). If one determines an absolute minimum number of individuals, based on duplication of teeth within age categories but not assessing details of antimeric morphology, one arrives at 17 individuals, 58.8% of whom are immature. Of these 47 immature specimens, 20 (or 42.6%) derive from juveniles (ages between 7 and 12 years).

As such, the Palomas dental sample has a substantially higher proportion of immature individuals than any of the comparative samples (Table 15.2). The Palomas MNI immature percentage is close to that for the Middle Pleistocene Atapuerca-SH sample (60.7% immature, n = 28) (Martinón-Torres et al. 2012). In addition, the probable proportion of immature individuals encompasses the proportion of immature individuals in the Krapina Neandertal sample (69.6%, n = 23), based on the maxillary or mandibular dentitions within age categories (Wolpoff 1979; see MNI sorting in Trinkaus 1995). However, both the Atapuerca-SH and Krapina samples are dominated by adolescent remains (Atapuerca-SH: 50.0%; Krapina: 43.5%), whereas the Palomas sample has an abundance of juvenile remains and only ≈18% adolescents. The Atapuerca-SH and Krapina samples also lack infant remains; excavation techniques of the 1890s may explain the absence of infants in the Krapina sample, but excavation procedures do not account for their absence from Atapuerca-SH. The one other large Neandertal sample, the fragmentary one from Hortus (de Lumley 1973), has a proportion of adults (56.5%, n = 23) closer to that of the overall fossil (and recent human) samples. It is nonetheless interesting that three of these four large samples (Atapuerca-SH, Krapina, and Palomas) are dominated by immature individuals.

The greater percentage of infant remains in the Upper Paleolithic and recent human samples is undoubtedly due to their occurrence in intentional burials (as with 86.6% of the Neandertal infants). The presence of infants in the Palomas sample does not reflect burial (given their occurrence as isolated mandibles or teeth), although one of the three possible burials at Palomas (Palomas 97) is juvenile. There have been suggestions of differential mature versus immature burial in the Upper Paleolithic (Zilhão and Trinkaus 2002b; Trinkaus et al. 2014a), but it does not seem to have affected the immature-mature proportion of the overall sample.

A Dearth of Older Adults

The Palomas sample mortality distribution is also notable for its dearth of older adult individuals, with only one of twelve dentally and pelvically aged individuals (not including Palomas 6) possibly having surpassed 40 years of age (Table 15.3). The one older adult is Palomas 1, who exhibits moderately advanced dental wear (chapters 6 and 10). Yet comparison with Neandertal and early modern human remains that provide nondental skeletal

indicators of their ages at death (Trinkaus 1995, 2011a; Hillson et al. 2006) suggests that the age at death of Palomas 1 may well have been within the fourth decade—younger than the younger to older adult age boundary of about 40 years. Therefore, placing Palomas 1 in the older adult category is conservative relative to recent humans.

It is possible that the dearth of older adults in the Palomas dental sample is due in part to the difficulties inherent in identifying human teeth that have largely or completely lost their crowns through wear, given the extensive abrasion evident on older Neandertal and early modern human dentitions (Heim 1976; Trinkaus 1983; Hillson 2006; Trinkaus et al. 2014a). Yet the inclusion in the Palomas sample of several partial roots (Palomas 33, 46, and 55) suggests that they would have been recognized as human unless markedly altered by pathological processes, such as hypercementosis.

The dearth of older adults, or the preponderance of prime-age adults, in the Palomas sample, however, is not unusual for Pleistocene human samples (Table 15.3), as has been previously noted (Trinkaus 1995, 2011a; Streeter et al. 2001; Trinkaus et al. 2014a). The Palomas sample, along with the small Middle Paleolithic modern human one, has the lowest proportion of older adults, but all of these samples have an overrepresentation of prime-age adults. Their older adult percentages are all below that of a pooled late Holocene archeological sample, some samples of which have been noted to have been demographically and socially challenging given a dearth of older adults (Howell 1982). Recent small-scale ethnographic samples all exhibit >50% older adults (pooled average: 61.9%; Trinkaus 1995).

Summary

The Palomas sample, therefore, however assessed with respect to its ages at death, is notable for the preponderance of immature individuals and for following the general Pleistocene pattern of few older adults. There are diverse behavioral and taphonomic (nondemographic) factors that could account for these distributions (Trinkaus 1995, 2011a; Bocquet-Appel and Arsuaga 1999), both at Palomas and among Pleistocene humans in general, as well as a generally high level of immature mortality and a low life expectancy among adults. Because it is unclear how most of the human remains came to be entombed within the Sima de las Palomas, it is difficult to parse out what combination of factors could have contributed to this unusual mortality profile.

The unusual mortality profile of the Palomas Neandertals is joined by their unusually high proportion of probably female adult remains, as indicated by the small body sizes of almost all of the remains (chapters 13 and 14). These two factors suggest unusual social patterns, whether in terms of differential age and sex mortality in the vicinity of the Sima de las Palomas or a socially determined composition of the individuals frequenting the locality on the side of the Cabezo Gordo. In any case, the age and sex composition of the overall sample, and probably of the stratigraphic subsets of it, is not a random sampling of an expected population of Neandertals.

TABLE 15.1. Distribution of the Palomas dental and mandibular remains by developmental stages and adult status, plus the Palomas 92, 96, and 97 associated skeletons

TOOTH	INFANT 0–2 YEARS	CHILD 3–6 YEARS	JUVENILE 7–12 YEARS	ADOLESCENT 13–18 YEARS	ADULT[a] 19+ YEARS
Maxilla					
di^1	SP39L				
di^2		SP85L			
dc^1					
dm^1	SP71L				
dm^2	SP36L				
I^1		SP37R	SP24L, SP34 L	SP79L, SP90L	SP73R
I^2				SP48R	
C^1			SP74L	SP35L	
P^3			SP60R	SP53L	
P^4			SP94L	SP68R	
M^1					
M^2					
M^3					SP51R
Mandible					
di_1					
di_2					
dc_1	SP95L	SP93R	SP31L		
dm_1	SP40L	SP25R	SP61R		
dm_2		SP69R, SP70R, SP83R, SP88L			
I_1			SP19L		SP21L
I_2			SP89R, SP91L		SP20L
C_1			SP18L, SP26R, SP82L	SP44R, SP54R	SP98R
P_3			SP45R, SP76R	SP75L	SP22L
P_4			SP57R, SP78L, SP87L		
M_1		SP84L			SP100L
M_2				SP29R	
M_3				SP58L	
Mandible		SP49RL, SP7L	SP80L		SP1RL, SP6L, SP23R, SP59L
Total[b]	5	11	19	11	11
MNI[b, c]	1	4	2	2	5
Skeletons	0	0	1	0	2

Note: See chapters 6 and 13 for detailed age-at-death assessments by specimen. Specimen number (SP) and side (R vs. L) are provided for each.

[a] All adult remains are together, although only Palomas 1 is possibly ≥40 years.

[b] The totals and the MNI (minimum number of individuals) are based only on the isolated teeth and mandibles and do not include the three associated skeletons.

[c] The MNI is the absolute minimum based on age intervals and side representation for individual teeth. Finer assessment of degrees of antimeric wear and morphology would increase the number of individuals represented. The mature mandibular molars are counted as 5, since Palomas 100 derives from a stratigraphically lower level than all but Palomas 1 and duplicates the Palomas 1 M_1s. The three associated skeletons are stratigraphically distinct from all of the dental and mandibular remains except Palomas 1, and it is not the same individual as either Palomas 92 or 96.

TABLE 15.2. Western Eurasian comparative age distributions for the Palomas dental and mandibular sample

GROUP	INFANT	CHILD	JUVENILE	ADOLESCENT	ADULT	N
Palomas[a]	8.3%	18.3%	33.3%	18.3%	21.7%	60
Mid. Pleist.[b]	0.0%	6.3%	4.3%	31.9%	57.4%	47
Neandertal[c]	6.3%	14.6%	13.8%	17.6%	47.7%	239
MPMH[d]	8.3%	16.7%	20.8%	8.3%	45.8%	24
E/MUP[e]	16.0%	8.5%	14.9%	13.8%	46.8%	94
Recent[f]	27.2%	12.8%	7.4%	10.5%	42.1%	3,197

Note: Use of the MNIs would reduce the preponderance of juveniles but maintain the dominance of immature individuals.
[a]The Palomas percentages are based on the isolated teeth and mandibles and three associated skeletons.
[b]Data from Atapuerca–Sima de los Huesos (n = 28), Arago, Aroeira, Aubesier, Biache, Boxgrove, La Chaise, Ehringsdorf, Mauer and Montmaurin (Lebel and Trinkaus 2011; Vlček 1993; Martinón-Torres et al. 2012; Trinkaus pers. obs.).
[c]Data from Trinkaus (1995, 2011, pers. obs.), Arsuaga et al. (2001), Akazawa and Muhesan (2002), Colombet et al. (2012).
[d]Middle Paleolithic modern humans; data from McCown and Keith (1939), Vandermeersch (1981), and Tillier (1999).
[e]Early/Mid Upper Paleolithic modern humans; data from Trinkaus (2011a) and Trinkaus et al. (2014a).
[f]Pooled recent human archeological samples, summarized in Trinkaus (1995).

TABLE 15.3. Younger (19–39 years) versus older (≥40 years) adults for Palomas and comparative Pleistocene and recent human archeological samples

GROUP	YOUNGER ADULT	OLDER ADULT	N
Palomas[a]	92.3%	7.7%	13
Middle Pleistocene	80.8%	19.2%	26
Neandertals	79.6%	20.4%	113
MPMH	90.9%	9.1%	11
E/MUP	70.5%	29.5%	44
Recent humans	63.4%	36.6%	1,345

Note: Palomas 1 is conservatively considered an older adult; specimens designated only as "adult" (for example, Palomas 6) are not included. Samples and sources are as in Table 15.2.
[a]The Palomas percentages are based on the number of isolated specimens (11) plus the two mature associated skeletons. Using MNIs would provide 7 adults, of which 85.7% were <40 years old.

16

The People of Palomas

ERIK TRINKAUS

THESE CONSIDERATIONS of the human fossil remains from the Sima de las Palomas, combining those scattered *ex situ* by nineteenth-century miners, those excavated from the relatively soft sediments of the Upper Cutting infilling, and the Palomas 92 partial postcrania extracted from the breccia of Conglomerate A, provide a wealth of information on the people who resided during the MIS 3 in and around the Cabezo Gordo. The bulk of the paleontological data is concerned with the abundant teeth from the site, both those still *in situ* in mandibles and maxillae and the numerous isolated ones. There are nonetheless insights from the cranial fragments, the partial mandibles, the isolated postcranial pieces, and the associated Palomas 92 (plus Palomas 96 and 97) partial skeletons.

The Palomas Humans as Neandertals

As discussed in chapter 3 and further documented in the subsequent chapters, these fossil remains do indeed represent Neandertals. The configurations of their supraorbital tori, mandibular symphyses, maxillary incisors, hamuli, distal manual phalanges, and femoral diaphyses, as well as their dental tissue proportions and lower limb articular to length proportions and some aspects of their upper limb remains, align them distinctly with the Neandertals as opposed to early modern humans. These aspects are joined by additional body proportions and further aspects of the shoulder anatomy, hand remains, and pelvis of Palomas 96. Yet many of their Neandertal features are shared with Middle Pleistocene archaic humans, and the gradual enrichment of the Middle Pleistocene human fossil record is increasing the list of skeletal aspects for which the Neandertals are ancestral as opposed to derived.

Although the Palomas remains are comfortably subsumed within the Neandertals *sensu lato*, they serve to extend the ranges of variation of these late archaic humans in a number of anatomical aspects. The supraorbital torus of Palomas 11 is unusually thin. The transverse sinus sulcus of Palomas 30 crosses the parietal bone. The mandibular corpus of Palomas 6 is unusually wide for its height. Palomas 6 and 23 have an unusual flange on the anterolateral basilar corpus that is either an exceptional anterior marginal tubercle or a unique feature. Palomas 80 has the most mesial position for a primary mental foramen.

Only one of the eight Palomas mandibular molars (the Palomas 29 M_2) has a midtrigonid crest, whereas the crest occurs in 97% of other Neandertal lower molars. Similarly, none of the four Palomas dm_2s present such a crest, whereas crests are present in 79% of Neandertal dm_2s. Even though the dental crowns of the Palomas teeth generally span the Neandertal size range, they tend toward the small size range; of the 54 permanent teeth providing crown diameters, half are within the Neandertal interquartile range, 35% are below the interquartile range, and 15% are below the previously known Neandertal range. In the context of reinforcing the tendency of Neandertal teeth to have relatively thin enamel, several of the Palomas ones have the thinnest or among the thinnest enamel caps. The Palomas 1 (C^1), 26 (C_1), 79 (I^1), and 96 (I_1) anterior teeth have unusually short roots for Neandertals. In a variety of other aspects of the dental crowns (discrete traits and contours), the Palomas teeth fit within Neandertal variation but are not distinct from those of early modern humans.

The Palomas postcrania follow the same pattern. The Palomas 92 and 96 humeri have only moderately thin pillars. The Palomas 64 radius has an anteromedial radial tuberosity, despite medial orientations for the Palomas 92 and 96 radii. The Palomas 92 ulnar trochlear notch is anteroproximally oriented. The Palomas 92 trapezium has the most dorsopalmarly convex metacarpal 1 facet known among archaic *Homo* remains, and the skeleton has a rather small trapezial tuberosity despite a large hamulus. The femoral diaphyses are characteristically archaic in lacking a pilaster, but Palomas 52 and 92 have anteroposteriorly deep midproximal diaphyses.

Added to these morphological and morphometric differences are the small dimensions of the Palomas adults (or near adults). The securely mature Palomas 28, 65, 77, 92, and 96 postcranial remains are at the lower limit of Neandertal body size indicators, and the Palomas 6 and 23 mandibles have among the shortest corpori. If the Palomas 16 and 52 diaphyses are mature, as their morphologies and cross-sectional proportions suggest, then those individuals were also quite small. Is the Palomas collection sampling a set of small females (only Palomas 96 is securely female), or was this a very small Neandertal population? As noted, other Iberian Neandertal postcrania are not diminutive.

These considerations combine to question the sources of this variation. The Palomas sample does span considerable time within early MIS 3, including among the most recent known Neandertals (those from the Upper Cutting infilling). It is therefore evident that at least a few populations through time are being sampled, especially if the *ex situ* remains came from deeper in the stratigraphic column. Is this variation due to pooling different populations? Yet the comparative Neandertal sample is substantially more temporally and geographically dispersed, extending from Atlantic Iberia to the Zagros Mountains and from the MIS 6/5 boundary to mid-MIS 3. Is it the result of Palomas being close to the southern end of the Iberian cul-de-sac and experiencing isolation by distance from other Neandertal populations? Yet considerations of other Iberian Neandertals, including those from Devil's Tower, Forbes' Quarry and Zafarraya (all further south) have not noted particularly non-Neandertal aspects in them. Could it be related to the relative recentness of the isolated remains from the upper infilling, given unusual features in the Initial Upper Paleolithic Saint-Césaire and Vindija Neandertals? Yet unusual features are also seen in the geologically older Palomas 92 (as well as Palomas 96) remains. Or are we just seeing what should be normal Neandertal variation, something obscured by limited samples elsewhere? Yet other large samples (for example, Shanidar), as well as regionally adjacent remains from sites that were similarly close in time (for example, La Chapelle-aux-Saints, La Ferrassie, La Quina, and the other Neandertal sites in southwestern France), do not appear to show the same variations in these individual traits. Only the Krapina sample appears somewhat variable, albeit in different traits than those of the Palomas remains.

The Palomas Neandertals as Humans

In addition to these affinity and variation issues, the Palomas Neandertal sample provides other, more paleobiological, insights. The sample is an exceptionally young one, with a dominance of immature individuals in the dental remains and a reasonable abundance of them in the skeletal remains. Moreover, among those immature individuals, juveniles appear to be the most numerous. Two other Neandertal sites (among those with multiple remains) have an abundance of immature individuals: La Ferrassie and Krapina (as well as the Middle Pleistocene Atapuerca-SH site). Krapina shares with Palomas the absence of older adults, but La Ferrassie has two of the more elderly Neandertal adults. The other Neandertal site with a number of individuals represented, Shanidar, has similar numbers of infants, young adults, and older adults. Palomas also reinforces the pan-Pleistocene pattern of a dearth of older (more than 40 years old) adults in the fossil record, since the probably oldest individual (based on dental wear: Palomas 1) is very unlikely to have lived past the fourth decade and may well have been considerably younger.

None of these mortality profiles (separately or pooled together) represent a paleodemographic profile of a real population. The absence of older adults may be related to a complex combination of demographic, mobility, mortuary, and other behavioral factors, and the scarcity of infants may reflect preservational issues. Yet the abundance of juveniles (or, more appropriately, post-weaning to adolescent remains, ≈70% of the sample) raises questions as to the spatial distribution of age-related mortality in the vicinity of the Sima de las Palomas. And why do the adults all appear likely to be female, even if only Palomas 96 is pelvically sexed?

The Palomas sample also has a dearth of evidence for pathological lesions. The absence of osteoarthritis and other degenerative conditions can be attributed to the young age profile of the sample combined with the fragmentary nature of all of the cranial and postcranial remains (and especially articulations). Yet the abundant dental remains have a low level of developmental defects of the enamel, or hypoplasias. Such indicators of developmental stress episodes are ubiquitous among the Neandertals (and Pleistocene humans in general), but they are scarce and present only as minor defects on the Palomas teeth. There are dental carious lesions on two individuals, Palomas 25 and 59, and such lesions are becoming increasingly documented among Mediterranean (if not more northern) Neandertals. None of the Palomas remains provide any evidence of trauma, even minor cranial vault lesions, whereas they are common among other archaic human samples.

The sufficiently preserved anterior teeth all show labial instrumental marks, indicating use of those teeth for holding items while slicing them. Yet the one associated anterior-posterior dentition (Palomas 1) does not have an advanced degree of differential anterior dental occlusal wear. Those cut marks all indicate right-handedness, for the five or six individuals represented, yet the one individual providing sufficient preserved bilateral arm bones (Palomas 96) was distinctly left-handed.

Other paleobiological aspects of the Palomas remains follow the general Neandertal (or Pleistocene human) pattern. Their limbs are generally robust, compared to those of most Holocene humans but not relative to early modern humans. The only contrasts with early modern humans are in their buccal postcanine wear patterns, pectoralis major and gluteus maximus insertions, hamulus projections, and distal phalangeal tuberosities. Palomas 92 also has broad pedal phalangeal diaphyses, as with other Middle Paleolithic (but not Upper Paleolithic) humans, suggesting infrequent use of protective footwear.

Summary

In conclusion, the Palomas human remains provide a window on the Neandertals and Neandertal biology from southeastern Iberia in the Interpleniglacial. They appear to have been relatively healthy (if dying early). They raise questions as to the behavioral patterns behind their unusual age (and probable sex) mortality profile and its possible implications for the use of the Cabezo Gordo in the vicinity of the Sima de las Palomas. As with most Pleistocene humans, they were strongly built, in a manner similar to other Neandertals and their Middle Pleistocene ancestors. Yet in a variety of features throughout the skeleton and dentition, they extend our appreciation of the extent of Neandertal morphological variation, and they thereby provide a changed perspective on aspects of their paleobiology.

References

Adams, D. C., F. J. Rohlf, and D. E. Slice. 2004. Geometric morphometrics: Ten years of progress following the "revolution." *Italian Journal of Zoology* 71:5–16.

Adler, C. J., K. Dobney, L. S. Weyrich, J. Kaidonis, A. W. Walker, W. Haak, C. J. Bradshaw, G. Townsend, A. Sołtysiak, K. W. Alt, J. Parkhill, and A. Cooper. 2013. Sequencing ancient calcified dental plaque shows changes in oral microbiota with dietary shifts of the Neolithic and Industrial revolutions. *Nature Genetics* 45:450–55.

Adler, D. S., G. Bar-Oz, A. Belfer-Cohen, and O. Bar-Yosef. 2006. Ahead of the game: Middle and Upper Palaeolithic hunting behaviors in the southern Caucasus. *Current Anthropology* 47:89–118.

Akazawa, T., and S. Muhesan, eds. 2002. *Neanderthal Burials. Excavations of the Dederiyeh Cave, Afrin, Syria.* Kyoto: International Research Center for Japanese Studies.

Alcázar de Velasco, A., J. L. Arsuaga, I. Martínez, and A. Bonmatí. 2011. Revisión de la mandibula humana de Bañolas, Gerona, España. *Boletín de la Real Sociedad Española de Historia Natural, Sección Geológica* 105:99–108.

AlQahtani, S. J., M. P. Hector, and M. H. Liversidge. 2010. The London atlas of human tooth development and eruption. *American Journal of Physical Anthropology* 142:481–90.

Alroussan, M. F. 2009. The Mesolithic-Neolithic Transition in the Near East: Biological implications of the shift in subsistence strategies through the analysis of dental morphology and dietary habits of human populations in the Mediterranean area 12,000–5,000 BP. PhD diss., Universitat de Barcelona.

Alvesalo, L., and P. M. A. Tigerstedt. 1974. Heritabilities of human tooth dimensions. *Hereditas* 77:311–18.

Ånerud, Å. G., H. Löe, H. Boysen, and M. Smith. 1979. The natural history of periodontal disease in man. *Journal of Periodontal Research* 14:526–40.

Angelucci, D. E., D. Anesin, D. Susini, V. Villaverde, J. Zapata, and J. Zilhão. 2013. Formation processes at a high resolution Middle Paleolithic site: Cueva Antón (Murcia, Spain). *Quaternary International* 315:24–41.

Antón, S. C. 1990. Neandertals and the anterior dental loading hypothesis: A biomechanical evaluation of bite force production. *Kroeber Anthropological Society Papers* 71–72:67–76.

———. 1994. Mechanical and other perspectives on Neandertal craniofacial morphology. In *Integrative Paths to the Past*, edited by R. S. Corruccini and R. L. Ciochon, 677–95. Englewood Cliffs, NJ: Prentice Hall.

———. 1996. Tendon-associated bone features of the masticatory system in Neandertals. *Journal of Human Evolution* 31:391–408.

Arensburg, B. 1977. New Upper Paleolithic human remains from Israel. *Eretz Israel* 13:208–15.

———. 1991. The vertebral column, thoracic cage, and hyoid bone. In *Le Squelette Moustérien de Kébara 2*, edited by O. Bar Yosef and B. Vandermeersch, 113–46. Paris: CNRS.

Armitage, G. C. 2004. Periodontal diagnoses and classification of periodontal diseases. *Periodontology 2000* 34:9–21.

Arsuaga, J. L., I. Martínez, V. Villaverde, C. Lorenzo, R. Quam, J. M. Carretero, and A. Gracia. 2001a. Fósiles humanos del país valenciano.

In *De Neandertales a Cromañones: El Inicio del Poblamiento Humano en las Tierras Valencianas*, edited by V. Villaverde, 265–322. Valencia: Universitat de València.

Arsuaga, J. L., I. Martínez, C. Lorenzo, R. Quam, J. M. Carretero, and A. Gracia. 2001b. Las estrías del incisivo de Cova Negra. In *De Neandertales a Cromañones: El Inicio del Poblamiento Humano en las Tierras Valencianas*, edited by V. Villaverde, 327–28. Valencia: Universitat de València.

Arsuaga, J. L., V. Villaverde, R. Quam, A. Gracia, C. Lorenzo, I. Martínez, and J. M. Carretero. 2002. The Gravettian occipital bone from the site of Malladetes (Barx, Valencia, Spain). *Journal of Human Evolution* 43:381–93.

Arsuaga, J. L., V. Villaverde, R. Quam, I. Martínez, J. M. Carretero, C. Lorenzo, and A. Gracia. 2007. New Neandertal remains from Cova Negra (Valencia, Spain). *Journal of Human Evolution* 52:31–58.

Arsuaga, J. L., I. Martínez, L. J. Arnold, A. Aranburu, A. Gracia-Téllez, W. D. Sharp, R. M. Quam, C. Falguères, A. Pantoja-Pérez, J. Bischoff, E. Poza-Rey, J. M. Parés, J. M. Carretero, M. Demuro, C. Lorenzo, N. Sala, M. Martinón-Torres, N. García, A. Alcázar de Velasco, G. Cuenca-Bescós, A. Gómez-Olivencia, D. Moreno, A. Pablos, C. C. Shen, L. Rodríguez, A. I. Ortega, R. García, A. Bonmatí, J. M. Bermúdez de Castro, and E. Carnonell. 2014. Neandertal roots: Cranial and chronological evidence from Sima de los Huesos. *Science* 344:1358–63.

Arsuaga, J. L., J. M. Carretero, C. Lorenzo, A. Gómez-Olivencia, A. Pablos, L. Rodríguez, R. García-González, A. Bonmatí, R. M. Quam, A. Pantoja-Pérez, I. Martínez, A. Aranburu, A. Gracia-Téllez, E. Poza-Rey, N. Sala, N. García, A. Alcázar de Velasco, G. Cuenca-Bescós, J. M. Bermúdez de Castro M, and E. Carbonell. 2015. Postcranial morphology of the Middle Pleistocene humans from Sima de los Huesos, Spain. *Proceedings of the National Academy of Sciences, USA* 112:11524–29.

Auerbach, B. M., and C. B. Ruff. 2004. Human body mass estimation: A comparison of "morphometric" and "mechanical" methods. *American Journal of Physical Anthropology* 125:331–42.

———. 2006. Limb bone bilateral asymmetry: Variability and commonality among modern humans. *Journal of Human Evolution* 50:203–18.

———. 2010. Stature estimation formulae for indigenous North American populations. *American Journal of Physical Anthropology* 141:190–207.

Avishai, G., R. Müller, Y. Gabet, I. Bab, U. Zilberman, and P. Smith. 2004. New approach to quantifying developmental variation in the dentition using serial microtomographic imaging. *Microscopy Research and Technique* 65:263–69.

Bailey, S. E. 2002a. Neandertal dental morphology: Implications for modern human origins. PhD diss., Arizona State University.

———. 2002b. A closer look at Neanderthal postcanine dental morphology: The mandibular dentition. *Anatomical Record* 269:148–56.

———. 2004a. Derived morphology in Neandertal maxillary molars: Insights from above. *American Journal of Physical Anthropology* 123:57–58.

———. 2004b. A morphometric analysis of maxillary molar crowns of Middle–Late Pleistocene hominins. *Journal of Human Evolution* 47:183–98.

———. 2005. Diagnostic dental differences between Neandertals and Upper Paleolithic modern humans: Getting to the root of the matter. In *Current Trends in Dental Morphology Research*, edited by E. Zadzinska, 201–10. Lodz: University of Lodz Press.

———. 2006. Beyond shovel-shaped incisors: Neandertal dental morphology in a comparative context. *Periodicum Biologorum* 108:253–67.

Bailey, S. E., and J. J. Hublin. 2006. Dental remains from the Grotte du Renne at Arcy-sur-Cure (Yonne). *Journal of Human Evolution* 50:485–508.

Bailey, S. E., and J. M. Lynch. 2005. Diagnostic differences in mandibular P4 shape between Neandertals and anatomically modern humans. *American Journal of Physical Anthropology* 126:268–77.

Bailey, S. E., and A. A. Zubov. 2014. Dental occlusal discrete morphology. In *The People of Sunghir: Burials, Bodies and Behavior in the Earlier Upper Paleolithic*, edited by E. Trinkaus, A. P. Buzhilova, M. B. Mednikova, and M. V. Dobrovolskaya, 151–54. New York: Oxford University Press.

Bailey, S. E., T. D. Weaver, and J. J. Hublin. 2009. Who made the early Aurignacian and other early Upper Paleolithic industries? *Journal of Human Evolution* 57:11–26.

Bailey, S. E., M. M. Skinner, and J. J. Hublin. 2011. What lies beneath? An evaluation of lower molar trigonid crest patterns based on both dentine and enamel expression. *American Journal of Physical Anthropology* 145:505–18.

Baker, B. J., T. L. Dupras, and M. W. Tocheri. 2005. *The Osteology of Infants and Children*. College Station: Texas A&M University Press.

Baker, G., L. H. P. Jones, Jand I. D. Wardrop. 1959. Cause of wear in sheep's teeth. *Nature* 184:1583–84.

Bang, G. 1964. A comparison between the incidence

of dental caries in typical coastal populations and inland populations with particular regard to the possible effect of a high intake of salt water fish. *Odontologie Tidskr* 72:12–43.

Barroso-Ruíz, C., M. A. de Lumley, M. Caparrós, and L. Verdú. 2003. Los restos humanos neandertalenses de la Cueva del Boquete de Zafarraya. In *El Pleistocene Superior del la Cueva del Boquete de Zafarraya*, edited by C. Barroso-Ruíz, 327–87. Sevilla: Consejería de Cultura, Junta de Andalucía.

Bar-Yosef, O. 2004. Eat what is there: Hunting and gathering in the world of Neanderthals and their neighbors. *International Journal of Osteoarchaeology* 14:333–42.

Bax, J. S., and P. S. Ungar. 1999. Incisor labial surface wear striations in modern humans and their implications for handedness in Middle and Late Pleistocene hominids. *International Journal of Osteoarchaeology* 9:189–98.

Bayle, P. 2008a. Analyses quantitatives par imagerie à haute résolution des séquences de maturation dentaire et des proportions des tissus des dents déciduales chez les Néanderthaliens et les hommes modernes. PhD diss., Université Toulouse III—Paul Sabatier.

———. 2008b. Proportions des tissus des dents déciduales chez deux individus de Dordogne (France): l'Enfant néanderthalien du Roc de Marsal et le spécimen du Paléolithique supérieur final de La Madeleine. *Bulletins et Mémoires de la Société d'Anthropologie de Paris* 20:151–63.

Bayle, P., J. Braga, A. Mazurier, and R. Macchiarelli. 2009a. Dental developmental pattern of the Neanderthal child from Roc de Marsal: A high-resolution 3D analysis. *Journal of Human Evolution* 56:66–75.

———. 2009b. High-resolution assessment of the dental developmental pattern and characterization of tooth tissue proportions in the late Upper Paleolithic child from La Madeleine, France. *American Journal of Physical Anthropology* 138:493–98.

Bayle, P., R. Macchiarelli, E. Trinkaus, C. Duarte, A. Mazurier, and J. Zilhão. 2010. Dental maturational sequence and dental tissue proportions in the early Upper Paleolithic child from Abrigo do Lagar Velho, Portugal. *Proceedings of the National Academy of Sciences, USA* 107:1338–42.

Bayle, P., A. Balzeau, and C. Zanolli. 2013. A microCT-based longitudinal study of the dental developmental pattern in the Neandertal child from Pech de l'Azé, France (abstract). *Proceedings of the European Society for the Study of Human Evolution* 2:40.

Belcastro, M. G., E. Rastelli, and V. Mariotti. 2008. Variation in the degree of sacral vertebral body fusion in adulthood in two European modern skeleton collections. *American Journal of Physical Anthropology* 135:149–60.

Benazzi, S., K. Douka, C. Fornai, C. Bauer, O. Kullmer, I. Pap, F. Mallegni, P. Bayle, M. Coquerelle, S. Condemi, A. Ronchitelli, K. Harvati, and G. W. Weber. 2011a. Early dispersal of modern humans in Europe and implications for Neanderthal behaviour. *Nature* 479:525–29.

Benazzi, S., C. Fornai, P. Bayle, M. Coquerelle, O. Kullmer, F. Mallegni, and G. W. Weber. 2011b. Comparison of dental measurement systems for taxonomic assignment of Neanderthal and modern human lower second deciduous molars. *Journal of Human Evolution* 61:320–26.

Benazzi, S., D. Panetta, C. Fornai, M. Toussaint, G. Gruppioni, and J. J. Hublin. 2014. Technical note: Guidelines for the digital computation of 2D and 3D enamel thickness in hominoid teeth. *American Journal of Physical Anthropology* 153:305–13.

Bermúdez de Castro, J. M. 1986. Dental remains from Atapuerca (Spain): I. Metrics. *Journal of Human Evolution* 15:265–87.

———. 1988. Dental remains from Atapuerca/Ibeas (Spain): II. Morphology. *Journal of Human Evolution* 17:279–304.

Bermúdez de Castro, J. M., and A. Rosas. 2001. Pattern of dental development in hominid XVIII from the Middle Pleistocene Atapuerca–Sima de los Huesos site (Spain). *American Journal of Physical Anthropology* 114:325–30.

Bermúdez de Castro, J. M., T. G. Bromage, and Y. Fernández-Jalvo. 1988. Buccal striations on fossil human anterior teeth: Evidence of handedness in the Middle and early Upper Pleistocene. *Journal of Human Evolution* 17:403–12.

Bermúdez de Castro, J. M., A. Rosas, and M. E. Nicolás. 1999. Dental remains from Atapuerca TD-6 (Gran Dolina Site, Burgos, Spain). *Journal of Human Evolution* 37:523–66

Bermúdez de Castro, J. M., M. Martinón-Torres, S. Sarmiento, M. Lozano, J. L. Arsuaga, and E. Carbonell. 2003. Rates of anterior tooth wear in Middle Pleistocene hominins from Sima de los Huesos (Sierra de Atapuerca, Spain). *Proceedings of the National Academy of Sciences, USA* 100:11992–96.

Bermúdez de Castro, J. M., J. M. Carretero, R. García-Gonzáles, L. Rodríguez-García, M. Martinón-Torres, J. Rosell, R. Blasco, L. Martín-Francés, M. Modesto, and E. Carbonell. 2012. Early Pleistocene human humeri from the Gran Dolina–TD6 site (Sierra de Atapuerca, Spain). *American Journal of Physical Anthropology* 147:604–7.

Bernal, V. 2007. Role of wild plant foods among late Holocene hunter-gatherers from Central and North Patagonia (South America): An approach from dental evidence. *American Journal of Physical Anthropology* 133:1047–59.

Bernal, V., S. I. Perez, P. N. Gonzalez, and J. A. F. Diniz-Filho. 2010. Ecological and evolutionary factors in dental morphological diversification among modern human populations from southern South America. *Procedings of the Royal Society* 277B:1107–12

Beynon, A. D. 1987. Replication technique for studying microstructure in fossil enamel. *Scanning Microscopy* 1(2):663–69.

Beynon, A. D., and B. A. Wood. 1986. Variations in enamel thickness and structure in East African Hominoids. *American Journal of Physical Anthropology* 70:177–93.

Bocherens, H., D. G. Drucker, D. Billiou, M. Patou-Mathis, and B. Vandermeersch. 2005. Isotopic evidence for diet and subsistence pattern of the Saint-Césaire I Neanderthal: Review and use of a multi-source mixing model. *Journal of Human Evolution* 49:71–87.

Bocquet-Appel, J. P., and J. L. Arsuaga. 1999. Age distributions of hominid samples at Atapuerca (SH) and Krapina could indicate accumulation by catastrophe. *Journal of Archaeological Science* 26:327–38.

Bogin, B. 1997. Evolutionary hypotheses for human childhood. *Yearbook of Physical Anthropology* 40:63–89.

Bondioli, L., P. Bayle, C. Dean, A. Mazurier, L. Puymerail, C. Ruff, J. T. Stock, V. Volpato, C. Zanolli, and R. Macchiarelli. 2010. Morphometric maps of long bone shafts and dental roots for imaging topographic thickness variation. *American Journal of Physical Anthropology* 142:328–34.

Bonfiglioli, B., V. Mariotti, F. Facchini, M. G. Belcastro, and S. Condemi. 2004. Masticatory and non-masticatory dental modifications in the Epipalaeolithic necropolis of Taforalt (Morocco). *International Journal of Osteoarchaeology* 14:448–56.

Bookstein, F. L. 1991. *Morphometric Tools for Landmark Data: Geometry and Biology*. Cambridge: Cambridge University Press.

———. 1997. Landmark methods for forms without landmarks: Morphometrics of group differences in outline shape. *Medical Image Analysis* 1:225–43.

Boule, M. 1911–13. L'homme fossile de La Chapelle-aux-Saints. *Annales de Paléontologie* 6:111–72; 7:21–56, 85–192; 8:1–70.

Brabant, H., and A. Sahly. 1964. Étude des dents néandertaliennes découvertes dans la Grotte du Portel, en Ariège (France). *Bulletin du Groupement International pour la Recherche Scientifique en Stomatologie* 7:237–54.

Brace, C. L. 1975. Comment on "Did La Ferrassie I use his teeth as tools?." *Current Anthropology* 16:396–97.

———. 1995. Biocultural interaction and the mechanism of mosaic evolution in the emergence of "modern" morphology. *American Anthropologist* 97:711–21.

Brace, C. L., A. S. Ryan, and B. H. Smith. 1981. Comment on "Tooth wear in La Ferrassie man." *Current Anthropology* 22:426–30.

Braga, J., F. Thackeray, G. Subsol, J. L. Kahn, D. Maret, J. Treil, and A. Beck. 2010. The enamel-dentine junction in the postcanine dentition of *Australopithecus africanus*: Intra individual metameric and antimeric variation. *Journal of Anatomy* 216:62–79.

Bräuer G. 1988. Osteometrie. In *Anthropologie I*, edited by R. Knussman, 160–232. Stuttgart: Fischer Verlag.

Brennan, M. U. 1991. Health and disease in the Middle and Upper Paleolithic of southwestern France: A bioarcheological study. PhD diss., New York University.

Broderick, A. H. 1948. *Early Man: A Survey of Human Origins*. London: Hutchinson.

Bromage, T. G., and A. Boyde. 1984. Microscopic criteria for the determination of directionality of cutmarks on bone. *American Journal of Physical Anthropology* 65(4):359–66.

Bromage, T., J. Bermúdez de Castro, and Y. Fernández-Jalvo. 1991. The SEM in taphonomic research and its application to studies of cutmarks generally and the determination of handedness specifically. *Anthropologie* 29:163–69.

Brose, D. S., and M. H. Wolpoff. 1971. Early Upper Paleolithic man and Late Middle Paleolithic tools. *American Anthropologist* 73:1156–94.

Brown, K., D. Fa, G. Finlayson, and C. Finlayson. 2011. Small game and marine resource exploitation by Neanderthals: The evidence from Gibraltar. In *Trekking the Shore: Changing Coastlines and the Antiquity of Coastal Settlement*, edited by N. F. Bicho, J. A. Haws, and L. G. Davis, 247–71. New York: Springer.

Bruner, E., and M. Lozano. 2014. Extended mind and visuo-spatial integration: Three hands for the Neandertal lineage. *Journal of Anthropological Sciences* 92:273–80.

Byers, S., K. Aloshima, and B. Curran. 1989.

Determination of stature from metatarsal length. *American Journal of Physical Anthropology* 79:275–79.

Carretero, J. M., J. L. Arsuaga, and C. Lorenzo. 1997. Clavicles, scapulae, and humeri from the Sima de los Huesos site (Sierra de Atapuerca, Spain). *Journal of Human Evolution* 33:357–408.

Carrión, J. S., E. I. Yll, M. J. Walker, A. J. Legaz, C. Chaín, and A. López. 2003. Glacial refugia of temperate, Mediterranean and Ibero–North African flora in south-eastern Spain: New evidence from cave pollen at two Neanderthal man sites. *Global Ecology and Biogeography* 12:119–29.

Carrión, J. S., E. I. Yll, C. Cháin, M. Dupré, M. J. Walker, A. Legaz, and A. López. 2005. Fitodiversidad arbórea en el litoral del sureste español durante el Pleistoceno Superior. In *Geomorfologia litoral i Quaternari, Homenatge al Profesor Vicenç Rosselló i Verger*, edited by E. Sanjaume and J. F. Mateu, 103–12. Valencia: Universitat de València.

Carter, J. T. 1928. The Teeth of Rhodesian Man. In *Rhodesian Man and Associated Remains*, edited by F. A. Bather, 64–65. London: British Museum (Natural History).

Cartmill, M., and F. H. Smith. 2009. *The Human Lineage*. New York: Wiley-Blackwell.

Caspari, R., and S. H. Lee. 2004. Older age becomes common late in human evolution. *Proceedings of the National Academy of Sciences, USA* 101:10895–900.

Catalano, S. A., P. A. Goloboff, and N. P. Giannini. 2010. Phylogenetic morphometrics (I): The use of landmark data in a phylogenetic framework. *Cladistics* 26:1–11.

Chase, P. G., and V. Teilhol. 2009. The fossil human remains. In *The Cave of Fontéchevade: Recent Excavations and Their Paleoanthropological Implications*, edited by P. G. Chase, A. Debénath, H. L. Dibble, and S. P. McPherron, 103–16. Cambridge: Cambridge University Press.

Chen, H., X. Zhou, H. Fujita, M. Onozuka, and K. Y. Kubo. 2013. Age-related changes in trabecular and cortical bone microstructure. *International Journal of Endocrinology* 2013:213234. doi:10.1155/2013/213234.

Chirchir, H., T. L. Kivell, C. B. Ruff, J. J. Hublin, K. J. Carlson, B. Zipfel, and B. G. Richmond. 2015. Recent origin of low trabecular bone density in modern humans. *Proceedings of the National Academy of Sciences, USA* 112:366–71.

Churchill, S. E., and V. Formicola. 1997. A case of marked bilateral asymmetry in the upper limbs of an Upper Palaeolithic male from Barma Grande (Liguria), Italy. *International Journal of Osteoarchaeology* 7:18–38.

Churchill, S. E., O. M. Pearson, F. E. Grine, E. Trinkaus, and T. W. Holliday. 1996. Morphological affinities of the proximal ulna from Klasies River Mouth Main Site: Archaic or modern? *Journal of Human Evolution* 31:213–37.

Clement, A. F., S. W. Hillson, and L. C. Aiello. 2012. Tooth wear, Neanderthal facial morphology, and the anterior dental loading hypothesis. *Journal of Human Evolution* 62(3):367–76.

Clement, A. J. 1956. Caries in the South African apeman: Some examples of undoubted pathological authenticity believed to be 800,000 years old. *British Dental Journal* 101:4–7.

Coleman, M. N., and M. W. Colbert. 2007. Technical note: CT thresholding protocols for taking measurements on three-dimensional models. *American Journal of Physical Anthropology* 133:723–25.

Colombet, P., P. Bayle, I. Crevecoeur, J. G. Ferrié, and B. Maureille. 2012. New Mousterian neonates from the southwest of France (Saint-Césaire, Charente-Maritime) (abstract). *European Society for the Study of Human Evolution* 2:38.

Compton, T., and C. Stringer. 2012. The human remains. In *Neanderthals in Wales: Pontnewydd and the Elwy Valley Caves*, edited by S. Aldhouse-Green, R. Peterson, and E. A. Walker, 118–230. Oxford: Oxbow Books.

Condemi, S., D. Tardivo, B. Foti, S. Ricci, P. Giunti, and L. Longo. 2012. A case of an osteolytic lesion on an Italian Neanderthal jaw. *Comptes Rendus Palevol* 11:79–83.

Constantino, P. J., J. J. W. Lee, H. Chai, B. Zipfel, C. Ziscovici, B. R. Lawn, and P. W. Lucas. 2010. Tooth chipping can reveal the diet and bite forces of fossil hominins. *Biology Letters* 6:826–29.

Coqueugniot, H. 1999. Le crâne d'*Homo sapiens* en Eurasie: Croissance et variation depuis 100.000 ans. *British Archaeological Reports International Series* 822:1–197.

Costa, R. L., Jr. 1982. Periodontal disease in the prehistoric Ipiutak and Tigara skeletal remains from Point Hope, Alaska. *American Journal of Physical Anthropology* 59:97–100.

Cowgill, L. W. 2010. The ontogeny of Holocene and Late Pleistocene human postcranial strength. *American Journal of Physical Anthropology* 141:16–37.

Cowgill, L. W., M. B. Mednikova, A. P. Buzhilova, and E. Trinkaus. 2015. The Sunghir 3 Upper Paleolithic juvenile: Pathology and persistence in the Paleolithic. *International Journal of Osteoarchaeology* 25:176–87.

Crevecoeur, I., P. Bayle, H. Rougier, B. Maureille, T. Higham, J. van der Plicht, N. De Clerck, and

P. Semal. 2010. The Spy VI child: A newly discovered Neandertal infant. *Journal of Human Evolution* 59:641–56.

Crevecoeur, I., M. Skinner, S. Bailey, P. Gunz, S. Bortoluzzi, A. Brooks, C. Burlet, E. Cornelissen, N. De Clerck, B. Maureille, P. Semal, Y. Vanbrabant, and B. Wood. 2014. First early hominin from Central Africa (Ishango, Democratic Republic of Congo). *PLoS ONE* 9:e84652.

Daegling, D. J. 2001. Biomechanical scaling of the hominoid mandibular symphysis. *Journal of Morphology* 250:12–23.

Dalton, D. 2011. The vertebral column. In *Joint Structure and Function*, 5th ed., edited by P. K. Levangie, 139–91. Philadelphia: F. A. Davis.

Daura, J., M. Sanz, M. E. Subirá, R. Quam, J. M. Fullola, and J. L. Arsuaga. 2005. A Neandertal mandible from the Cova del Gegant (Sitges, Barcelona, Spain). *Journal of Human Evolution* 49:56–70.

Davies, T. G. H., and P. O. Pedersen. 1955. The degree of attrition of the deciduous teeth and the first permanent molars of primitive and urbanised Greenland natives. *British Dental Journal* 99:35–43.

Dean, M. C. 2006. Tooth microstructure tracks the pace of human life-history evolution. *Proceedings of the Royal Society* 273B:2799–2808.

Dean, M. C., and T. J. Cole. 2013. Human life history evolution explains dissociation between the timing of tooth eruption and peak rates of root growth. *PLoS ONE* 8:e54534.

Dean, M. C., and B. Wood 2013. A digital radiographic atlas of great apes skull and dentition. In *Digital Archives of Human Paleobiology*, edited by L. Bondioli and R. Macchiarelli, CD-ROM. Milan: ADS Solutions.

Dean, M. C., A. Rosas, A. Estalrrich, A. García-Tabernero, R. Huguet, C. Lalueza-Fox, M. Bastir, and M. de la Rasilla. 2013. Longstanding dental pathology in Neandertals from El Sidrón (Asturias, Spain) with a probable familial basis. *Journal of Human Evolution* 64:678–86.

Defleur, A. 1993. *Les Sépultures Moustériennes*, Paris: CNRS Éditions.

de Lumley, M. A. 1973. Anténéandertaliens et Néandertaliens du bassin méditerranéen occidental européen. *Études du Quaternaires* 2:1–626.

Demirjian, A., H. Goldstein, and J. M. Tanner. 1973. A new system of dental age assessment. *Human Biology* 45:211–27.

DeNiro, M. J. 1985. Post-mortem preservation and alteration of *in vivo* bone collagen isotope ratios in relation to paleodietary reconstruction. *Nature* 317:806–9.

Deter, C. A. 2009. Gradients of occlusal wear in hunter-gatherers and agriculturalists. *American Journal of Physical Anthropology* 138:247–54.

Dias, G., and N. Tayles. 1997. "Abscess cavity"—a misnomer. *International Journal of Osteoarchaeology* 7:548–54.

Dias, G. J., K. Prasad, and A. L. Santos. 2007. Pathogenesis of apical periodontal cysts: Guidelines for diagnosis in palaeopathology. *International Journal of Osteoarchaeology* 17:619–26.

Dinnis, R., S. M. Bello, A. T. Chamberlain, C. Coleman, and C. Stringer. 2014. A cut-marked Neolithic human tooth from Ash Tree Shelter, Derbyshire, UK. *Cave and Karst Science* 41:114–17.

Doboş, A., A. Soficaru, and E. Trinkaus. 2010. The prehistory and paleontology of the Peştera Muierii, Romania. *Études et Recherches Archéologiques de l'Université de Liège* 124:1–122.

Dobson, S. D., and E. Trinkaus. 2002. Cross-sectional geometry and morphology of the mandibular symphysis in Middle and Late Pleistocene *Homo*. *Journal of Human Evolution* 43:67–87.

Doube, M., M. M. Kłosowski, I. Arganda-Carreras, F. P. Cordelières, R. P. Dougherty, J. S. Jackson, B. Schmid, J. R. Hutchinson, and S. J. Shefelbine. 2010. BoneJ: Free and extensible bone image analysis in ImageJ. *Bone* 47:1076–79.

Dunning, J. M. 1953. The influence of latitude and distance from seacoast on dental disease. *Journal of Dental Research* 32:811–29.

Eke, P. I., B. A. Dye, L. Wei, G. O. Thornton-Evans, and R. J. Genco. 2012. Prevalence of periodontitis in adults in the United States, 2009 and 2010. *Journal of Dental Research* 91:914–20.

El-Zaatari, S., F. E. Grine, P. S. Ungar, and J. J. Hublin. 2011. Ecogeographic variation in Neandertal dietary habits: Evidence from occlusal molar microwear texture analysis. *Journal of Human Evolution* 61:411–24.

Enlow, D., and M. G. Hans. 1996. *Essentials of Facial Growth*. Philadelphia: W. B. Saunders.

Eschman, P. N. 1992. *SLCOMM Version 1.6*. Albuquerque: Eschman Archeological Services.

Estalrrich, A., and A. Rosas. 2013. Handedness in Neandertals from the El Sidrón (Asturias, Spain): Evidence from instrumental striations with ontogenetic inferences. *PLoS ONE* 8:e62797.

———. 2015. Division of labor by sex and age in Neandertals: An approach through the study of activity-related dental wear. *Journal of Human Evolution* 80:51–63.

Estalrrich, A., A. Rosas, S. García-Vargas, A. García-Tabernero, D. Santamaria, and M. de la

Rasilla. 2011. Subvertical grooves on interproximal wear facets from the El Sidrón (Asturias, Spain) Neandertal dental sample. *American Journal of Physical Anthropology* 144:154–61.

Estebaranz, F., L. M. Martínez, O. Hiraldo, V. Espurz, A. Bonnin, M. Farrés, and A. Pérez-Pérez. 2004. Tooth crown size and dentine exposure in *Australopithecus* and early *Homo*: Testing hypothesis of dietary related selective pressures. *Anthropologie* 42:59–63.

Estebaranz, F., L. M. Martínez, J. Galbany, D. Turbón, and A. Pérez-Pérez. 2009. Testing hypotheses of dietary reconstruction from buccal dental microwear in *Australopithecus afarensis*. *Journal of Human Evolution* 57:739–50.

Estebaranz, F., J. Galbany, L. M. Martínez, D. Turbón, and A. Pérez-Pérez. 2012. Buccal dental microwear analyses support greater specialization in consumption of hard foodstuffs for *Australopithecus anamensis*. *Journal of Anthropological Science* 90:163–85

Feeney, R. N. M., J. P. Zermeno, D. J. Reid, S. Nakashima, H. Sano, A. Bahar, J. J. Hublin, and T. M. Smith. 2010. Enamel thickness in Asian human canines and premolars. *Anthropological Science* 118:191–98.

Fiore, I., L. Bondioli, J. Radovčić, D. W. Frayer. 2015. Handedness in the Krapina Neandertals: A re-evaluation. *PaleoAnthropology* 2015:19–36.

Fiorenza, L., and O. Kullmer. 2013. Dental wear and cultural behavior in Middle Paleolithic humans from the Near East. *American Journal of Physical Anthropology* 152:107–17.

Fiorenza, L., S. Benazzi, and O. Kullmer. 2009. Morphology, wear, and 3D digital surface models: Materials and techniques to create high-resolution replicas of teeth. *Journal of Anthropological Science* 87:211–18.

Fiorenza, L., S. Benazzi, J. Tausch, O. Kullmer, T. G. Bromage, and F. Schrenk. 2011. Molar macrowear reveals Neanderthal eco-geographic dietary variation. *PLoS ONE* 6:e14769.

Fischer, E. 1906. Die Variationen an Radius und Ulna des Menschen. *Zeitschrift für Morphologie und Anthropologie* 9:147–246.

Fleitmann, D., H. Cheng, S. Badertscher, R. L. Edwards, M. Mudelsee, O. M. Göktürk, A. Fankhauser, R. Pickering, C. C. Raible, A. Matter, J. Kramers, and O. Tüysüz. 2009. Timing and climatic impact of Greenland interstadials recorded in stalagmites from northern Turkey. *Geophysical Research Letters* 36:L19707. doi:10.1029/2009GL040050.

Fornai, C., S. Benazzi, J. Svoboda, I. Pap, K. Harvati, and G. W. Weber. 2014. Enamel thickness variation of deciduous first and second upper molars in modern humans and Neanderthals. *Journal of Human Evolution* 76:83–91.

Franciscus, R. G. 2003. Internal nasal floor configuration in *Homo* with special reference to the evolution of Neandertal facial form. *Journal of Human Evolution* 44:701–29.

Franciscus, R. G., and E. Trinkaus. 1988. The Neandertal nose (abstract). *American Journal of Physical Anthropology* 75:209–10.

Frayer, D. W. 1977. Metric dental change in the European Upper Paleolithic and Mesolithic. *American Journal of Physical Anthropology* 46:109–20

———. 1978. Evolution of the dentition in Upper Paleolithic and Mesolithic Europe. *University of Kansas Publications in Anthropology* 10:1–201.

———. 1989. Oral pathologies in the European Upper Paleolithic and Mesolithic. In *People and Culture in Change: Upper Palaeolithic, Mesolithic and Neolithic Populations of Europe and the Mediterranean Basin*, edited by I. Hershkovitz. *British Archaeological Reports* S508:255–82.

Frayer, D. W., I. Fiore, C. Lalueza-Fox, J. Radovčić, and L. Bondioli. 2010. Right handed Neandertals: Vindija and beyond. *Journal of Anthropological Science* 88:113–27.

Frayer, D. W., M. Lozano, J. M. Bermúdez de Castro, E. Carbonell, J. L. Arsuaga, J. Radovčić, I. Fiore, and L. Bondioli. 2012. More than 500,000 years of right-handedness in Europe. *Laterality* 17:51–69.

Friedlaender, J. S., and H. L. Bailit. 1969. Eruption times of the deciduous and permanent teeth of natives on Bougainville Island, Territory of New Guinea: A study of racial variation. *Human Biology* 41:51–65.

Fu, Q., M. Hajdinjak, O. T. Moldovan, S. Constantin, S. Mallick, P. Skoglund, N. Patterson, N. Rohland, I. Lazaridis, B. Nickel, B. Viola, K. Prüfer, M. Meyer, J. Kelso, D. Reich, and S. Pääbo. 2015. An early modern human from Romania with a recent Neanderthal ancestor. *Nature* 524:216–19.

Fujita, H. 2012. Periodontal Diseases in Anthropology. In *Periodontal Diseases: A Clinician's Guide*, edited by K. Manaki, 279–94. Rijeka: INTECH Open Access Publishers. doi:10.5772/27146.

Fukase, H., O. Kondo, and H. Ishida. 2015. Size and placement of developing anterior teeth in immature Neanderthal mandibles from Dederiyeh Cave, Syria: Implications for emergence of the modern human chin. *American Journal of Physical Anthropology* 156:482–88.

Galbany, J., and A. Pérez-Pérez. 2006. Tamaño dental, desgaste oclusal y microestriación dentaria en primates Hominoidea. *Revista Española de Antropología Biológica* 26:11–18.

Galbany, J., L. M. Martínez, O. Hiraldo, V. Espurz, F. Estebaranz, M. Sousa, H. Martínez-López-Amor, A. M. Medina, M. Farrés, A. Bonnin, C. Bernis, D. Turbon, and A. Pérez-Pérez. 2004a. *Teeth: Catálogo de los moldes de dientes de homínidos de la Universitat de Barcelona*. Barcelona: Universitat de Barcelona.

Galbany, J., L. M. Martínez, and A. Pérez-Pérez. 2004b. Tooth replication techniques, SEM imaging, and microwear analysis in primates: Methodological obstacles. *Anthropologie* 41:5–12.

Galbany, J., L. M. Martínez, H. M. López-Amor, V. Espurz, O. Hiraldo, A. Romero, J. De Juan, and A. Pérez-Pérez. 2005. Error rates in dental buccal microwear quantification using Scanning Electron Microscopy. *Scanning* 27:23–29.

Galbany, J., J. Altmann, A. Pérez-Pérez, and S. C. Alberts. 2011. Age and individual foraging behavior predict tooth wear in Amboseli baboons. *American Journal of Physical Anthropology* 44:51–59.

Galván, B., C. M. Hernández, C. Mallol, N. Mercier, A. Sistiaga, and V. Soler. 2014. New evidence of early Neanderthal disappearance in the Iberian Peninsula. *Journal of Human Evolution* 75:16–27.

Gambier, D. H., B. Maureille, and R. White. 2004. Vestiges humains des niveaux de l'Aurignacien ancien du site de Brassempouy (Landes). *Bulletins et Mémoires de la Société d'Anthropologie de Paris* 16:49–88.

Garralda, M. D. 2006. Los Neandertales en la Península Ibérica. *Munibe* 57:289–314.

Garralda, M. D., B. Galván, C. M. Hernández, C. Mallol, J. A. Gómez, and B. Maureille. 2014. Neanderthals from El Salt (Alcoy, Spain) in the context of the latest Middle Palaeolithic populations from the southeast of the Iberian Peninsula. *Journal of Human Evolution* 75:1–15.

Genovés, S. 1967. Proportionality of the long bones and their relation to stature among Mesoamericans. *American Journal of Physical Anthropology* 26:67–78.

Geraads, D., and E. Tchernov. 1983. Fémurs humains du pléistocène moyen de Gesher Benot Ya'acov (Israël). *L'Anthropologie* 87:138–41.

Gibert, J., M. J. Walker, A. Malgosa, F. Sánchez, P. J. Pomery, D. Hunter, A. Arribas, and A. Maillo. 1994. Hominids in Spain: Ice Age Neanderthals from Cabezo Gordo. *Research & Exploration* 10:120–23.

Gómez-Olivencia, A., C. Couture-Veschambre, S. Madelaine, and B. Maureille, B. 2013a. The vertebral column of the Regourdou Neandertal. *Journal of Human Evolution* 64:582–607.

Gómez-Olivencia, A., E. Been, J. L. Arsuaga, and J. T. Stock. 2013b. The Neandertal vertebral column 1: The cervical spine. *Journal of Human Evolution* 64:608–30.

Gómez-Robles, A. 2010. Análisis de la forma dental en la filogenia humana: Tendencias y modelos evolutivos basados en métodos de morfometría geométrica. PhD diss., University of Granada.

Gómez-Robles, A., M. Martinón-Torres, J. M. Bermúdez de Castro, A. Margvelashvili, M. Bastir, A. Arsuaga, A. Pérez-Pérez, F. Estebaranz, and L. Martínez. 2007. A geometric morphometric analysis of hominin upper first molar shape. *Journal of Human Evolution* 53:272–85.

Gómez-Robles, A., M. Martinón-Torres, J. M. Bermúdez de Castro, L. Prado, S. Sarmiento, and J. L. Arsuaga. 2008. Geometric morphometric analysis of the crown morphology of the lower first premolar of hominins with special attention to Pleistocene *Homo*. *Journal of Human Evolution* 55:627–38.

Gómez-Robles, A., A. J. Olejniczak, M. Martinón-Torres, L. Prado-Simón, and J. M. Bermúdez de Castro. 2011a. Evolutionary novelties and losses in geometric morphometrics: A practical approach through hominin molar morphology. *Evolution.* 65:1772–90.

Gómez-Robles, A., M. Martinón-Torres, J. M. Bermúdez de Castro, L. Prado-Simón, and J. L. Arsuaga. 2011b. A geometric morphometric analysis of hominin upper premolars. Shape variation and morphological integration. *Journal of Human Evolution* 61:688–702.

Goodman, A. H., and J. C. Rose. 1990. Assessment of systemic physiological perturbations from dental enamel hypoplasias and associated histological structures. *American Journal of Physical Anthropology* 33:59–110.

Goose, D. H. 1967. Preliminary study of tooth size in families. *Journal of Dental Research* 46:959–62.

Grayson, D. K., and F. Delpech. 2003. Ungulates and the Middle-to-Upper Paleolithic transition at Grotte XVI (Dordogne, France). *Journal of Archaeological Science* 30:1633–48.

Grimaud-Hervé, D. 1997. *L'Évolution de l'Encéphale chez* Homo erectus *et* Homo sapiens: *Exemples de l'Asie et de l'Europe*. Paris: CNRS Éditions.

Grine, F. E. 2005. Enamel thickness of deciduous and permanent molars in modern *Homo sapiens*. *American Journal of Physical Anthropology* 126:14–31.

Grine, F. E., A. J. Gwinnett, and J. H. Oaks. 1990. Early hominid dental pathology: Interproximal caries in 1.5 million-year-old *Paranthropus robustus* from Swartkrans. *Archives Oral Biology* 35:381–86.

Grine, F., W. L. Jungers, P. V. Tobias, and O. M. Pearson. 1995. Fossil *Homo* femur from Berg Aukas, northern Namibia. *American Journal of Physical Anthropology* 97:151–85.

Grine, F. E., P. S. Ungar, and M. F. Teaford. 2002. Error rates in dental microwear quantification using scanning electron microscopy. *Scanning* 24:144–53.

Grine, F. E., H. F. Smith, C. P. Hessy, and E. J. Smith. 2009. Phenetic affinities of Plio-Pleistocene *Homo* fossils from South Africa: Molar cusp proportions. In *The First Humans: Origin and Early Evolution of the Genus* Homo, edited by F. E. Grine, J. G. Fleagle, and R. E. Leakey, 49–64. Dordrecht: Springer.

Guatelli-Steinberg, D. 2009. Recent studies of dental development in Neandertals: Implications for Neandertal life histories. *Evolutionary Anthropology* 18:9–20.

Guatelli-Steinberg, D., C. S. Larsen, and D. L. Hutchinson. 2004. Prevalence and the duration of linear enamel hypoplasia: A comparative study of Neandertals and Inuit foragers. *Journal of Human Evolution* 47:65–84.

Gügel, I. L., G. Grupe, and K. H. Kunzelmann. 2001. Simulation of dental microwear: Characteristic traces by opal phytoliths give clues to ancient human dietary behavior. *American Journal of Physical Anthropology* 114:124–38.

Gustafsson, A., and P. Lindenfors. 2009. Latitudinal patterns in human stature and sexual stature dimorphism. *Annals of Human Biology* 36:74–87.

Hardy, B. L. 2010. Climatic variability and plant food distribution in Pleistocene Europe: Implications for Neanderthal diet and subsistence. *Quaternary Science Reviews* 29:662–79.

Hardy, B. L., and M. H. Moncel. 2011. Neanderthal use of fish, mammals, birds, starchy plants and wood 125–250,000 years ago. *PLoS ONE* 6:e23768.

Hardy, B. L., M. Kay, A. E. Marks, and K. Monigal. 2001. Stone tool function at the Paleolithic sites of Starosele and Buran Kaya III, Crimea: Behavioral implications. *Proceedings of the National Academy of Sciences, USA* 98:10972–77.

Hauser, G., and C. F. DeStefano. 1989. *Epigenetic Variants of the Human Skull*. Stuttgart: E. Schweizerbart'sche Verlagsbuchhandlung.

Heim, J. L. 1976. Les hommes fossiles de La Ferrassie I: Le gisement. Les squelettes adultes (crâne et squelette du tronc). *Archives de l'Institut de Paléontologie Humaine* 35:1–331.

Heim, J. L., and J. Granat. 1995. La mandibule de l'enfant néandertalien de Malarnaud (Ariége): Une nouvelle approche anthropologique par la radiographie et la tomodensitométrie. *Anthropologie et Préhistoire* 106:79–96.

Hemming, S. R. 2004. Heinrich Events: Massive Late Pleistocene detritus layers of the North Atlantic and the global climate imprint. *Review of Geophysics* 42:RG1005.

Henry, A. G. 2010. Plant foods and the dietary ecology of Neandertals and modern humans. PhD diss., George Washington University.

Henry, A. G., K. D. Gordon, E. Trinkaus, and A. S. Brooks. 2006. Teeth as tools? A comparison of Neanderthal and early modern human incisor microwear (abstract). *PaleoAnthropology* 2006: A88.

Henry, A. G., A. S. Brooks, and D. R. Piperno. 2011. Microfossils in calculus demonstrate consumption of plants and cooked foods in Neanderthal diets (Shanidar III, Iraq; Spy I and II, Belgium). *Proceedings of the National Academy of Sciences, USA* 108:486–91.

Henry, A. G., A. S. Brooks, and D. R. Piperno. 2014. Plant foods and the dietary ecology of Neanderthals and early modern humans. *Journal of Human Evolution* 69:44–54.

Hershkovitz, I., P. Smith, R. Sarig, R. Quam, L. Rodríguez, R. García, J. L. Arsuaga, R. Barkai, and A. Gopher. 2011. Middle Pleistocene dental remains from Qesem Cave (Israel). *American Journal of Physical Anthropology* 144:575–92.

Hildebrand, T., and P. Rüegsegger. 1997. A new method for the model-independent assessment of thickness in three-dimensional images. *Journal of Microscopy* 185:67–75.

Hillson, S. W. 1992. Impression and replica methods for studying hypoplasia and perikymata on human tooth crown surfaces from archaeological sites. *International Journal of Osteoarchaeology* 2:65–78.

———. 1996. *Dental Anthropology*. Cambridge: Cambridge University Press.

———. 2001. Recording dental caries in archaeological human remains. *International Journal of Osteoarchaeology* 11:249–89.

———. 2006. Dental morphology, proportions, and attrition. In *Early Modern Human Evolution in Central Europe: The People of Dolní Věstonice and*

Pavlov, edited by E. Trinkaus and J. A. Svoboda, 179–223. New York: Oxford University Press.

———. 2008. The current state of dental decay. In *Technique and Application in Dental Anthropology* (Vol. 53), edited by J. D. Irish and G. C. Nelson, 111–35. Cambridge: Cambridge University Press.

Hillson, S. W., and S. Bond. 1997. Relationship of enamel hypoplasia to the pattern of tooth crown growth: A discussion. *American Journal of Physical Anthropology* 104:89–104.

Hillson, S. W., and E. Trinkaus. 2002. Comparative dental crown metrics. In *Portrait of the Artist as a Child. The Gravettian Human Skeleton from the Abrigo do Lagar Velho and Its Archeological Context*, edited by J. Zilhão and E. Trinkaus. *Trabalhos de Arqueologia* 22:356–64.

Hillson, S. W., R. G. Franciscus, T. W. Holliday, and E. Trinkaus. 2006. The ages-at-death. *In Early Modern Human Evolution in Central Europe: The People of Dolní Věstonice and Pavlov*, edited by E. Trinkaus and J. A. Svoboda, 31–45. New York: Oxford University Press.

Hillson, S. W., S. A. Parfitt, S. M. Bello, M. B. Roberts, and C. B. Stringer. 2010. Two hominin incisor teeth from the Middle Pleistocene site of Boxgrove, Sussex, England. *Journal of Human Evolution* 59:493–503.

Hlusko, L. J. 2015. Elucidating the evolution of hominid dentition in the age of phenomics, modularity, and quantitative genetics. *Annals of Anatomy–Anatomischer Anzeiger* 203:3–11. doi:10.1016/j.anat.2015.05.001.

Hlusko, L. J., J. P. Carlson, D. Guatelli-Steinberg, K. L. Krueger, B. Mersey, P. S. Ungar, and A. Defleur. 2013. Neanderthal teeth from Moula-Guercy, Ardèche, France. *American Journal of Physical Anthropology* 151(3):477–91.

Hoffmann, D. L., A. W. G. Pike, K. Wainer, and J. Zilhão. 2013. New U-series results for the speleogenesis and the Palaeolithic archaeology of the Almonda karstic system (Torres Novas, Portugal). *Quaternary International* 294:168–82.

Holliday, T. W. 1995. Body size and proportions in the Late Pleistocene Western Old World and the origins of modern humans. PhD diss., University of New Mexico.

———. 2006. Neanderthals and modern humans: An example of a mammalian syngameon? In *Neanderthals Revisited: New Approaches and Perspectives*, edited by K. Harvati and T. Harrison, 289–306. New York: Springer. .

Holliday, T. W., J. R. Gautney, and L. Friedl. 2014. Right for the wrong reasons: Reflections on modern human origins in the post-Neanderthal genome era. *Current Anthropology* 55:696–724.

Holt, B. M. 2003. Mobility in Upper Paleolithic and Mesolithic Europe: Evidence from the lower limb. *American Journal of Physical Anthropology* 122:200–215.

Horvath, J. E., G. L. Ramachandran, O. Fedrigo, W. J. Nielsen, C. C. Babbitt, E. M. St. Clair, L. W. Pfefferle, J. Jernvall, G. A. Wray, and C. E. Wall. 2014. Genetic comparisons yield insight into the evolution of enamel thickness during human evolution. *Journal of Human Evolution* 73:75–87.

Howell, N. 1982. Village composition implied by paleodemographic life table: The Libben site. *American Journal of Physical Anthropology* 59:263–69.

Humphrey, L. T., I. De Groote, J. Morales, N. Barton, S. Collcutt, C. B. Ramsey, and A. Bouzouggar. 2014. Earliest evidence for caries and exploitation of starchy plant foods in Pleistocene hunter-gatherers from Morocco. *Proceedings of the National Academy of Science, USA* 111:954–59.

Jidoi, K., T. Nara, and Y. Dodo. 2000. Bony bridging of the mylohyoid groove of the human mandible. *Anthropological Science* 108:345–70.

Johansson, A. K., R. Sorvari, D. Birkhed, and J. H. Meurman. 2001. Dental erosion in deciduous teeth: An in vivo and in vitro study. *Journal of Dentistry* 29:333–40.

Jolly, C. J. 2001. A proper study of mankind: Analogies from the Papionin monkeys and their implications for human evolution. *Yearbook of Physical Anthropology* 44:177–204.

Kaifu, Y., K. Kazutaka, G. C. Townsend, and L. C. Richards. 2003. Tooth wear and the "design" of the human dentition: A perspective from evolutionary medicine. *Yearbook Physical Anthropology* 46:47–61.

Kallay. J. 1970. Komparativne napomene o čeljustima Krapinskih praljudi obzirom na položaj među Hominidima (Comparative observations on the mandible of a Krapina Neanderthal man in respect to its position among Hominids). In *Krapina 1899–1969*, edited by M. Malez, 153–64. Zagreb: Jugoslavenske akademije znanosti I umjetmosti.

Kay, R. F. 1975. The functional adaptation of primate teeth. *American Journal of Physical Anthropology* 43:195–216.

Kay, R. F., and K. M. Hiiemae. 1974. Jaw movement and tooth use in recent and fossil primates. *American Journal of Physical Anthropology* 40:227–56.

King, T., P. Andrews, and B. Boz. 1999. Effect of

taphonomic processes on dental microwear. *American Journal of Physical Anthropology* 108:359–73.

Kirschbaum, M. U. F. 2004. Direct and indirect climate change effects on photosynthesis and transpiration. *Plant Biology* 6:242–53.

Kivell, T. L., A. S. Deane, M. W. Tocheri, C. M. Orr, P. Schmid, J. Hawks, L. R. Berger, and S. E. Churchill. 2015. The hand of *Homo naledi*. *Nature Commununications* 6:8431.

Koby, F. E. 1956. Une incisive néandertalienne trouvée en Suisse. *Verhandlungen der Nuturforschers Gesellschaft in Basel* 67:1–15.

Kono, R. T. 2004. Molar enamel thickness and distribution patterns in extant great apes and humans: New insights based on a 3-dimensional whole crown perspective. *Anthropological Science* 112:121–46.

Krueger, K. L. 2011. Dietary and behavioral strategies of Neandertals and anatomically modern humans: Evidence from anterior dental microwear texture analysis. PhD diss., University of Arkansas.

Krueger K. L., and P. S. Ungar. 2012. Anterior dental microwear texture analysis of the Krapina Neandertals. *Central European Journal of Geosciences* 4(4):651–62.

Kupczik, K., and J. J. Hublin. 2010. Mandibular molar root morphology in Neanderthals and Late Pleistocene and recent *Homo sapiens*. *Journal of Human Evolution* 59:525–41.

Lacy, S. A. 2014a. The oral pathological conditions in the Broken Hill (Kabwe) cranium. *International Journal of Paleopathology* 7:57–63.

———. 2014b. Oral health and its implications in Late Pleistocene Western Eurasian humans. PhD diss., Washington University.

Lacy, S. A., X. J. Wu, C. Z. Jin, D. G. Qin, Y. J. Cai, and E. Trinkaus. 2012. Dentoalveolar paleopathology of the early modern humans from Zhirendong, South China. *International Journal of Paleopathology* 2:10–18.

Lalueza, C. 1992. Information obtained from the microscopic examination of cultural striations in human dentition. *International Journal of Osteoarchaeology* 2:155–69.

Lalueza, C., and D. W. Frayer. 1997. Nondietary marks in the anterior dentition of the Krapina Neanderthals. *International Journal of Osteoarchaeology* 7:133–49.

Lalueza, C., and A. Pérez-Pérez. 1994. Dietary information through the examination of plant phytoliths on the enamel surface of human dentition *Journal of Archaeological Science* 21:29–34

Lalueza, C., A. Pérez-Pérez, E. Chimenos, J. Maroto, and D. Turbón. 1993a. Estudi radiogràfic i microscòpic de la mandíbula de Banyoles: Patologies i estat de conservació. In *La Mandíbula de Banyoles en el Context dels Fossils Humans del Pleistocè*, edited by J. Maroto, 135–44. Girona: Centre d'Investigacions Arqueológiques.

Lalueza, C., A. Pérez-Pérez, and D. Turbon. 1993b. Microscopic study of the Banyoles mandible (Girona, Spain): Diet, cultural activity, and toothpick use. *Journal of Human Evolution* 24:281–300.

———. 1996. Dietary inferences through buccal microwear analysis of Middle and Upper Pleistocene human fossils. *American Journal of Physical Anthropology* 100:367–87.

Lanfranco, L. P., and S. Eggers. 2012. Caries through time: An anthropological overview. In *Contemporary Approach to Dental Caries*, edited by M. Y. Li, 3–34. Rijeka: INTECH Open Access Publishers. doi:10.5772/38059.

Lavigne, S. E., and J. E. Molto. 1995. System of measurement of the severity of periodontal disease in past populations. *International Journal of Osteoarchaeology* 5:265–73.

Lebel, S., and E. Trinkaus. 2002a. A carious Neandertal molar from the Bau de l'Aubesier, Vaucluse, France. *Journal of Archaeological Science* 29:555–57.

———. 2002b. Middle Pleistocene human remains from the Bau de l'Aubesier. *Journal of Human Evolution* 43:659–85.

Le Cabec, A., P. Gunz, K. Kupczik, J. Braga, and J. J. Hublin. 2013. Anterior tooth root morphology and size in Neanderthals: Taxonomic and functional implications. *Journal of Human Evolution* 64:169–93.

Le Luyer, M. 2016. Évolution de l'architecture interne des dents humaines à la fin du Pléistocène et au début de l'Holocène. PhD diss., Université de Bordeaux.

Le Luyer, M., S. Rottier, and P. Bayle. 2014. Comparative patterns of enamel thickness topography and oblique molar wear in two Early Neolithic and Medieval populations. *American Journal of Physical Anthropology* 155:162–72.

Liu, W., L. A. Schepartz, S. Xing, S. Miller-Antonio, X. J. Wu, E. Trinkaus, and M. Martinón-Torres. 2013. Late Middle Pleistocene hominin teeth from Paxian Dadong, south China. *Journal of Human Evolution* 64:337–55.

Liu, W., M. Martinón-Torres, Y. J. Cai, H. W. Tong, S. W. Pei, M. J. Sier, X. H. Wu, R. L. Edwards, H. Cheng, Y. Y. Li, X. X. Yang, J. M. Bermúdez de Castro, and X. J. Wu. 2015. The earliest unequivocally modern humans in southern China. *Nature* 526:696–99.

Liversidge, H. M., and T. Molleson. 2004. Variation in crown and root formation and eruption of human deciduous teeth. *American Journal of Physical Anthropology* 123:172–80.

Löe, H., Å. G. Ånerud, H. Boysen, and E. Morrison. 1986. Natural history of periodontal disease in man. *Journal of Clinical Periodontology* 13:431–40.

Lorenzo, C. 2007. Evolución de la mano en los homínidos: Análisis morfológico de los fósiles de la Sierra de Atapuerca. PhD diss., Universidad Complutense de Madrid.

Lorenzo, C., J. M. Carretero, J. L. Arsuaga, I. Martínez, A. Gracia, and R. Quam. 2012. Hands, laterality and language: Hand morphology in the Sima de los Huesos site (Sierra de Atapuerca, Spain) (abstract). *American Journal of Physical Anthropology Supplement* 54:195–96.

Lorenzo, J. I., and L. Montes. 2001. Restes néandertaliens de la Grotte de "Los Moros de Gabasa" (Huesca, Espagna). In *Les premiers hommes modernes de la Péninsule Ibérique*, edited by J. Zilhão, T. Aubry, and A. F. Carvalho. *Trabalhos de Arqueologia* 17:77–86.

Lozano, M., J. M. Bermúdez de Castro, E. Carbonell, and J. L. Arsuaga. 2008. Nonmasticatory uses of anterior teeth of Sima de los Huesos individuals (Sierra de Atapuerca, Spain). *Journal of Human Evolution* 55:713–28.

Lozano, M., M. Mosquera, J. M. Bermúdez de Castro, J. L. Arsuaga, and E. Carbonell. 2009. Right handedness of *Homo heidelbergensis* from Sima de los Huesos (Atapuerca, Spain) 500,000 years ago. *Evolution and Human Behavior* 30(5):369–76.

Lozano-Ruiz, M., J. M. Bermúdez de Castro, M. Martinón-Torres, and S. Sarmiento. 2004. Cutmarks on fossil human anterior teeth of the Sima de los Huesos Site (Atapuerca, Spain). *Journal of Archaeological Science* 31:1127–35.

Lukacs, J. R., and J. N. Pal. 1993. Mesolithic subsistence in north India: Inferences from dental attributes. *Current Anthropology* 34:745–65.

Lunt, R. C., and B. D. Law. 1974. A review of the chronology of calcification of deciduous teeth. *Journal of the American Dental Association* 89:599–606.

Macchiarelli, R., L. Bondioli, A. Débénath, A. Mazurier, J. F. Tournepiche, W. Birch, and C. Dean. 2006. How Neanderthal molar teeth grew. *Nature* 444:748–51.

Macchiarelli, R., L. Bondioli, and A. Mazurier. 2008. Virtual dentitions: Touching the hidden evidence. In *Technique and Application in Dental Anthropology*, edited by J. D. Irish and G. C. Nelson, 426–48. Cambridge: Cambridge University Press.

Macchiarelli, R. P. Bayle, L. Bondioli, A. Mazurier, and C. Zanolli. 2013. From outer to inner structural morphology in dental anthropology. The integration of the third dimension in the visualization and quantitative analysis of fossil remains. In *Anthropological perspectives on tooth morphology: genetics, evoltion, variation*, edited by R. G. Scott and J. D. Irish, 250–277. Cambridge: Cambridge University Press.

Macho, G. A., and I. R. Spears. 1999. The effects of loading on the biomechanical behavior of molars of *Homo*, *Pan* and *Pongo*. *American Journal of Physical Anthropology* 109:211–77.

Mahoney, P. 2006. Dental microwear from the Natufian hunter-gatherers and early Neolithic farmers: Comparisons within and between samples. *American Journal of Physical Anthropology* 130:308–19.

———. 2010. Two-dimensional patterns of human enamel thickness on deciduous (dm1, dm2) and permanent first (M1) mandibular molars. *Archives of Oral Biology* 55:115–26.

———. 2013. Testing functional and morphological interpretations of enamel thickness along the deciduous tooth row in human children. *American Journal of Physical Anthropology* 151:518–25.

Mallegni, F., and E. Trinkaus. 1997. A reconsideration of the Archi 1 Neandertal mandible. *Journal of Human Evolution* 33:651–68.

Marshall, J., and R. Gardner. 1957. *The Hunters.* !Kung Series (film). Watertown, MA: Documentary Educational Resources.

Martin, H. 1923. *l'Homme fossile de la Quina.* Paris: Doin.

Martin, L. 1985. Significance of enamel thickness in hominoid evolution. *Nature* 324:260–63.

Martínez, L. M., and A. Pérez-Pérez. 2004. Post-mortem wear as indicator of taphonomic processes affecting enamel surfaces of hominin teeth from Laetoli and Olduvai (Tanzania): Implications to dietary interpretations. *Anthropologie* 42:37–42.

Martinón-Torres, M. 2006. Evolución del aparato dental en homínidos: estudio de los dientes humanos del Pleistoceno de la Sierra de Atapuerca (Burgos). PhD diss., Universidad de Santiago de Compostela.

Martinón-Torres, M., J. M. Bermúdez de Castro, A. Gómez-Robles, L. Prado-Simón, and J. L. Arsuaga. 2012. Morphological description and comparison of the dental remains from Atapuerca–Sima de los Huesos site (Spain). *Journal of Human Evolution* 62:7–58.

Martinón-Torres, M., M. Martínez de Pinillos, M. M. Skinner, L. Martín-Francés, A. Gracia-Téllez, I. Martínez, J. L. Arsuaga, and J. M. Bermúdez de Castro. 2014. Talonid crests expression at the enamel–dentine junction of hominin lower permanent and deciduous molars. *Comptes Rendus Palevol* 13:223–34.

Marzke, M. W., M. W. Tocheri, B. Steinberg, J. D. Femiani, S. P. Reece, R. L. Linscheid, C. M. Orr, and R. F. Marzke. 2010. Comparative 3D quantitative analyses of trapeziometacarpal joint surface curvatures among living catarrhines and fossil hominins. *American Journal of Physical Anthropology* 141:38–51.

Matiegka, J. 1934. *Homo předmostensis: Fosilní člověk z Předmostí na Moravě I, Lebky.* Prague: Česká Akademie Věd a Umění.

Maureille, B., H. Rougier, F. Houët, and B. Vandermeersch. 2001. Les dents inférieures du Néandertalien Regourdou 1 (site de Regourdou, commune de Montignac, Dordogne): Analyses métriques et comparatives. *Paléo* 13:183–200.

McCown, T. D., and A. Keith. 1939. *The Stone Age of Mount Carmel II: The Fossil Human Remains from the Levalloiso-Mousterian.* Oxford: Clarendon Press.

McHenry, H. M. 1994. Early hominid postcrania: Phylogeny and function. In *Integrative Paths to the Past*, edited by R. S. Corruccini and R. L. Ciochon, 168–251. Englewood Cliffs, NJ: Prentice Hall.

McHenry, H. M., R. S. Corruccini, and F. C. Howell. 1976. Analysis of an early hominid ulna from the Omo Basin, Ethiopia. *American Journal of Physical Anthropology* 44:295–304.

McKern, T. W., and T. D. Stewart. 1957. *Skeletal Age Changes in Young American Males.* Headquarters Quartermaster Research and Development Command, Technical Report EP-45:1–179. Natick, MA: Quartermaster Research and Development Center.

Meadows, L., and R. Jantz. 1992. Estimation of stature from metacarpal lengths. *Journal of Forensic Sciences* 37:147–54.

Mednikova, M. B. 2011. A proximal pedal phalanx of a Paleolithic hominin from Denisova Cave, Altai. *Archaeology, Ethnology, and Anthropology of Eurasia* 39:129–38.

Meiklejohn, C., J. Baldwin, and C. Schentag. 1988. Caries as a probable dietary marker in the Western European Mesolithic. In *Diet and Subsistence: Current Archaeological Perspectives*, edited by B. V. Kennedy and G. M. Lemoine, 273–79. Calgary: Archaeological Association of the University of Calgary.

Merbs, C. F. 1983. *Patterns of Activity-Induced Pathology in a Canadian Inuit Population.* Ottawa: National Museums of Canada.

Miles, A. E. W. 1962. Assessment of the ages of a population of Anglo-Saxons from their dentitions. *Proceedings of the Royal Society of Medicine* 55:881–86.

Mittra, E. S., M. F. Smith, P. Lemelin, and W. L. Jungers. 2007. Comparative morphometrics of the primate apical tuft. *American Journal of Physical Anthropology* 134:449–59.

Molnar, P. 2008. Dental wear and oral pathology: Possible evidence and consequences of habitual use of teeth in a Swedish Neolithic sample. *American Journal of Physical Anthropology* 136(4):423–31.

Molnar, S. 1971. Human tooth wear tooth function and cultural variability. *American Journal of Physical Anthropology* 34:175–90.

Molnar, S., and D. G. Gantt. 1977. Functional implications of primate enamel thickness. *American Journal of Physical Anthropology* 46:447–54.

Molnar, S., and I. M. Molnar. 1985. The incidence of enamel hypoplasia among the Krapina Neandertals. *American Anthropologist* 87:536–49.

Molnar, S., and S. C. Ward. 1977. On the Hominid masticatory complex: Biomechanical and evolutionary perspectives. *Journal of Human Evolution* 6:557–68.

Molnar, S., C. Hildebolt, I. M. Molnar, J. Radovčić, and M. Gravier. 1993. Hominid enamel thickness: I. The Krapina Neanderthals. *American Journal of Physical Anthropology* 92:131–38.

Moorrees, C. F. A. 1957. *The Aleut Dentition.* Cambridge, MA: Harvard University Press.

Moorrees, C. F. A., E. A. Fanning, and E. E. Hunt. 1963a. Formation and resorption of three deciduous teeth in children. *American Journal of Physical Anthropology* 19:99–108.

———. 1963b. Age variation of formation stages for ten permanent teeth. *Journal of Dental Research* 42:1490–1502.

Morton, D. J. 1922–24. Evolution of the human foot. *American Journal of Physical Anthropology* 5:305–36; 7:1–52.

Musgrave, J. H. 1973. The phalanges of Neanderthal and Upper Palaeolithic hands. In *Human Evolution*, edited by M. H. Day, 59–85. London: Taylor and Francis.

Musgrave, J. H., and N. K. Harneja. 1978. The estimation of adult stature from metacarpal bone length. *American Journal of Physical Anthropology* 48:113–19.

Nagurka, M. L., and W. C. Hayes. 1980. An interactive graphics package for calculating cross-sectional properties of complex shapes. *Journal of Biomechanics* 13:59–64.

Neely, A. L., T. R. Holford, H. Löe, Å. G. Ånerud, and H. Boysen. 2001. The natural history of periodontal disease in man: Risk factors for progression of attachment loss in individuals receiving no oral health care. *Journal of Periodontology* 72:1006–15.

———. 2005. The natural history of periodontal disease in humans: Risk factors for tooth loss in caries free subjects receiving no oral health care. *Journal of Clinical Periodontology* 32:984–93.

NESPOS database. 2015. Neanderthal Studies Professional Online Service. http://www.nespos.org.

Niewoehner, W. A. 2001. Behavioral inferences from the Skhul/Qafzeh early modern human hand remains. *Proceedings of the National Academy of Sciences, USA* 98:2979–84.

———. 2006. Neanderthal hands in their proper perspective. In *Neanderthals Revisited: New Approaches and Perspectives*, edited by K. Harvati and T. Harrison, 157–90. New York: Springer.

Niewoehner, W. A., A. H. Weaver, and E. Trinkaus. 1997. Neandertal capitate-metacarpal articular morphology. *American Journal of Physical Anthropology* 103:219–33.

O'Connor, C. F., R. G. Franciscus, and N. E. Holton. 2005. Bite force production capability and efficiency in Neandertals and modern humans. *American Journal of Physical Anthropology* 127(2):129–51.

Odgaard, A., J. Kabel, B. van Rietbergen, M. Dalstra, and R. Huiskes. 1997. Fabric and elastic principal directions of cancellous bone are closely related. *Journal of Biomechanics* 30:487–95.

Ogilvie, M. D., B. K. Curran, and E. Trinkaus. 1989. Incidence and patterning of dental enamel hypoplasia among the Neandertals. *American Journal of Physical Anthropology* 79:25–41.

O'Higgins, P. 2000. The study of morphological variation in the hominid fossil record: Biology, landmarks, and geometry. *Journal of Anatomy* 197:103–20.

Olejniczak, A. J., and F. E. Grine. 2006. Assessment of the accuracy of dental enamel thickness measurements using microfocal x-ray computed tomography. *Anatomical Record* 288A:263–75.

Olejniczak, A. J., P. Tafforeau, T. M. Smith, H. Temming, and J. J. Hublin. 2007. Compatibility of microtomographic imaging systems for dental measurements. *American Journal of Physical Anthropology* 134:130–34.

Olejniczak, A. J., T. M. Smith, R. N. Feeney, R. Macchiarelli, A. Mazurier, L. Bondioli, A. Rosas, J. Fortea, M. de la Rasilla, A. Garcia-Tabernero, J. Radovcic, M. M. Skinner, M. Toussaint, and J. J. Hublin. 2008a. Dental tissue proportions and enamel thickness in Neandertal and modern human molars. *Journal of Human Evolution* 55:12–23.

Olejniczak, A. J., P. Tafforeau, R. N. M. Feeney, and L. B. Martin. 2008b. Three-dimensional primate molar enamel thickness. *Journal of Human Evolution* 54:187–95.

O'Neill, M. C., and C. B. Ruff. 2004. Estimating human long bone cross-sectional geometric properties: A comparison of noninvasive methods. *Journal of Human Evolution* 47:221–35.

Oztunc, H., O. Yoldas, and E. Nalbantoglu. 2006. The periodontal disease status of the historical population of Assos. *International Journal of Osteoarchaeology* 16:76–81.

Pablos, A., A. Gómez-Olivencia, A. García-Pérez, I. Martínez, C. Lorenzo, and J. L. Arsuaga. 2013. From toe to head: Use of robust regression methods in stature estimation based on foot remains. *Forensic Science International* 266:299e1–299e7.

Pampush, J. D., A. C. Duque, B. R. Burrows, D. J. Daegling, W. F. Kenney, and W. S. McGraw. 2013. Homoplasy and thick enamel in primates. *Journal of Human Evolution* 64:216–24.

Patte, E. 1960. Découverte d'un Néandertalien dans la Vienne. *L'Anthropologie* 64:512–17.

Pearson, O. M., and F. E. Grine. 1997. Reanalysis of the hominid radii from Cave of Hearths and Klasies River Mouth, South Africa. *Journal of Human Evolution* 32:577–92.

Pérez, P. J., J. L. Arsuaga, and J. M. Bermúdez de Castro. 1982. Atypical tooth wear in fossil man. *Paleopathology Newsletter* 39:11–13.

Pérez, S. I., V. Bernal, and P. N. Gonzalez. 2006. Differences between sliding semi-landmark methods in geometric morphometrics, with an application to human craniofacial and dental variation. *Journal of Anatomy* 208:769–84.

Pérez-Pérez, A., C. Lalueza, and D. Turbón. 1994. Intradividual and intragroup variability of buccal tooth striation pattern. *American Journal of Physical Anthropology* 94:175–87.

Pérez-Pérez, A., J. M. Bermúdez de Castro, and J. Arsuaga. 1999. Non-occlusal dental microwear analysis of 300,000-year-old *Homo heidelbergensis* teeth from Sima de Los Huesos (Sierra de Atapuerca, Spain). *American Journal of Physical Anthropology* 108:433–57.

Pérez-Pérez, A., V. Espurz, J. M. Bermúdez de

Castro, M. A. de Lumley, and D. Turbón. 2003. Non-occlusal dental microwear variability in a sample of Middle and Late Pleistocene human populations from Europe and the Near East. *Journal of Human Evolution* 44:497–513.

Peters, C. 1982. Electron-optical microscope study of incipient dental microdamage from experimental seed and bone crushing. *American Journal of Physical Anthropology* 57:283–301.

Pfeiffer, S., and P. King. 1983. Cortical bone formation and diet among protohistoric Iroquoians. *American Journal of Physical Anthropology* 60:23–28.

Pinilla, B. 2012. Dieta y adaptaciones ecológicas de las poblaciones humanas del Pleistoceno Medio y Superior. PhD diss., Universitat de Barcelona.

Pinilla, B., A. Romero, and A. Pérez-Pérez. 2011. Age-related variability in buccal dental microwear in Middle and Upper Pleistocene human populations. *Anthropological Review* 74:25–37.

Pinilla, B., A. Romero, A. Pérez-Pérez, and E. Trinkaus. 2014. Buccal dental microwear and diet of the Sunghir Upper Paleolithic modern humans (in Russian). *Archaeology, Ethnology and Anthropology in Eurasia* 58:131–42.

Piperno, D. 1988. *Phytolith Analysis: An Archaeological and Geological Perspective*. San Diego: Academic Press.

Piperno, D., E. Weiss, I. Holst, and D. Nadel. 2004. Processing of wild cereal grains in the Upper Palaeolithic revealed by starch grain analysis. *Nature* 430:670–73.

Plavcan, J. M., V. Meyer, A. S. Hammond, C. Couture, S. Madelaine, T. W. Holliday, B. Maureille, C. V. Ward, and E. Trinkaus. 2014. The Regourdou Neandertal body size. *Comptes Rendus Palevol* 13:747–54.

Ponce de León, M. S., and C. P. Zollikofer. 1999. New evidence from Le Moustier 1: Computer assisted reconstruction and morphometry of the skull. *Anatomical Record* 254:474–89.

Puech, P. F. 1979. The diet of early man: Evidence from abrasion of teeth and tools. *Current Anthropology* 20:590–92.

Puech, P. F., H. Albertini, and N. T. W. Mills. 1980. Dental destruction in Broken-Hill man. *Journal of Human Evolution* 9:33–39.

Puech, P. F., A. Prone, and H. Albertini. 1981. Reproduction expérimentale des processus d'altération de la surface dentaire par friction non abrasive et non adhésive: application à l'alimentation de l'homme fossile. *Comptes Rendus de l'Académie des Sciences Paris, Ser. II* 293:729–734.

Puech, P. F., A. Prone, and R. Kraatz. 1982. Microscopie de l'usure dentaire chez l'homme fossile: Bol alimentaire et environnement. *Comptes Rendus de l'Académie des Sciences Paris* 290D:1413–16.

Puech, P. F., H. Albertini, and C. Serratrice. 1983. Tooth microwear and dietary pattern in early hominids from Laetoli, Hadar, and Olduvai. *Journal of Human Evolution* 12:721–29.

Puech, P. F., F. Cianfarani, and H. Albertini. 1986. Dental microwear features as an indicator for plant food in early hominids: A preliminary study of enamel. *Human Evolution* 1:507–15.

Puech, P., S. Puech, F. Cianfarani, and H. Albertini. 1987. Tooth wear and dexterity in *Homo erectus*. In *Hominidae*, edited by G. Giacobini, 247–50. Milan: JACA.

Pyle, S. I., A. M. Waterhouse, and W. W. Greulich. 1971. *A Radiographic Standard of Reference for the Growing Hand and Wrist*. Cleveland: Case Western Reserve University Press.

Quam, R. M., J. L. Arsuaga, J. M. Bermúdez de Castro, J. C. Díez, C. Lorenzo, J. M. Carretero, N. García, A. I. Ortega. 2001. Human remains from Valdegoba Cave (Huérmeces, Burgos, Spain). *Journal of Human Evolution* 41:385–435.

Rak, Y. 1986. The Neanderthal: A new look at an old face. *Journal of Human Evolution* 15:151–64.

———. 1991. The pelvis. In *Le Squelette Moustérien de Kébara 2*, edited by O. Bar Yosef and B. Vandermeersch, 147–56. Paris: CNRS.

Ramsey, C. B., T. Higham, A. Bowles, and R. Hedges. 2004. Improvements to the pretreatment of bone at Oxford. *Radiocarbon* 46:155–63.

Raxter, M. H., C. B. Ruff, A. Azab, M. Erfan, M. Soliman, and A. El-Sawaf. 2008. Stature estimation in ancient Egyptians: A new technique based on anatomical reconstruction of stature. *American Journal of Physical Anthropology* 136:147–55.

Rhodes, S. E., M. J. Walker, M. López-Martinez, M. Haber-Uniarte, A. López-Jiménez, A. T. Buitrago-López, and G. Dewar. 2013. Analysis of *Hystrix* specimens recovered from Sima de las Palomas, Murcia, Spain: Identification and Paleoenvironmental revision. *Program with Abstracts, 41st Annual Meeting of the Canadian Association of Physical Anthropology*, 47–48. http://blog.utsc.utoronto.ca/capa/files/2013/09/CAPA-2013-Program-Book1.pdf.

Richards, G. D., R. S. Jabbour, and J. Y. Anderson. 2003. Medial mandibular ramus: Ontogenetic, idiosyncratic, and geographic variation in recent *Homo*, great apes, and fossil hominids. *British Archaeological Reports: International Series* 1138:1–113.

Richards, M. P., and E. Trinkaus. 2009. Isotopic evidence for the diets of European Neandertals and early modern humans. *Proceedings of the National Academy of Sciences USA* 106:16034–39.

Rightmire, G. P., and H. J. Deacon. 1991. Comparative studies of Late Pleistocene human remains from Klasies River Mouth, South Africa. *Journal of Human Evolution* 20:131–56.

Robinson, D. L., P. G. Blackwell, E. C. Stillman, and A. H. Brook. 2002. Impact of landmark reliability on the planar Procrustes analysis of tooth shape. *Archives of Oral Biology* 47:545–54.

Robinson, J. T. 1952. Some hominid features of the ape-man dentition. *Journal of the Dental Association of South Africa* 7:102–13.

Robson Brown, K., S. Tarsuslugil, V. N. Wijayathunga, and R. K. Wilcox. 2014. Comparative finite-element analysis: A single computational modelling method can estimate the mechanical properties of porcine and human vertebrae. *Journal of the Royal Society Interface* 11:20140186. doi:10.1098/rsif.2014.0186.

Robu, M., J. K. Fortin, M. P. Richards, C. C. Schwartz, J. Wynn, C. T. Robbins, and E. Trinkaus. 2013. Isotopic evidence for dietary flexibility among European Late Pleistocene cave bears (*Ursus spelaeus*). *Canadian Journal of Zoology* 91:227–34.

Rodríguez-Vidal, J., F. d'Errico, F. G. Pacheco, R. Blasco, J. Rosell, R. P. Jennings, A. Queffeler, G. Finlayson, D. A. Fa, J. M. G. López, J. S. Carrión, J. J. Negro, S. Finlayson, L. M. Caceres, M. A. Bernal, S. F. Jiménez, and C. Finlayson. 2014. A rock engraving made by Neanderthals in Gibraltar. *Proceedings of the National Academy of Sciences, USA* 111:13301–6.

Rohlf, F. J. 1998. Thin plate spline software, tpsUtil; tpsRelw. Morphometrics at SUNY Stony Brook, Department of Ecology and Evolution, University of Stony Brook, New York.

———. 2005. Thin plate spline software, tpsDig2. Morphometrics at SUNY Stony Brook, Department of Ecology and Evolution, University of Stony Brook, New York.

———. 2007. Thin plate spline software, tpsRelw. Morphometrics at SUNY Stony Brook, Department of Ecology and Evolution, University of Stony Brook, New York.

Ronderos, M., B. L. Pihlstrom, and J. S. Hodges. 2001. Periodontal disease among indigenous people in the Amazon rain forest. *Journal of Clinical Periodontology* 28:995–1003.

Rosas, A. 2001. Occurrence of Neanderthal features in mandibles from the Atapuerca-SH site. *American Journal of Physical Anthropology* 114:74–91.

Rosas, A., and M. Bastir. 2004. Geometric morphometric analysis of allometric variation in the mandibular morphology of the hominids of Atapuerca, Sima de los Huesos site. *Anatomical Record* 278A:551–60.

Rosas, A., C. Martínez-Maza, M. Bastir, A. García-Tabernero, C. Lalueza-Fox, R. Huguet, J. E. Ortiz, R. Julià, V. Soler, T. de Torres, E. Martínez, J. C. Cañaveras, S. Sánchez-Moral, S. Cuezva, J. Lario, D. Santamaría, M. de la Rasilla, and J. Fortea. 2006. Paleobiology and comparative morphology of a late Neandertal sample from El Sidrón, Asturias, Spain. *Proceedings of the National Academy of Sciences, USA* 103:19266–71.

Rose, J. C., K. W. Condon, and A. H. Goodman. 1985. Diet and dentition: Developmental disturbances. In *The Analysis of Prehistoric Diets*, edited by R. I. Gilbert and J. H. Mielke, 281–305. Orlando: Academic Press.

Rougier, H. 2003. *Étude Descriptive et Comparative de Biache-Saint-Vaast 1 (Biache-Saint-Vaast, Pas-de-Calais, France)*. Thèse de Doctorat, Université de Bordeaux 1.

Rougier, H., and E. Trinkaus. 2013. The human cranium from the Peştera cu Oase, Oase 2. In *Life and Death in the Peştera cu Oase: A Setting for Modern Human Emergence in Europe*, edited by E. Trinkaus, S. Constantin, and J. Zilhão, 257–320. New York: Oxford University Press.

Roy, T. A., C. B. Ruff, and C. C. Plato. 1994. Hand dominance and bilateral asymmetry in the structure of the second metacarpal. *American Journal of Physical Anthropology* 94:203–11.

Ruff, C. B. 2000. Body size, body shape, and long bone strength in modern humans. *Journal of Human Evolution* 38:269–90.

———. 2010. Body size and body shape in early hominins: Implications of the Gona pelvis. *Journal of Human Evolution* 58:166–78.

Ruff, C. B., and W. C. Hayes. 1983. Cross-sectional geometry of Pecos Pueblo femora and tibiae—a biomechanical investigation: I. Method and general patterns of variation. *American Journal of Physical Anthropology* 60:259–381.

Ruff, C. B., W. W. Scott, and A. Y. C. Liu. 1991. Articular and diaphyseal remodeling of the proximal femur with changes in body mass in adults. *American Journal of Physical Anthropology* 86:397–413.

Ruff, C. B., A. Walker, and E. Trinkaus. 1994. Postcranial robusticity in *Homo*, III: Ontogeny. *American Journal of Physical Anthropology* 93:35–54.

Ruff, C. B., B. M. Holt, M. Niskanan, V. Sládek, M. Berner, E. Garofalo, H. M. Garvin, M. Hora, H. Maijanen, S. Ninimäki, K. Salo, E. Schuplerová, and D. Tompkins. 2012. Stature and body mass estimation from skeletal remains in the European Holocene. *American Journal of Physical Anthropology* 148:601–17.

Ryan, A. S. 1980. Anterior dental microwear in hominid evolution: Comparisons with humans and nonhuman primates. PhD diss., University of Michigan.

Sakura, H. 1970. Dentition of the Amud man. In *The Amud Man and his Cave Site*, edited by H. Suzuki and F. Takai, 207–29. Tokyo: Therapeia.

Salazar-García, D. C., R. C. Power, A. S. Serra, V. Villaverde, M. J. Walker, and A. G. Henry. 2013. Neanderthal diets in central and southeastern Mediterranean Iberia. *Quaternary International* 318:3–18.

Sánchez-Cabeza, J. A., J. García-Orellana, and L. Gibert. 1999. Uranium-thorium dating of natural carbonates: Application to the Cabezo-Gordo site (Murcia, Spain). In *Los hominidos y su Entorno en el Pleistoceno inferior y medio de Eurasia*, edited by J. Gibert, F. Sánchez, L. Gibert, and F. Ribot, 261–68. Orce: Museo de Prehistoria y Paleontología "J. Gibert."

Sanson, G. D., S. A. Kerr, and K. A. Gordon. 2007. Do silica phytoliths really wear mammalian teeth? *Journal of Archaeological Science* 34:526–31.

Santamaría, D., and M. de la Rasilla. 2013. Datando el final del Paleolítico medio en la Península Ibérica: Problemas metodológicos y límites de la interpretación. *Trabajos de Prehistoria* 70:241–63.

Sarrión, I. 2006. Hallazgo de un parietal humano del tránsito Pleistoceno medio-superior procedente de la Cova del Bolomor, Tavernes de la Valldigna, Valencia. *Archivo de Prehistoria Levantina* 26:11–23.

Scheuer, L., S. Black, and A. Christie. 2000. *Developmental Juvenile Osteology*. San Diego: Academic Press.

Schwartz, G. T. 2000. Taxonomic and functional aspects of the patterning of enamel thickness distribution in extant large bodied hominoids. *American Journal of Physical Anthropology* 111:221–44.

Schwartz, G. T., and M. C. Dean. 2008. Charting the chronology of developing dentitions. In *Technique and application in dental anthropology*, edited by J. D. Irish and G. C. Nelson, 219–33. Cambridge: Cambridge University Press.

Scolan, H., F. Santos, A. M. Tillier, B. Maureille, and A. Quintard. 2012. Des nouveaux vestiges néanderthaliens à Las Pélénos (Monsempron-Libos, Lot-et-Garonne, France). *Bulletins et Mémoires de la Société d'Anthropologie de Paris* 24:69–95.

Scott, G. R., and C. G. Turner II. 1997. *The Anthropology of Modern Human Teeth: Dental Morphology and Its Variation in Recent Human Populations.* Cambridge: Cambridge University Press.

Scott, G. R., and J. R. Winn. 2011. Dental chipping: Contrasting patterns of microtrauma in Inuit and European populations. *International Journal of Osteoarchaeology* 21(6):723–31.

Semal, P., H. Rougier, I. Crevecoeur, C. Jungels, D. Flas, A. Hauzeur, B. Maureille, M. Germonpré, H. Bocherens, S. Pirson, L. Cammaert, N. De Clerck, A. Hambucken, T. Higham, M. Toussaint, and J. van der Plicht. 2009. New data on the late Neandertals: Direct dating of the Belgian Spy fossils. *American Journal of Physical Anthropology* 138:421–28.

Semenov, S. A. 1964. *Prehistoric Technology: An Experimental Study of the Oldest Tools and Artefacts from Traces of Manufacture and Wear.* London: Cory, Adams, and Mackay.

Sen, J., T. Kanchan, A. Ghosh, M. Nitish, and K. Krishan. 2014. Estimation of stature from lengths of index and ring fingers in a northeastern Indian population. *Journal of Forensic and Legal Medicine* 22:10–15.

Senut, B. 1981. *L'Humérus et ses Articulations chez les Hominidés Plio-Pleistocènes.* Paris: CNRS.

Shah, A. A., C. Elcock, and A. H. Brook. 2005. Posterior tooth morphology and lower incisor crowding. *Dental Anthropology* 18:37–42.

Shang, H., and E. Trinkaus. 2010. *The Early Modern Human from Tianyuan Cave, China.* College Station: Texas A&M University Press.

Shaw, C. N., C. L. Hofmann, M. D. Petraglia, J. T. Stock, and J. S. Gottschall. 2012. Neandertal humeri may reflect adaptation to scraping tasks, but not spear thrusting. *PLoS ONE* 7(7):e40349.

Sheets, H. D. 2001. IMP (Integrated Morphometric Package). Morphometrics software, Department of Physics, Canisius College. http://www.canisius.edu/~sheets/morphsoft.html.

Skinner, M. F. 1997. Dental wear in immature Late Pleistocene European hominines. *Journal of Archaeological Science* 24:677–700.

Skinner, M. M., Z. Alemseged, C. Gaunitz, and J. J. Hublin. 2015. Enamel thickness trends in Plio-Pleistocene hominin mandibular molars. *Journal of Human Evolution* 85:35–45.

Skinner, M. M., B. Wood, C. Boesch, A. J. Olejniczak, A. Rosas, T. M. Smith, and J. J. Hublin. 2008. Dental trait expression at the enamel-dentine junction of lower molars in extant and fossil hominoids. *Journal of Human Evolution* 54:173–86.

Slice, D. E., ed. 2005. *Modern Morphometrics in Physical Anthropology.* New York: Kluwer Academic/ Plenum Publishers.

Smith, B. H. 1984. Patterns of molar wear in hunter-gatherers and agriculturalists. *American Journal of Physical Anthropology* 63:39–56.

———. 1991. Standards of human tooth formation and dental age assessment. In *Advances in Dental Anthropology,* edited by M. A. Kelley and C. S. Larsen, 143–68. New York: Wiley-Liss.

Smith, F. H. 1976. *The Neandertal remains from Krapina: A descriptive and comparative study.* Report of Investigations 15. Knoxville: Department of Anthropology, University of Tennessee.

———. 1978. Evolutionary significance of the mandibular foramen area in Neandertals. *American Journal of Physical Anthropology* 48:523–32.

———. 1983. Behavioral interpretations of changes in craniofacial morphology across the archaic/ modern *Homo sapiens* transition. In *The Mousterian Legacy: Human Biocultural Change in the Upper Pleistocene,* edited by E. Trinkaus, 137–209. British Archaeological Reports International Series 164. Oxford: British Archaeological Reports.

———. 1993. Models and realities in modern human origins: The African fossil evidence. In *The Origin of Modern Humans and the Impact of Chronometric Dating,* edited by M. J. Aitken, C. B. Stringer, and P. A. Mellars, 234–48. Princeton, NJ: Princeton University Press.

Smith, F. H, and S. P. Paquette. 1989. The adaptive basis of Neandertal facial form, with some thoughts on the nature of modern human origins. In *The Emergence of Modern Humans: Biocultural Adaptations in the Later Pleistocene,* edited by E. Trinkaus, 181–210. Cambridge: Cambridge University Press.

Smith, F. H., and G. C. Ranyard. 1980. Evolution of the supraorbital region in Upper Pleistocene fossil hominids from south-central Europe. *American Journal of Physical Anthropology* 53:589–610.

Smith, P. 1977a. Selective pressures and dental evolution in hominids. *American Journal of Physical Anthropology* 47:453–58.

———. 1977b. Regional variation in tooth size and pathology in fossil hominids. *American Journal of Physical Anthropology* 47:459–66.

———. 1978. Evolutionary changes in the deciduous dentition of Near Eastern populations. *Journal of Human Evolution* 7:401–8.

Smith, T. M. 2008. Incremental dental development: Methods and applications in hominoid evolutionary studies. *Journal of Human Evolution* 54:205–24.

———. 2013. Teeth and human life-history evolution. *Annual Review of Anthropology* 42:191–208.

Smith, T. M., and P. Tafforeau. 2008. New visions of dental tissue research: Tooth development, chemistry, and structure. *Evolutionary Anthropology* 17:213–26.

Smith, T. M., A. J. Olejniczak, J. P. Zermeno, P. Tafforeau, M. M. Skinner, A. Hoffmann, J. Radovčić, M. Toussaint, R. Kruszynski, C. Menter, J. Moggi-Cecchi, U. A. Glasmacher, O. Kullmer, F. Schrenk, C. Stringer, and J. J. Hublin. 2012. Variation in enamel thickness within the genus *Homo. Journal of Human Evolution* 62:395–411.

Smith, T. M., P. Tafforeau, A. Le Cabec, A. Bonnin, A. Houssaye, J. Pouech, J. Moggi-Cecchi, F. Manthi, C. Ward, M. Makaremi, and C. G. Menter. 2015. Dental ontogeny in Pliocene and early Pleistocene hominins. *PLoS ONE* 10:e0118118.

Sognnaes, R. F. 1956. Histological evidence of developmental lesions in teeth originating from Paleolithic, prehistoric, and ancient man. *American Journal of Pathology* 32:547–77.

Solan, M., and M. H. Day. 1992. The Baringo (Kapthurin) ulna. *Journal of Human Evolution* 22:307–13.

Soressi, M., S. P. McPherron, M. Lenoir, T. Dogandžić, P. Goldberg, Z. Jacobs, Y. Maigrot, N. L. Martisius, C. E. Miller, W. Rendu, M. Richards, M. M. Skinner, T. E. Steele, S. Talamo, and J. P. Texier. 2013. Neandertals made the first specialized bone tools in Europe. *Proceedings of the National Academy of Sciences, USA* 110:14186–90.

Spencer, M. A., and B. Demes. 1993. Biomechanical analysis of masticatory system configuration in Neandertals and Inuits. *American Journal of Physical Anthropology* 91(1):1–20.

Spoor, C. F., F. W. Zonneveld, and G. A. Macho. 1993. Linear measurements of cortical bone and dental enamel by computed tomography: Applications and problems. *American Journal of Physical Anthropology* 91:469–84.

Stefan, V. H., and E. Trinkaus. 1998. Discrete trait and dental morphometric affinities of the Tabun 2 mandible. *Journal of Human Evolution* 34:443–68.

Stevens, R. E., and R. E. M. Hedges. 2004. Carbon and nitrogen stable isotope analysis of northwest European horse bone and tooth collagen, 40,000 BP–present: Palaeoclimatic interpretations. *Quaternary Science Reviews* 23:977–91.

Stewart, T. D. 1962. Neanderthal cervical vertebrae with special attention to the Shanidar

Neanderthals from Iraq. *Bibliotheca Primatologica* 1:130–54.

Stott, J. R. R., W. C. Hutton, and I. A. F. Stokes. 1973. Forces under the foot. *Journal of Bone and Joint Surgery* 55B:335–44.

Streeter, M., S. D. Stout, E. Trinkaus, C. B. Stringer, M. B. Roberts, and S. A. Parfitt. 2001. Histomorphometric age assessment of the Boxgrove 1 tibial diaphysis. *Journal of Human Evolution* 40:331–38.

Stringer, C. B., J. C. Finlayson, R. N. E. Bartond, Y. Fernández-Jalvo, I. Cáceres, R. C. Sabin, E. J. Rhodes, A. P. Currant, J. Rodríguez-Vidal, F. Giles-Pacheco, and J. A. Riquelme-Cantal. 2008. Neanderthal exploitation of marine mammals in Gibraltar. *Proceedings of the National Academy of Sciences, USA* 105:14319–24.

Ströhle, A., and A. Hahn. 2011. Diets of modern hunter-gatherers vary substantially in their carbohydrate content depending on ecoenvironments: Results from ethnographic analysis. *Nutrition Research* 31:429–35.

Susman, R. L. 1998. Hand function and tool behavior in early hominids. *Journal of Human Evolution* 35:23–46.

Susman, R. L., D. de Ruiter, and C. K. Brain. 2001. Recently identified postcranial remains of *Paranthropus* and early *Homo* from Swartkrans Cave, South Africa. *Journal of Human Evolution* 41:607–29.

Suwa, G., and R. T. Kono. 2005. A micro-CT based study of linear enamel thickness in the mesial cusp section of human molars: Reevaluation of methodology and assessment of within-tooth, serial, and individual variation. *Anthropological Science* 113:273–89.

Suwa, G., B. A. Wood, and T. D. White. 1994. Further analysis of mandibular molar crown and cusp areas in Pliocene and early Pleistocene hominids. *American Journal of Physical Anthropology* 93:407–26.

Suwa, G., T. D. White, and F. Howell. 1996. Mandibular postcanine dentition from the Shungura formation, Ethiopia: Crown morphology, taxonomic allocations, and Plio-Pleistocene hominid evolution. *American Journal of Physical Anthropology* 101:247–82.

Suwa, G., B. Asfaw, R. T. Kono, D. Kubo, C. O. Lovejoy, and T. D. White. 2009. The *Ardipithecus ramidus* skull and its implications for Hominid origins. *Science* 326:94–99.

Svensson, A., K. K. Andersen, M. Bigler, H. B. Clausen, D. Dahl-Jensen, S. M. Davies, S. J. Johnsen, R. Muschler, F. Parrenin, S. O. Rasmussen, R. Röthlisberger, I. Seierstad, J. P. Steffensen, and B. M. Vinther. 2008. A 60,000 year Greenland stratigraphic ice core chronology. *Climate of the Past Discussions* 4:47–57.

Teaford, M. F., and J. Lytle. 1996. Diet-induced changes in rates of human tooth microwear: A case study involving stone-ground maize. *American Journal of Physical Anthropology* 100:143–47.

Tillier, A. M. 1979. La dentition de l'enfant moustérien Châteauneuf 2 découverte à l'Abri de Hauteroche (Charente). *L'Anthropologie* 83:417–38.

———. 1999. *Les Enfants Moustériens de Qafzeh: Interprétation Phylogénétique et Paléoauxologique.* Paris: CNRS Éditions.

Tillier, A. M., and E. Genet-Varcin. 1980. La plus ancienne mandibule d'enfant découverte en France dans le gisement de La Chaise de Vouthon (Abri Suard) en Charente. *Zeitschrift für Morphologie und Anthropologie* 71:196–214.

Tillier, A. M., B. Arensburg, and H. Duday. 1989. La mandibule et les dents du Néanderthalien de Kébara (Homo 2), Mont Carmel, Israel. *Paléorient* 15:39–58.

Tillier, A. M., B. Arensburg, Y. Rak, and B. Vandermeersch. 1995. Middle Palaeolithic dental caries: New evidence from Kebara (Mount Carmel, Israel). *Journal of Human Evolution* 29:189–92.

Toussaint, M., A. J. Olejniczak, S. El Zaatari, P. Cattelain, D. Flas, C. Letourneux, and S. Pirson. 2010. The Neandertal lower right deciduous second molar from Trou de l'Abîme at Couvin, Belgium. *Journal of Human Evolution* 58:56–67.

Townsend, G. C., and T. Brown. 1978. Heritability of permanent tooth size. *American Journal of Physical Anthropology* 49:497–503.

Trinkaus, E. 1975. A functional analysis of the Neandertal foot. PhD diss., University of Pennsylvania.

———. 1980. Sexual differences in Neanderthal limb bones. *Journal of Human Evolution* 9:377–97.

———. 1981. Neanderthal limb proportions and cold adaptation. In *Aspects of Human Evolution*, edited by C. B. Stringer, 187–224. London: Taylor & Francis.

———. 1983. *The Shanidar Neandertals*. New York: Academic Press.

———. 1985. Pathology and the posture of the La Chapelle-aux-Saints Neandertal. *American Journal of Physical Anthropology* 67:19–41.

———. 1987. The Neandertal face: Evolutionary and functional perspectives on a recent hominid face. *Journal of Human Evolution* 16(5):429–43.

———. 1989. Olduvai Hominid 7 trapezial

metacarpal 1 articular morphology: Contrasts with recent humans. *American Journal of Physical Anthropology* 80:411–16.

———. 1992. Morphological contrasts between the Near Eastern Qafzeh-Skhul and late archaic human samples: Grounds for a behavioral difference? In *The Evolution and Dispersal of Modern Humans in Asia*, edited by T. Akazawa, K. Aoki, and T. Kimura, 277–94. Tokyo: Hokusen-Sha Publishing Company.

———. 1993. Variability in the position of the mandibular mental foramen and the identification of Neandertal apomorphies. *Rivista di Antropologia* 71:259–74.

———. 1995. Neanderthal mortality patterns. *Journal of Archaeological Science* 22:121–42.

———. 2002. The mandibular morphology. In *Portrait of the Artist as a Child. The Gravettian Human Skeleton from the Abrigo do Lagar Velho and Its Archeological Context*, edited by J. Zilhão and E. Trinkaus, 312–25. *Trabalhos de Arqueologia 22.*

———. 2003. Neandertal faces were not long; Modern human faces are short. *Proceedings of the National Academy of Sciences, USA* 100:8142–45.

———. 2004. Dental crown dimensions of Middle and Late Pleistocene European humans. In *Miscalánea en Homenaje a Emiliano Aguirre III: Paleoantropología*, edited by S. Rubio, 393–98. Alcalá de Henares: Museo Arqueológico Regional.

———. 2005. Anatomical evidence for the antiquity of human footwear use. *Journal of Archaeological Science* 32:1515–26.

———. 2006a. Modern human versus Neandertal evolutionary distinctiveness. *Current Anthropology* 47:597–620.

———. 2006b. The lower limb remains. In *Early Modern Human Evolution in Central Europe: The People of Dolní Věstonice and Pavlov*, edited by E. Trinkaus and J. A. Svoboda, 380–418. New York: Oxford University Press.

———. 2007. European early modern humans and the fate of the Neandertals. *Proceedings of the National Academy of Sciences, USA* 104:7367–72.

———. 2011a. Late Pleistocene adult mortality patterns and modern human establishment. *Proceedings of the National Academy of Sciences, USA* 108:1267–71.

———. 2011b. The postcranial dimensions of the La Chapelle-aux-Saints 1 Neandertal. *American Journal of Physical Anthropology* 145:461–68.

———. 2012. The human humerus from the Broken Hill Mine, Kabwe, Zambia. *American Journal of Physical Anthropology* 149:312–17.

———. 2013. The paleobiology of modern human emergence. In *Origins of Modern Humans: Biology Reconsidered*, 2nd ed., edited by F. H. Smith and J. C. M. Ahern, 393–434. Hoboken: John Wiley & Sons.

———. 2015. The appendicular skeletal remains of Oberkassel 1 and 2. In *The Late Glacial Burial from Oberkassel Revisited*, edited by L. Giemsch and R. W. Schmitz, 75–132. Darmstadt: Verlag Phillip von Zabern.

———. 2016a. The evolution of the hand in Pleistocene *Homo*. In *The Evolution of the Primate Hand: Perspectives from Anatomical, Developmental, Functional, and Paleontological Evidence*, edited by T. L. Kivell, P. Lemelin, B. G. Richmond, and D. Schmitt, 545–71. Dordrecht: Springer.

———. 2016b. *The Krapina Human Postcranial Remains: Morphology, Morphometrics, and Paleopathology.* Zagreb: FF Press.

———. 2016c. The sexual attribution of the La Quina 5 Neandertal. *Bulletins et Mémoires de la Société d'Anthropologie de Paris* 28:111–7.

Trinkaus, E., and S. E. Churchill. 1988. Neandertal radial tuberosity orientation. *American Journal of Physical Anthropology* 75:15–21.

———. 1999. Diaphyseal cross-sectional geometry of Near Eastern Middle Palaeolithic humans: The humerus. *Journal of Archaeological Science* 26(2):173–84.

Trinkaus, E., and C. E. Hilton. 1996. Neandertal pedal proximal phalanges: Diaphyseal loading patterns. *Journal of Human Evolution* 30:399–425.

Trinkaus, E., and B. Pinilla. 2009. Dental caries in the Qafzeh 3 Middle Paleolithic modern human. *Paléorient* 35:69–76.

Trinkaus, E., and M. P. Richards. 2013. Stable isotopes and dietary patterns of the faunal species from the Peştera cu Oase. In *Life and Death at the Peştera cu Oase: A Setting for Modern Human Emergence in Europe*, edited by E. Trinkaus, S. Constantin, and J. Zilhão, 211–26. New York: Oxford University Press.

Trinkaus, E., and H. Rougier. 2013. The human mandible from the Peştera cu Oase, Oase 1. In *Life and Death in the Peştera cu Oase: A Setting for Modern Human Emergence in Europe*, edited by E. Trinkaus, S. Constantin, and J. Zilhão, 234–56. New York: Oxford University Press.

Trinkaus, E., and C. B. Ruff. 1999. Diaphyseal cross-sectional geometry of Near Eastern Middle Paleolithic humans: The femur. *Journal of Archaeological Science* 26:409–24.

———. 2000. Comment on O. M. Pearson, "Activity,

climate, and postcranial robusticity: Implications for modern human origins and scenarios of adaptive change." *Current Anthropology* 41:598.

———. 2012. Femoral and tibial diaphyseal cross-sectional geometry in Pleistocene *Homo*. *PaleoAnthropology* 2012:13–62.

Trinkaus, E., and H. Shang. 2008. Anatomical evidence for the antiquity of human footwear: Tianyuan and Sunghir. *Journal of Archaeological Science* 35:1928–33.

Trinkaus, E., and P. Shipman. 1993. *The Neandertals: Changing the Image of Mankind*. New York: Alfred A. Knopf.

Trinkaus, E., and J. Zilhão. 2013. Paleoanthropological implications of the Peştera cu Oase and its contents. In *Life and Death at the Peştera cu Oase: A Setting for Modern Human Emergence in Europe*, edited by E. Trinkaus, S. Constantin, and J. Zilhão, 389–400. New York: Oxford University Press.

Trinkaus, E., S. E. Churchill, and C. B. Ruff. 1994. Postcranial robusticity in *Homo*, II: Humeral bilateral asymmetry and bone plasticity. *American Journal of Physical Anthropology* 93:1–34.

Trinkaus, E., R. J. Smith, and S. Lebel. 2000a. Dental caries in the Aubesier 5 Neandertal primary molar. *Journal of Archaeological Science* 27:1017–21.

Trinkaus, E., J. Svoboda, D. L. West, V. Sládek, S. W. Hillson, E. Drozdová, and M. Fišáková. 2000b. Human remains from the Moravian Gravettian: Morphology and taphonomy of isolated elements from the Dolní Věstonice II site. *Journal of Archaeological Science* 27:1115–32.

Trinkaus, E., A. E. Marks, J. P. Brugal, S. E. Bailey, W. J. Rink, and D. Richter. 2003. Later Middle Pleistocene human remains from the Almonda karstic system, Torres Novas, Portugal. *Journal of Human Evolution* 45, 219–26.

Trinkaus, E., S. W. Hillson, R. G. Franciscus, and T. W. Holliday. 2006. Skeletal and dental paleopathology. In *Early Modern Human Evolution in Central Europe: The People of Dolní Věstonice and Pavlov*, edited by E. Trinkaus and J. A. Svoboda, 419–58. New York: Oxford University Press.

Trinkaus, E., J. Maki, and J. Zilhão. 2007. Middle Paleolithic human remains from the Gruta da Oliveira (Torres Novas), Portugal. *American Journal of Physical Anthropology* 134:263–73.

Trinkaus, E., J. Svoboda, P. Wojtal, M. Nývltová Fišáková, and J. Wilczyński. 2010. Human remains from the Moravian Gravettian: Morphology and taphonomy of additional elements from Dolní Věstonice II and Pavlov I. *International Journal of Osteoarchaeology* 20:645–69.

Trinkaus, E., S. E. Bailey, and H. Rougier. 2013. The dental and alveolar remains of Oase 1 and 2. In *Life and Death at the Peştera cu Oase: A Setting for Modern Human Emergence in Europe*, edited by E. Trinkaus, S. Constantin, and J. Zilhão, 348–74. New York: Oxford University Press.

Trinkaus, E., A. P. Buzhilova, M. B. Mednikova, and M. V. Dobrovolskaya. 2014a. *The People of Sunghir: Burials, Bodies and Behavior in the Earlier Upper Paleolithic*. New York: Oxford University Press.

Trinkaus, E., T. W. Holliday, and B. M. Auerbach. 2014b. Neandertal clavicle length. *Proceedings of the National Academy of Sciences, USA* 111:4438–42.

Trotter, M., and G. C. Gleser. 1952. Estimation of stature from long bones of American whites and negroes. *American Journal of Physical Anthropology* 10:463–514.

Turner, C. G. II, C. R. Nichol, and G. R. Scott. 1991. Scoring procedures for key morphological traits of the permanent dentition: The Arizona State University dental anthropology system. In *Advances in Dental Anthropology*, edited by M. A. Kelley and C. S. Larsen, 13–31. New York: Wiley-Liss.

Twiesselmann, F. 1973. Évolution des dimensions et de la forme de la mandibule, du palais et des dents de l'homme. *Annales de Paléontologie (Vertébrés)* 59:173–277.

Underhill, T. E., J. O. Katz, T. L. Pope Jr., and C. L. Dunlap. 1992. Radiologic findings of diseases involving the maxilla and mandible. *American Journal of Roentgenology* 159:345–50.

Uomini, N. T. 2008. In the knapper's hands: Identifying handedness from lithic production and use. In *"Prehistoric Technology" 40 Years Later: Functional Studies and the Russian Legacy*, edited by L. Longo and N. Skakun, 51–62. Oxford: BAR International Series.

———. 2011. Handedness in Neanderthals. In *Neanderthal Lifeways, Subsistence, and Technology: One Hundred Fifty Years of Neanderthal Study*, edited by N. J. Conard and J. Richter, 139–58. New York: Springer.

Vandermeersch, B. 1981. *Les Hommes Fossiles de Qafzeh (Israël)*. Paris: CNRS.

———. 1991. La ceinture scapulaire et les membres supérieures. In *Le Squelette Moustérien de Kébara 2*, edited by O. Bar Yosef and B. Vandermeersch, 157–78. Paris: CNRS.

Verna, C. 2006. *Les Restes Humains Moustériens de la Station Amont de La Quina (Charente, France)*. Thèse de Doctorat, Université de Bordeaux 1.

Verna, C., V. Dujardin, and E. Trinkaus. 2012. The

Aurignacian human remains from La Quina-Aval (France). *Journal of Human Evolution* 62:605–17.

Villa, P., and W. Roebroeks. 2014. Neandertal demise: An archeological analysis of the modern human superiority complex. *PLoS ONE* 9(4):e96424.

Villemeur, I. 1994. *La Main des Néandertaliens: Comparaison avec la Main des Hommes de Type Moderne—Morphologie et Mécanique.* Paris: CNRS.

Vlček, E. 1967. Die Sinus frontales bei europäischen Neandertalern. *Anthropologischer Anzeiger* 30:166–89.

———. 1969. *Neandertaler der Tschechoslowakei.* Prague: Academia Prag.

———. 1993. *Fossile Menschenfunde von Weimar-Ehringsdorf.* Stuttgart: Konrad Theiss.

Volpato, V., R. Macchiarelli, D. Guatelli-Steinberg, I. Fiore, L. Bondioli, and D. W. Frayer. 2012. Hand to mouth in a Neandertal: Right-handedness in Regourdou 1. *PLoS ONE* 7(8):e43949.

Walker, M. J. 2001. Excavations at Cueva Negra del Estrecho del Río Quípar and Sima de las Palomas del Cabezo Gordo: Two sites in Murcia (southeast Spain) with Neanderthal skeletal remains, Mousterian assemblages, and late Middle to early Upper Pleistocene fauna. In *A Very Remote Period Indeed*, edited by S. Milliken and J. Cook, 153–59. Oxford: Oxbow Books.

———. 2009. La Sima de las Palomas del Cabezo Gordo en Torre Pacheco y la Cueva Negra des Estrecho del Río Quípar en Caravaca de la Cruz: Dos ventanas sobre la vida y la muerte del hombre fósil en Murcia. In *Darwin y la Evolución Humana*, ed. T. F. Verdú and F. A. Martínez, 71–96. Murcia: Caja Mediterráneo.

Walker, M. J., and J. Gibert. 1999. Dos yacimientos murcianos con restos neandertalenses: La Sima de las Palomas del Cabezo Gordo y la Cueva Negra del Estrecho del Quípar de La Encarnación. In *Actas del XXIX Congreso Nacional de Arqueología, Cartagena*, 1:299–310. Murcia: Instituto de Patrimonio Histórico.

Walker, M. J., J. Gibert, F. Sánchez, A. V. Lombardi, I. Serrano, A. Eastham, F. Ribot, A. Arribas, A. Cuenca, J. A. Sánchez, J. García, L. Gibert, S. Albadalejo, and J. A. Andreu. 1998. Two SE Spanish Middle Palaeolithic sites with Neanderthal remains: Sima de la Palomas del Cabezo Gordo and Cueva Negra del Estrecho del Río Quípar (Murcia Province). Internet Archaeology 5. http://intarch.ac.uk/journal/issue5/walker.

Walker, M. J., J. Gibert, F. Sánchez, A. V. Lombardi, I. Serrano, A. Gómez, A. Eastham, F. Ribot, A. Arribas, A. Cuenca, L. Gibert, S. Albadelejo, and J. A.

Andreu. 1999. Excavation of new sites of early man in Murcia: Sima de las Palomas del Cabezo Gordo and Cueva Negra del Estrecho del Río Quípar de La Encarnación. *Human Evolution* 14:99–123.

Walker, M. J., J. Gibert, M. V. López, A. V. Lombardi, A. Pérez-Pérez, J. Zapata, J. Ortega, T. Higham, A. Pike, J. L. Schwenninger, J. Zilhão, and E. Trinkaus. 2008. Late Neandertals in southeastern Iberia: Sima de las Palomas del Cabezo Gordo, Murcia, Spain. *Proceedings of the National Academy of Sciences, USA* 105:20631–36.

Walker, M. J., A. V. Lombardi, J. Zapata, and E. Trinkaus. 2010. Neandertal mandibles from the Sima de la Palomas del Cabezo Gordo, Murcia, southeastern Spain. *American Journal of Physical Anthropology* 142:261–72.

Walker, M. J., J. Ortega, M. V. López, K. Parmová, and E. Trinkaus. 2011a. Neandertal postcranial remains from the Sima de las Palomas del Cabezo Gordo, Murcia, southeastern Spain. *American Journal of Physical Anthropology* 144:505–15.

Walker, M. J., J. Ortega, K. Parmová, M. V. López, and E. Trinkaus. 2011b. Morphology, body proportions, and postcranial hypertrophy of a female Neandertal from the Sima de las Palomas, southeastern Spain. *Proceedings of the National Academy of Sciences, USA* 108:10087–91.

Walker, M. J., J. Zapata, A. V. Lombardi, and E. Trinkaus. 2011c. New evidence of dental pathology in 40,000 year old Neandertals. *Journal of Dental Research* 90:428–32.

Walker, M. J., M. V. López-Martínez, J. Ortega-Rodrigáñez, M. Haber-Uriarte, A. López-Jiménez, A. Avilés-Fernández, J. L. Polo-Camacho, M. Campillo-Boj, J. García-Torres, J. S. Carrión-García, M. San Nicolás-del Toro, and T. Rodríguez-Estrella. 2012. The excavation of buried articulated Neanderthal skeletons at Sima de las Palomas (Murcia, SE Spain). *Quaternary International* 259:7–21.

Walker, M. J., J. Ortega-Rodrigáñez, A. Agut-Giménez, M. Soler-Laguía, C. P. E. Zollikofer, and M. Ponce de León. 2013. The Sima de las Palomas Neanderthal skeletons: First steps towards "virtual" reconstruction (abstract). *Proceedings of the European Society for the Study of Human Evolution* 1:191.

Wang, Y. J., H. Cheng, R. L. Edwards, Z. S. An, J. Y. Wu, C. C. Shen, and J. A. Dorale. 2001. A high-resolution absolute-dated Late Pleistocene monsoon record from Hulu Cave, China. *Science* 294:2345–48.

Ward, C. V., W. H. Kimbel, and D. C. Johanson. 2011. Complete metatarsal and arches in the foot of *Australopithecus afarensis*. *Science* 331:750–53.

Ward, C. V., M. W. Tocheri, J. M. Plavcan, F. H. Brown, and F. K. Manthi. 2014. Early Pleistocene third metacarpal from Kenya and the evolution of modern human-like hand morphology. *Proceedings of the National Academy of Sciences, USA* 111:121–24.

Warinner, C., J. F. Matias Rodrigues, R. Vyas, C. Trachsel, N. Shved, J. Grossmann, A. Radini, Y. Hancock, R. Y. Tito, S. Fiddyment, C. Speller, J. Hendy, S. Charlton, H. U. Luder, D. C. Salazar-Garcia, E. Eppler, R. Seiler, L. H. Hansen, J. A. Samaniego Castuita, S. Barkow-Oesterreicher, K. Y. Teoh, C. D. Kelstrup, J. V. Olsen, P. Nanni, T. Kawai, E. Willersley, C. von Mering, C. M. Lewis, M. J. Collins, M. T. P. Gilbert, F. Rühli, and E. Cappellini. 2014. Pathogens and host immunity in the ancient human oral cavity. *Nature Genetics* 46:336–44.

Weaver, A., T. Holliday, C. Ruff, and E. Trinkaus. 2001. The fossil evidence for the evolution of human intelligence in Pleistocene *Homo*. In *In the Mind's Eye: Multidisciplinary Perspectives on the Evolution of Human Cognition*, edited by A. Nowell, 154–72. Ann Arbor: International Monographs in Prehistory Archaeological Series.

Weaver, T. D. 2009. The meaning of Neandertal skeletal morphology. *Proceedings of the National Academy of Sciences, USA* 106:16028–33.

Webb, P. A. O., and J. M. Suchey. 1985. Epiphyseal union of the anterior iliac crest and medial clavicle in a modern multiracial sample of American males and females. *American Journal of Physical Anthropology* 68:457–66.

Weidenreich, F. 1937. The dentition of *Sinanthropus pekinensis*: A comparative odontography of the Hominids. *Palaeontologia Sinica* D1:1–180.

Weyer, E. 1959. *Primitive Peoples Today*. London: Hamish Hamilton.

Willman, J. C. 2016. The Non-Masticatory Use of the Anterior Teeth among Late Pleistocene Humans. PhD diss., Washington University.

Willman, J. C., J. Maki, P. Bayle, E. Trinkaus, and J. Zilhão. 2012. Middle Paleolithic human remains from the Gruta da Oliveira (Torres Nova), Portugal. *American Journal of Physical Anthropology* 149:39–51.

Wolff, E. W., J. Chappellaz, T. Blunier, S. O. Rasmussen, and A. Svensson. 2010. Millennial-scale variability during the last glacial: The ice core record. *Quaternary Science Reviews* 29:2828–38.

Wolpoff, M. H. 1971. Metric trends in hominid dental evolution. *Case Western Reserve University Studies in Anthropology* 2:1–244.

———. 1979. The Krapina dental remains. *American Journal of Physical Anthropology* 50:67–114.

Wood, B. A., and S. A. Abbott. 1983. Analysis of dental morphology of Plio-Pleistocene hominids I: Mandibular molars: Crown area measurements and morphological traits. *Journal of Anatomy* 136:197–219.

Wood, R. A., C. Barroso-Ruíz, M. Caparrós, J. F. Jordá Pardo, B. Galván Santos, and T. F. G. Higham. 2013. Radiocarbon dating casts doubt on the late chronology of the Middle to Upper Palaeolithic transition in southern Iberia. *Proceedings of the National Academy of Sciences, USA* 110:2781–86.

Woodburn, J., and S. Hudson. 1966. *The Hadza: The Food Quest of a Hunting and Gathering Tribe of Tanzania* (16 mm film). London: London School of Economics.

Wu, X. J., and E. Trinkaus. 2014. The Xujiayao 14 mandibular ramus and Pleistocene *Homo* mandibular variation. *Comptes Rendus Palevol* 13:333–41.

———. 2015. Neurocranial trauma on the late archaic human remains from Xujiayao, northern China. *International Journal of Osteoarchaeology* 25:245–52.

Wu, X. J., L. A. Schepartz, W. Liu, and E. Trinkaus. 2011. Antemortem trauma and survival in the Late Middle Pleistocene human cranium from Maba, south China. *Proceedings of the National Academy of Sciences, USA* 108:19558–62.

Wu, X. J., S. D. Maddux, L. Pan, and E. Trinkaus. 2012. Nasal floor variation among eastern Eurasian Pleistocene *Homo*. *Anthropological Science* 120:217–26.

Wu, X. J., S. Xing, and E. Trinkaus. 2013. An enlarged parietal foramen in the late archaic Xujiayao 11 neurocranium from northern China, and rare anomalies among Pleistocene *Homo*. *PLoS ONE* 8:e59587.

Wu, X. J., I. Crevecoeur, W. Liu, S. Xing, and E. Trinkaus. 2014. Temporal labyrinths of eastern Eurasian Pleistocene humans. *Proceeding of the National Academy of Sciences, USA* 111:10509–13.

Xia, J., J. Zheng, D. D. Huang, Z. R. Tian, L. Chen, Z. R. Zhou, P. S. Ungar, and L. M. Qian. 2015. A new model to explain tooth wear with implications for microwear formation and diet reconstruction. *Proceedings of the National Academy of Sciences, USA* 112:10669–72.

Yokley, T. R., and S. E. Churchill. 2006. Archaic and modern human distal humeral morphology. *Journal of Human Evolution* 51:603–16.

Zanolli, C., and A. Mazurier. 2013. Endostructural characterization of the *H. heidelbergensis* dental remains from the early Middle Pleistocene site of Tighenif, Algeria. *Comptes Rendus Palevol* 12:293–304.

Zanolli, C., P. Bayle, and R. Macchiarelli. 2010. Tissue proportions and enamel thickness distribution in the early Middle Pleistocene human deciduous molars from Tighenif, Algeria. *Comptes Rendus Palevol* 9:341–48.

Zanolli, C., L. Bondioli, A. Coppa, M. C. Dean, P. Bayle, F. Candilio, S. Capuani, D. Dreossi, I. Fiore, D. W. Frayer, Y. Libsekal, L. Mancini, L. Rook, T. M. Tekle, C. Tuniz, and R. Macchiarelli. 2014. The late Early Pleistocene human dental remains from Uadi Aalad and Mulhuli-Amo (Buia) Eritrean Danakil: Macromorphology and microstructure. *Journal of Human Evolution* 74:96–113.

Zelditch, M. L., D. L. Swiderski, H. D. Sheets, and W. L. Fink. 2004. *Geometric Morphometrics for Biologists: A Primer*. San Diego: Elsevier.

Zheng, J., B. Yang, S. Tuomasjukka, S. Ou, and H. Kallio. 2009. Effects of latitude and weather conditions on contents of sugars, fruit acids, and ascorbic acid in black currant (*Ribes nigrum* L.) juice. *Journal of Agricultural Food Chemistry* 57:2977–87.

Zilberman, U., and P. Smith. 1992. A comparison of tooth structure in Neanderthals and early *Homo sapiens sapiens*: A radiographic study. *Journal of Anatomy* 180:387–93.

Zilberman, U., M. Skinner, and P. Smith. 1992. Tooth components of mandibular deciduous molars of *Homo sapiens sapiens* and *Homo sapiens neanderthalensis*: A radiographic study. *American Journal of Physical Anthropology* 87:255–62.

Zilhão, J. 2006. Chronostratigraphy of the Middle-to-Upper Paleolithic transition in the Iberian peninsula. *Pyrenae* 37:7–84.

Zilhão, J., and E. Trinkaus,. eds. 2002a. *Portrait of the Artist as a Child: The Gravettian Human Skeleton from the Abrigo do Lagar Velho and Its Archeological Context*. Trabalhos de Arqueologia 22. Lisboa: Instituto Português de Arqueologia.

———. 2002b. Social implications. In *Portrait of the Artist as a Child: The Gravettian Human Skeleton from the Abrigo do Lagar Velho and Its Archeological Context*, edited by J. Zilhão and E. Trinkaus, 519–42. *Trabalhos de Arqueologia* 22.

Zilhão, J., D. E. Angelucci, E. Badal-García, F. d'Errico, F. Daniel, L. Dayet, K. Douka, T. F. G. Higham, M. J. Martínez-Sánchez, R. Montes-Bernárdez, S. Murcia-Mascarós, C. Pérez-Sirvent, C. Roldán-García, M. Vanhaeren, V. Villaverde, R. Wood, and J. Zapata. 2010. Symbolic use of marine shells and mineral pigments by Iberian Neandertals. *Proceedings of the National Academy of Sciences, USA* 107:1023–28.

Zollikofer, C. P. E., M. Ponce de León, R. W. Schmitz, and C. B. Stringer. 2008. New insights into Mid-Late Pleistocene fossil hominin paranasal sinus morphology. *Anatomical Record* 291:1506–16.

Zubov, A. A. 1992. The epicristid or middle trigonid crest defined. *Dental Anthropology* 6:9–10.

Contributors

MIGUEL ALCARAZ

Departamento de Dermatología, Estomatología, Radiología y Medicina Física, Facultad de Medicina, Universidad de Murcia, Campus Universitario de Espinardo, 30100 Murcia, Spain.
Email: mab@um.es

PRISCILLA BAYLE

UMR-5199 PACEA, Bâtiment B8, Université de Bordeaux, Allée Geoffroy Saint-Hilaire, 33615 Pessac cedex, France.
Email: priscilla.bayle@u-bordeaux.fr

MARIA HABER

Área de Prehistoria, Departamento de Prehistoria, Arqueología, Historia Antigua, Historia Medieval y Ciencias y Técnicas Historiográficas, Facultad de Letras, C/ Santo Cristo, 1, Campus de La Merced, Universidad de Murcia, 30001 Murcia, Spain.
Email: mariahaber@um.es

SARAH A. LACY

Department of Anthropology and Archaeology, 518 Clark Hall, University of Missouri–St. Louis, One University Boulevard, St. Louis, MO 63121, USA.
Email: lacysa@umsl.edu

MONA LE LUYER

UMR-5199 PACEA, Bâtiment B8, Université de Bordeaux, Allée Geoffroy Saint-Hilaire, 33615 Pessac cedex, France.
Email: mona.le-luyer@u-bordeaux.fr

A.VINCENT LOMBARDI

2584 Blossom Lane, New Castle PA 16105, USA.
Email: avincelombardi@gmail.com

MARIANO V. LÓPEZ

Calle Pintor Joaquín 10–4°-1, 30009 Murcia, Spain.
Email: marianlopez@hotmail.com

JON ORTEGA

Departamento de Zoología y Antropología Física, Facultad de Biología, Campus Universitario de Espinardo, Edifico 20, Universidad de Murcia, 30100 Murcia, Spain.
Email: ventepalportus@yahoo.es

KLÁRA KARAKOVÁ

International Clinical Research Center, St. Anne's University Hospital Brno, Pekarska 53, 656 91 Brno, Czech Republic.
Email: klara.karakova@fnusa.cz

ALEJANDRO PÉREZ-PÉREZ

Sección de Antropología, Departamento de Biología Animal, Facultad de Biología, Universitat de Barcelona, Edificio Margalef, 2a planra, Avenida Diagonal 645, 08028 Barcelona, Spain.
Email: martinez.perez-perez@ub.edu

BEATRIZ PINILLA

Sección de Antropología, Departamento de Biología Animal, Facultad de Biología, Universitat de Barcelona, Edificio Margalef, 2a planra, Avenida Diagonal 645, 08028 Barcelona, Spain.
Email: beatriz.pinilla@gmail.com

KATHARINE A. ROBSON BROWN

Centre for Human Evolutionary Research, University of Bristol, Bristol BS8 1UU, UK.
Email: kate.robson-brown@bristol.ac.uk

ALEJANDRO ROMERO
Departamento de Biotecnología, Facultad de Ciencias, Universidad de Alicante, Ap.C.99, E-03080 Alicante, Spain.
Email: arr@ua.es.

ERIK TRINKAUS
Department of Anthropology, Washington University, Saint Louis, MO 63130, USA.
Email: trinkaus@wustl.edu

MICHAEL J. WALKER
Departamento de Zoología y Antropología Física, Facultad de Biología, Campus Universitario de Espinardo, Edifico 20, Universidad de Murcia, 30100 Murcia, Spain.
Email: walker@um.es, mjwalke@gmail.com

JOHN C. WILLMAN
Department of Anthropology, Washington University, Saint Louis, MO 63130, USA.
Email: jcwillman@wustl.edu

JOSEFINA ZAPATA
Departamento de Zoología y Antropología Física, Facultad de Biología, Campus Universitario de Espinardo Edifico 20, Universidad de Murcia, 30100 Murcia, Spain.
Email: jzapata@um.es

Index

Palomas Fossils Index